ENCYCLOPEDIA OF PHYSICS

EDITOR IN CHIEF
S. FLÜGGE

VOLUME XLIX/6

GEOPHYSICS III

PART VI

BY

G.M. NIKOL'SKIJ · K. RAWER · P. STUBBE
L. THOMAS · T. YONEZAWA

EDITOR
K. RAWER

WITH 124 FIGURES

SPRINGER-VERLAG
BERLIN HEIDELBERG NEW YORK
1982

HANDBUCH DER PHYSIK

HERAUSGEGEBEN VON

S. FLÜGGE

BAND XLIX/6

GEOPHYSIK III

TEIL VI

VON

G.M. NIKOL'SKIJ · K. RAWER · P. STUBBE
L. THOMAS · T. YONEZAWA

BANDHERAUSGEBER

K. RAWER

MIT 124 FIGUREN

SPRINGER-VERLAG
BERLIN HEIDELBERG NEW YORK
1982

Professor Dr. SIEGFRIED FLÜGGE

Physikalisches Institut der Universität, D-7800 Freiburg i.Br.

Professor Dr. KARL RAWER

D-7801 March-Hugstetten

Universität Freiburg, D-7800 Freiburg i.Br.

ISBN 3-540-07080-X Springer-Verlag Berlin Heidelberg New York
ISBN 0-387-07080-X Springer-Verlag New York Heidelberg Berlin

Library of Congress Catalog Card Number A 56-2942.
Printed in Germany.

Satz, Druck und Bindearbeiten: Universitätsdruckerei H. Stürtz AG, 8700 Würzburg.
2153/3130-543210

Contents

Introduction: A Summary Description of Aeronomy

K. Rawer

In November 1647 Blaise Pascal, mathematician and, later, famous philosopher and religious thinker, asked his brother-in-law to walk up mount Puy de Dôme carrying with him what we now call a mercury barometer. His "great experiment" might be considered as the starting point of atmospheric science[1]. While the hypothesis of hydrostatic equilibrium was fundamental for Pascal's proposal, S.P. Laplace[2] later added the structure equation of ideal gases and so founded the barometric formula, a relation between pressure and altitude[3].

For a long time no other experimental evidence was available from higher altitudes, so that it became current practice to extrapolate the barometric formula. In doing so, one obtained an extremely rapid decrease of pressure with increasing height. Until the twentieth century nobody expected special conditions to exist at higher altitudes because the atmosphere was considered to be a purely terrestrial phenomenon.

We know now that there is in fact a very considerable outside influence arising from solar radiation[4], and this influence[5] produces what is called the upper atmosphere. Explaining this connection is the principal subject of this volume. Apart from high latitudes where particular conditions are found, solar wave radiation is the most important factor producing heating, dissociation, and even ionization at certain pressure levels. The first two influences can very considerably increase the scale height[3] and so extend the upper part of the atmosphere to much larger heights than expected from the barometric formula. Thus, the very existence of the upper atmosphere is due to the Sun's influence. Its physics is, therefore, particularly linked with the well-known variations of solar activity[4, 5].

The first discovery indicating some absorption of specific solar radiation by the atmosphere dates from 1878[6]. An explanation was given only three years

---

[1] Pascal, B. (1648): Récit de la grande Expérience de l'Equilibre dans les Liqueurs. Paris: Ch. Saureux. It was the main aim of P.'s experiment to prove a rational theory of the vacuum phenomenon appearing in such instruments, and disguise the earlier idea of the so-called horror vacui

[2] Laplace, S.P. (1805): Traité de mécanique céleste, tome 4. Paris: Courcier (Livre X, Chap. IV)

[3] Rawer, K.: This Encyclopedia, vol. 49/7, his Sect. 1

[4] Nikol'skij, G.M.: Contribution in this volume, p. 309

[5] Schmidtke, G.: This Encyclopedia, vol. 49/7

[6] Cornu, A. (1879): C.R. Acad. Sci. Paris *88*, 1101, *89*, 808. – Chappuis, J. (1880): C.R. Acad. Sci. Paris *91*, 985; J. Phys. Paris (2) *1*, 494

later by HARTLEY[7], who showed that *ozone* is the absorbing medium. This was the first proof that the upper atmosphere contains some constituents which do not occur in the troposphere. The most brilliant proof, which included an in-situ shape measurement of the lower part of the ozone profile (up to 34 km), was given much later by E. and V.H. REGENER[8]. Ozone is a minor constituent in the atmosphere and as such it cannot greatly change the structure relation. However, it absorbs solar radiation in spectral ranges that are of longer wavelengths than those producing ozone. Since the intensity in the solar spectrum increases with increasing wavelength, this absorbed energy flux is large enough to provoke an important temperature increase[9], with a maximum around 50 km, thus extending the atmosphere upwards. This range is now called the *mesosphere*[3].

Ozone formation in the atmosphere was the first subject of aeronomic theory[10]. *Aeronomy*[10-12] is the discipline of atmospheric science which tries to explain the special features of the upper atmosphere, in particular the profiles of density, constituents, and temperatures. It is, of course, important to distinguish between major and minor constituents. However, even very minor constituents may play a great rôle in the *chemical reactions* which are now considered to be of much greater importance in aeronomy than previously assumed. This is particularly true for the ionospheric *ion chemistry*, which is discussed rather extensively in the contribution by L. THOMAS in this volume.

Above the mesosphere, the structure of the upper atmosphere is largely controlled by the appearance of atomic oxygen[13] as a product of *photo dissociation*, a feature which is inherent in all aeronomic theories explaining the formation of ozone[10]. Unlike ozone in the mesosphere, atomic oxygen becomes a major constituent in the *thermosphere*[3]. Due to diffusion it is even the most prominent (99%) constituent in a large height range[14] with a lower limit between 220 and 580 km and an upper one a few 100 km higher up. All this is described in T. YONEZAWA's contribution to this volume.

Full diffusive separation leads to a monotonic decrease with increasing height of the mean molecular weight $\bar{m}$; the value of atomic oxygen (16 au) is then reached somewhere between 250 and 700 km (according to the exospheric temperature[14]), but the decrease continues and $\bar{m}$ decreases more and

[7] Hartley, W.N. (1881): J. Chem. Soc. *39*, 57 and 111

[8] Regener, E., Regener, V.H. (1934): Phys. Z. *35*, 788; Nature *134*, 380. These authors used a balloon-borne quartz spectrograph with an ingenious indirect illumination device[9]

[9] Paetzold, H.-K., Regener, E. (1957): Contribution in this Encyclopedia, vol. 48, p. 370. The record of the 1934 experiment is shown there as Fig. 12

[10] Fabry, C. (1929): Gerlands Beitr. Geophys. *24*, 1. – Mecke, R. (1930): Z. wiss. Photogr. *29*, 27; (1931): Trans. Faraday Soc. *27*, 375. – Chapman, S. (1929): Gerlands Beitr. Geophys. *24*, 66; (1930): Mem. R. Meteorol. Soc. *3*, 103. – Götz, E.W.P. (1931): Ergeb. Kosm. Phys. *1*, 180

[11] Chapman, S. (1960): in [12], p. 1

[12] Ratcliffe, J.A. (ed.) (1960): Physics of the upper atmosphere. New York: Academic Press

[13] Chapman, S. (1930): Phil. Mag. *10*, 345. – Flory, P.J. (1936): J. Chem. Phys. *4*, 23. – Penndorf, R. (1936): Beiträge zum Ozonproblem. Leipzig: Geophys. Inst. Universität (Ser. 2), *8*, 181. – Wulf, O.R., Deming, L.S. (1936): Terr. Magn. Atmos. Electr. *41*, 299. – Vegard, L. (1938): Ergeb. Exakten Naturwiss. *17*

[14] Committee on Space Research (1972): CIRA 1972, COSPAR International Reference Atmosphere 1972. Berlin: Akademie-Verlag

more due to helium and hydrogen, the relative abundance of which becomes important at higher levels[14].

The neutral atmosphere has a rather well-defined height limit, because at some level of extremely low density, upward travelling molecules no longer find collision partners, and thus they describe ballistic orbits in the Earth's gravity field. This height range is called the *exosphere*. Part of the molecules, namely those which in the thermal velocity distribution have obtained a vertical velocity component above the "escape velocity" (about $11\,\mathrm{km\,s^{-1}}$), in fact do escape into interplanetary space[15]. This phenomenon is also discussed by T. YONEZAWA in Chap. V of his contribution to this volume.

Ionization of atmospheric molecules, mainly by solar XUV light[16], is the source of the *ionosphere*[17,18]. Almost up to exospheric heights, the ionospheric plasma has low numerical density compared with that of the neutrals. Thus, as explained by K. RAWER in his contribution in the subsequent volume, the ionized constituents are minor ones even above the peak altitude of absolute plasma density, which appears at a height somewhere between 250 and 500 km[17,18]. However, since in the absence of collisions they are attached to the magnetic field of the Earth, the terrestrial ionosphere extends well above the exosphere. In fact, strictly speaking, the plasma in the *magnetosphere*[19], which extends up to at least 10 Earth radii, must be considered as to be the outermost part of the terrestrial atmosphere.

In the height range between about 60 km and the outermost levels of the exosphere, the ionized constituents play a non-negligible rôle in aeronomy. Originally, only one interaction with neutrals was considered, namely collisional damping of radio waves. Collision is discussed in K. SUCHY's contribution to this volume. Apart from this, the ionization balance between production and loss was formulated almost independently from the neutrals, or more precisely speaking, their influence was parameterized by using descriptive, height-dependent loss coefficients[20].

In the following we shall consider separately production and loss of ionization. The most important source of ionization is *solar wave radiation*. The fundamental ideas concerning this mechanism have not been seriously changed since the pioneering work of PANNEKOCK[21] and LENARD[22]. The shape of the production versus height profile was calculated for the most simple case quite

---

[15] Jeans, J.H. (1904): The dynamical theory of gases. Cambridge: University Press

[16] This is the current designation for the wavelength range between limits of about 0.05 and 200 nm

[17] Rawer, K., Ramakrishnan, S., Bilitza, D. (1978): URSI-COSPAR International Reference Ionosphere 1978. Brussels: International Union of Radio Science. – Rawer, K. (1981): International Reference Ionosphere IRI 79, 2nd edn. Boulder (Colo., USA): World Data Center A for Solar-Terrestrial Physics

[18] Rawer, K., Suchy, K. (1967): Contribution in this Encyclopedia, vol. 49/2, p. 1

[19] Poeverlein, H. (1972): Contribution in this Encyclopedia, vol. 49/4, p. 7

[20] Majumdar, R.C. (1937): Z. Phys. *107*, 599; (1938): Ind. J. Phys. *12*, 75

[21] Pannekoek, A. (1926): C.R. Acad. Amsterdam *29*, 1165. This author applied the theory of Saha, M.N. (1920): Phil. Mag. *1*, 1025 to the particular conditions in the ionosphere

[22] Lenard, P. (1900): Ann. Phys. *1*, 486. – Lenard, P., Ramsauer, C. (1910): Sitzungsber. Heidelb. Akad. Wiss. (28), 3; (1911): ibidem (24), 3

early by ELIAS[23]. However, in these early days the intensity distribution in the solar XUV spectrum[16] could not be measured since this light does not penetrate to the surface. Estimates of the intensity distribution during this time were very far from reality[24]. Only with the advent of rockets and satellites could direct measurements in space and thus outside the dense atmosphere be made in this spectral range. It took a lot of time and effort before such measurements were reliable (see G.M. NIKOL'SKIJ's contribution to this volume).

Though quite a few problems are still open, important progress has been accomplished with satellites launched in the last decade. It now appears feasible to specify spectral intensities that have still rather coarse spectral resolution but should be sufficient for aeronomic purposes (see G. SCHMIDTKE's contribution in volume 49/7). In the same contribution, wavelength-dependent ionization cross sections are also given. In this problem area too, important progress has been made during the last decade, mainly by laboratory measurements in different countries.

If the solar XUV spectrum and the relevant cross sections are known, one still needs a model of the neutral atmosphere (see L. THOMAS' contribution to this volume, Chap. F) in order to compute the primary production of ionization[25], see also G. SCHMIDTKE's contribution[25] in volume 49/7[26].

We now come to the loss term. Originally, "recombination" and "attachment" coefficients were used as height-dependent parameters to be interpreted in terms of atomic physics[27]. The complex nature of the actual physics was first taken into account by BATES and MASSEY[28] in 1947. They replaced the assumption of direct (radiative) recombination by "*dissociative recombination*". This is discussed in detail in L. THOMAS' contribution in this volume. The now generally accepted reaction is in fact a chain of successive reactions, in the simplest case a charge-transfer process followed by dissociative recombination, i.e. recombination with an electron with simultaneous dissociation of the original *molecular* ion. Thus, instead of the traditionally assumed simple reactions of recombination or attachment, we have now a complicated system. While this was a step towards better physical understanding, the application of this model needs information about the rôle of the neutrals in the charge-exchange reaction such that a kind of coupling between ionized and neutral atmospheres must be assumed. During the last two decades the chemical aspect has been developed more and more, so that today rather complex sets of chemical reaction equations are considered. Since, unfortunately, most of the reaction coefficients are temperature dependent, one has to take account of the heat balance also. A summary account can be found in K. RAWER's and – for certain aspects – in G. SCHMIDTKE's contributions in volume 49/7; details, in particular about chemistry, are given by L. THOMAS in this volume.

[23] Elias, G.J. (1923): Tydskr. Nederl. Radio Gen. *2*, 1. The more general case of including grazing incidence on a curved, rotating Earth was resolved by Chapman, S. (1931): Proc. Phys. Soc. London *43*, 26 and 483; Proc. R. Soc. London *A 132*, 353

[24] Rawer, K. (1981): Adv. Space Res. *1*, 87

[25] See also: Thomas, L.: Contribution in this volume, Chapt. B

[26] See further: Rawer, K.: This Encyclopedia, vol. 49/7, Sects. 10, 11 and 12

[27] Massey, H.S.W. (1937): Proc. R. Soc. London *A 163*, 542

[28] Bates, D.R., Massey, H.S.W. (1947): Proc. R. Soc. London *A 192*, 1

Finally, *plasma motions* with non-vanishing divergence are of considerable importance, in particular at altitudes above about 250 km where the ionization, when produced, lasts for several hours. Quite generally, plasma density gradients provoke *diffusion*, mainly in the vertical direction. This phenomenon is very important for the formation of the final shape of the ionized layers, which can differ considerably from the production profile. This subject is dealt with by T. YONEZAWA in his contribution in this volume.

Apart from this always present influence, the discovery of a strong coupling between neutral and plasma motions by KOHL and KING[29] has finally allowed a determination of the (field parallel) plasma motions induced by neutral winds and by this an understanding of a large part of the apparent irregularities in the temporal behaviour of the ionospheric F2 layer. Another influence that is essential at low and at high latitudes is that of electric fields which provoke motions perpendicular to the magnetic field lines. The influence of motions on the plasma density is the subject of P. STUBBE's contribution in this volume.

The present situation of aeronomy may be characterized by saying that interactions between the different constituents, and thermal influences too, were found to be much more important than expected. Thus a full theory should consider all of them together. On the other hand, the solution for such complex systems becomes so much involved that even with advanced numerical methods, convergence problems may show up that are sometimes so difficult that some simplifying assumptions need to be introduced (see K. RAWER's contribution in volume 49/7).

For this reason *semi-empirical methods* have been quite often applied during the last years, in particular in view of the conditions at the limits. One typical example which is discussed in volume 49/7[30,5] supposes measured data at some critical height level to be available and so avoids assumptions which are often uncertain. On the other hand, these theories consider a larger geographic area in a realistic manner (including heat deposit by corpuscular influences). In fact, aeronomic theory should tend to produce world-wide solutions, but we are still rather far from this goal.

---

[29] Kohl, H., King, J.W. (1965): Nature *206*, 699. - Risbeth, H., Megill, L.R., Cahn, J.H. (1965): Annls. Géophys. *21*, 235

[30] Rawer, K.: This Encyclopedia, vol. 49/7, Sect. 17

The Neutral and Ion Chemistry of the Upper Atmosphere

By

L. THOMAS

With 50 Figures

1. Introduction. The lower atmosphere of the Earth has been studied extensively for many years. The emphasis has in the main been given to the dynamical processes involved in meteorology although recent studies of polluted atmospheres have demonstrated the scientific interest and practical importance of particular constituents at the lowest altitudes. At greater heights the studies of chemical reactions form an essential part of the study of Aeronomy which is directed towards an improved understanding of the chemical and physical processes which govern the distributions of the neutral and ionized constituents [*1*]. For the present purposes the upper atmosphere is taken as the region above about 60 km where free electrons exist in sufficient concentrations to affect the propagation of radio waves. Because of this characteristic the region is referred to as the ionosphere and much of the initiative for studying both the ionized and neutral constituents at these heights arose from the need to understand ionospheric observations. It has become clear that the neutral and ionized chemistries cannot be realistically separated. For instance, nitric oxide represents a critical constituent in the ion chemistry of the lower ionosphere but its production is itself dependent on the formation of $N(^2D)$ atoms, largely during ionization processes.

The purpose of the present article is to describe the chemical processes which are believed to determine the neutral and ion composition. The details of the height distributions of neutral and ionized constituents over the height range of interest are not established, and the horizontal and temporal variations have not been clearly identified to date. It is, therefore, considered sufficient to describe in Chap. A what are believed to be the main features of the structure and composition of the neutral and ionized atmosphere between 60 and 1,000 km. These features are determined by the interactions between solar radiations and atmospheric constituents. In order to understand the details of these interactions it is necessary to have a knowledge of the relevant solar fluxes and the efficiencies of the primary interaction processes. These topics are considered in Chap. B. The products of these interactions can undergo collisions with themselves or other atmospheric and ionospheric constituents, some of these collisions being accompanied by chemical change, as outlined in Chap. C. The present understanding of such chemical reactions has arisen chiefly from laboratory studies carried out in recent years and it is appropriate to indicate in Chap. D the techniques which have made the greatest contribution in this area. Information on chemical reactions has also been derived from space-borne measurements and, indeed, the study of such reactions in the height range 100 to 250 km was a primary objective of the Atmospheric Explorer series of satellites [1,2]. Chapter E is concerned with space-borne and ground-based methods of measuring neutral and ion composition. On the basis of the laboratory results, theoretical models of neutral and ion composition have been developed to provide an understanding of the atmospheric data. These models will be outlined in Chaps. F and G, respectively. In describing these models emphasis has been given to the chemistry operating and little attention has been given to horizontal transport processes. Finally, the formation of excited species, which give rise to airglow emissions, and their roles in atmospheric and ionospheric processes are outlined since it seems likely that such species will be paid increased attention in the future.

The article relates in the main to normal conditions at middle latitudes and does not consider the effects associated with precipitated high-energy particles and the general interaction at high latitudes with the magnetosphere, the completely ionized region extending out to several earth radii.

A. Neutral and ionized atmospheric structure and composition

2. Neutral atmosphere. It is customary to classify the different regions of the neutral atmosphere on the basis of the height variation of temperature, as illustrated in Fig. 1 [*2*]. The troposphere is characterised by a reduction in temperature with height up to the minimum at the tropopause; at greater heights the temperature rises to a maximum at the stratopause which corresponds approximately to the lower limit of the height region of interest in the present review. Above the stratopause the temperature decreases through the mesosphere to a minimum value at the mesopause, and at lower thermospheric heights it increases and finally assumes an approximately constant value above about 300 km. The values of temperature vary markedly with time of day, season, position and the level of solar activity, especially at the greater heights; those shown in Fig. 1 are from the U.S. Standard Atmosphere [*3*], and correspond to average daytime conditions at middle latitudes.

[1] Dalgarno, A., Hanson, W.B., Spencer, N.W., Schmerling, E.R. (1973): Radio Sci. *8*, 263

[2] The National Bureau of Standards (Washington, DC20234, USA) by its National Standard Reference Data System provides critically evaluated numerical data, e.g. NBS Spec. Publ. 513 (Reaction Rate and Photochemical Data for Atmospheric Chemistry – 1977)

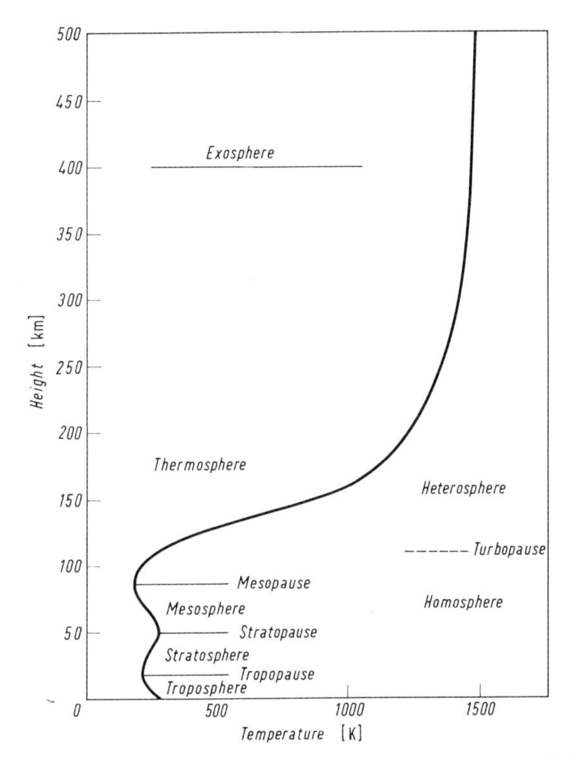

Fig. 1. The vertical distribution of neutral gas temperature taken from the U.S. Standard Atmosphere [*3*]

A second basis for classification arises from the dynamical processes which control the height distributions of the major atmospheric constituents. In the homosphere, up to about 105 km, mixing of the constituents by winds, turbulence, and the general circulation is sufficiently rapid to maintain almost constant proportions of the major constituents, molecular nitrogen and oxygen, and carbon dioxide, as shown by the full lines in Fig. 2, after THOMAS[1]. In contrast, in-situ mixing processes do not affect significantly the height variations of constituents in the heterosphere, above about 105 km, each assuming a distribution determined by its molecular weight and the atmospheric temperature, as illustrated in the CIRA model atmosphere [*4*] shown in Fig. 3. At still greater heights, in the exosphere, atomic hydrogen and helium become major constituents, and the atoms of these gases have sufficient kinetic energy to escape from the terrestrial atmosphere.

At heights below the exosphere the height distributions of gases can be considered on the basis of a hydrostatic model. To a first approximation the variation of atmospheric density is related to the atmospheric pressure gradient by:

$$g\rho = -\mathrm{d}p/\mathrm{d}h, \tag{2.1}$$

[1] Thomas, L. (1976): J. Atmos. Terr. Phys. *38*, 61

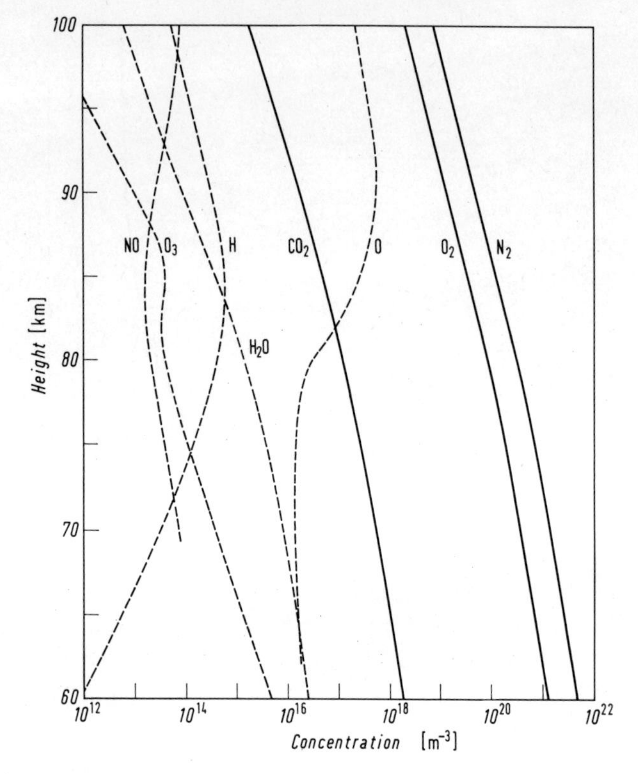

Fig. 2. The vertical distributions from 60 to 100 km of major neutral constituents (full curves) and chemically active minor neutral constituents (broken curves), after THOMAS[1]

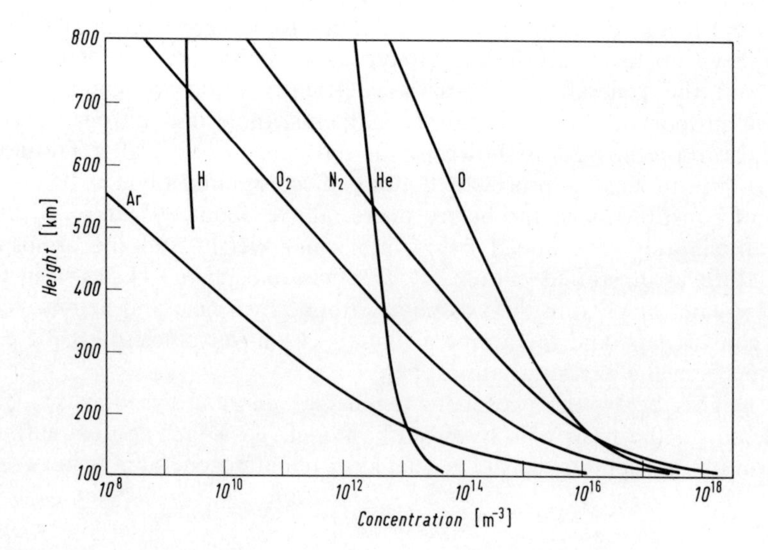

Fig. 3. The vertical distributions from 100 to 800 km of neutral constituents, as represented in the CIRA model for an exospheric temperature of 1,500 K [*4*]

where ρ represents the density, g the acceleration of gravity at height h, and p the pressure at that height.

This relation, together with the ideal gas equation, leads to:

$$\frac{\mathrm{d}p}{p} = -\frac{\mathrm{d}h}{kT/\bar{m}g}, \tag{2.2}$$

where k denotes Boltzmann's constant, T the atmospheric temperature and $\bar{m}$ the mean molecular mass.

The quantity $H = kT/\bar{m}g$ is known as the atmospheric scale height. If it is assumed to be constant, then

$$p = p_0 \exp(-h'/H), \tag{2.3}$$

where h' represents the height above the level of the reference pressure P_0.

It is evident that in the homosphere, where there is no major change in composition, the mean molecular mass $\bar{m}$ remains constant. In this situation the measurement of scale height will yield a measurement of atmospheric temperature. However, in the heterosphere the relative proportions of light constituents increase and the mean molecular mass decreases with increasing height. Consequently, the interpretation of the height variation of pressure or density in this part of the atmosphere is dependent on a knowledge of temperature or mean molecular mass.

For chemically active constituents the height distributions are strongly influenced by photochemical processes. The combined effect of such processes and mixing gives rise to the height distributions of water vapour, atomic hydrogen, atomic oxygen, ozone, and nitric oxide shown by the broken curves in Fig. 2; the first four of these have been deduced from a theoretical model, as described in Sect. 23α, and the fifth has been derived from observations of nitric oxide fluorescence in the dayglow by MEIRA[2] (Sect. 18). It is expected that the atomic oxygen and nitric oxide show marked reductions at the lower altitudes during nighttime.

3. Ionized atmosphere. Although the early radio-wave experiments used for sounding the ionosphere indicated distinct layers, subsequent measurements with rocket-borne devices have revealed that the electron concentration increases monotonically with height, but shows ledges located at 70 to 90 km, 100 to 120 km and 150 to 180 km, and a maximum in the range 200 to 300 km [5]. According to a convention originated by Appleton the letters D, E, F1 and F2 are used to denote these successive features in the height distribution of electron concentration, as illustrated in Fig. 4 taken from the International Reference Ionosphere [6]: an additional distinct layer sometimes observed near 65 km has been called the C layer. The full curve represents noon, high solar activity conditions at mid-latitudes during summer. The corresponding distribution for nighttime is represented by the broken curve and shows much reduced concentrations below about 300 km, a pronounced minimum near

[2] Meira, L.G. (1971): J. Geophys. Res. *76*, 202

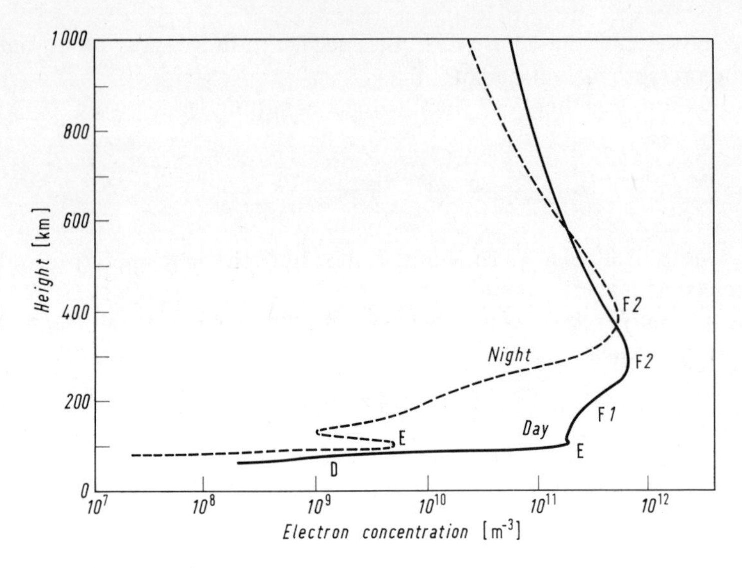

Fig. 4. The vertical distributions of electron concentration representative of high solar activity summer conditions at mid-latitudes, corresponding to the International Reference Ionosphere for a Covington 10.7 cm solar flux index of 140 [*6*]

140 km, and a sharp decrease below 90 km. The electron concentration is reduced to about 10^{10} to $10^{11}\,m^{-3}$ near 1,000 km and at greater heights the F2 region merges into the magnetosphere.

In addition to the day-to-night changes mentioned above, the ionospheric regions show marked variations with season and latitude. The dependence on solar zenith angle found in the changes of D, E and F1 layers with time of day and with season noted in early observations demonstrated the solar control of these regions; at night these three layers virtually disappear, as illustrated in Fig. 4. Subsequent rocket-borne observations of the attenuation of solar radiations confirmed that the ionization is produced by the absorption of solar extreme ultra-violet and X-radiation. The F2 layer does not vary regularly, although there is a day-to-night change as shown in Fig. 4. The generally anomalous character of this uppermost layer can be explained when account is taken of the transport of ionization in the presence of the geomagnetic field in addition to production by photoionization and chemical loss.

The thermal balance of the ionosphere has been subject to considerable study since the early rocket-borne observations revealed a lack of thermal equilibrium between electrons, ions and neutral gases at heights above about 150 km. This effect is known to be a common feature of both the daytime and nighttime ionosphere, and has been well demonstrated by in-situ sampling with rockets and satellites and also by ground-based measurements using the incoherent radar scatter technique (Sect. 20), as shown by the results of EVANS[1] in Fig. 5. It is known to arise from heating of the electron gas by electrons

[1] Evans, J.V. (1975): Proc. IEEE *63*, 1636

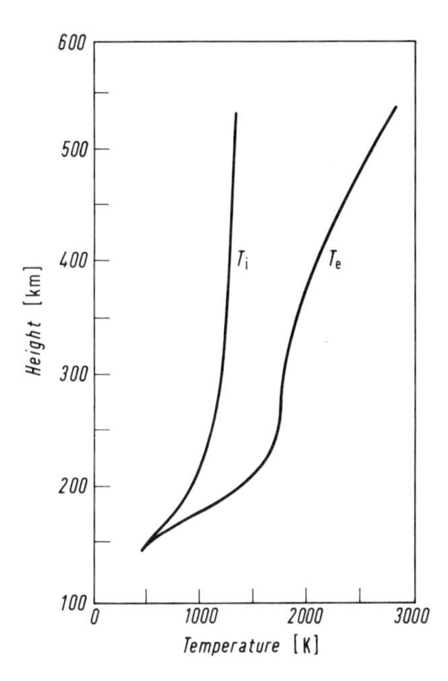

Fig. 5. The vertical distributions of electron and ion temperatures determined by incoherent-scatter measurements at 09.32 EST on 3 July 1970, at Millstone Hill, Massachusetts, after EVANS[1]

released with excess energy during the photoionization process. The ion temperature is expected to increase with height from that of the neutral gas near 200 to 300 km to that of the electrons at great heights. It is well known that a knowledge of both electron and ion temperatures is required to evaluate plasma transport effects. In addition, the importance of reactant temperatures has attracted increasing attention in considerations of the electron and ion chemistry of the E and F layers.

The ion composition of the E and F layers derived from mass-spectrometer measurements during daytime conditions near the minimum of solar activity is illustrated in Fig. 6, after JOHNSON[2]. Attention is drawn to the major importance of O_2^+ and NO^+ ions at E-region heights, the predominance of O^+ above 160 km and, finally, the change from O^+ to H^+ and He^+ ions at the greatest heights. These results are reasonably representative of the conditions stated but marked variations are often observed. Of particular interest in the present study is the increased concentration of NO^+ relative to O_2^+ at E-region heights during conditions of increased solar activity and during nighttime.

The extension of rocket-borne measurements of ion composition to lower heights has shown that NO^+ and O_2^+ remain the major ions down to about 82 km in daytime and 86 km at nighttime. In addition, a broad layer of metal ions generally occurs near 93 km and more variable layers occur at greater heights, in association with sporadic-E layers. However, the most striking

---

[2] Johnson, C.Y. (1966): J. Geophys. Res. *71*, 330

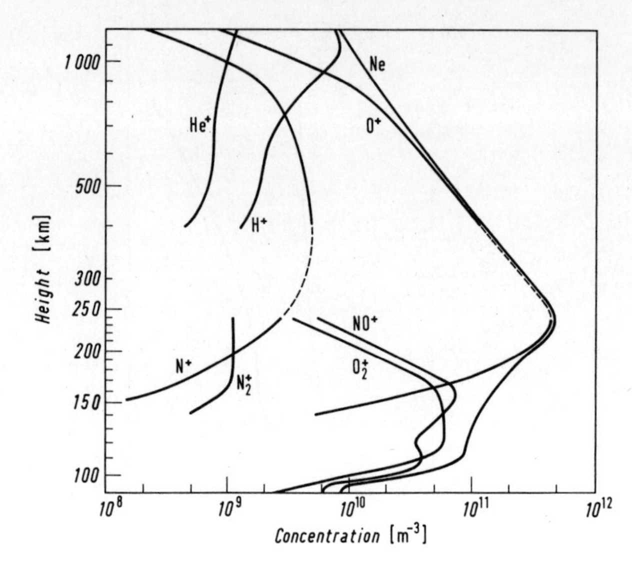

Fig. 6. The positive-ion composition and vertical distribution of electron concentration in the E and F regions (N_e) representative of solar minimum daytime conditions, as derived from mass-spectrometer and radio-wave measurements, after JOHNSON[2]

feature of these measurements at lower heights is that water cluster ions, $H^+ \cdot (H_2O)_n$, dominate the ion composition below about 82 km in daytime and below about 86 km at nighttime over a wide range of latitudes and geophysical conditions. The presence of the lower metal-ion layer and of the water cluster ions is illustrated in Fig. 7 which reproduces the results of measurements carried out by NARCISI et al.[3] during daytime. As will be explained in Section 19α, there is still some doubt whether the lower-order cluster ions $H^+ \cdot (H_2O)_2$ and H_3O^+ are representative of ambient conditions, because of possible fragmentation of higher-order weakly-bound ions caused by the supersonic velocity of the rocket or through collisions during sampling. However, it is not disputed that water cluster ions are a common feature of the middle and lower D layer during normal conditions.

It is well known that the relatively large ambient pressures encountered at D-region heights permit the formation of negative ions by the three-body attachment of electrons to neutral atoms and molecules. However, the information available on negative-ion composition, derived from two sets of mass-spectrometer measurements, has shown a serious discrepancy, as illustrated in Fig. 8. This shows measurements carried out at a high-latitude site by ARNOLD et al.[4] during nighttime under weak auroral conditions and at a mid-latitude site by NARCISI et al.[3] near totality of a solar eclipse. The results of ARNOLD et al. indicate that CO_3^- was the dominant ion present and that significant

[3] Narcisi, R.S., Bailey, A.D., Wlodyka, L.E., Philbrick, C.R. (1972): J. Atmos. Terr. Phys. *34*, 647

[4] Arnold, F., Kissel, J., Krankowsky, D., Wieder, H., Zahringer, J. (1971): J. Atmos. Terr. Phys. *33*, 1169

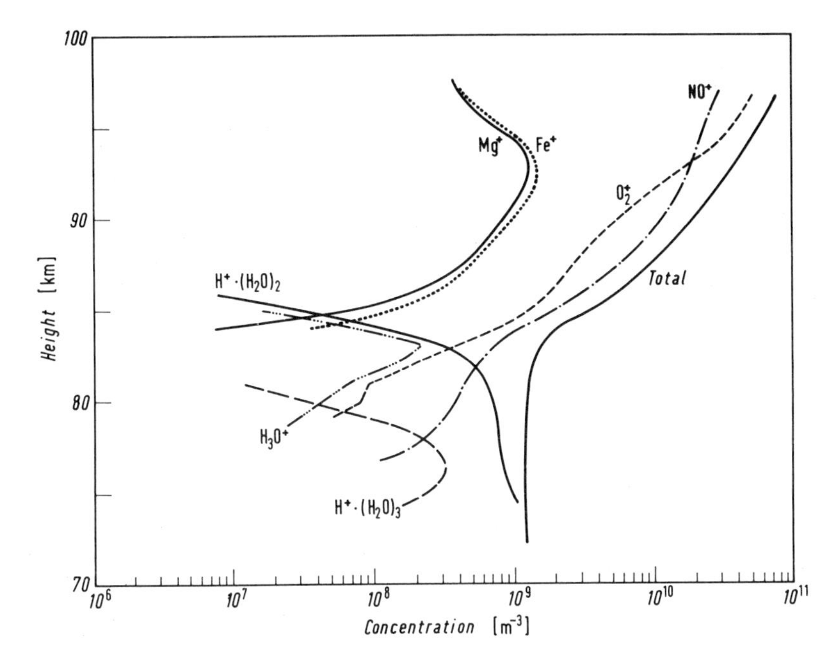

Fig. 7. The positive-ion composition of the D region derived from mass-spectrometer measurements at a solar zenith angle of 20° during November 1966 at Cassino, Brazil, after NARCISI et al.[3]

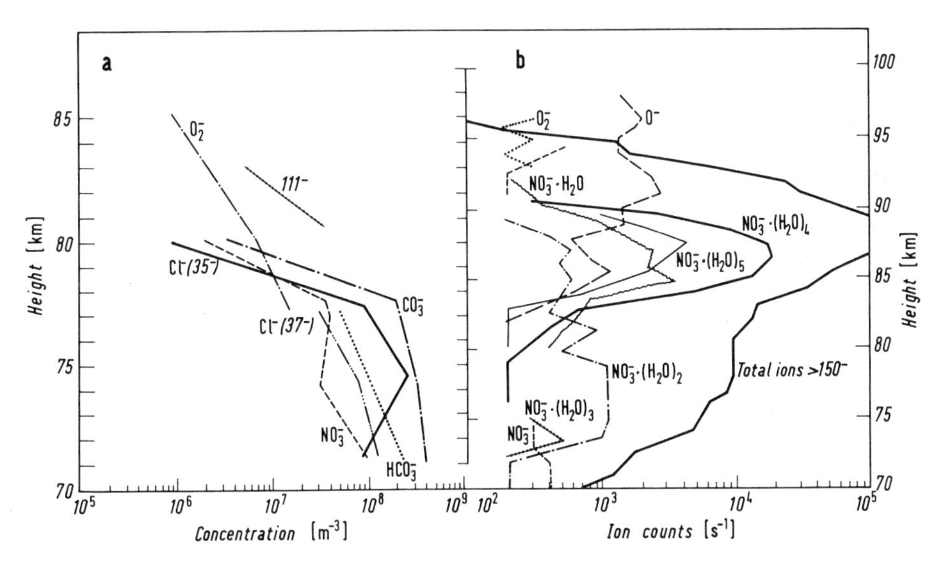

Fig. 8a, b. The negative-ion composition of the D region as derived from mass-spectrometer measurements (**a**) during rocket ascent and descent at 23.58 LT on 23 March 1970 at Andøya, Norway, by ARNOLD et al.[4], and (**b**) during rocket ascent near totality of a solar eclipse at 13.39 LT on 7 March 1970 at Wallops Island, Virginia, USA, by NARCISI et al.[3]

concentrations of HCO_3^- and NO_3^- were also present between about 70 and 80 km; in addition substantial concentrations of Cl^- ions, of isotopic masses 35^- and 37^-, were observed. The measurements of NARCISI et al. were dominated by cluster ions of masses consistent with $NO_3^- \cdot (H_2O)_n$ and possible admixtures of $CO_3^- \cdot (H_2O)_n$, where $n=0$ to 5; in addition a layer of ions of masses in excess of 150^- was observed near 88 km. It will also be seen in Sect. 28 that neither of these sets of observations is consistent with what is expected from laboratory measurements of negative-ion reactions.

B. Interactions of solar radiations with atmospheric gases

4. Absorption of radiations

α) Introduction. In the consideration of the passage of solar ultra-violet and X-radiation through the upper atmosphere, scattering and emission can be neglected and the equation of radiative transfer assumes a simple form. The change in intensity from I to $I-\Delta I$ due to absorption by a particular constituent when the radiation traverses a distance Δs is then given by:

$$\Delta I = -I \sigma_T N \Delta s, \tag{4.1}$$

where σ_T represents the total absorption cross-section and N the concentration of the constituent.

Integration gives

$$I = I_0 \exp\left(-\int_{\infty}^{s} \sigma_T N \, ds\right), \tag{4.2}$$

where I_0 represents the intensity when $s=\infty$, i.e. at the top of the absorbing atmosphere.

The quantity $\int_{\infty}^{s} \sigma_T N \, ds$ is customarily referred to as the optical depth.

In many applications, the absorption coefficient k is adopted, this being related to the cross-section by:

$$k = \sigma_T N_L, \tag{4.3}$$

where N_L is Loschmidt's number ($2.69 \cdot 10^{25}\ m^{-3}$).

Although resonance transitions giving rise to sharp atomic absorption lines or molecular bands do occur, processes involving photons sufficiently energetic to cause dissociation or ionization play the major roles in the interactions of solar radiations with the Earth's upper atmosphere. The absorption cross-section generally increases with increase in photon energy to a maximum value near the threshold of the particular process involved, and then decreases slowly as the photon energy increases further.

In general, measurements of partial absorption cross-sections have not been made. Instead, the experiments provide the ratios of these quantities to the total cross-section. In particular, the measurements of photoionization cross-section σ_I usually involve the determination of the photoionization yield η, the number of ions produced for each photon absorbed, given by:

$$\sigma_I = \eta \sigma_T. \tag{4.4}$$

In considering the penetration of solar radiations through the atmosphere, the absorption by all constituents needs to be included. Allowance for the zenith angle of the sun is straightforward except for very large values when account has to be taken of the curvature of the Earth's surface.

The cross-sections needed for investigations of photodissociation and photoionization rates are of two classes. Those for molecular oxygen and nitrogen determine the intensities of solar radiations that are important at any height. The second class refers to minor gaseous constituents which are optically thin down to the lowest height of interest. It is convenient to consider the absorption cross-section data for these two types of constituents in the wavelength region of interest i.e. from about 350 nm downwards [7].

β) Absorption cross-sections for different spectral regions

(i) *Above 175 nm*

Molecular oxygen O_2. For the atmospheric region of interest, this constituent imposes the greatest limit on penetration, the effects of molecular nitrogen being negligible. The absorption features consist of the weak Herzberg ($A^3\Sigma_u^+ - X^3\Sigma_g^-$) continuum which extends from 185 nm to 242 nm, and the absorption bands of the Schumann-Runge ($B^3\Sigma_u^- - X^3\Sigma_g^-$) system between 175 nm and 205 nm (the corresponding continuum being found below 175 nm).

The cross-section data obtained for the Herzberg continuum by DITCHBURN and YOUNG[1] vary from about $1.3 \cdot 10^{-27}\,m^2$ at 200 nm to $10^{-28}\,m^2$ near the dissociation limit.

Measurements of absorption cross-sections in the region of the Schumann-Runge band system carried out over a period of about 50 years have shown large variations. Recent high-resolution measurements by ACKERMAN and BIAUME[2] and HUDSON and MAHLE[3] have shown large linewidths, indicating that predissociation occurs. This is illustrated in the potential energy diagram of Fig. 9. It can be seen that the curve for the unstable $^3\Pi_u$ state crosses that of the $B\,^3\Sigma_u^-$ state, reportedly in the region of vibrational quantum number $v' = 3$ or 4. It is, therefore, possible for a radiationless transition to occur, so that the molecule moves from the stable part of the $^3\Sigma_u^-$ state to the unstable $^3\Pi_u$ condition and is dissociated.

[1] Ditchburn, R.W., Young, P.A. (1962): J. Atmos. Terr. Phys. *24*, 127
[2] Ackerman, M., Biaume, F. (1970): J. Mol. Spectrosc. *35*, 73
[3] Hudson, R.D., Mahle, S.H. (1972): J. Geophys. Res. *77*, 2902

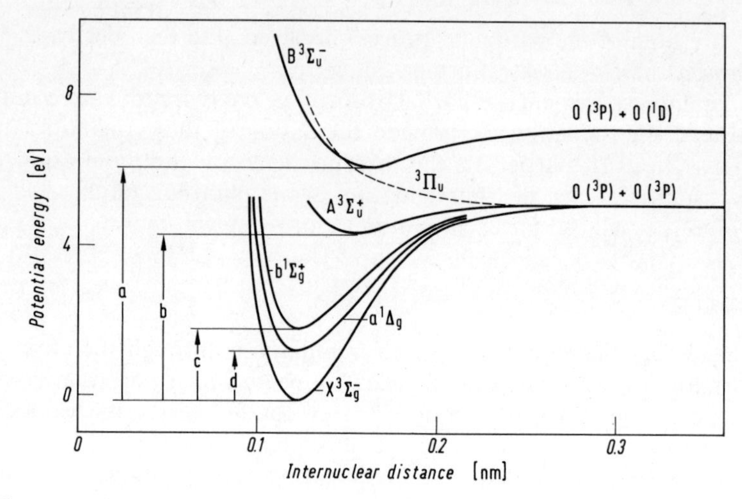

Fig. 9. The potential energy level diagram for molecular oxygen. The transitions shown correspond to the following band system: *a* Schumann-Runge; *b* Herzberg; *c* Atmospheric; *d* Infra-red Atmospheric

Detailed theoretical calculations for the Schumann-Runge bands by ACKERMAN et al.[4] have demonstrated the marked changes of cross-section which occur with temperature. With high-resolution values, variations by a factor of two can occur for a temperature change of 100 K but the effects are still noticeable for values integrated over large wavelength intervals, as illustrated in the results presented by KOCKARTS[5] shown in Table 1; account is also taken of the Herzberg continuum.

Table 1. Average cross-sections for molecular oxygen (after KOCKARTS[5])

Wavelength range/nm	Cross-section/m^2	
	$T=300$ K	$T=160$ K
175.44–176.99	$1.28\cdot10^{-23}$	$1.57\cdot10^{-23}$
176.99–178.57	1.18	1.18
178.57–180.18	$7.37\cdot10^{-24}$	$6.06\cdot10^{-24}$
180.18–181.82	4.77	5.21
181.82–183.49	3.16	2.94
183.49–185.18	1.61	1.33
185.18–186.92	$8.74\cdot10^{-25}$	$7.25\cdot10^{-25}$
186.92–188.68	4.19	3.40
188.68–190.48	1.90	1.37
190.48–192.31	$9.48\cdot10^{-26}$	$4.84\cdot10^{-26}$
192.31–194.18	6.24	5.72
194.18–196.08	2.15	1.87
196.08–198.02	$7.56\cdot10^{-27}$	$5.42\cdot10^{-27}$
198.02–200.00	3.06	1.77
200.00–202.02	1.94	1.49

[4] Ackerman, M., Biaume, F., Kockarts, G. (1970): Planet. Space Sci. *18*, 1639

[5] Kockarts, G. (1971) in: Mesospheric models and related experiments. Fiocco, G. (ed), p. 160. Dordrecht: Reidel

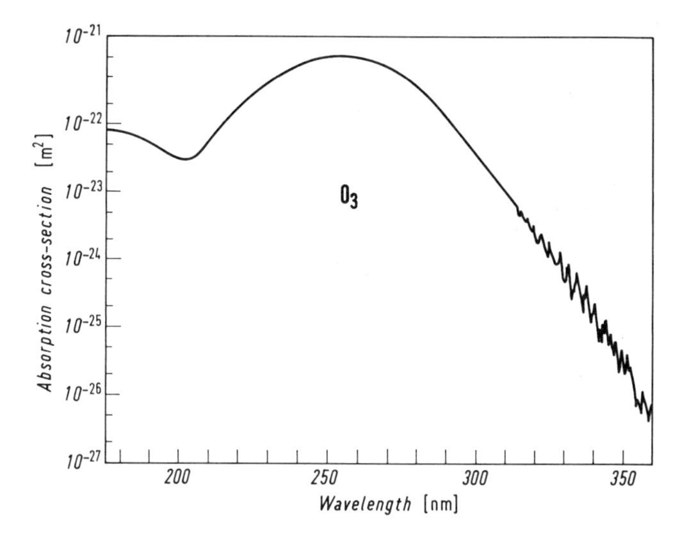

Fig. 10. The variation of the absorption cross-section of ozone with wavelength in the Hartley and Huggins bands, as presented by ACKERMAN[6]

For detailed calculations of absorption in the thermosphere and mesosphere, temperature-dependent cross-sections should be adopted.

Ozone O_3. The absorption of solar radiation by this minor constituent begins in the infra-red region, the cross-section increasing with decreasing wavelength to a maximum value of about $5 \cdot 10^{-25}\,\mathrm{m}^2$ near 600 nm in the Chappuis bands. Of greater importance for atmospheric considerations is the region below 350 nm, consisting of the Huggins bands down to 300 nm and the Hartley bands between 300 nm and 200 nm. Cross-section measurements for the wavelength region 200 nm to 350 nm have been carried out by several groups, but shorter wavelengths have attracted less attention. A summary of the data available for the range 175 nm to 350 nm, as given by ACKERMAN[6], is shown in Fig. 10.

Nitric Oxide NO. Measurements of the absorption spectrum of nitric oxide in the 175 to 235 nm spectral region by THOMPSON et al.[7] and MARMO[8] have shown a strong band structure which is strongly bandwidth dependent. The results of THOMPSON et al. were obtained at 6 mm ($800\,\mathrm{Nm}^{-2}$) pressure to avoid effects of dimerization and peak values of about $2.7 \cdot 10^{-23}\,\mathrm{m}^2$ were obtained near 190 nm.

Metallic atoms. The absorptions of solar radiations of wavelengths greater than 175 nm by atoms of metals such as sodium, potassium and calcium represent the only important photoionization processes for this spectral region,

---

[6] Ackerman, M. (1971) in: Mesospheric models and related experiments. Fiocco, G. (ed), p. 149. Dordrecht: Reidel

[7] Thompson, B.A., Harteck, P., Reeves, R.R. (1963): J. Geophys. Res. *68*, 6431

[8] Marmo, M.M. (1953): J. Opt. Soc. Am. *43*, 1186

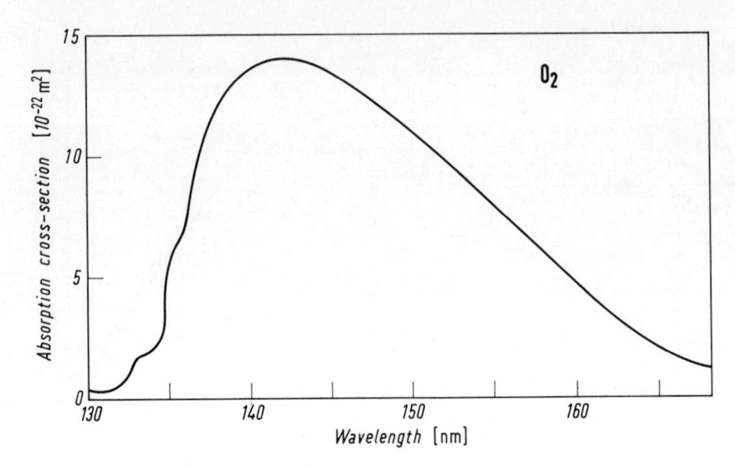

Fig. 11. The variation of the absorption cross-section for molecular oxygen in the Schumann-Runge continuum region, after METZER and COOK[10]

the threshold for these atoms being 241.2 nm (5.1 eV), 285.6 nm (4.3 eV) and 202.8 nm (6.1 eV) for Na, K and Ca, respectively[9]. Although photoionization cross-sections for radiations greater than 200 nm are less than $4 \cdot 10^{-23}\,m^2$, the ionization coefficients can be significant at certain times of day.

(ii) *From 175 nm to 100 nm*

Molecular oxygen O_2. The absorption spectrum in the region 170 nm to 130 nm has been studied extensively and consists of the strong Schumann-Runge continuum and three dissociation continua near 135.2 nm, 133.2 nm and 129.3 nm. The results obtained by METZER and COOK[10] are shown in Fig. 11. At still shorter wavelengths, between 130 nm and 102.5 nm, the absorption spectrum consists of a complexity of bands with the possibility of a weak underlying continuum. The absorption cross-section of the molecule in this region oscillates between values of about $10^{-24}\,m^2$ to greater than $10^{-21}\,m^2$. One of the windows corresponds to Lyman-α, at 121.6 nm, and is of particular interest since the solar line has an energy of several $mW\,m^{-2}$ and is important for certain photodissociation and photoionization processes. Near the limit of this spectral region is the ionization threshold of the molecule at 102.7 nm (12.1 eV).

Molecular nitrogen N_2. The absorption in this spectral region arises chiefly from the Lyman-Birge-Hopfield ($a\,^1\Pi_g - X\,^1\Sigma_g^+$) bands, Fig. 12. These are forbidden transitions and the absorption is very weak, WATANABE[11] reporting maximum values of $7 \cdot 10^{-25}\,m^2$ for the bands and upper limits of $3 \cdot 10^{-26}\,m^2$ between the bands.

[9] Swider, W. (1969): Planet. Space Sci. *17*, 1233
[10] Metzer, P.H., Cook, G.R. (1964): J. Quant. Spectrosc. Radiat. Transfer *4*, 107
[11] Watanabe, K. (1958): Adv. Geophys. *5*, 153

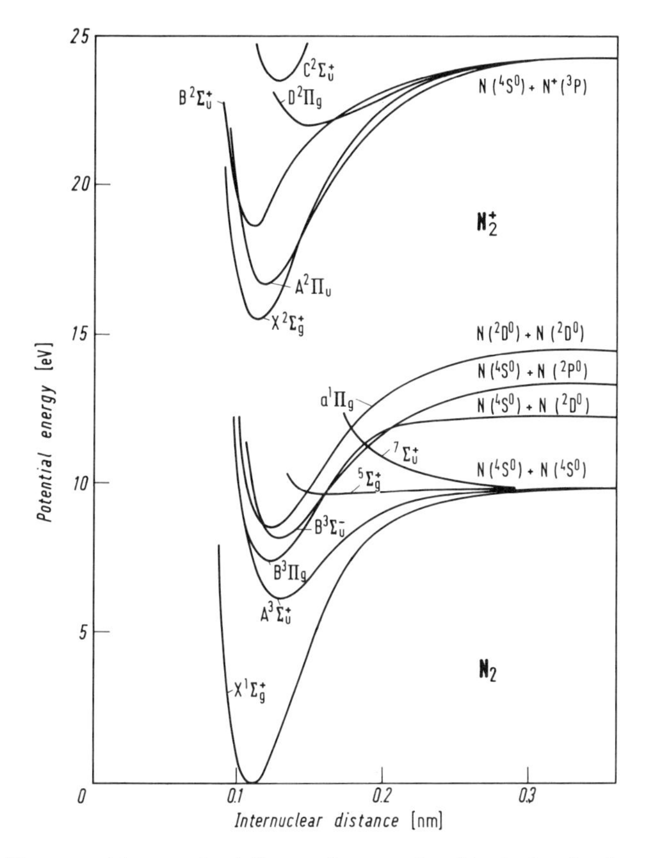

Fig. 12. The potential energy level diagram for neutral and ionized molecular nitrogen

Nitric oxide NO. The absorption by this molecule is of interest since the first ionization potential is 134.3 nm (9.3 eV) and, therefore, radiations in the spectral windows at shorter wavelengths, particularly Lyman-α, are capable of producing ionization. The absorption spectrum of the molecule is very complex, the ionization continuum being superposed by a number of diffuse bands. The ionization cross-section increases with decreasing wavelength to a value of about $2 \cdot 10^{-22}\,m^2$ near the wavelength of Lyman-α.

Water vapour H_2O. The absorption spectrum in this spectral region consists of a continuous absorption down to 145 nm and diffuse bands at shorter wavelengths. The measurements of WATANABE and ZELIKOFF[12] are shown in Fig. 13.

(iii) *From 100 nm to 10 nm*

At heights where the main ionization processes occur, solar radiations encounter an atmosphere composed primarily of molecular and atomic ni-

[12] Watanabe, K., Zelikoff, M. (1953): J. Opt. Soc. Am. *43*, 753

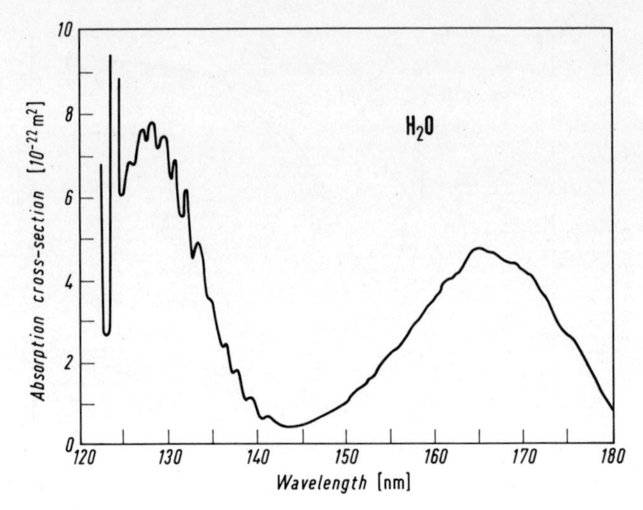

Fig. 13. The variation of the absorption cross-section of water vapour with wavelength, after WATANABE and ZELIKOFF[12]

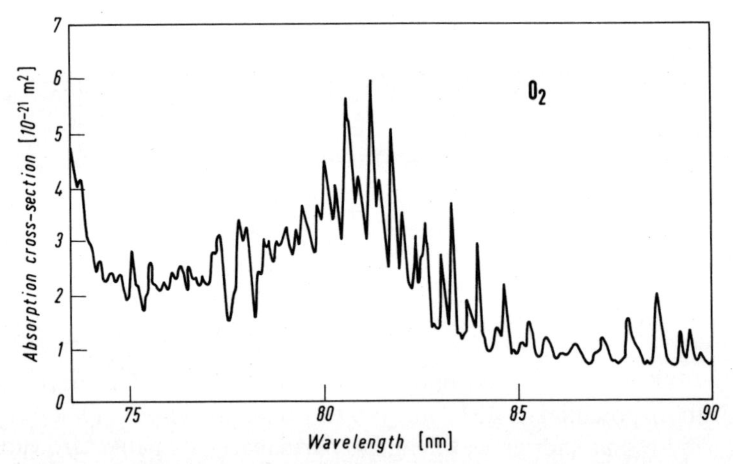

Fig. 14. The variation of the total absorption cross-section for molecular oxygen with wavelength, as presented by MARR[13]

trogen and oxygen. The photoionization of molecules such as ozone, water vapour and methane can be neglected since no radiations capable of ionizing them are available at heights where their concentrations are significant.

Molecular oxygen O_2. It is believed that for this region preionization, analogous to predissociation described above, to the state corresponding to the first ionization potential is important. As a result, diffuse bands occur in the absorption spectrum. A complex system of bands assigned to a Rydberg series has been observed down to 60 nm, as illustrated in the total absorption cross-section data presented by MARR[13] and shown in Fig. 14. Measurements of the

[13] Marr, G.V. (1965): Proc. R. Soc. London, Ser. A *288*, 531

photoionization yield also show considerable structure, the values varying from about 0.6 to 0.3 between 90 nm and 80 nm and increasing to about 0.7 near 70 nm.

Values of total absorption cross-section below 60 nm show a general decrease with decreasing wavelength, from about $2.8 \cdot 10^{-21}\,m^2$ at 60 nm to $1.0 \cdot 10^{-21}\,m^2$ near 20 nm. The results also indicate that the photoionization efficiency is 100% below 60 nm, and the values of absorption cross-section are therefore representative of photoionization also.

A variety of ionization processes are possible, as shown in Table 2.

Table 2. Ionization thesholds of molecular oxygen

Wavelength (nm)	Energy (eV)	O_2^+ product ion state
102.7	12.1	$X\,^2\Pi_g$
77.0	16.1	$a\,^4\Pi_u$
73.7	16.8	$A\,^2\Pi_u$
68.2	18.2	$b\,^4\Sigma_g^-$
61.4	20.3	$c\,^2\Sigma_g^-$
50.5	24.6	$c\,^4\Sigma_u^-$

Little information is available about the relative importance of the different ion states in the upper atmosphere.

Observations of DIBELER and WALKER[14] also showed that dissociative ionization of molecular oxygen can occur for wavelengths shorter than 71.9 nm (17.2 eV), and subsequent measurements by COMES et al.[15] have yielded values of cross-sections of about $3 \cdot 10^{-22}\,m^2$ between 61 nm and 65 nm. DOOLITTLE et al.[16] have also shown that electronically, excited atomic oxygen ions are produced with a wavelength of 58.4 nm.

Molecular nitrogen N_2. The absorption spectrum shows a complex system of bands which have been ordered into a Rydberg series and an ionization continuum below the first ionization limit at 79.6 nm (15.6 eV). In view of the particularly marked structure in the spectral region down to this limit, high-resolution measurements are required for quantitative studies of absorption by this molecule. Although measurements using line emission sources are available for the Carbon III line at 97.7 nm, it is customary to adopt mean values for shorter wavelengths down to the first ionization limit.

Measurements by various workers of total absorption cross-section at shorter wavelengths, down to about 60 nm, do not show good agreement, the mean value between 65 nm and 70 nm being about $3 \cdot 10^{-21}\,m^2$. Improved agreement is shown between the measurements at still shorter wavelengths, the values varying continuously from about $2.5 \cdot 10^{-21}\,m^2$ near 54 nm to about

[14] Dibeler, V.H., Walker, J.A. (1967): J. Opt. Soc. Am. *57*, 1007

[15] Comes, F.J., Speier, F., Elzer, A. (1968): Z. Naturforsch. Teil A *23*, 125

[16] Doolittle, P.H., Schoen, R.I., Schubert, K.E. (1968): J. Chem. Phys. *49*, 5108

$1.0 \cdot 10^{-21}\,m^2$ near 20 nm. Measurements of the ionization yield indicate values of about 0.95 at 69 nm and 1.0 for the continuum region from 65 nm to 1 nm.

As with molecular oxygen, various ionization states of molecular nitrogen ions are possible, Fig. 12, as shown in Table 3.

Table 3. Ionization thresholds of molecular nitrogen

Wavelength (nm)	Energy (eV)	N_2^+ product ion state
79.6	15.6	$X\,^2\Sigma_g^+$
74.3	16.7	$A\,^2\Pi_u$
66.1	18.8	$B\,^2\Sigma_u^+$
56.4	22.0	$D\,^2\Pi_g$
52.6	23.6	$C\,^2\Sigma_u^+$

Alternatively, dissociative photoionization can occur and measurements near 51 nm have indicated a cross-section about 0.01 of the total photoionization value[17].

Atomic oxygen O. The wavelength limit for photoionization of ground-state atomic oxygen is 91 nm, corresponding to an ionization potential of 13.6 eV; the ion produced is $O^+(^4S)$. The cross-section data for absorption of, and photoionization by, shorter wavelengths are largely based on theoretical predictions, although some observational information is available. The most recent theoretical results of HENRY[18] show good agreement with the experimental values of COMES et al.[15], each being about $3 \cdot 10^{-22}\,m^2$ from the ionization limit down to 77 nm; the results found experimentally by CAIRNS and SAMSON[19] were slightly larger, increasing from about $5 \cdot 10^{-22}\,m^2$ to $8 \cdot 10^{-22}\,m^2$ over the same wavelength range. At still shorter wavelengths, in the continuum below 66.5 nm, the cross-sections approach the value $10^{-21}\,m^2$. For the region below about 40 nm, experimental values can be estimated from the measurements for molecular oxygen by SAMSON and CAIRNS[20]; the atomic cross-sections decrease from about $9.0 \cdot 10^{-22}\,m^2$ at 40 nm to $4.5 \cdot 10^{-22}\,m^2$ at 21 nm.

The production of excited states of the atomic ion during photoionization has attracted attention in the last decade. Of particular interest are the metastable ions ^2D and ^2P corresponding to threshold wavelengths near 73.2 nm (16.9 eV) and 66.5 nm (18.6 eV), respectively. From theoretical values of the photoionization cross-sections to the ^2D and ^2P states relative to those for the ground state[21], and information on the transitions to the ground state, it has been estimated that ^2D ions constitute more than 50% of the total atomic oxygen ions produced during photoionization.

[17] Weissler, G.L., Samson, J.A.R., Ogawa, M., Cook, G.R. (1959): J. Opt. Soc. Am. *49*, 338
[18] Henry, R.J.W. (1968): J. Chem. Phys. *48*, 3635
[19] Cairns, R.B., Samson, J.A.R. (1965): Phys. Rev. A *139*, 1403
[20] Samson, J.A.R., Cairns, R.B. (1965): J. Opt. Soc. Am. *55*, 1035
[21] Henry, R.J.W. (1967): Planet. Space Sci. *15*, 1747

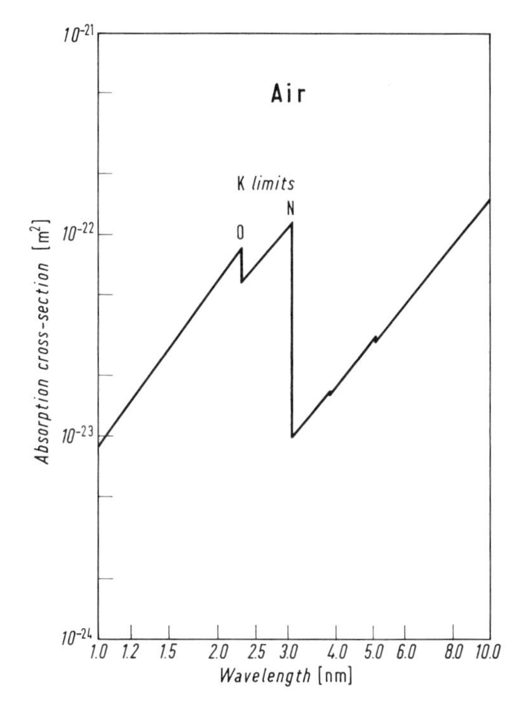

Fig. 15. The absorption cross-section of air for X-rays of wavelength 1.0 to 10 nm, as presented by BANKS and KOCKARTS [8]

Atomic nitrogen N. Both the theoretical calculations by HENRY[18] and the experimental values of COMES and ELZER[22] show values of about $1.0 \cdot 10^{-21}\,\mathrm{m}^2$ for the atomic nitrogen cross-section near the ionization limit of the ground state at 85.2 nm (14.5 eV). At shorter wavelengths, the experimental values are rather smaller than those derived theoretically, but the additional values deduced from the SAMSON and CAIRNS[20] data for molecular nitrogen for wavelengths below 55 nm show better agreement with theory. As for atomic oxygen, the possibility of various ionization states exists, the ionization limits near 61 nm (20.3 eV) and 36.7 nm (33.8 eV) corresponding to ions $N^+(^5S)$ and $N^+(^3S)$, respectively. It appears, however, that photoionization leads predominantly to the ground-state ion $N^+(^3P)$.

(iv) *Shorter than 10 nm*

The absorption of radiations with wavelengths less than about 20 nm (62.0 eV) involves inner-shell electrons and is, consequently, atomic in nature. The atmospheric absorption is, therefore, not dependent on the molecular form of the atoms. The variation of absorption cross-section for air in the wavelength range 0.1–10 nm is represented in Fig. 15, after BANKS and KOCKARTS [8]. For this range ionization by the photoelectric effect predominates and it is only for

[22] Comes, F.J., Elzer, A. (1968): Z. Naturforsch. Teil A *23*, 133

wavelengths below about 0.05 nm that Compton scattering becomes important. The K-shell ionization limits for atomic oxygen and nitrogen are shown at 2.3 nm (539 eV) and 3.1 nm (400 eV), respectively, the cross-sections being about $1.0 \cdot 10^{-22}\,m^2$.

For considerations of atmospheric chemistry, absorption of solar radiations giving rise to ionization or dissociation is most significant. For convenience, the first energy thresholds for ionization and dissociation are presented for the most important constituents in Table 4.

Table 4. Ionization and dissociation thresholds

Constituent	Ionization/eV	Dissociation/eV
O	13.6	
N	14.5	
H	13.6	
Na	5.1	
K	4.3	
Ca	6.1	
O_2	12.1	5.1
N_2	15.6	9.9
O_3	12.8	1.1
NO	9.3	6.5
H_2O	12.6	5.0
$O_2(a\,^1\Delta_g)$	11.1	5.1

γ) Penetration of solar radiations into the atmosphere. The available information on absorption cross-sections combined with data on atmospheric structure and composition enables an examination to be made of the atmospheric penetration of solar radiations over the entire wavelength range of interest. The results can be conveniently summarised in terms of the heights at which the solar fluxes incident vertically are reduced by $\exp(-1)$ from the free space values, i.e. the heights of unit optical depth. Fig. 16 shows the variation of these heights for wavelengths up to 310 nm. Also shown marked are the wavelengths corresponding to the ionization thresholds of ground-state atomic oxygen, molecular nitrogen and nitric oxide, and of the ground-state and first electronically, excited state of molecular oxygen, and also to the dissociation thresholds of nitric oxide, water vapour and molecular oxygen.

For wavelengths down to about 210 nm the atmospheric penetration is limited by absorption in ozone; for the 200 to 100 nm region molecular oxygen is the primary absorbing gas. For radiations in the region of 140 to 150 nm, near the peak of the Schumann-Runge continuum, the maximum absorption occurs at about 110 km. Deep narrow windows are shown on the shorter wavelength side, one coinciding almost exactly with the wavelength of Lyman-α, permitting this radiation to penetrate to heights below 80 km. For ultraviolet wavelengths shorter than 100 nm, absorption by all the major atmospheric constituents occurs and the height of maximum absorption is greater,

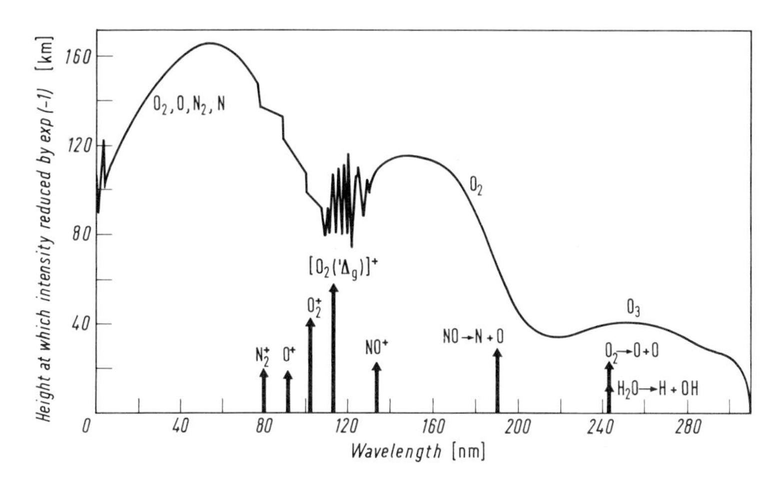

Fig. 16. The variation with wavelength of the height at which solar radiations incident vertically are reduced by exp(−1). The wavelengths corresponding to certain ionization and dissociation thresholds are indicated

being as high as 170 km. The softer X-rays, of wavelengths between 10 nm and 1 nm, again penetrate down to heights of about 120 to 100 km, whereas still shorter wavelength X-rays are absorbed below 100 km.

5. Relevant solar-flux intensities

α) Continuum and monochromatic radiation. The effects of heating, ionization and dissociation of the upper atmospheric constituents under normal conditions can be largely attributed to the absorption of solar radiations. The spectral region of primary interest in the present study is that below about 350 nm for which information on intensities has been derived from rocket, satellite and balloon observations [9]. The solar radiation is represented by a continuum corresponding to a blackbody of temperature 5,500 to 6,000 K down to about 300 nm. For shorter wavelengths the effective temperature drops to a value of 5,000 K near 240 nm and remains at about 4,750 K over the range 200 to 150 nm. For still smaller wavelengths the continuum is characterised by a higher temperature but the radiation from monochromatic emission lines becomes relatively more important and dominates the solar spectrum below about 123 nm. More details may be found in the contributions by NIKOL'SKIJ in this volume and by SCHMIDTKE in volume 49/7.

In considering the solar-flux intensities for the region below 350 nm it is convenient to consider a division near 130 nm. Although certain metal atoms do have ionization thresholds at longer wavelengths, this value represents the upper limit for the photoionization of most atmospheric constituents. Furthermore, the major photodissociation processes occur at wavelengths above this value[0].

---

[0] Nikol'skij, G.M.: Contribution in this volume, p. 309

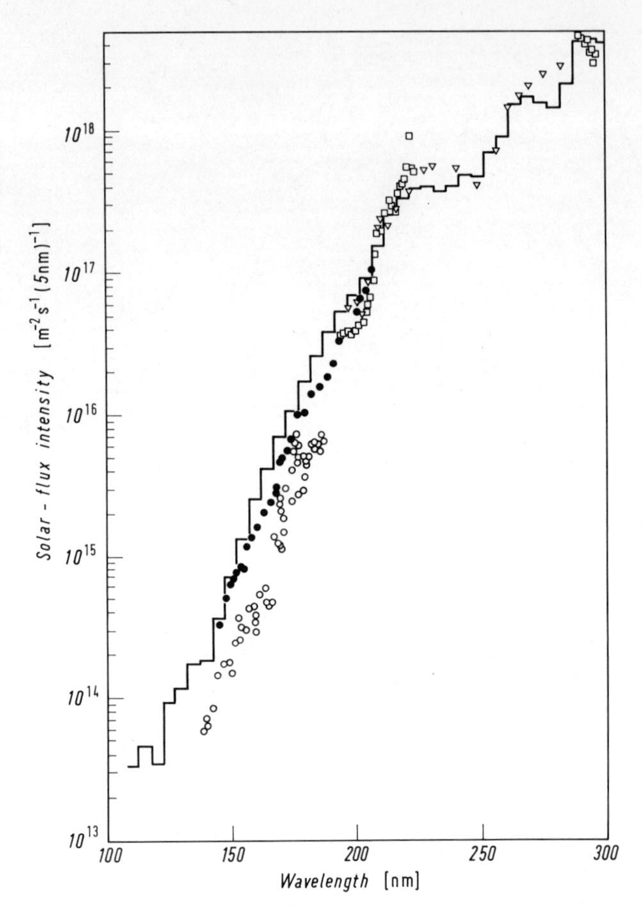

Fig. 17. Measurements of solar-flux intensities between 100 and 300 nm, from ACKERMAN[1]: ⎍ DETWILER et al.[2]; □ ACKERMAN et al.[3]; ▽ BONNET[4]; ○ PARKINSON and REEVES[5]; ● WIDING et al.[6]

β) Fluxes in the spectral range 300 nm to 130 nm. Until recently this part of the solar spectrum has been rather neglected in spite of its importance in the photodissociation of molecular oxygen, water and nitric oxide in the mesosphere and thermosphere, and in the thermal balance of these two regions of the atmosphere.

Figure 17 shows a summary from ACKERMAN[1] of most of the measurements made in this part of the spectral region prior to 1970[2-6]. Of particular interest

---

[1] Ackerman, M. (1971) in: Mesospheric models and related experiments. Fiocco, G. (ed), p. 149. Dordrecht: Reidel

[2] Detwiler, C.R., Garrett, D.L., Purcell, J.D., Tousey, R. (1961): Ann. Geophys. *17*, 263

[3] Ackerman, M., Frimont, D., Pastiels, R. (1968): Ciel Terre *84*, 408

[4] Bonnet, R.M. (1968): Space Res. 7, 458

[5] Parkinson, W.H., Reeves, E.M. (1969): Sol. Phys. *10*, 342

[6] Widing, K.G., Purcell, J.D., Sandlin, G.D. (1970): Sol. Phys. *12*, 52

is the difference in solar fluxes in the 150 to 200 nm region between the more recent measurements and the early measurements of DETWILER et al.[2]. Reviews by SIMON[7] and in [9] describe more recent observations and confirm the smaller values of fluxes in the 130 to 180 nm range. It can be shown that the adoption of the smaller absolute intensities has a significant influence on the photodissociation rate of molecular oxygen and, thereby, on the composition of the mesosphere and thermosphere, and also on the energy input into the lower thermosphere. A variation of intensity with solar activity has been reported from satellite-borne observations carried out during 1969 to 1972, the effect being most marked at the shorter wavelenghts, amounting to about 50% near 175 nm[8]. However, this has not been confirmed in more recent measurements[9].

γ) Fluxes in the spectral range 130 nm to 10 nm. Since the ionization thresholds for ground-state constituents correspond to wavelengths of 102.7 nm for molecular oxygen, 79.6 nm for molecular nitrogen, 91.0 nm for atomic oxygen and 134.3 nm for nitric oxide, it is seen that this part of the spectral region is responsible for the major photoionization processes [*10*]. For such processes important lines occur at 121.6 nm (Lyman-α), 102.6 nm (Lyman-β), and 97.7 nm (Carbon III). For still shorter wavelengths, between about 80 and 90 nm, the solar emission is characterised by the Lyman continuum in which the band structure complicates the calculation of atmospheric absorption and ionization. The lines of multi-ionized atoms become relatively more important at still shorter wavelengths.

The major difficulty in the estimation of absolute solar fluxes in this spectral region arises from the use of standards for laboratory calibration of rocket or satellite equipment and from the changes in performance of critical components between the times of calibration and observation in space[0]. The majority of results have been derived from rocket-borne measurements, although OSO 3 satellite also yielded data and the Atmospheric Explorer series of satellites has provided information on temporal changes. The integrated values of solar flux between 122.0 nm and 5.2 nm emitted from the whole solar disc, as derived from the flight of a rocket spectrometer by HEROUX et al.[10] on 23 August 1972, when the solar zenith angle was 27°, are summarised in Table 5; these values have been corrected for atmospheric absorption. The authors estimate that the errors for the intensities are ±30% in the wavelength range 125.0 to 15.0 nm and ±50% between 15.0 nm and 5.2 nm. It is to be noted that these integrated values include, and are often dominated by, the intensities of dominant lines. For example, in the range 102.7 to 122.0 nm the major contribution arises from the Lyman-α flux which amounts to $2.89 \cdot 10^{15}$ photons $m^{-2} s^{-1}$, and for the range 30.0 to 50.5 nm the line of ionized helium near 30.4 nm accounts for $7.2 \cdot 10^{13}$ photons $m^{-2} s^{-1}$. It is considered that the values of solar intensities are representative of a moderately quiet day since the Covington Index deduced from the 10.7 cm solar emission was 120.

[7] Simon, P.C. (1978): Planet. Space Sci. *26*, 355

[8] Heath, D.F. (1973): J. Geophys. Res. *78*, 2779

[9] Hinteregger, H.E. (1976): J. Atmos. Terr. Phys. *38*, 791

[10] Heroux, L., Cohen, M., Higgins, J.E. (1974): J. Geophys. Res. *79*, 5237

Table 5. Solar intensities between 122 and 5.2 nm (after HEROUX et al.[10])

Wavelength range (nm)	Intensity at top of atmosphere	
	$/10^{13}$ photons $m^{-2} s^{-1}$	$/mW m^{-2}$
122.0–102.7	297.5	4.87
102.7–91.0	9.62	0.20
91.0–80.0	8.00	0.18
80.0–63.0	<2.28	<0.059
63.0–50.4	5.65	0.20
50.4–30.0	14.65	0.836
30.0–20.0	9.96	0.784
20.0–10.0	5.15	0.546
10.0–5.2	1.68	0.466

A comparison of measurements during six rocket flights[11] has shown that the solar flux between 30 and 122 nm shows relatively little change with solar activity but the flux at wavelengths shorter than 30 nm increases appreciably with increasing activity.

δ) Fluxes in the spectral region below 10 nm. The chief characteristic of this spectral region, which represents an important source of ionization in the lower ionosphere, is the marked dependence of intensities on the level of solar activity[0]. The effect is most marked for X-rays of wavelengths shorter than 1 nm and becomes more pronounced with decrease in wavelength, as illustrated by the representative values of BANKS and KOCKARTS [8] given in Table 6, which also indicate the enhancements during solar flares.

Table 6. Dependence of intensities of short X-rays on solar activity (after BANKS and KOCKARTS [8])

Solar conditions	Intensity/mW m^{-2}		
	0.2 nm	0.4 nm	0.6 nm
Completely quiet	10^{-8}	10^{-7}	10^{-6}
Lightly distributed	10^{-6}	10^{-5}	10^{-4}
Disturbed	10^{-5}	10^{-4}	10^{-3}
Strong solar flares	10^{-3}	10^{-2}	10^{-1}

For the softer X-rays between about 1 and 10 nm rather smaller variations have been reported[12]. Thus for the region between 4 and 6 nm a change by a factor of 7 was observed between the epochs of solar minimum and maximum, the corresponding changes for wavelengths below 0.8 nm being by a factor of approximately 600.

[11] Heroux, L., Higgins, J.E. (1977): J. Geophys. Res. *82*, 3307
[12] Kreplin, R.W. (1961): Ann. Geophys. *17*, 151

The measurements of HEROUX et al.[10] presented in Table 5 extended down to a wavelength of 5.2 nm. The integrated flux below 10 nm amounted to about 0.5 mW m^{-2} but as already noted this could be subject to an error of about $\pm 50\%$. Within this uncertainty the value is reasonably consistent with the moderately quiet character of the day of observation.

6. Photodissociation rates

α) Molecular oxygen O_2. The dissociation of molecular oxygen can be conveniently considered in terms of the first and second dissociation limits:

$$O_2 + h\nu(\lambda < 243\,\text{nm}) = O(^3P) + O(^3P) - 5.11\,\text{eV} \tag{6.1}$$

$$O_2 + h\nu(\lambda < 175\,\text{nm}) = O(^3P) + O(^1D) - 7.08\,\text{eV}. \tag{6.2}$$

As indicated in Sect. 4β, the absorption cross-section for the Herzberg continuum is rather small and the photodissociation rate is almost constant for heights above about 50 km.

The predissociation which occurs for $v' = 3$ or 4 of the upper state of the Schumann-Runge bands (Sect. 4β) provides an additional source of $O(^3P)$ atoms in the Herzberg continuum region. Calculations of absorption cross-sections at very small intervals have been possible over the whole Schumann-Runge system (ACKERMAN et al.[1]), using the results of absorption cross-section measurements and the determination of the structure of certain Schumann-Runge bands. These absorption cross-sections have been used by NICOLET[2] to deduce the photodissociation coefficients for the Schumann-Runge bands in the mesosphere and lower thermosphere. The results derived for overhead sun using the atmospheric model given in Table 4.3d of BANKS and KOCKARTS [8] (NICOLET, pivate communication) are presented in Fig. 18 together with the contributions of Lyman-α, the Herzberg continuum and the Schumann-Runge continuum. This diagram shows that the Schumann-Runge continuum dominates the molecular oxygen dissociation in the thermosphere but is superseded by the Schumann-Runge bands in the middle mesosphere, and that the Herzberg continuum is the major cause of dissociation at still lower heights.

β) Ozone O_3. Although photodissociation of ozone, to give atomic and molecular oxygen, is energetically possible for wavelengths shorter than 1,180 nm (1.05 eV), the greater contribution for the height region of interest is made by ultra-violet wavelengths.

The photodissociation products of ozone play an important role in atmospheric chemical processes. The first electronically-excited state of molecular oxygen ($O_2(^1\Delta_g)$) is particularly important, the processes responsible for its production being as follows:

$$O_3 + h\nu(\lambda \leqq 611.2\,\text{nm}) \rightarrow O_2(^1\Delta_g) + O(^3P) - 2.03\,\text{eV} \tag{6.3}$$

[1] Ackerman, M., Biaume, F., Kockarts, G. (1970): Planet. Space Sci. *18*, 1639

[2] Nicolet, M. (1971) in: Mesospheric models and related experiments. Fiocco, G. (ed), p. 1. Dordrecht: Reidel

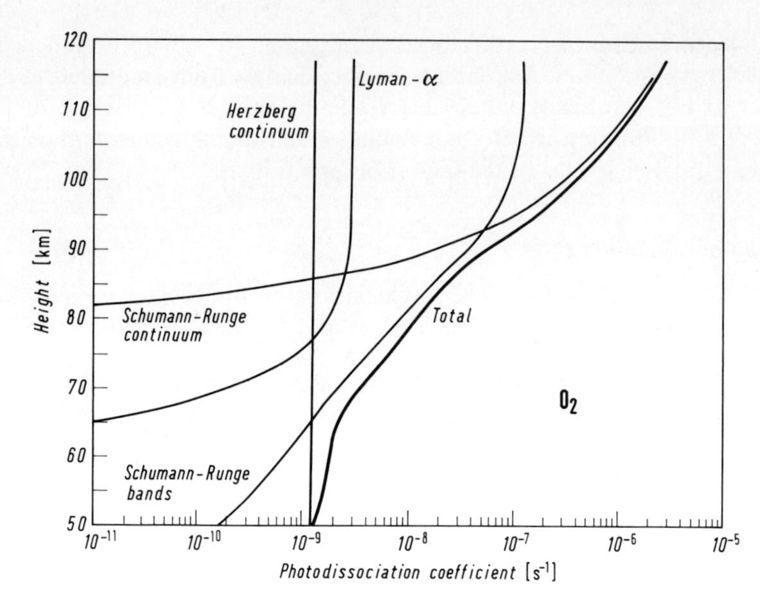

Fig. 18. The photodissociation coefficient of molecular oxygen in the mesosphere and lower thermosphere for overhead sun conditions, as computed by NICOLET[2] using the model given in Table 4.3d of [8]

$$O_3 + h\nu(\lambda \leqq 310.3\,\text{nm}) \rightarrow O_2(^1\Delta_g) + O(^1D) - 3.99\,\text{eV} \tag{6.4}$$

$$O_3 + h\nu(\lambda \leqq 199.4\,\text{nm}) \rightarrow O_2(^1\Delta_g) + O(^1S) - 6.22\,\text{eV}. \tag{6.5}$$

The respective values of solar photon-flux intensities at these wavelengths are approximately $3 \cdot 10^{19}$, $6 \cdot 10^{18}$ and $4 \cdot 10^{16}\,\text{m}^{-2}\,\text{s}^{-1}$ for 5 nm wavelength intervals, and these values and the corresponding absorption cross-sections imply that the region near 310 nm is probably the most effective. The photolysis of ozone in the Hartley bands is now recognised as the primary process for the formation of $O_2(^1\Delta_g)$ molecules which show a peak concentration near 50 km in daytime and make an important contribution to the near infra-red dayglow.

Calculation shows that the photodissociation coefficient corresponding to (6.4) is near $10^{-2}\,\text{s}^{-1}$ throughout the height region of interest, even for solar zenith angles approaching 90°. Because of the variation of ozone absorption cross-section with wavelength, this dissociation arises largely from absorption of radiation in excess of 230 nm. It follows, therefore, that the photodissociation rate for ozone in the Hartley bands is not affected by absorption of solar radiation by molecular oxygen and, in particular, by the complicated nature of absorption in the Schumann-Runge bands.

γ) Water vapour, H_2O, and other hydrogen-oxygen constituents. The photodissociation of water vapour is possible for all solar radiations of wavelengths less than 247 nm (5.02 eV). However, the major contributions in the upper mesosphere and lower thermosphere are from Lyman-α radiation and the

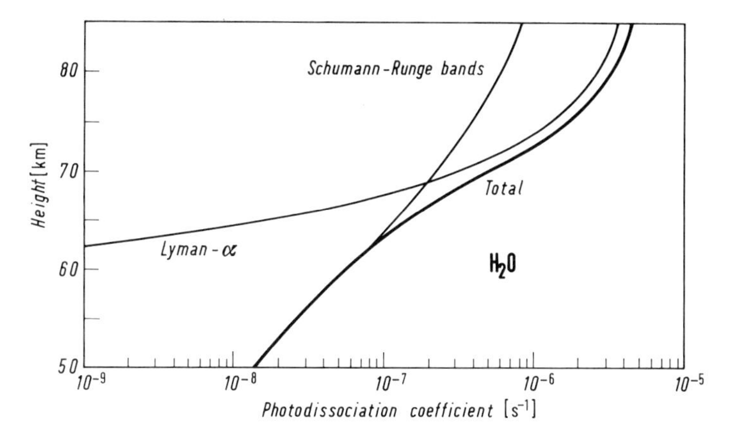

Fig. 19. The photodissociation coefficient of water vapour in the mesosphere for overhead sun conditions, as computed by NICOLET [2]

Schumann-Runge continuum, the Schumann-Runge bands playing the major role in the lower mesosphere. The principal dissociation mechanism is believed to be

$$H_2O + h\nu(\lambda < 242\,\text{nm}) \rightarrow OH(X\,^2\Pi) + H(^2S) - 5.12\,\text{eV}. \qquad (6.6)$$

In the case of Lyman-α the H atoms will be produced with high kinetic energy.

Alternative routes, involving the production of molecular hydrogen and atomic oxygen, are also possible:

$$H_2O + h\nu(\lambda < 247\,\text{nm}) \rightarrow H_2 + O - 5.02\,\text{eV}, \qquad (6.7)$$

where for Lyman-α the atom is in the $O(^1D)$ state.

Another possibility for Lyman-α is the production of the OH radical in the electronically-excited state, $OH(^2\Sigma^+)$, the energy yield being such that vibrational excitation corresponding to $v' \leqq 2$ is possible.

The photodissociation coefficients for Lyman-α and the Schumann-Runge bands as computed for overhead sun conditions by NICOLET [2], are shown in Fig. 19. In the calculations, account was taken of the fine structure of the molecular oxygen absorption cross-section in the Schumann-Runge band region.

The corresponding photodissociation of hydrogen peroxide yields two hydroxyl radicals:

$$H_2O_2 + h\nu(\lambda < 577\,\text{nm}) = 2\,OH - 2.15\,\text{eV}. \qquad (6.8)$$

The photodissociation coefficient is found to be about $10^{-4}\,s^{-1}$ throughout the mesosphere and thermosphere.

A third hydrogen-oxygen constituent which is of interest in the mesosphere is the hydroperoxyl radical, HO_2. For the dissociation of this it is customary to assume similar values of absorption coefficient, and therefore dissociation rates, as for hydrogen peroxide.

δ) Nitric oxide NO. The dissociation limit for nitric oxide corresponds to a wavelength of 190 nm and for this and shorter wavelengths the following process occurs:

$$NO(X^2\Pi) + h\nu(\lambda \leqq 190\,\text{nm}) \rightarrow N + O - 6.51\,\text{eV}. \tag{6.9}$$

In the actual dissociation of the molecule, predissociation has been observed in the γ, β, δ and ε bands, corresponding to transitions to the $A\,^2\Sigma^+$, $B\,^2\Pi$, $C\,^2\Pi$ and $D\,^2\Sigma^+$ states, respectively.

In considering the dissociation of the molecule in the mesosphere and stratosphere, it is to be noted that the absorption spectrum extends into the region of the Schumann-Runge bands of molecular oxygen. Hence, detailed calculations of nitric oxide photodissociation rates need to take account of the fine structure in the variation of the molecular oxygen absorption cross-section in this spectral region, including the effect of the temperature dependence of the cross-section. From such a treatment CIESLIK and NICOLET[3] have concluded that the major contribution to the dissociation of nitric oxide in the mesosphere and stratosphere is the predissociation of the δ(0–0) and δ(1–0) bands. The coefficients for overhead Sun conditions for 80 km and 60 km are $6.80 \cdot 10^{-6}\,\text{s}^{-1}$ and $4.44 \cdot 10^{-6}\,\text{s}^{-1}$ for the (0–0) band, and $6.82 \cdot 10^{-6}\,\text{s}^{-1}$ and $2.18 \cdot 10^{-6}\,\text{s}^{-1}$ for the (1–0) band.

7. Photoionization rates

α) General. As indicated in Sect. 5γ, solar radiations of wavelengths less than 79.6 nm (15.6 eV) are capable of ionizing the major atmospheric constituents. In addition, particular minor constituents can be ionized by longer wavelengths, the chief example being nitric oxide whose ionization by Lyman-α will be shown to provide a major contribution to the D region. For this part of the ionosphere, ionization by cosmic rays is also important below about 70 km, especially at middle and high latitudes. More generally, corpuscular radiation can provide an important but irregular source of ionization at middle and high latitudes at all ionospheric heights.

In considering photoionization it is convenient to work in terms of the photon flux φ, and it follows from relation (4.2) that:

$$\varphi(\lambda) = \varphi_0(\lambda) \exp\left(-\int_{\infty}^{s} \sigma_T(\lambda)\, N\, \mathrm{d}s\right), \tag{7.1}$$

where φ_0 represents the photon flux outside the atmosphere.

For solar zenith angles up to about 80° it is acceptable to consider a plane stratified atmosphere for which case the optical path ds and the height interval dh are related by $\mathrm{d}s = -\sec\chi\, \mathrm{d}h$, where χ represents the zenith angle.

[3] Cieslik, S., Nicolet, M. (1973): Aeronom. Acta A *107*, 1

The photoionization rate at a height h for a given constituent i is then given by:

$$Q_i(h,\lambda)=N_i(h)\,\sigma_{\mathrm{I}}^i(\lambda)\,\varphi_0(\lambda)\exp\left(-\int_h^\infty N_i(h)\,\sigma_{\mathrm{T}}^i(\lambda)\sec\chi\,\mathrm{d}h\right). \tag{7.2}$$

ELIAS[0] and later CHAPMAN [*11*] examined the rate of production of ionization Q as a function of height for an isothermal atmosphere made up of a single constituent distributed with height according to the relation $N(h)=N_0\exp(-h/H)$, where H is the scale height (Sect. 2). CHAPMAN showed that if z represents the height measured from the level of maximum production in units of H, then Q is given at all heights by:

$$Q(z)=Q_{\mathrm{m}}\exp[1-z-\exp(-z)], \tag{7.3}$$

where Q_{m} represents the maximum production rate and is given by:

$$Q_{\mathrm{m}}=\frac{\varphi_0\cos\chi}{eH}\,\frac{\sigma_{\mathrm{I}}}{\sigma_{\mathrm{T}}}. \tag{7.4}$$

Furthermore, this maximum production rate occurs at the level of unit optical depth, i.e. where

$$N\sigma_{\mathrm{T}}H\sec\chi=1. \tag{7.5}$$

The general problem of calculating the total rate of production $Q(h)$ at each level for a wide range of wavelengths λ_1 to λ_2 incident on an atmosphere containing many constituents involves a complicated form of expression (7.2):

$$Q(h)=\sum_i N_i(h)\int_{\lambda_1}^{\lambda_2}\sigma_{\mathrm{I}}^i(\lambda)\,\varphi_0(\lambda)\exp-\left[\sum_i\sigma_{\mathrm{T}}^i(\lambda)\sec\chi\int_h^\infty N_i(h)\,\mathrm{d}h\right]\mathrm{d}\lambda. \tag{7.6}$$

Numerical methods are normally adopted for computing this function over appropriate ranges of wavelengths [*12*]. Such ranges of wavelengths can be chosen on the basis of information on atmospheric penetration depths outlined in Sect. 4γ. It can be seen from Fig. 16 that a division between radiations absorbed above and below 100 km is convenient, and it will be found that a natural boundary also exists between the E and F regions of the ionosphere occurring above this level and the D region located below. The major sources of ionization on either side of this level will, therefore, be considered in turn.

β) E and F regions. The photoionization rates in the height region 100 to 300 km calculated by HEROUX et al.[1] for different spectral regions are shown in Fig. 20. These calculations are based on the solar-flux intensities above the atmosphere shown in Table 5 and refer to a solar zenith angle of 27°, corresponding to that for the time of the rocket flight on which the fluxes were

[0] Elias, G.J. (1923): Tijdschr. Ned. Radio. Gen. *2*, 1

[1] Heroux, L., Cohen, M., Higgins, J.E. (1974): J. Geophys. Res. *79*, 5237

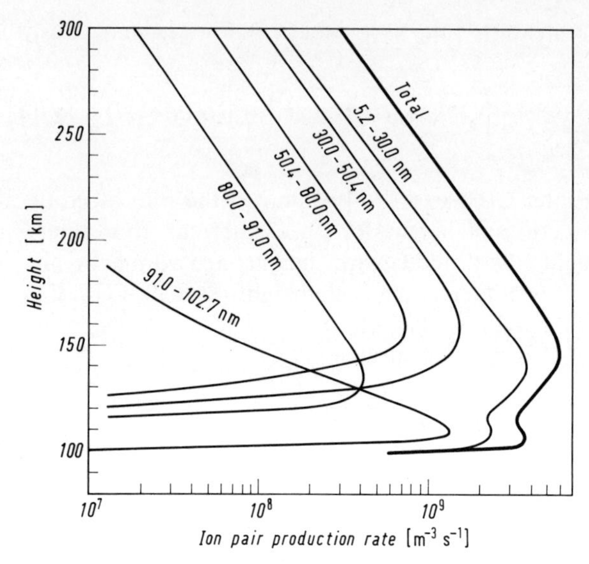

Fig. 20. The photoionization rates for different wavelength regions as calculated by HEROUX et al.[1] from solar intensity measurements on 23 August 1972 for a zenith angle of 27°, using the JACCIA[2] atmospheric model for this date

measured; the appropriate atmospheric model[2] was adopted. The photoionization cross-sections of HUFFMAN[3] were adopted in the computations, these values taking into account secondary-electron ion production at wavelengths shorter than 17 nm. It is seen that for the F region the radiations of special importance are those between 5.2 nm and 80 nm. A closer examination shows that for the wavelength range 5.2 to 30 nm, wavelengths longer than about 10 nm contribute to the F-region ionization; the radiations of wavelengths less than 10 nm contribute to the E region and here the wavelength ranges 91 to 102.7 nm and 80 to 91 nm also represent major sources of ionization. In the first of these two latter wavelength ranges, Lyman-β at 102.6 nm and the Carbon III line at 97.7 nm are of particular importance.

The corresponding production rates for the three ion species O^+, O_2^+ and N_2^+ are shown in Fig. 21. Attention is drawn to the dominance of O^+ at F-region heights; by contrast, in the E region O_2^+ and N_2^+ represent the major primary ions. As already mentioned in Sect. 4β(iii), significant proportions of these primary ions might be produced in excited states. Furthermore, photoelectrons are produced during the photoionization process with mean energies of several eV at high altitudes. In addition to raising the temperature of the ambient electrons, these photoelectrons may have sufficient energy to excite electronic states, such as the ^1D state of atomic oxygen or even the ^3S state of helium atoms, as well as vibrational or rotational states of molecules, such as N_2.

---

[2] Jacchia, L.G. (1971): Revised static models of the thermosphere and exosphere with empirical temperature profiles. Special Report 332, Smithson Astrophys. Observ. Cambridge, Mass.

[3] Huffman, R.E. (1969): Can. J. Chem. *47*, 1823

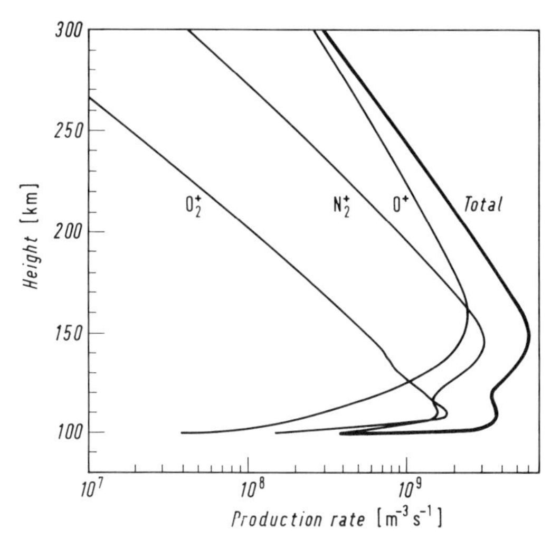

Fig. 21. The production rates of O^+, N_2^+ and O_2^+ ions as calculated by HEROUX et al.[1] from solar intensity measurements on 23 August 1972 for a zenith angle of 27°, using the JACCHIA[2] atmospheric model for this date

γ) D region. It has been seen that Lyman-α radiation corresponds to an absorption window of molecular oxygen and penetrates to heights below 80 km; unit optical depth for overhead sun is near 75 km. This radiation is capable of ionizing nitric oxide, whose ionization threshold is near 134 nm, and this represents a major source of ionization in the D region[4]. The production rates deduced for a solar zenith anglo of 60° with a Lyman-α flux of $4.3\,\mathrm{mW\,m^{-2}}$ and the nitric oxide distribution deduced by MEIRA[5] are shown in Fig. 22. Also shown is the production rate for the ionization of $O_2(^1\Delta_g)$ by radiations from 102.7 nm to the threshold at 111.8 nm, as deduced by HUFFMAN et al.[6] after allowance was made for absorption by atmospheric carbon dioxide. The corresponding results for 0.2 to 0.8 nm X-rays, soft X-rays, Lyman-β and other extreme ultra-violet radiations have been calculated by BOURDEAU et al.[7]. The production rate resulting from the entry of galactic cosmic rays was deduced for a geomagnetic latitude of 50° by WEBBER[8]. It is to be noted from these results, which refer to solar minimum conditions, that the maximum production of ionization in the middle D region results from the photoionization of nitric oxide by Lyman-α radiation. There is, however, considerably uncertainty about the nitric oxide distribution and its changes with time. For illustrative purposes, the broken curve represents the production rates for 0.2 to 0.8 nm X-rays expected for disturbed solar conditions. Comparison with the results of BOURDEAU et al.[7] indicates increases of production rates by a factor of about 40 in the middle D region and by a larger factor at lower heights.

---

[4] Nicolet, M. (1945): Mem. Inst. R. met. Belg. *19*, 1
[5] Meira, L.G. (1971): J. Geophys. Res. *76*, 202
[6] Huffman, R.E., Paulsen, D.E., Larrabee, J.C., Cairns, R.B. (1971): J. Geophys. Res. *76*, 1028
[7] Bourdeau, R.E., Aikin, A.C., Donley, J.L. (1966): J. Geophys. Res. *71*, 727
[8] Webber, W. (1962): J. Geophys. Res. *67*, 5091

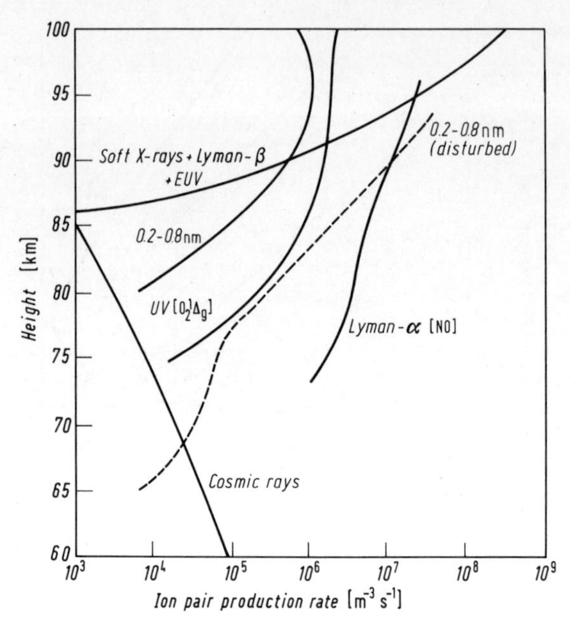

Fig. 22. The ionization rates in the daytime D region for solar minimum conditions and a solar zenith angle of 60°. The rates for nitric oxide are based on the height distribution reported by MEIRA[5] and a Lyman-α flux of 4.3 mW m^{-2}; the rates for $O_2(^1\Delta_g)$ are from HUFFMAN et al.[6]; the results for 0.2–0.8 nm, soft X-rays, Lyman-β and other extreme ultra-violet radiations are from BOURDEAU et al.[7]; the rates for galactic cosmic rays are from WEBER[8]. The production rates for 0.2–0.8 nm radiations expected for disturbed solar conditions are represented by the broken curve

BOURDEAU et al. state that the integrated flux adopted in their calculations was less than $2.5 \cdot 10^{-4}$ mW m^{-2} and on the basis of the values given in Table 6 it seems possible that conditions could have been lightly disturbed. From an analysis of measurements of X-ray data, POPPOFF et al.[9] have shown that the production rates due to hard X-rays can vary by more than three orders of magnitude over the eleven-year solar cycle for non-flare conditions. For the ultra-violet radiations, a corresponding increase by a factor of 1.5 has been suggested for Lyman-α flux by WEEKS[10] and by a factor of 1.7 for the Oxygen VI line at 103.2 nm by HALL et al.[11], but the simultaneous changes in NO and $O_2(^1\Delta_g)$ concentrations could be more important. For the production rates due to cosmic rays, WEBBER[8] has suggested that the solar maximum values will be reduced by a factor of 2 relative to those shown in Fig. 22.

8. Fluorescence and resonance scattering. In addition to photodissociation and ionization, excitation of atmospheric constituents can result from the absorption of solar radiations. The excited atom or molecule will return directly to the ground state, or to an intermediate excited state, with the

[9] Poppoff, I.G., Whitten, R.C., Edmonds, R.S. (1964): J. Geophys. Res. *69*, 4081
[10] Weeks, L.H. (1967): Astrophys. J. *147*, 1203
[11] Hall, L.A., Higgins, J.E., Chagnon, C.W., Hinteregger, H.E. (1969): J. Geophys. Res. *74*, 4181

emission of characteristic radiations. In the case when the excitation and emission wavelength are identical the process is referred to as resonance scattering, and when the emitted wavelength is different from that absorbed the process is described as fluorescence.

It can be shown that for an incident parallel solar beam of flux $\pi\varphi$ per unit frequency interval at a wavelength corresponding to a transition from level 1 to level 2 of a constituent, the number of photons resonantly scattered per second per unit atom is given by:

$$g=\pi\varphi\frac{A_{21}}{\sum_j A_{2j}}\int\sigma_{\mathrm{T}}(\nu)\,\mathrm{d}\nu=\pi\varphi\frac{\pi q^2}{mc}f_{2j}\frac{A_{21}}{\sum_j A_{2j}},\tag{8.1}$$

where $\sigma_{\mathrm{T}}(\nu)$ represents the cross-section of an atom at frequency ν, A_{2j} are the Einstein coefficients for transitions from level 2, f_{2j} are the oscillator strengths of the multiplet arising from transitions from level 2, q and m are the charge and mass of the electron, respectively, and c the velocity of light.

The mean radiative lifetime, given by $\frac{1}{\sum_j A_{2j}}$, is typically in the range 10^{-7} – 10^{-9} s and collisional deactivation can lead to a reduction in the intensity of the scattered radiation.

Resonance scattering and photoelectron excitation are major processes for the generation of dayglow emissions. Such scattering is observed from neutral atomic and molecular constituents and also from ions. Thus excitation of the oxygen triplet $\mathrm{O}(^3\mathrm{S}_1-{}^3\mathrm{P}_{2,1,0})$ gives rise to emissions observed at wavelengths of 130.2, 130.4 and 130.6 nm, especially from heights above 400 km; the nitric oxide emissions dominate the dayglow between 200 nm and 300 nm and observations of the $\gamma(\mathrm{A}\,^2\Sigma^+-\mathrm{X}\,^2\Pi)$ bands have provided useful estimates of the molecular concentrations at mesospheric and thermospheric heights; finally, emissions at 393.3 and 396.8 nm during twilight periods have been attributed to the resonance excitation of the calcium ion $\mathrm{Ca}^+(^2\mathrm{P})$ and those at 391.4 nm are attributed to the First Negative band of $\mathrm{N}_2^+(\mathrm{B}\,^2\Sigma_u^+-\mathrm{X}\,^2\Sigma_g^+)$, Fig. 12.

C. Collision processes and chemical reactions

9. Elastic and inelastic collisions. The study of collisions of various types [*13*] is fundamental to an understanding of many aspects of the physics and chemistry of the natural and ionized atmosphere. In one class of interaction the collisions are elastic in that the linear and angular momenta are conserved and also the total kinetic energy of the system of colliding particles remains unchanged, i.e. there are no permanent changes in the internal excitation energies of the particles. An example is found in the diffusion of gases. In the inelastic class of collisions, momentum is still conserved but the total kinetic energy of the system changes as a result of the excitation or de-excitation of one or both of the particles. Such collisions are involved in the cooling of the ionospheric electron gas, the excitation of airglow and auroral emissions, and the chemical reactions of both neutral and ionized constituents with which we are primarily concerned in this chapter. In these reactions, transfer of particles can occur between the colliding molecules or ions and the most common types of these rearrangement collisions will be treated in Sect. 12. However, it is instructive to consider first the characteristics of collisions in which there is no change in chemical identity and then to indicate the type of complications arising from chemical changes.

10. Collision parameters[1]

α) In the simplest situation it can be assumed that the interaction between particles depends only on the relative speed with which they approach. In this approach the "kinetic collision frequency", i.e. the number of collisions per unit time of one particle of type 1 with particles of type 2, is given by:

$$\nu_{12}=N_2\, v_{12}\, \sigma(v_{12}), \tag{10.1}$$

where v_{12} represents the relative velocity, N_2 the concentration of particle 2 and $\sigma(v_{12})$ the cross-section.

The average collision frequency is then given by the integral over the particle velocity distributions

$$\bar{\nu}_{12}=\frac{1}{N_1}\int v_{12}\,\sigma(v_{12})\,f(v_1)\,f(v_2)\,dv_1\,dv_2 \tag{10.2}$$

and the characteristics of the interactions, the mean time between collisions, τ_{12}, and the mean free path between collisions, λ_{12}, are given by:

$$\bar{\tau}_{12}=1/\bar{\nu}_{12} \tag{10.3}$$

$$\lambda_{12}=\bar{v}_{12}/\bar{\nu}_{12}. \tag{10.4}$$

Under the assumption that the collisions occur between rigid spheres of masses m_1, m_2, the temperatures of these components being T_1, T_2 (each having a Maxwellian velocity distribution), it can be shown that

$$\bar{\nu}_{12}=N_2\left(\frac{8k}{\pi}\right)^{\frac{1}{2}}\left[\frac{T_1}{m_1}+\frac{T_2}{m_2}\right]^{\frac{1}{2}}\sigma_0, \tag{10.5}$$

where k represents Boltzmann's constant and σ_0 is given by the area of a circle of radius equal to the sum of the radii of the two types of sphere.

[1] Suchy, K.: More detailed contribution in this Encyclopedia, vol. 49/7

β) For several problems of aeronomic interest, however, the cross-section appropriate to momentum transfer is relevant. This is defined in terms of the number of particles scattered into an element of solid angle $d\Omega$ orientated in the spherical coordinate system characterised by the polar angle θ and azimuth Φ:

$$\sigma_M(v_{12}) = \int_0^{2\pi}\int_0^{\pi} \sigma(\theta, \Phi, v_{12})(1-\cos\theta)\sin\theta \, d\theta \, d\Phi, \tag{10.6}$$

where $\sigma(\theta, \Phi, v_{12})$ represents the differential cross-section. In Suchy's contribution[2] an expression is given for the momentum transport collision frequency from which σ_0 in equation (10.5) is replaced by:

$$\frac{4}{3}\frac{1}{C^6}\int_0^{\infty} v_{12}^5 \sigma_M(v_{12}) \exp\left(-\frac{v_{12}^2}{C^2}\right) dv_{12}, \tag{10.7}$$

where

$$C^2 = \left(\frac{2kT_1}{m_1} + \frac{2kT_2}{m_2}\right).$$

Note that C is a representative average velocity. This type of formulation finds application in a range of aeronomic studies involving elastic collisions of different types of particle. Thus a momentum transfer cross-section and the corresponding "transfer collision frequency" arise in considerations of the collisional momentum exchange between different gaseous species, the energy exchange between gases of different temperatures, and the elastic collisions of electrons with neutral atoms or molecules[1].

γ) In the case of collisions between ions and neutral atoms or molecules, an induced dipole attraction dominates the interaction at low temperatures. The potential function for an ion of charge Zq and an atom of polarizability α, separated by a distance d, is then given by[3]:

$$V_p(d) = \frac{C_4}{d^4}$$

with

$$C_4 = -\frac{1}{2}\frac{u(Zq)^2}{4\pi\varepsilon_0}\frac{\alpha u}{4\pi\varepsilon_0}, \tag{10.8}$$

where ε_0 is the dielectric permittivity of free space ($=8.854\cdot 10^{-12}\,\mathrm{As\,V^{-1}\,m^{-1}}$). $u=1$ in rationalized systems of units (e.g. SI) but $=4\pi$ in non-rationalized systems of units.

For higher temperatures, above about 300 K, a short-range quantum mechanical repulsion becomes important, this more than compensating for the dipole attraction. However, with the limited experimental data available for such temperatures, it is customary to adopt the collision frequency deduced for

[2] Suchy, K.: This Encyclopedia, vol. 49/7, Eq. (19.3)
[3] Suchy, K.: This Encyclopedia, vol. 49/7, Eq. (9.2)

the polarization force over the whole range of ion and neutral particle temperatures. DALGARNO et al.[4] have shown that the ion-neutral velocity dependent momentum transfer cross-section for single charged ions is given by:

$$\sigma_M(v_{12}) = 2.2\pi \frac{Zq}{v_{12}} \frac{u}{4\pi\varepsilon_0} \left(\frac{\alpha}{\mu_A}\right)^{\frac{1}{2}}, \tag{10.9}$$

where μ_A represents the ion-neutral reduced mass

$$\mu_A = \frac{m^+ m}{m^+ + m}$$

m^+ and m representing the ion and neutral masses.

For ion-molecule collisions, this corresponds to the Langevin cross-section for ions reacting with non-polar molecules. Equation (10.9) represents the special case of Suchy's general Eq. (6.19)[1] for $l=1$ and $n=4$, when the dimensionless factor $Q^{(1)*}$ assumes the value 2.2092.

Substituting for $\sigma_M(v_{12})$ in Eq. (10.7), SUCHY[5] shows that the averaged ion-neutral momentum transfer collision frequency is given by:

$$\nu_{in} = N_n \pi |C_4|^{\frac{1}{2}} Q^{(1)*} \left(\frac{2}{\mu_A}\right)^{\frac{1}{2}} \tag{10.10}$$

which may be written for convenience as

$$\frac{\nu_{in}}{s^{-1}} = 2.6 \times 10^{-15} \frac{N_n}{m^{-3}} \left(\frac{\alpha u/4\pi\varepsilon_0 \cdot 10^{-30} m^3}{\mu_A/u_A}\right)^{\frac{1}{2}}, \tag{10.11}$$

where N_n represents the neutral gas concentration, α the reduced atomic polarizability, and u_A the atomic mass unit. Table 7, compiled from BANKS and KOCKARTS [8], shows the polarizabilities of certain neutral gases and the ion-neutral momentum transfer collision frequencies obtained from experiments or from Eqs. (10.5), (10.7) for temperatures of 300 K and 1,000 K.

Table 7. Polarizabilities, α, and averaged ion-neutral momentum transfer collision frequencies ν_{in}

Ion	Neutral gas	$\frac{\alpha u/4\pi\varepsilon_0}{10^{-30} m^3}$	$10^{16} \frac{\nu_{in}}{s^{-1} N_n/m^{-3}}$	
			300 K	1,000 K
O^+	N_2	1.76	9.2	9.5±2.0
O_2^+			7.6	5.9
N^+			11.4	12.1
N_2^+	O_2	1.59	7.2	5.7
O^+			10.2	10.8
NO	O	0.79±0.02	6.0	4.7
O_2^+			5.9	4.6

[4] Dalgarno, A., McDowell, M.R.C., Williams, W. (1958): Phil. Trans. R. Soc. London, Ser. A *250*, 411

[5] Suchy, K.: This Encyclopedia, vol. 49/7, Eq. (18.10b)

δ) Ion-neutral collisions which are accompanied by a transfer of charge[6] are important for considerations of ion energy transfer and ion diffusion. It is found that in such collisions the original kinetic energy of both particles tends to be retained and the interaction is essentially elastic although the identity of the ion is changed. A particular example of such processes is that of resonance charge transfer which occurs between an ion and its parent atom where the energy defect is small. In addition, certain ion-neutral pairs exhibit an accidental coincidence in ionization energies, a resonance process then permitting a rapid charge exchange at thermal energies. It will be seen in Sect. 32 that such an exchange between atomic oxygen ions and hydrogen atoms represents the main source of atomic hydrogen ions in the upper atmosphere.

ε) The inelastic collision of precipitated electrons and protons with neutral atoms or molecules can lead to ionization if the energy of the charged particle is sufficient; these collision processes represent the most important source of ionization at high latitudes [*1*]. For lower energies the collisions result in the excitation of the neutral particles, thereby leading to auroral and airglow emissions. Attention will be drawn to such inelastic processes where they relate to the neutral or ion chemistry of the upper atmosphere. However, in the space available no detailed treatment of such processes is possible.

ζ) As for collisions between charged particles, so-called COULOMB-collisions, see K. SUCHY in volume 49/7, his Sect. 12.

11. Collisions leading to chemical change. The chemistry of neutral and ionized atoms and molecules in the upper atmosphere [*14*] involves a number of collision processes between particles which are approximately in chemical equilibrium. A two-body process involving reactants A and B can be represented by:

$$\mathrm{A + B \rightarrow Products}. \tag{11.1}$$

The specific loss rate for reactant A is then given by k_2 [B] where k_2 represents the rate coefficient of the reaction concerned and [B] the number density (concentration) of constituent B. By analogy with the kinematic interactions considered in equation (10.1), it can be seen that k_2 represents the product of an effective collision cross-section and the relative velocity of constituents A and B. For illustrative purposes we can consider the collision of rigid spheres of diameters $3.6 \cdot 10^{-10}$ m and mass $5.6 \cdot 10^{-26}$ kg, corresponding to oxygen molecules, at a temperature of 300 K:

$$v_{12} = \left(\frac{8k}{\pi}\right)^{\frac{1}{2}} \left(\frac{2T}{m}\right)^{\frac{1}{2}} = 630\ \mathrm{m\,s^{-1}} \tag{11.2}$$

$$\sigma_0 = 4.1 \cdot 10^{-19}\ \mathrm{m}^2 \tag{11.3}$$

and

$$\bar{\nu}_{12} = 2.6 \cdot 10^{-16}\ \mathrm{s}^{-1}\ [\mathrm{B}]. \tag{11.4}$$

[6] Compare: Suchy, K., this Encyclopedia, vol. 49/7, his Sect. 15

In similar fashion a three-body process can be considered:

$$A + B + C \rightarrow \text{Products.} \tag{11.5}$$

In this case the specific loss rate for constituent A is given by k_3 [B] [C], where k_3 represents the rate coefficient. Again by considering the nearly simultaneous collision of the three particles, it is found that for rigid spheres of the same characteristics as before

$$\bar{v}_{123} = 7.6 \cdot 10^{-44}\,\text{s}^{-1}[\text{B}]\,[\text{C}]. \tag{11.6}$$

These values of $\bar{v}_{12}$ and $\bar{v}_{123}$ represent gas-kinetic type of collision frequencies. The factors then correspond to gas-kinetic rate coefficients, although in actual reactions a factor of probability that a reaction will occur is usually involved. This is expressed in the ARRHENIUS[1] relation in which the rate constant is set equal to the collision frequency times the fraction of activated molecules:

$$k = v \exp\left(-\frac{\mathscr{E}}{kT}\right), \tag{11.7}$$

where k is Boltzmann's constant. The activation energy $\mathscr{E}$ refers to one collision; it can be considered as the energy barrier that must be overcome in the reaction.

In the case of ion-molecule reactions the strong polarization forces that operate can distort the binding fields so that the activation energy for an ion-atom interchange reaction is lowered to a negligible value.

12. Types of neutral and ion chemical reactions

α) Recombination of atoms. The recombination of two atoms to form a molecule, with the excess energy being transferred into radiation, is of some importance in connection with the excitation of certain airglow emissions. However, the rate coefficients for radiative associations are less than $10^{-21}\,\text{m}^3\,\text{s}^{-1}$, even when allowed transitions are involved, and these processes do not play an important role in the neutral atmosphere chemistry.

Three-body recombination of atoms is also possible:

$$A + B + M \rightarrow AB + M. \tag{12.1}$$

Here the third body M serves to conserve both momentum and energy. In some cases the molecule AB is initially in an excited state during the association but a quenching collision with M serves to stabilise it.

This type of reaction is effective at heights below about 120 km where the concentration of the third body is sufficient, and the following represent important examples of three-body recombination processes in this part of the

[1] Arrhenius, S. (1889): Z. Phys. Chem. *4*, 226

upper atmosphere:

$$O + O + O \rightarrow O_2 + O(^1S) + 0.93\ \text{eV} \tag{12.2}$$

$$O + O_2 + M \rightarrow O_3 + M + 1.05\ \text{eV} \tag{12.3}$$

$$H + O_2 + M \rightarrow HO_2 + M + 1.99\ \text{eV}. \tag{12.4}$$

It has been considered that reaction (12.2) is the principal excitation mechanism for the $O(^1S - {}^1D)$ airglow emission at 557.7 nm, the value for the rate coefficient reported by FELDER and YOUNG[0], $5.0 \cdot 10^{-45}\ m^6\ s^{-1}$, seeming to be adequate to explain the observed intensities. However, later measurements have cast doubt on this conclusion[1]. There is some uncertainty about the rate coefficient for reaction (12.3) which is important in the production of ozone at mesospheric and lower heights, recent measurements reviewed by NIKI[2] indicating a value near $5.0 \cdot 10^{-46}\ m^6\ s^{-1}$ at temperatures near 300 K. Reaction (12.4) is the major production mechanism for the HO_2 radical, and an important loss process for H atoms in the mesosphere, the measured rate coefficients reported by DAVIS[3] being consistent with $2.0 \cdot 10^{-44}\ m^6\ s^{-1}$ at 300 K. It is of interest to compare these three-body rate coefficients with that deduced in Sect. 11 for the three-body collisions of oxygen molecules considered as spherical hard spheres (relation (11.6)).

β) Atom-interchange reactions. A class of reactions which is known to play an important role, particularly in the chemistry of hydrogen and oxygen compounds in the mesosphere, is represented by:

$$A + BC \rightarrow AB + C. \tag{12.5}$$

As an example, a reaction between atomic oxygen and ozone provides an important loss process for these constituents in the mesosphere:

$$O + O_3 \rightarrow 2O_2 + 4.06\ \text{eV}. \tag{12.6}$$

There is some uncertainty about the rate coefficient but recent information[3] has indicated a value of $1.5 \cdot 10^{-17} \exp\left(-\frac{2{,}200}{T/K}\right) m^3\ s^{-1}$.

Another example involving oxygen atoms is the reaction with hydroxyl radicals which provides a loss process for both constituents and a source of hydrogen atoms in the mesosphere:

$$OH + O \rightarrow O_2 + H + 0.72\ \text{eV}. \tag{12.7}$$

---

[0] Felder, W., Young, R.A. (1972): J. Chem. Phys. *56*, 6028

[1] Slanger, T.G., Black, G. (1977): Planet. Space Sci. *25*, 79

[2] Niki, H. (1974): Can. J. Chem. *52*, 1397

[3] Davis, D.D. (1974): Can. J. Chem. *52*, 1405

From an examination of recent measurements tabulated by DAVIS[3] the rate coefficient for this reaction is seen to be about $4.0 \cdot 10^{-17}\,m^3\,s^{-1}$.

The atom-interchange reaction between hydrogen atoms and ozone molecules is believed to give rise to vibrationally-excited hydroxyl radicals in the mesosphere:

$$H \rightarrow O_3 \rightarrow (OH)^* + O_2 + 3.34\ eV. \tag{12.8}$$

The energy yield is considered to be sufficient to populate the radical up to and including the vibrational level $v''=9$ and, thereby, provide the source of the Meinel bands in the airglow.

An atom-interchange reaction involving an electronically-excited atom has been identified as a major source of nitric oxide molecules in the lower thermosphere:

$$N(^2D) + O_2 \rightarrow NO + O + 3.77\ eV. \tag{12.9}$$

The rate coefficient of $1.4 \cdot 10^{-17}\,m^3\,s^{-1}$ deduced from laboratory measurements by SLANGER et al.[4] is several orders of magnitude larger than the corresponding value found for the ground-state atom, $N(^4S)$, at temperatures representative of the lower thermosphere.

γ) Recombination of positive ions with electrons

(i) *Radiative.* The charge neutralization of an electron-ion pair can occur by radiative recombination in which energy is released as a photon and internal excitation of the atom:

$$A^+ + e^- \rightarrow A^* + h\nu. \tag{12.10}$$

Theoretical estimates of rate coefficients[5] have yielded values of $3.7 \cdot 10^{-18}\,m^3\,s^{-1}$ for atomic oxygen ions at 250 K and have also predicted a $T_e^{-0.7}$ variation with electron temperature. In view of the resulting small values of rate coefficients over the range of electron temperature encountered in the upper atmosphere, it is considered that radiative recombination plays little part in the disappearance of either electrons or atomic ions. Loss of these constituents normally occurs through reactions of the atomic ion with molecules (Subsect. δ) and the subsequent dissociative recombination of the molecular ions formed.

(ii) *Dissociative.* The main loss of electrons by recombination throughout the ionosphere is by the dissociative recombination of molecular ions:

$$AB^+ + e^- \rightarrow A^* + B^*. \tag{12.11}$$

Either or both of the products A and B can be left in an excited state depending on the energy yield of the reaction.

It is found that the rate coefficients observed are about five orders of magnitude larger than for radiative recombination of atomic ions.

---

[4] Slanger, T.G., Wood, B.J., Black, G. (1971): J. Geophys. Res. *76*, 8430
[5] Massey, H.S.W., Bates, D.R. (1943): Rep. Prog. Phys. *9*, 62

It can be shown theoretically that the average collision frequency between electrons and ions decreases as the electron temperature increases[6]. The variations of the dissociative recombination of the atmospheric ions with electron temperature observed in the laboratory are complicated (Sect. 16) but for O_2^+ and N_2^+ are reasonably consistent with the expressions:

$$\alpha_D(O_2^+) = 2.0 \cdot 10^{-13} \left(\frac{300}{T_e/K}\right)^{0.7} \mathrm{m^3\,s^{-1}} \quad \text{(Mehr and Biondi}^{7}\text{)} \tag{12.12}$$

$$\alpha_D(N_2^+) = 1.8 \cdot 10^{-13} \left(\frac{300}{T_e/K}\right)^{0.39} \mathrm{m^3\,s^{-1}} \quad \text{(Mehr and Biondi}^{7}\text{).} \tag{12.13}$$

For NO^+, different dependences on electron temperature have been shown in two sets of measurements:

$$\begin{aligned} \alpha_D(NO^+) &= 4.4 \cdot 10^{-13} \left(\frac{300}{T_e/K}\right)^{1.2} \mathrm{m^3\,s^{-1}} \quad \text{for } T_e < 400\,\mathrm{K} \\ &= 2.9 \cdot 10^{-13} \left(\frac{500}{T_e/K}\right)^{0.37} \mathrm{m^3\,s^{-1}} \quad \text{for } T_e > 500\,\mathrm{K} \\ &\text{(Huang et al.}^{8}\text{)} \\ \alpha_D(NO^+) &= 4.3 \cdot 10^{-13} \left(\frac{300}{T_e/K}\right)^{0.83} \mathrm{m^3\,s^{-1}} \quad \text{for } T_e > 300\,\mathrm{K} \\ &\text{(Walls and Dunn}^{9}\text{).} \end{aligned} \tag{12.14}$$

Values of these three rate coefficients have also been deduced in analyses of Atmospheric Explorer satellite data, as reviewed by Torr and Torr [*15*]; the values for O_2^+ are in agreement with (12.12); those for N_2^+ are larger than expected from (12.13); and those for NO^+ show a $T_e^{-0.85}$ dependence at temperatures above 500 K, in agreement with the second set of results in (12.14).

These values refer to the total recombination and little information is available about the yield of the various excited atomic states. The possible states are illustrated in Fig. 23 which shows the energy level diagrams of the lowest states of atomic nitrogen and oxygen.

In the case of NO^+ and N_2^+ ions, the possibility of producing $N(^2D)$ atoms is of particular interest:

$$NO^+(^1\Sigma^+) + e^- \rightarrow N(^2D) + O(^3P) + 0.38\,\mathrm{eV} \tag{12.15}$$

$$N_2^+(^2\Sigma_g^+) + e^- \rightarrow N(^2D) + N(^4S) + 3.44\,\mathrm{eV} \tag{12.16}$$

$$N_2^+(^2\Sigma_g^+) + e^- \rightarrow N(^2D) + N(^2D) + 1.06\,\mathrm{eV.} \tag{12.17}$$

---

[6] See Rawer, K., Suchy, K., this Encyclopedia, vol. 49/2, Sect. 3

[7] Mehr, F.J., Biondi, M.A. (1969): Phys. Rev. *181*, 264

[8] Huang, C.M., Biondi, M.A., Johnsen, R. (1975): Phys. Rev. A *11*, 901

[9] Walls, F.L., Dunn, G.H. (1974): J. Geophys. Res. *79*, 1911

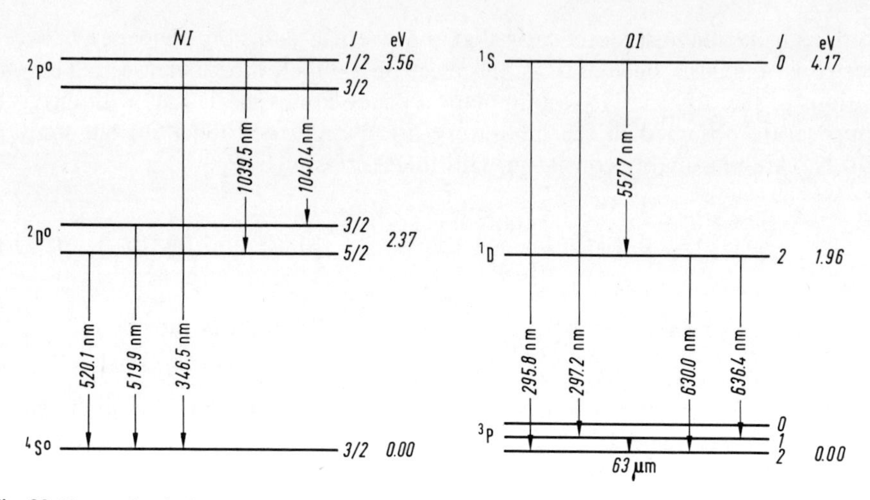

Fig. 23. Energy level diagrams of the lowest states of atomic nitrogen and oxygen; the wavelengths of the forbidden transitions are shown

The $N(^2D)$ atoms are involved in the excitation of the 520 nm emission in the airglow, $N(^2D-^4S)$, and as mentioned previously in the production of nitric oxide by reacting with molecular oxygen. However, alternative recombination reactions to (12.15), (12.16) and (12.17) are possible which yield ground-state atoms, $N(^4S)$, instead.

The productions of $O(^1S)$ and $O(^1D)$ atoms from the dissociative recombination of $O_2^+(^2\Pi_g)$ ions have been used in the interpretation of airglow emission at 557.7 nm and 630 nm in terms of the transitions $O(^1S-^1D)$ and $O(^1D-^3P_2)$, respectively. In the recombination of the O_2^+ ion all combinations of pairs of atoms from $O(^3P)$, $O(^1D)$ and $O(^1S)$ are possible, excepting the case of two $O(^1S)$ atoms. There is still some doubt about the relative yields of the five reactions possible. However, it is well established that the production of $O(^1D)$ atoms by this process is responsible for the excitation of one feature of the 630 nm airglow, namely the arcs observed about 20° north and south of the magnetic equator.

The other dissociative recombination processes of primary concern in the present study are those of the water cluster ion series, $H^+ \cdot (H_2O)_n$, and of the metallic oxide ions, each being of interest in the lower ionosphere. The laboratory measurements by LEU et al.[10] for the water cluster ions at different gas temperatures are summarised in Table 8.

A comparison of the value for H_3O^+ with previous measurements at higher temperatures reveals a substantial temperature dependence, as expected for dissociative recombination. However, the values shown for the ion $H^+ \cdot (H_2O)_3$ shows no such dependence, thus indicating that a modified recombination of this ion is probably taking place. The table suggests an increase in recombination coefficient with the increasing number of water molecules.

[10] Leu, M.T., Biondi, M.A., Johnsen, R. (1973): Phys. Rev. A 7, 292

Table 8. Electron-ion recombination coefficients of $H^+ \cdot (H_2O)_n$ ions, in units of $10^{-12}\,m^3\,s^{-1}$ (after LEU et al.[10])

T/K	H_3O^+ (19⁺)	$H^+ \cdot (H_2O)_2$ (37⁺)	$H^+ \cdot (H_2O)_3$ (55⁺)	$H^+ \cdot (H_2O)_4$ (73⁺)	$H^+ \cdot (H_2O)_5$ (91⁺)	$H^+ \cdot (H_2O)_6$ (109⁺)	$H^+ \cdot (H_2O)_7$ (127⁺)
546	1.0	2.0	4.0				
415		2.2	4.2				
300			3.8	4.9			
245					6.0	7.5	≤ 10

δ) Positive ion-neutral reactions. The importance of ion-neutral reactions in the F region of the ionosphere was recognised in the early years of aeronomic studies by BATES and MASSEY [*16*] in their consideration of the formation of molecular ions from the dominant ion O^+. It has subsequently been realised that binary reactions dominate the positive-ion chemistry of the E and F regions whilst both binary and three-body reactions are important in the D region.

(i) *O^+ reactions.* The reactions of O^+ ions with molecular nitrogen and oxygen:

$$O^+ + N_2 \rightarrow NO^+ + N + 1.10\,\text{eV} \tag{12.18}$$

$$O^+ + O_2 \rightarrow O_2^+ + O + 1.55\,\text{eV} \tag{12.19}$$

play a central role in the F region since they convert the slowly recombining atomic ion into molecular ions whose recombination rates with electrons are some five orders of magnitude larger. Recent laboratory measurements of these reactions have concentrated on their temperature or energy dependences. The results for these and other ion-neutral reactions deduced from observations with flowing-afterglow systems (Sect. 15β) as presented by LINDINGER et al.[11] are shown in Fig. 24; good agreement is found between these results and those deduced from ionospheric data [*15*]. The laboratory data do not extend to the temperatures of about 2,000 K sometimes expected at F-region heights but it seems certain from related measurements of reactions (12.18) and (12.19) at varying ion kinetic energy that the rate coefficients will have minimum values at certain temperatures. A marked dependence of the rate coefficient of reaction (12.18) on vibrational excitation of molecular nitrogen has also been observed by SCHMELTEKOPF et al.[12]. For vibrational temperatures above about 1,200 K a rapid increase occurs, the values changing by about an order of magnitude as the temperature increases up to 3,000 K. It seems likely that this dependence on vibrational excitation of molecular nitrogen could result in a very rapid formation of NO^+ ions during disturbed conditions and, thereby, large electron loss rates.

---

[11] Lindinger, W., Fehsenfeld, F.C., Schmeltekopf, A.L., Ferguson, E.E. (1974): J. Geophys. Res. *79*, 4753

[12] Schmeltekopf, A.L., Ferguson, E.E., Fehsenfeld, F.C. (1968): J. Chem. Phys. *48*, 2966

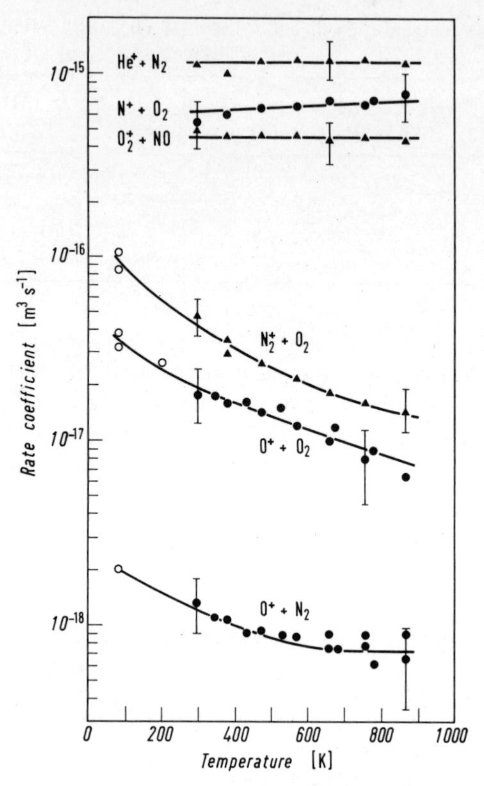

Fig. 24. Measurements of the rate coefficients of ion-molecule reactions as a function of temperature, as presented by LINDINGER et al.[11]

(ii) *He^+ reactions.* The main loss process for He^+ ions at high altitudes is through reactions with molecular nitrogen:

$$He^+ + N_2 \rightarrow N^+ + N + He + 0.28\ \text{eV} \tag{12.20}$$

$$He^+ + N_2 \rightarrow N_2^+ + He + 8.99\ \text{eV}. \tag{12.21}$$

The total rate constant for these reactions has been well established by several laboratory measurements, those of LINDINGER et al.[11] in Fig. 24 showing little evidence of a variation with temperature below 900 K. Little dependence has been found on the nitrogen vibrational temperature although the branching ratio k_{20}/k_{21} increases with increasing temperature[12].

(iii) *O_2^+ reactions.* Besides dissociative recombination, the major loss processes for this ion involve reactions with minor neutral constituents, such as NO:

$$O_2^+ + NO \rightarrow NO^+ + O_2 + 2.80\ \text{eV}. \tag{12.22}$$

The value of the rate coefficient for this reaction and the negligible temperature dependence are illustrated in Fig. 24.

A similar reaction occurs with atomic nitrogen:

$$O_2^+ + N \rightarrow NO^+ + O + 4.19\,\text{eV}. \tag{12.23}$$

A value of $1.8 \cdot 10^{-16}\,\text{m}^3\,\text{s}^{-1}$ has been reported for this reaction by FITE[13].

The reaction with the major constituent N_2 to yield NO^+ and NO is also exothermic. However, the simultaneous breaking of two bonds renders it very slow and, in fact, no firm figure for the rate coefficient has been obtained from laboratory measurements.

In addition, three-body attachment to molecular oxygen and water vapour has been observed to occur sufficiently rapidly to play important roles in the D region (Sects. 15ζ, 27).

(iv) *N_2^+ reactions.* The following fast reaction limits the concentration of N_2^+ ions at heights below about 110 km:

$$N_2^+ + O_2 \rightarrow O_2^+ + N_2 + 3.52\,\text{eV}. \tag{12.24}$$

The temperature dependence of the reaction relevant to this range of heights is shown in Fig. 24.

At greater heights, reactions with atomic oxygen become more important:

$$N_2^+ + O \rightarrow NO^+ + N + 3.06\,\text{eV} \tag{12.25}$$

$$N_2^+ + O \rightarrow O^+ + N_2 + 1.96\,\text{eV}. \tag{12.26}$$

Reaction (12.25) is sufficiently exothermic for $N(^2D)$ atoms to be produced. Measurements of the total rate coefficient and of the branching ratio have been carried out over a range of ion energy, as reviewed by MCFARLAND et al.[14]. The rate coefficient is $1.5 \cdot 10^{-16}\,\text{m}^3\,\text{s}^{-1}$ at 300 K (0.04 eV energy) and more than 90% of the reactions follow channel (12.25), this proportion decreasing with increasing energy. Similar results have been obtained from analyses of ionospheric data from the Atmospheric Explorer satellites [*15*].

(v) *N^+ reactions.* Ion-atom interchange and charge-transfer reactions with molecular oxygen represent the main loss for N^+ ions:

$$N^+ + O_2 \rightarrow NO^+ + O + 6.66\,\text{eV} \tag{12.27}$$

$$N^+ + O_2 \rightarrow O_2^+ + N + 2.47\,\text{eV}. \tag{12.28}$$

The temperature dependence of the total rate coefficient is shown in Fig. 24 but there is still considerable uncertainty about the branching ratio.

(vi) *NO^+ reactions.* As illustrated in Fig. 25, NO^+ ions are formed by a number of ion-molecule reactions involving ions with higher ionization potentials. Laboratory measurements have shown that the NO^+ ions themselves show very slow binary reactions with the major atmospheric constituents.

---

[13] Fite, W.L. (1969): Can. J. Chem. *47*, 1797

[14] McFarland, M., Albritton, D.L., Fehsenfeld, F.C., Ferguson, E.E., Schmeltekopf, A.L. (1974): J. Geophys. Res. *79*, 2925

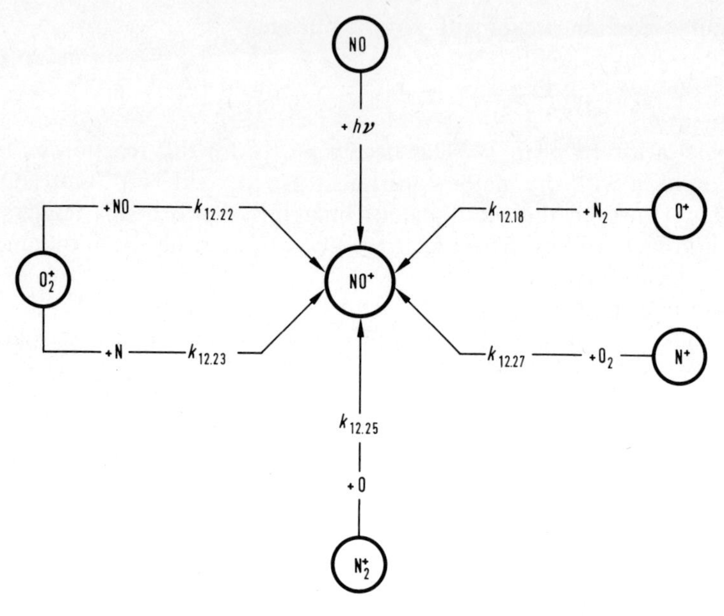

Fig. 25. Principal reactions for the formation of NO^+

However, three-body processes have been observed, these being of considerable importance in consideration of the D-region positive-ion chemistry. Specifically, such processes are thought to be involved in the conversion of NO^+ into the water cluster ions $H^+\cdot(H_2O)_n$, Sect. 27. The formation of $NO^+\cdot H_2O$ is believed to be the initial stage in this conversion. Although three-body association with water vapour is known to occur with a rate coefficient of about $10^{-40}\,m^6\,s^{-1}$, a more rapid loss of NO^+ occurs by the corresponding reactions with the more abundant CO_2 or N_2 constituents:

$$NO^+ + CO_2 + M \rightleftarrows NO^+\cdot CO_2 + M \qquad (12.29)$$

$$NO^+ + N_2 + M \rightleftarrows NO^+\cdot N_2 + M. \qquad (12.30)$$

These reactions are then believed to be followed by rapid switching reactions

$$NO^+\cdot N_2 + CO_2 \rightarrow NO^+\cdot CO_2 + N_2 \qquad (12.31)$$

or

$$NO^+\cdot N_2 + H_2O \rightarrow NO^+\cdot H_2O + N_2 \qquad (12.32)$$

and

$$NO^+\cdot CO_2 + H_2O \rightarrow NO^+\cdot H_2O + CO_2. \qquad (12.33)$$

Lack of information prevents the accurate estimation of the heats of reaction.

Important recent measurements by JOHNSEN et al.[15] have revealed a marked

[15] Johnsen, R., Huang, C.M., Biondi, M.A. (1975): J. Chem. Phys. *63*, 3374

temperature dependence for the rate coefficient of the forward reaction in (12.30), an estimated binding energy of only 0.18 eV for the ion $NO^+ \cdot N_2$, and a correspondingly pronounced temperature dependence for the collisional break-up of this ion. The details of the current ideas on D-region positive-ion chemistry based on these and subsequent reactions will be described in Sect. 27.

ε) Negative-ion reactions. The formation of negative ions at D-region heights, arising from the relatively high gaseous densities which permit three-body attachment of electrons, provides the lowest part of the ionosphere with a unique character. The most important of the electron attachment processes is that involving molecular oxygen:

$$e^- + O_2 + O_2 \rightarrow O_2^- + O_2 + 0.43\,\text{eV} \tag{12.34}$$

which has a rate coefficient of $1.6 \cdot 10^{-42}\,\text{m}^6\,\text{s}^{-1}$ at room temperatures (CHANIN et al.[16]).

Laboratory measurements carried out in the past decade by FEHSENFELD et al.[17,18] have shown that the O_2^- ions formed by reaction (12.34) can either charge transfer with ozone to form O_3^- or undergo a three-body association with molecular oxygen to form O_4^-. Subsequent reactions with minor atmospheric constituents are then thought to occur, giving rise to ions such as CO_3^-, CO_4^-, NO_2^- and NO_3^- and hydrates of at least some of these ions. Alternatively, the O_2^- can undergo very rapid associative detachment with atomic oxygen[17]:

$$O_2^- + O \rightarrow O_3 + e^- + 0.62\,\text{eV}. \tag{12.35}$$

The large rate coefficient, $3.3 \cdot 10^{-16}\,\text{m}^3\,\text{s}^{-1}$, reported for this reaction implies that it can serve to circumvent the reaction scheme outlined above.

Loss of negative ions can also arise from recombination with positive ions (Sect. 17) or by photodetachment (Sect. 28).

ζ) Quenching of electronically-excited species. Attention has been drawn above to the possibility of producing excited species in exothermic chemical reactions, and the importance of solar radiations as a source of energy for such species was also indicated in Sect. 8. These species can play an important role in atmospheric processes if their excitation energy is transferred to other constituents during collisions, resulting either in the excitation of these constituents or in the initiation of a chemical reaction. The greater values of chemical rate coefficients for excited species compared with those for the corresponding ground-state species is largely responsible for the atmospheric or ionospheric importance of excited species. Whether or not these species enter into such reactions depends in part on their radiative lifetimes. These are shown for certain atmospheric species in Table 9.

[16] Chanin, L.M., Phelps, A.V., Biondi, M.A. (1959): Phys. Rev. Lett. *2*, 344

[17] Fehsenfeld, F.C., Schmeltekopf, A.L., Schiff, H.I., Ferguson, E.E. (1967): Planet. Space Sci. *15*, 373

[18] Fehsenfeld, F.C., Ferguson, E.E., Bohme, D.K. (1969) Planet. Space Sci. *17*, 1759

Table 9. Radiative lifetimes of excited species

Excited state	Ground state	Radiative lifetime(s)
$He(2\,^3S)$	$He(1\,^1S)$	$4.5\cdot10^4$
$O(^1D)$	$O(^3P)$	110
$O(^1S)$		0.74
$N(^2D)$	$N(^4S)$	$9.4\cdot10^3$
$O_2(a\,^1\Delta_g)$	$O_2(X\,^3\Sigma_g^-)$	$2.7\cdot10^3$
$O_2(b\,^1\Sigma_g^+)$		12
$O_2(A\,^3\Sigma_u^+)$		~ 1
$OH(X\,^2\Pi)_{Vibr.}$	$OH(X\,^2\Pi)$	$6\cdot10^{-2}$
$O^+(^2D)$	$O^+(^4S)$	$1.3\cdot10^4$
$O^+(^2P)$		~ 5
$O_2^+(a\,^4\Pi_u)$	$O_2^+(X\,^2\Pi_g)$	Long

The collisional deactivation, or quenching, of the initially excited species can occur by different paths but most laboratory studies have given information only on the rate of the overall process.

(i) *$O(^1D)$ and $O(^1S)$ reactions.* $O(^1D)$ atoms are very effectively quenched by molecular nitrogen and oxygen, the rate coefficients based on various laboratory measurements being about $6\cdot10^{-17}\,m^3\,s^{-1}$ and $3\cdot10^{-17}\,m^3\,s^{-1}$, respectively. The reaction with N_2 is believed to be one of the main sources of vibrationally-excited nitrogen:

$$O(^1D) + N_2 \rightarrow (N_2)^* + O(^3P) + 1.97\text{ eV} \tag{12.36}$$

in which excitation occurs for 33% of the collisions (SLANGER and BLACK[19]). Collisions with oxygen are capable of vibrationally exciting the molecule or producing electronically-excited states:

$$O(^1D) + O_2(X\,^3\Sigma_g^-) \rightarrow O(^3P) + [O_2(X\,^3\Sigma_g^-)]^* + 1.97\text{ eV} \tag{12.37}$$

$$O(^3P) + O_2(a\,^1\Delta_g) + 0.99\text{ eV} \tag{12.38}$$

$$O(^3P) + O_2(b\,^1\Sigma_g^+) + 0.34\text{ eV}. \tag{12.39}$$

The shorter radiative lifetime of $O(^1S)$ implies that quenching is competitive as a loss process only at lower heights, below about 95 km. In contrast to $O(^1D)$, quenching by molecular nitrogen is unimportant and the rate coefficient for molecular oxygen is also smaller by about two orders of magnitude. In fact, measurements by ATKINSON and WELGE[20] have shown that the coefficient is temperature dependent, with a value of $4.9\cdot10^{-18}\exp\left(\frac{-860}{T/K}\right)$ $m^3\,s^{-1}$. This implies that for mesospheric heights, quenching by both molecular oxygen and $O(^3P)$ atoms is important, since a rate coefficient of $7.5\cdot10^{-18}\,m^3\,s^{-1}$ has been reported for the atomic collision by FELDER and YOUNG[21].

[19] Slanger, T.G., Black, G. (1974): J. Chem. Phys. *60*, 468
[20] Atkinson, R., Welge, K.H. (1972): J. Chem. Phys. *57*, 3689
[21] Felder, W., Young, R.A. (1972): J. Chem. Phys. *56*, 6028

(ii) *Metastable molecular oxygen reactions.* Although $O_2(^1\Delta_g)$, the upper state for transitions responsible for the Infra-red Atmospheric bands, shows a relatively long lifetime, Table 9, quenching becomes important only at low mesospheric heights owing to the small values of rate coefficients for quenching by molecular oxygen[22] and nitrogen[23]. In fact, it is found that the radiative and quenching rates are equal at about 70 km.

The rate coefficients for quenching of $O_2(b\,^1\Sigma_g^+)$, the upper state of the Atmospheric bands, by molecular oxygen and nitrogen are several orders of magnitude larger than for $O_2(^1\Delta_g)$.

For $O_2(A\,^3\Sigma_u^+)$, the upper state of the Herzberg band system, there are very few laboratory measurements of the rate coefficients for quenching processes, but the data available and estimates from atmospheric data seem to indicate an effective value of about $10^{-21}\,m^3\,s^{-1}$.

(iii) *Metastable molecular and atomic oxygen ion reactions.* The long radiative lifetimes of these ions imply that deactivation by collisions is important. Of particular interest are the ion-transfer reactions with molecular nitrogen:

$$O^+(^2D) + N_2 \rightarrow O + N_2^+ + 1.34\,eV \tag{12.40}$$

$$O^+(^2P) + N_2 \rightarrow O + N_2^+ + 3.04\,eV \tag{12.41}$$

$$O_2^+(a\,^4\Pi_u) + N_2 \rightarrow O_2 + N_2^+ + 0.48\,eV. \tag{12.42}$$

Laboratory measurements[24] with a mixture of ^2D and ^2P ions have shown a rate coefficient for (12.40) and (12.41) together of $1.5 \cdot 10^{-16}\,m^3\,s^{-1}$, with a 90% yield of N_2^+ ions and 10% of the reactions producing NO^+, as with ground-state O^+ ions, reaction (12.18). Analyses of satellite measurements [*15*] have provided estimates of the coefficients of (12.40) and (12.41) separately, these being more than three-times larger than the combined laboratory value given above. Large values, in excess of $2 \cdot 10^{-16}\,m^3\,s^{-1}$, have also been shown by two sets of laboratory measurements[24,25] for reaction (12.42).

D. Laboratory measurements of relevant rate coefficients

13. Requirements for measurements of neutral and ion reactions. Although studies of neutral and ion reactions have formed part of an active research area for several decades, it is only in the last 15 years that determined efforts have been

[22] Becker, K.H., Groth, W., Schurath, U. (1971): Chem. Phys. Lett. *8*, 259

[23] Schiff, H.I. (1972): Ann. Geophys. *28*, 67

[24] Glosik, J., Rakshit, A.B., Twiddy, N.D., Adams, N.G., Smith, D. (1978): J. Phys. B *11*, 3365

[25] Lindinger, W., Albritton, D.L., McFarland, M., Fehsenfeld, F.C., Schmeltekopf, A.L., Ferguson, E.E. (1975): J. Chem. Phys. *62*, 4101

applied to studies of elementary reactions relevant to the neutral and ion chemistry of the upper atmosphere. The upsurge of interest was chiefly inspired by the results of in-situ measurements of neutral and ion composition, and also of ground-based and rocket-borne studies of airglow emissions. More recently, the areas of gas kinetics and photochemistry have benefited from increased support initiated by the needs identified in studies of environmental problems, but much of the new information obtained relates to processes confined to the stratospheric and tropospheric regions. For the purpose of the present article it is useful to examine the types of experiments which have provided information on reactions involved in the neutral and ion chemistry of the region above 60 km.

Many of the requirements of laboratory measurements are common to neutral and ion reactions. Thus a knowledge of the reaction products and their energy states, and of branching ratios if relevant, is essential in the application of the rate-coefficient data to theoretical models. For the three-body reactions involved in the positive-ion, negative-ion and neutral chemistry of the D region and mesosphere, measurements at temperatures down to about 120 K are required. On the other hand, for reactions relevant to the ion chemistry of the E and F regions, measurements over the temperature range 300 to 2,000 K are needed to represent ionospheric conditions, and the state of non-thermal equilibrium between electrons and ions or neutral particles also needs to be reproduced. The variety of laboratory techniques that have been developed satisfy some of these requirements. Brief descriptions of the techniques employed in neutral gas and ion reactions are given below; it has been considered appropriate to pay the greatest attention to experiments relating to the ion chemistry.

14. Neutral gas reactions. In comparison with the measurements of reactions involving ions and electrons, neutral gas reaction studies suffer from a generally greater complexity, smaller cross-sections, greater energy barriers and less predictable intermediate states. The information available on neutral reaction kinetics has then been derived from various methods which can be conveniently classed as thermal or photochemical. The former include observations associated with explosions and other moderately high-temperature conditions, studies of flame processes, and shock-tube measurements. The earlier photochemical methods involved the measurement of the production or destruction of reactants for known intensities of relatively weak irradiation, usually interpreted on the basis of assumed steady-state conditions. Recently, more direct approaches have been devised, employing flash photolysis in stationary reactants or discharge excitation in flowing reactants, and these two techniques now form the basis of systems directed towards the measurement of reaction coefficients of atmospheric interest.

α) Flash-photolysis methods

In the flash-photolysis approach, initiated by NORRISH and PORTER[1], a photochemically active reactant is illuminated with a flash lamp and the concentration of reactant product or intermediate

[1] Norrish, R.G.W., Porter, G. (1949): Nature *164*, 658

is monitored as a function of time; resonance fluorescence or absorption of a continuously operated source provides a convenient means of measuring the concentration of the atom or radical of interest. The use of *Q*-switched Ruby lasers and second-harmonic generation with non-linear crystals have made it possible to use ultra-violet radiation pulses of nanosecond duration, and mode-locked operation of Ruby or Neodymium lasers will enable the techniques to be extended into the picosecond range.

The illumination cell can be operated over a wide range of temperatures, and the method also offers the advantages of being capable of operation at moderately high pressures, being free of surface effects, and of high sensitivity. However, it suffers from lack of versatility, the difficulty of generating some species, absorption of the illuminating radiation by various constituents present, and the need to produce the mixture of gases with the reacting species being present in such concentrations as to provide a convenient decay rate.

In order to demonstrate the general principle of this type of experiment it is useful to consider that used by SLANGER and BLACK [2] for observations of the electronic to vibrational energy transfer between $O(^1D)$ and N_2 and between $O(^1D)$ and CO. The basis of the method was that by observing resonance fluorescence of the $(0-1)$ band at 159.7 nm of the $CO(A\,^1\Pi - X\,^1\Sigma^+)$ system the ambient concentration of CO molecules in the $v''=1$ level of the $X\,^1\Sigma_g^+$ ground state could be monitored. Thus by modulating the source of $O(^1D)$ atoms and observing the decay of the emission from $CO^*(v''=1)$ it was possible to estimate the contribution to the concentration in this level of the reaction:

$$O(^1D) + CO \rightarrow O(^3P) + CO^*(v''=1). \tag{14.1}$$

The effect of N_2 addition to the $O(^1D)-CO$ system could be interpreted to estimate the efficiency of the corresponding reaction of $O(^1D)$ and N_2. Vibrationally-excited N_2 does not radiate and its principal loss process was by energy transfer to CO:

$$N_2^*(v''=1) + CO(v''=0) \rightleftarrows N_2(v''=0) + CO^*(v''=1). \tag{14.2}$$

The forward reaction was more rapid than the observed loss rate of $CO^*(v''=1)$ and an increase in the $CO^*(v''=1)$ decay time was expected. Measurements for different pressures of N_2 allowed the efficiency of the electronic to vibrational energy transfer for the $O(^1D)-N_2$ system to be found relative to that for $O(^1D)-CO$ system, this being derived by relating the $CO^*(v''=1)$ concentration to the $O(^1D)$ production rate in the absence of N_2.

The experimental arrangement is shown in Fig. 26. The source of the $(0-1)$ band of $CO(A\,^1\Pi - X\,^1\Sigma^+)$ was the lamp situated below the reaction chamber and operating with a 1% mixture of CO_2 in argon. The source of $O(^1D)$ was the photodissociation of O_2 by the 147 nm radiation from a Xenon resonance lamp located above the reaction chamber. The procedure was to pulse the Xenon lamp, whose turn-off time was 30 μs, and to observe the decay of the continually-excited signal associated with the $CO^*(v''=1)$ molecules, which had lifetimes in the range 30 to 400 ms. This signal was monitored by a solar blind photomultiplier, which was protected from light from the Xenon lamp by a mixture of 0.2% of nitrous oxide in argon in attenuation chamber 3. Attenuation chambers 1 and 2 contained, respectively, a 20% mixture of ethane in helium at atmospheric pressure and a 20% mixture of CO in argon at atmospheric pressure and were designed to eliminate the $(0-0)$ band emission of $CO(A\,^1\Pi - X\,^1\Sigma^+)$ from the reaction chamber

[2] Slanger, T.G., Black, G. (1974): J. Chem. Phys. *60*, 468

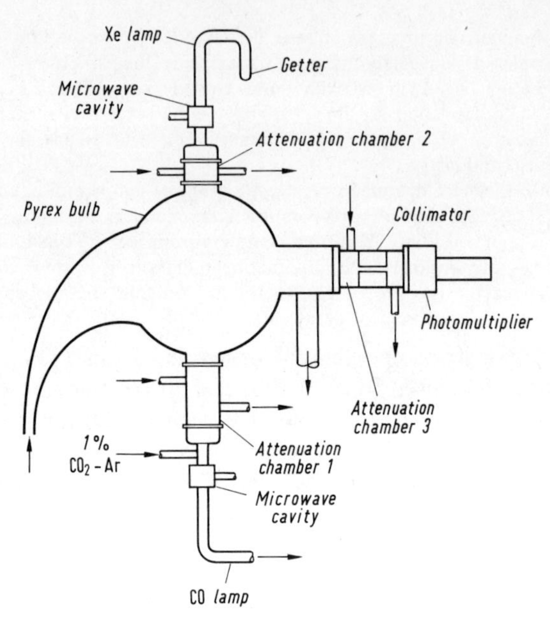

Fig. 26. The reaction cell and lamps used in the flash-photolysis experiment of SLANGER and BLACK[2]

and also radiation from the Xenon lamp capable of exciting vibrational states of ground-state CO. It is of interest to note that as the CO pressures used in the reaction chamber were generally in excess of 130 $\mathrm{N\,m^{-2}}$ (1 torr), resonance trapping of the 4.7 μm fundamental band occurred. Since the emission of this radiation represented the primary loss process of $CO^*(v''=1)$, the lifetime was substantially longer than the natural radiative lifetime and the concentration of the excited molecule was therefore increased.

The efficiency found for the transfer of electronic energy of $O(^1D)$ to vibrational energy of nitrogen, $33 \pm 10\%$, was considerably larger than theoretical estimates predicted. From a comparison of the energy transferred to N_2 vibration by this reaction with that resulting from the reaction of $N(^4S)$ with NO, SLANGER and BLACK[2] concluded that the quenching of $O(^1D)$ represents the dominant source of vibrationally-excited molecular nitrogen at altitudes below 300 km.

β) Discharge-flow methods. The discharge-flow technique makes use of the same principle as the flowing-afterglow method considered for ion-molecule reactions (Sect. 15β) in that the region in which the reactants are generated is separated from the reaction zone; the two gas streams transporting the relevant reactants are brought together immediately upstream of the reaction zone. The different applications of the technique have made use of various methods for the formation of the reactive species: glow discharges, thermal dissociation, photolysis and chemical reactions. Similarly, a variety of detection arrangements are possible: mass spectrometry, resonance fluorescence or absorption, chemiluminescent emission, catalytic probe calorimetry, electron spin resonance, and chemical analysis. The choice of source and detector depends in

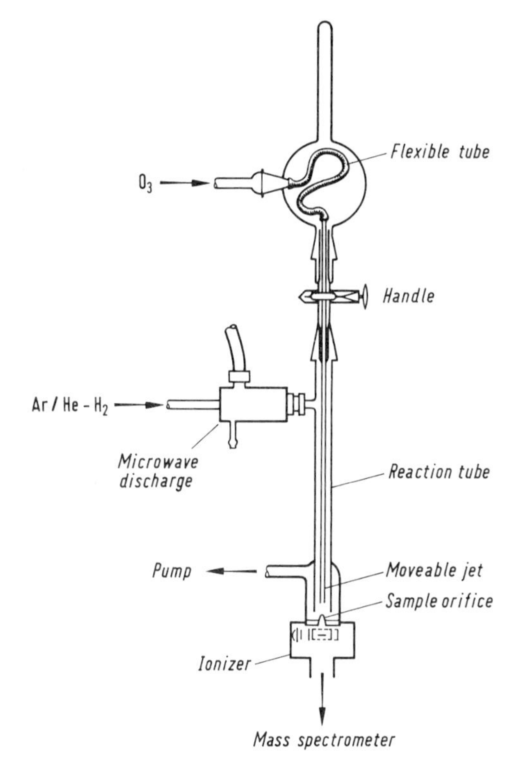

Fig. 27. The discharge-flow method used in the study of the reaction of hydrogen atoms with ozone by PHILLIPS and SCHIFF [3]

part on the experiment configuration chosen, particularly the extent to which the source, reactant inlet, or detector are movable. The main drawbacks of the discharge-flow approach to the study of neutral reactions are associated with the uncertainties arising from measurements of flow characteristics and reactions at walls.

An arrangement making use of a movable reactant jet is shown in Fig. 27 which represents the equipment employed by PHILLIPS and SCHIFF [3] in their measurement of the rate coefficient of the reaction between hydrogen atoms and ozone (reaction (12.8)). In this experiment the H atoms were generated by passing a stream of argon or helium containing H_2 through a microwave discharge. This gas passed down the reaction tube and was removed by a pump. The ozone molecules were introduced into the reaction tube at a jet which could be moved within the reaction tube by means of the handle shown, thereby altering the reaction time before the products passed through the sample orifice to the mass spectrometer. Very few measurements have been carried out on this reaction, which represents an important source of OH in the mesosphere, is responsible of the Meinel hydroxyl bands in the airglow (Fig. 50), and is a major loss process for H atoms in the mesosphere. The value obtained with the early form of equipment shown in Fig. 27, $2.6 \cdot 10^{-17}\,\mathrm{m^3\,s^{-1}}$, is still accepted as the most reliable.

15. Ion-molecule reactions. The study of ion-molecule reactions [*17*] has grown rapidly since the mid 1950's, much of the work having been inspired by the

[3] Phillips, L.F., Schiff, H.I. (1962): J. Chem. Phys. *37*, 1233

requirements of aeronomical investigations. Both positive and negative ion-molecule reactions have been involved and a great variety of experimental techniques have been applied [*18*]. Methods of sampling afterglows have been employed over the past two decades; although stationary afterglows have been widely used, the introduction of the flowing-afterglow system has greatly extended the numbers of reactions studied. The observation of reactions in drift tubes offers the advantage of providing data over a wide range of ion translational energies, and measurements at energies in excess of thermal have been derived from beam experiments which offer a great versatility in the range of reactions that can be examined. The combination of features of the flowing-afterglow and drift-tube techniques promises to offer particular advantages, and the use of mass-spectrometer experiments at high pressures has also provided important information relating to reactions occurring in the ionospheric D region.

α) Stationary pulsed afterglows. In the stationary-afterglow approach, measurements are confined to the period after the completion of energy input into the gas, when the gas returns to an equilibrium condition. It is considered that electron excitation processes are not present, since the electrons are thermalised by collisions with the neutral gas particles in time scales shorter than the reaction times being examined. As a result, the situation is not so complicated as in active discharges, and afterglows lend themselves more readily to quantitative measurements.

The pulsed-afterglow technique has been pioneered by Sayers and collaborators [1,2] who used a mass spectrometer to measure the ion composition as a function of time in the afterglow. Early measurements were made of the reactions of atomic oxygen ions with molecular nitrogen and oxygen (reactions (12.18) and (12.19)). In their analysis they examined the decay of O^+ ions resulting from the reactions and did not observe the product ions. Measurements for the reaction with molecular oxygen by SMITH and FOURACRE [3] over the temperature range 185 to 576 K suggested an inverse square-root temperature dependence for the rate coefficient.

The technique was also exploited by FITE and colleagues [4,5] who arranged to minimise the unknown effects of electron-ion recombination and of changes in ambipolar diffusion, the other loss mechanisms for the reactant ions.

The uncertainty about the states of the reactant gases is a major limitation of the stationary-afterglow technique; the ionizing discharge can cause excitation, and also dissociation. The use of photoionization sources has served to reduce the problem. Such a source has been used by LINEBURGER, PUCKETT and TEAGUE [6,7] in their measurements of NO^+ reactions which are of particular relevance to the D-region positive-ion chemistry. PUCKETT and TEAGUE [7] made use of $NO-H_2O$ gas mixtures to examine reactions which could produce water cluster ions from the NO^+ precursor ion. It was arranged that this primary reactant ion was produced by the photoionization of nitric oxide with krypton resonance radiation at 123.6 and 116.5 nm, a MgF_2 window providing a cut-off at 115 nm to preclude the direct photoionization of water vapour by radiations

---

[1] Dickinson, P.H.G., Sayers, J. (1960): Proc. Phys. Soc. London, *76*, 137

[2] Batey, P.H., Court, G.R., Sayers, J. (1965): Planet. Space Sci. *13*, 911

[3] Smith, D., Fouracre, R.A. (1968): Planet. Space Sci. *16*, 243

[4] Fite, W.L., Rutherford, J.A., Snow, W.R., van Lint, V.A.J. (1962): Discuss. Faraday Soc. *33*, 264

[5] Fite, W.L., Rutherford, J.A. (1964): Discuss. Faraday Soc. *37*, 192

[6] Lineburger, W.C., Puckett, L.J. (1969): Phys. Rev. *187*, 286

[7] Puckett, L.J., Teague, M.W. (1971): J. Chem. Phys. *54*, 2564

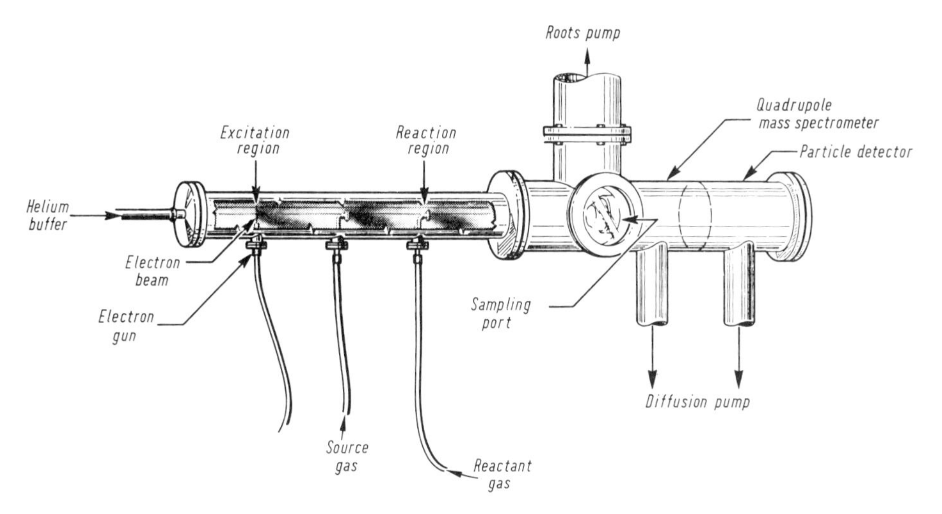

Fig. 28. The flowing-afterglow system used for studies of ion-molecule reactions by FERGUSON et al.[8]

shorter than 98.4 nm. These measurements yielded values of rate coefficients and/or equilibrium constants of reactions leading to the hydrates of NO^+ and water cluster ions.

β) Flowing afterglow. The flowing-afterglow technique introduced in 1964 by FERGUSON et al.[8] at Boulder has proved to be more versatile than any other method, and has made the greatest contribution to our knowledge of ion-molecule reaction rate coefficients.

The general arrangement of the system is shown in Fig. 28. In this form, ions were produced by an electron gun at the end of a metre-long tube of 8 cm inner diameter, although earlier versions made use of electrodeless microwave discharges or discharge electrodes consisting of a cylindrical cathode and small wire anode. Helium gas was passed as a buffer down the tube at a flow velocity of about $100\ \mathrm{m\,s^{-1}}$, this flow being maintained by a Roots-type pump at pressures between about 13 to $1{,}300\ \mathrm{N\,m^{-2}}$. The helium was partly ionized and the ions, together with triplet helium metastable atoms, flowed past a small jet by which the source gas was introduced into the system. These reactants could be ionized by either the He^+ ions alone, as in the production of N^+ and N_2^+ from N_2, or by He^+ and $He(2\,^3S)$, as in the production of O^+ and O_2^+ ions from O_2; if necessary, steps could be taken to remove He^+ or $He(2\,^3S)$ selectively. The ions produced then reacted with the relevant gas introduced at a second jet. The ion composition of the plasma was monitored at the end of the tube by a quadrupole mass spectrometer (Sect. 19α (iii)), pumped by a diffusion pump, followed by a particle detector. Since the neutral reactants were normally present in excess over the primary ion, the concentration of the primary ion decreased exponentially with time. The measurement then involved the recording of the decrease in primary-ion signal and the increase in the product-ion signal as a function of added neutral reactant. Both pulsed and steady-state modes of operation have been used.

The versatility of the technique permitted by the possible addition of various gases downstream from the active discharge probably represents the major advantage of the flowing-afterglow approach. In particular, measurements involving neutral constituents of low ionization potential or great chemical activity are made possible, and arrangements can also be made for introducing known concentrations of reactants. A second advantage relates to the state of the reactants. Since the neutral constituents are introduced downstream of the discharge they are usually in the ground-electronic and vibrational state. In addition, it is believed that superelastic collisions between

[8] Ferguson, E.E., Fehsenfeld, F.C., Schmeltekopf, A.L. (1969): Adv. At. Mol. Phys. 5, 1

electrons and ions almost certainly ensure that the ions are in the ground-electronic state before reaching the reaction zone.

Attention has already been drawn in Sect. 12δ to the results of measurements with the flowing-afterglow system for a number of positive-ion charge-transfer and ion-atom interchange reactions, Fig. 24. It is to be noted that these measurements extended to 900 K, the highest temperature at which ion-molecule reactions have been measured to date, and illustrate that fast reactions show little temperature dependence. A particularly important result obtained related to the reaction between atomic oxygen ions and molecular nitrogen (reaction (12.18)). This is one of the slowest exothermic ion-molecule reactions, the rate coefficient being about $1.5 \cdot 10^{-18}\,m^3\,s^{-1}$ at 300 K and decreases to about $8 \cdot 10^{-19}\,m^3\,s^{-1}$ near 900 K. However, vibrational excitation of the nitrogen by a discharge before it entered the flowing afterglow was accompanied by a large increase in the rate coefficient[9], as pointed out in Sect. 12δ. Spectroscopic determination of the N_2 vibrational distribution has permitted the examination of the reaction coefficient as a functional of vibrational temperature; an increase to about $5 \cdot 10^{-17}\,m^3\,s^{-1}$ was found for the upper limit of N_2 vibrational temperature considered, 6,000 K.

Another type of positive-ion reaction examined with the flowing-afterglow method has been the rapid conversion of metallic oxide ions, MgO^+ and SiO^+, to the metal ions by reaction with oxygen atoms[10,11]. These reactions are of interest in the interpretation of the concentrations of metal ions at E-region heights as observed by rocket-borne mass spectrometers.

The flowing-afterglow technique, in parallel with the mass-spectrometer ion source and drift-tube methods and the stationary-afterglow approach, has been applied to the measurement of reactions believed to be important in the formation of water cluster ions $H^+ \cdot (H_2O)_n$ in the D region from O_2^+ and NO^+, the primary ions formed during photoionization. The details of these reactions will be presented in Sect. 27 but it is of particular interest to note that the flowing-afterglow method has provided relevant information on the three-body attachment of NO^+ to CO_2 (DUNKUN et al.[12]), believed to be involved in the formation of $NO^+ \cdot H_2O$, and more recently on the reactions of this hydrate with H atoms and the OH radical proposed to produce the lowest-order water cluster ion H_3O^+ (FEHSENFELD et al.[13]).

Since negative ions can also be withdrawn and monitored from the flowing-afterglow system, studies of negative-ion reactions have also been possible. As indicated in Sect. 12ε, several measurements of negative-ion neutral reactions of relevance to the ionospheric D region have been measured[14]. It will be shown in Sect. 28 that these results form the basis of present theoretical models of D-region negative-ion chemistry. The extension of this work to include hydrated

[9] Schmeltekopf, A.L., Ferguson, E.E., Fehsenfeld, F.C. (1968): J. Chem. Phys. *48*, 2966

[10] Ferguson, E.E., Fehsenfeld, F.C. (1968): J. Geophys. Res. *73*, 6215

[11] Ferguson, E.E. (1972): Radio Sci. *7*, 397

[12] Dunkin, D.B., Fehsenfeld, F.C., Schmeltekopf, A.L., Ferguson, E.E. (1971): J. Chem. Phys. *54*, 3817

[13] Fehsenfeld, F.C., Howard, C.J., Harrop, W.J., Ferguson, E.E. (1975): J. Geophys. Res. *80*, 2229

[14] Fehsenfeld, F.C., Ferguson, E.E., Bohme, D.K. (1969): Planet. Space Sci. *17*, 1759

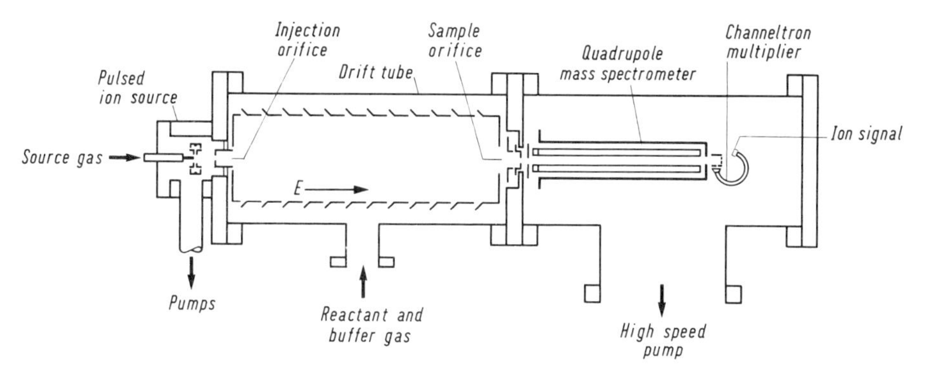

Fig. 29. The drift-tube system and mass-spectrometer arrangement used by JOHNSON and BIONDI[17]

negative ions by FEHSENFELD and FERGUSON[15] underlines the versatility of the flowing-afterglow technique.

ADAMS and SMITH[16] have recently developed an apparatus termed SIFT (selected ion flow tube) which has many of the features of the flowing-afterglow experiment but makes use of a mass-analysed beam of ions injected into the gas flow; these ions originate in a low-pressure microwave discharge in the appropriate gas and are extracted through a quadrupole mass filter. This SIFT technique offers the advantage that a single ion species can be introduced into the gas flow in the absence of excited species, electrons and photons. In their first measurements at 300 K with this apparatus Adams and Smith have shown that in the reactions of N^+ ions with O_2 molecules, more than 90% of the interactions follow channels (12.27) and (12.28) but dissociative charge transfer, yielding O^+, occurs in about 6% of the interactions.

γ) Drift tube. In the drift-tube technique, ions drift under the action of a longitudinal electrical field in an enclosure containing gas at a relatively high pressure. The temporal decay of a pulse of ions in the presence of the reactant gas is measured as a function of the ratio of electric field to gas pressure which controls the ion energy distribution.

Up until about 1960 this technique was applied in the main to measure ion mobilities. However, the inclusion of mass-spectrometer devices for the sampling of ion species at the end of their path through the drift tube has led to its increasing application in studies of ion-molecule reactions. Nevertheless, the interpretation of the measurements in terms of reaction data requires an analysis in which account is also taken of drift and diffusion effects.

Fig. 29 shows the arrangement used by JOHNSON and BIONDI[17] in their measurements of the reactions (12.18) and (12.19). The gas to be ionized, in this case oxygen, was admitted to a differentially-pumped electron-bombardment ion source. The sign and operation of the source was such that the oxygen did not enter the drift/reaction region. In the drift tube itself the uniform axial electric field E was generated by voltages applied to eleven cylindrical, overlapping guard rings and two end plates. At the end of the drift tube the ions passed through a small orifice into a quadrupole mass spectrometer and after mass analysis were detected by a channeltron electron multiplier.

The results obtained with this equipment over the O^+ ion energy range 0.04 to 0.1 eV showed reasonable agreement with those provided by the flowing-afterglow technique over the temperature range 300 to 600 K (Fig. 24).

---

[15] Fehsenfeld, F.C., Ferguson, E.E. (1974): J. Chem. Phys. *61*, 3181
[16] Adams, N.G., Smith, D. (1976): J. Phys. B *9*, 1439
[17] Johnsen, R., Biondi, M.A. (1973): J. Chem. Phys. *59*, 3504

However, at energies of about 0.15 eV for the N_2 reaction and 0.25 eV for the O_2 reaction the rate coefficients increased rapidly with increasing energies. JOHNSON and BIONDI have noted that the increase in the coefficient for the N_2 reaction is qualitatively similar to the change with N_2 vibrational temperature found by SCHMELTEKOPF et al.[9] and pointed out that both effects might occur in the F region, giving rise to enhanced values of this rate coefficient.

JOHNSON et al.[18] reported measurements of reactions between N^+ and N_2^+ ions with O_2 molecules using this type of apparatus, and more recently the equipment shown in Fig. 29 and a revised form capable of operating over a temperature range of about 130 to 500 K have been used[19] to determine as a function of temperature the equilibrium constant and the forward reaction coefficient of the three-body association of molecular nitrogen to nitric oxide ions (reaction (12.30)). Reference to this measurement and to its significance in studies of the D region was made in Sect. 12δ.

Another measurement of ionospheric interest carried out with the drift-tube technique has been the association of Na^+ ions with carbon dioxide (KELLER and BEYER[20]). It is considered that this might be relevant to the chemistry of metallic-ion layers in the D and E regions, as indicated in Sect. 30.

δ) Flow-drift tube. It can be seen from the previous discussion that although the flowing-afterglow technique offers a great versatility in measurement, it is restricted to the energy range from 0.01 to 0.11 eV, corresponding to temperatures of about 80 to 900 K. The drift method, although capable of measurement over a wide energy range, is limited in its versatility. The beam technique again offers chemical versatility but is restricted to energies greater than thermal, usually in excess of 1 eV ion kinetic energy. A development by MCFARLAND et al.[21] combined a flowing-afterglow system and drift tube, and thereby bridged the gap between the energy ranges covered by these two techniques and also provided chemical versatility.

A buffer gas, usually helium, flowed through a tube of length 1.25 m and inner diameter 8 cm under the action of a Roots pump backed by a mechanical pump; tube pressures of 25 to 150 $N m^{-2}$ were maintained. Ions were produced at one end of the tube by electron impact and were carried by a gas flow to a section where an electric field served to separate negative and positive ions. For reaction-rate measurements the ions passed to a drift-reaction section where the reactant gas was added, and the reduced ion concentration was monitored at the end of the tube by a mass spectrometer. The measurement of the depletion of the reactant ion as a function of the reactant-gas concentration provided the rate coefficient at the ion energy determined by the ratio of electric field to buffer-gas density in the drift-reaction section. Variation of this ratio provided the measurement of the particular rate coefficient as a function of the kinetic energy of the ion.

This technique has been applied to the measurement of both positive and negative-ion reactions, and also of mobilities by incorporating arrangements for pulses of ions to enter the drift-reaction section. For atmospheric considerations results have been obtained[22,23] for reactions of N^+ with O_2

---

[18] Johnsen, R., Brown, H.L., Biondi, M.A. (1970): J. Chem. Phys. *52*, 5080

[19] Johnsen, R., Huang, C.M., Biondi, M.A. (1975): J. Chem. Phys. *63*, 3374

[20] Keller, G.E., Beyer, R.A. (1971): J. Geophys. Res. *76*, 289

[21] McFarland, M., Albritton, D.L., Fehsenfeld, F.C., Ferguson, E.E., Schmeltekopf, A.L. (1973): J. Chem. Phys. *59*, 6610

[22] McFarland, M., Albritton, D.L., Fehsenfeld, F.C., Ferguson, E.E., Schmeltekopf, A.L. (1973): J. Chem. Phys. *59*, 6620

[23] McFarland, M., Albritton, D.L., Fehsenfeld, F.C., Ferguson, E.E., Schmeltekopf, A.L. (1974): J. Geophys. Res. *79*, 2925

(reactions (12.27), (12.28)), N_2^+ ions with O_2 (reaction (12.24)), N_2^+ ions with O (reactions (12.25), (12.26)), and of O^+ ions with N_2 and O_2 (reactions (12.18), (12.19)).

The measurements show good agreement with the values obtained for thermal energies from flowing-afterglow studies, with the drift-tube results at intermediate energies, and with beam data at the highest energies. In addition, the values and variations with temperature for reactions (12.18) and (12.19) have been confirmed by information derived from satellite measurements [*15*]. The overall laboratory data show that except for the N^+ reactions with O_2, the variation of rate coefficients with energy show minima, these occurring at energies of about 0.4 eV for (12.24), 0.1 eV for (12.18), 0.2 eV for (12.19) and 0.3 eV for (12.25), (12.26). McFarland et al.[22] indicate the possible importance of the energy dependences to studies of reaction mechanisms and, thereby, emphasise the need for measurements over wide energy ranges.

ε) Beam experiments. The principle of beam experiments is different from those described previously since it involves the use of nearly monoenergetic ions of a single mass reacting with reactant gases, the primary and secondary ions then being analysed with regard to mass, energy and angular distribution.

The separation of the ion source and reaction region by a mass selector allows time for the short-lived excited states to decay and the interacting beam only contains ions in the ground or long-lifetime excited states (i.e. greater than about 10^{-5} s). The reactant gas, which forms the target for the ion beam, is either contained in an enclosed collision chamber or is in the form of a beam that is intersected in an open evacuated region. The single-beam technique permits the use of larger neutral gas densities and is well suited to the measurement of total cross-sections. However, collisions occur over the whole length of the primary-ion path within the chamber and the effective size of the reaction volume will vary with scattering angle. In the case of the crossed-beam approach the reaction volume is small and nearly point-source scattering occurs, and this type of experiment is capable of producing more detailed information than the single-beam approach. The relatively lower pressure of the reactant gas, 10^{-5} to $10^{-4}\,\mathrm{N\,m^{-2}}$ compared with 10^{-3} to $10^{-1}\,\mathrm{N\,m^{-2}}$ in the reaction chamber of the single-beam experiment, does not lend itself to direct measurement and usually needs to be calculated.

The mass selection of the primary ions, the possible control of both ion and neutral beams, and the knowledge of the relative directions of incident beams and angular distribution of scattered ions represent unique features of the beam technique. These are of particular importance in relation to investigations of molecular reaction dynamics. As such, the technique offers the greatest information on basic data relevant to reaction mechanisms and energetics. The main disadvantage of the beam technique is its restriction to higher ion energies, and, in particular, its inability to operate effectively at thermal or near-thermal energies. This aspect is particularly serious in view of the evidence for the change in character of reaction mechanisms at a few tenths of an electron volt, as mentioned in relation to the comparison above of the results derived with the flow-drift and other techniques. However, the development of the merged-beam method, in which the ions and molecular beams are made to intersect collinearly, has enabled an extension to lower energies in some ion-molecule reactions.

The single and crossed-beam experiments have been applied successively to the measurement of a number of reactions[24, 25], generally at ion energies above 1 eV, including some involving excited states, such as reaction (12.40) and the following[26]:

$$O_2^+(a\,^4\Pi_u) + Ca \rightarrow Ca^+ + O_2. \qquad (15.1)$$

[24] Rutherford, J.A., Vroom, D.A. (1971): J. Chem. Phys. *55*, 5622

[25] Rutherford, J.A., Vroom, D.A. (1974): J. Chem. Phys. *61*, 2514

[26] Rutherford, J.A., Mathis, R.F., Turner, B.R., Vroom, D.A. (1972): J. Chem. Phys. *57*, 3087

In this connection, the crossed-beam technique has provided the most effective and, in most cases, the only measurements of the charge transfer between ions O^+, O_2^+, N^+, N_2^+ and NO^+ with metal atoms such as Mg and Fe[27,28].

ζ) Mass-spectrometer ion source. The use of a mass spectrometer in the analysis of the primary and secondary ions from an ion source of suitable design probably represented the earliest approach to the investigation of ion-molecule reactions. For the direct measurement of thermal reaction coefficients, pulse-operated ion sources offer advantages over d.c. ion sources.

KEBARLE and collaborators at the University of Alberta have successfully applied a pulsed electron-beam mass spectrometer to the measurement of clustering reactions believed to be important in the D region. The use of a relatively high pressure in the ion source (100 to 500 $N\,m^{-2}$) has avoided the possible production of ions in excited states. A 4 kV electron beam entered the ion source through a narrow slit and in the centre of the source, about 10^{-3} m below the electron beam, was an ion exist slit. The field-free character of the ion source allowed the ions to diffuse from the electron beam, to escape through the exist slit and be analysed with a quadrupole mass spectrometer.

The experiments of primary interest here are those involving reactions with water. In order to introduce a known small concentration (0.04 to 1.3 $N\,m^{-2}$) in a large concentration (100 to 500 $N\,m^{-2}$) of oxygen or nitrogen, a flow system was used in which the water vapour was introduced at a controlled rate into the oxygen or carrier gas stream. These experiments[29,30] provided the first measurements of reactions:

$$O_2^+ + O_2 + O_2 \rightarrow O_4^+ + O_2 \tag{15.2}$$

$$O_2^+ + H_2O + O_2 \rightarrow O_2^+ \cdot (H_2O) + O_2 \tag{15.3}$$

$$O_4^+ + H_2O \rightarrow O_2^+ \cdot (H_2O) + O_2 \tag{15.4}$$

$$O_2^+ \cdot (H_2O) + H_2O \rightarrow H_3O^+ \cdot OH + O_2 \tag{15.5}$$

$$\rightarrow H_3O^+ + OH + O_2 \tag{15.6}$$

$$H_3O^+ \cdot OH + H_2O \rightarrow H^+ \cdot (H_2O)_2 + OH \tag{15.7}$$

$$H_3O^+ + H_2O + M \rightarrow H^+ \cdot (H_2O)_2 + M \tag{15.8}$$

and successive three-body attachment of water molecules to produce higher-order water cluster ions.

It will be seen in Sect. 27 that these reactions form the basis of the reaction scheme for the formation of water clusters in the D region from O_2^+ ions as precursor. In addition, previous measurements by KEBARLE et al.[31] using α-particle and proton-beam spectrometers yielded values of the equilibrium constants for the formation and thermal decomposition of the water cluster ions of different order which led to estimates of the corresponding heats and entropies of solvation.

---

[27] Rutherford, J.A., Mathis, R.F., Turner, B.R., Vroom, D.A. (1971): J. Chem. Phys. *55*, 3785
[28] Rutherford, J.A., Vroom, D.A. (1972): J. Chem. Phys. *57*, 3091
[29] Good, A., Durden, D.A., Kebarle, P. (1970): J. Chem. Phys. *52*, 212
[30] Good, A., Durden, D.A., Kebarle, P. (1970): J. Chem. Phys. *52*, 222
[31] Kebarle, P., Searles, S.K., Zolla, A., Scarborough, J., Arshadi, M. (1967): J. Am. Chem. Soc. *89*, 6393

16. Dissociative recombination. The dissociative recombination of molecular ions represents the major chemical loss process for electrons at all heights in the ionosphere. The atmospheric conditions vary markedly over the height range of interest and it is important that the laboratory measurements relate as closely as possible to these conditions. In view of the marked increase of temperature with height between about 90 km and 250 km, and the lack of thermal equilibrium between electrons and ions or neutral particles at the upper levels, the dependence of recombination rate coefficients on neutral and electron temperatures is particularly important. In addition, the electronic and vibrational state of the ion should be known if possible.

Early measurements for oxygen and nitrogen ions did not employ mass analysis and it is not possible to assess the reliability of those measurements. Some of the results for nitrogen showed a substantial pressure dependence of the inferred values of recombination coefficient which suggested a change in ion composition, perhaps from N_2^+ to N_4^+. Furthermore, discrepancies between the results reported for O_2^+ and NO^+ with more recent results could be interpreted in terms of the presence of O_3^+ or O_4^+ and $(NO^+)_2$, respectively, which would not have been recognised without careful mass analysis.

In the measurements for NO^+ various methods have been used for the production of the ion. A repetitively-pulsed electrical discharge was used to ionize NO gas by MENTZONI and DONOHUE[1] and the electron density decay was monitored with a microwave technique. In view of the possible production of several ionized species in an electrical discharge in NO, the absence of arrangements for mass analysis was a serious drawback. Chemi-ionization in mixtures of NO and N_2 was used to produce NO^+ in the experiment of YOUNG and ST. JOHN[2]; currents arriving at pulsed collector electrodes were used to measure the rate of growth and equilibrium value of charge density. However, no mass analysis was used to identify the dominant ion. GUNTON and SHAW[3] photoionized purified NO or NO – Ne mixtures with single pulses of Lyman-α radiation within a 10 GHz microwave cavity, and used a microwave-cavity reflection technique to measure the electron concentration decay. A differentially-pumped mass spectrometer was incorporated but this did not operate effectively at the high gas pressures required in the measurements of recombination coefficient. They reported a $T^{-1.2}$ dependence on temperature. WELLER and BIONDI[4] also made use of photoionization as a source of NO^+ in a microwave-afterglow apparatus represented in the simplified diagram of Fig. 30. Highly purified nitric oxide was contained with neon buffer gas in a cylindrical copper microwave cavity which formed part of a bakeable ultra-high vacuum system. The pressures chosen for NO($1\,\mathrm{N\,m^{-2}}$) and neon (about $650\,\mathrm{N\,m^{-2}}$) avoided complex-ion formation and ensured recombination-controlled conditions. The use of single ionizing pulses from a Lyman-α lamp avoided the build up of negative ions. Measurements of the resonant frequency of the microwave cavity provided the decay of electron concentration, and simultaneous measurements were made of mass-identified ions diffusing to the cavity walls and passing through the orifice to the quadrupole mass spectrometer. Only measurements in which the change in NO^+ wall current followed the decay of electron concentration were accepted in the deduction of rate coefficients. This deduction involved comparison of the measurements with solutions of the continuity equation for electrons. Values of $7.4 \cdot 10^{-13}$, $4.1 \cdot 10^{-13}$, and $3.1 \cdot 10^{-13}\,\mathrm{m^3\,s^{-1}}$ were obtained at 200, 300 and 450 K, respectively, that for 300 K agreeing well with the value found previously by GUNTON and SHAW[3]. In addition, operation of the equipment at higher NO densities, when $(NO^+)_2$ was the dominant ion, provided a measurement of $1.7 \cdot 10^{-12}\,\mathrm{m^3\,s^{-1}}$ for this dimer ion at 300 K. All these measurements were carried out under conditions when the

---

[1] Mentzoni, M., Donohue, J. (1967): Can. J. Phys. *45*, 1565

[2] Young, R.A., St. John, G. (1966): Phys. Rev. *152*, 25

[3] Gunton, R.C., Shaw, T.M. (1965): Phys. Rev. *140*, A756

[4] Weller, C.S., Biondi, M.A. (1968): Phys. Rev. *172*, 198

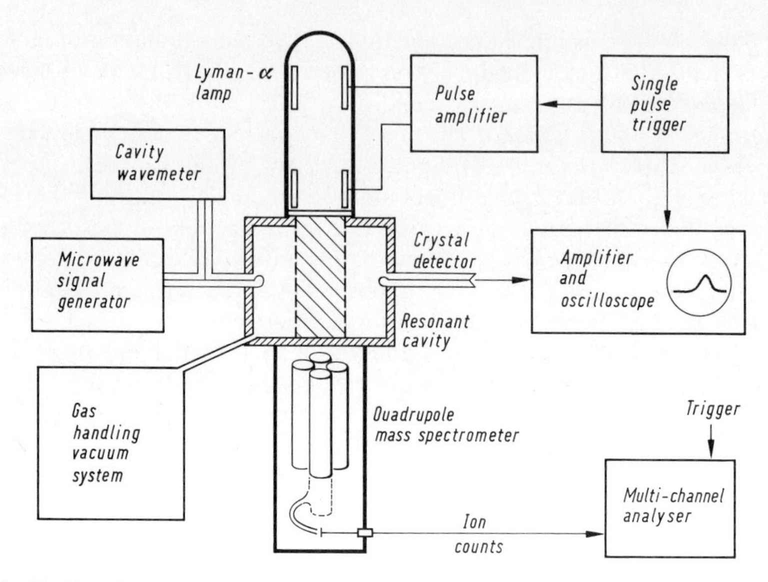

Fig. 30. The microwave mass-spectrometer afterglow apparatus of WELLER and BIONDI[4]

electron, ion and gas temperatures were similar. More recent measurements by HUANG et al.[5] have shown a $T_e^{-1.2}$ dependence for the recombination coefficient of NO^+ on electron temperature below 400 K and a $T_e^{-0.37}$ dependence above 500 K. These values referred to the ground-electronic state and probably the $v=1$ or $v=0$ vibrational level. A $T_e^{-0.83}$ dependence was found by WALLS and DUNN[6] from the depletion of trapped NO^+ ions when an electron beam of known energy was introduced; the storage time was sufficient to ensure that the ions were in the ground-electronic and vibrational state. As mentioned in Sect. 12γ(ii), this temperature dependence was similar to that found from ionospheric data [*15*].

KASNER and BIONDI[7] made use of a similar apparatus to that shown in Fig. 30 for the measurement of the rate coefficient of O_2^+, except that the ions were produced by a single microwave discharge. In the mixture of $O_2(0.1\,N\,m^{-2})$, krypton (or argon) $(130\,N\,m^{-2})$ and neon $(2{,}600\,N\,m^{-2})$ used, the O_2^+ ions were thought to be produced by charge transfer from krypton (or argon), itself ionized by a Penning ionization reaction with neon metastables. O_2^+ ions in the ground-electronic and low-vibrational state were expected to be produced. The measurements carried out over the temperature range 200 to 700 K indicated a T^{-1} dependence and a value of $2.2\cdot 10^{-13}\,m^3\,s^{-1}$ at 300 K.

Measurements of electron concentration changes carried out over the temperature range 180 to 630 K by SMITH and GOODALL[8] in helium-oxygen afterglows yielded values of O_2^+ recombination coefficients in good agreement with the results of KASNER and BIONDI[7], except at the lowest temperature. KASNER and BIONDI[7] also found a value of $2.3\cdot 10^{-12}\,m^3\,s^{-1}$ for the dimer ion O_4^+ at 205 K. All these values, like those from the early NO^+ measurements, referred to conditions when the electron, ion and gas temperatures were identical. MEHR and BIONDI[9] have used an apparatus similar to that shown in

---

[5] Huang, C.M., Biondi, M.A., Johnsen, R. (1973): Phys. Rev. A 11, *901*
[6] Walls, F.L., Dunn, G.H. (1974): J. Geophys. Res. *79*, 1911
[7] Kasner, W.H., Biondi, M.A. (1968): Phys. Rev. *174*, 139
[8] Smith, D., Goodall, C.V. (1968): Planet. Space Sci. *16*, 1177
[9] Mehr, F.J., Biondi, M.A. (1969): Phys. Rev. *181*, 264

Fig. 30, but modified to include arrangements for controlled electron heating during the afterglow, to examine the dependence of the dissociative recombination of O_2^+ and N_2^+ on electron temperature. The value found for O_2^+ at an electron temperature of 300 K, $1.95 \cdot 10^{-13}\,m^3\,s^{-1}$, was in good agreement with that found by KASNER and BIONDI[7] for similar electron, ion and neutral temperatures. In addition, the values varied as $T_e^{-0.70}$ over the range 300 to 1,200 K and as $T_e^{-0.56}$ between 1,200 and 5,000 K; the ionospheric determination of the rate coefficient showed good agreement with these high-temperature values [*15*]. In contrast to the general agreement found between the two sets of laboratory measurements for O_2^+ up to temperatures of 700 K (MEHR and BIONDI[9]), the results found by MEHR and BIONDI for N_2^+ showed significant differences from the measurement of KASNER[10] in which the electron, ion and gas temperatures were similar. The value found by MEHR and BIONDI for an electron temperature of 300 K, $1.8 \cdot 10^{-13}\,m^3\,s^{-1}$ was substantially smaller than the corresponding value reported by KASNER, $2.7 \cdot 10^{-13}\,m^3\,s^{-1}$. In addition, whereas MEHR and BIONDI fround a $T_e^{-0.39}$ dependence, KASNER found that the rate coefficient was essentially independent of temperature. The N_2^+ ions in both experiments should have been in the same state (ground-electronic and low-vibrational) since in each case they were produced by Penning ionization of N_2 by neon metastables. MEHR and BIONDI suggested that the increasing populations of excited vibrational states of N_2^+ as the neutral gas temperature is increased leads to enhancements of recombination coefficient which could compensate for the decrease of coefficient with increasing T_e for an ion in a given vibrational state. Vibrational excitation of the ions has also been invoked to explain the large values of coefficient deduced from ionospheric data [*15*].

A particularly interesting set of measurements carried out by LEU et al.[11] with the microwave-afterglow/mass-spectrometer apparatus depicted in Fig. 30 relate to the recombination of water cluster ions $H^+ \cdot (H_2O)_n$. The losses of ions and electrons were examined in the afterglow phase of a plasma generated by a microwave pulse in a helium-water vapour mixture. It appears that following the formation of metastable helium atoms by electron impact, Penning ionization of water vapour occurred and this was followed by a rapid two-body reaction forming H_3O^+ and a sequence of three-body clustering processes giving the higher-order cluster ions. By varying the gas temperature and the partial pressure of water vapour, it was arranged that different groups of the water cluster ion series dominated the afterglow. The results of their recombination measurements have been summarised in Table 8.

17. Positive-ion and negative-ion recombination. The measurement of the rate coefficients for positive-ion negative-ion mutual neutralisation (recombination) reactions at thermal energies has attracted considerably less attention than the corresponding ion-molecule and electron-ion interactions. The earliest studies were carried out under high-pressure conditions and related to three-body recombination. However, even for D-region conditions the measurements for the three-body processes represented effective two-body rate coefficients substantially smaller than true two-body recombinations.

---

[10] Kasner, W.H. (1967): Phys. Rev. *164*, 194
[11] Leu, M.T., Biondi, M.A., Johnsen, R. (1973): Phys. Rev. A 7, 292

In the measurements of MAHAN and PERSON[1], ions were produced by vacuum ultra-violet photolysis in gas mixtures made up of relatively small concentrations of NO and NO_2 in high pressures ($1.3 \cdot 10^3$ to $9.1 \cdot 10^4\,N\,m^{-2}$) of inert gases. Charge collection in a parallel-plate ionization chamber but without mass identification was used to determine two-body rate coefficients, by extrapolating the high-pressure data to the limit of zero pressure. A value of $2.1 \cdot 10^{-13}\,m^3\,s^{-1}$ was estimated for the two-body recombination of both NO^+ with NO_2^- and NO^+ with NO_3^-. EISNER and HIRSH[2] found values of $1.75 \cdot 10^{-13}\,m^3\,s^{-1}$ and $3.4 \cdot 10^{-14}\,m^3\,s^{-1}$ for these two reactions, respectively, by observing the decay of a thermal plasma produced from bombardment by 1 MeV electrons of nitrogen-oxygen mixtures at 300 to 3,000 $N\,m^{-2}$ pressure. In this case, mass identification was included and the ionization density was measured with a planar radio-frequency conductance probe.

Until recently the most fruitful attack on the measurement of ion-ion recombination coefficients has been carried out with the merged-beam approach.

This involved the merging, by means of an electromagnet, of two beams of particles and observing the interactions over paths of known lengths. In the experiment of ABERTH and PETERSON[3] the beams had kilovolt energies, but relative energies as low as 0.1 eV, travelled together for 0.33 m and were collected in Faraday cups after separation by electrostatic deflection. The neutral particles formed by the neutralisation of the positive and negative ions continued along the direction of the superimposed beams and were detected by secondary-electron emission.

In the application of this technique by PETERSON et al.[4], values of rate coefficients were inferred for reactions of $O^+ + O^-$, $N^+ + O^-$, $O_2^+ + O_2^-$, $N_2^+ + O_2^-$, $NO^+ + NO_2^-$, and $O_2^+ + NO_2^-$, these ranging from $1.6 \cdot 10^{-13}\,m^3\,s^{-1}$ to $5.1 \cdot 10^{-13}\,m^3\,s^{-1}$. In this merged-beam approach, control of the identity of ions and of their relative energies is ensured but there is some uncertainty about the internal states of the reactants, particularly for species having metastable electronic states. Thus if the interacting beams contain a substantial fraction of excited states, the rate coefficients found could be larger than those for ground-state reactants.

In this connection, the values of ion-ion recombination coefficients provided by the merged-beam technique for reactions involving molecular negative ions are significantly larger than the theoretical estimates of OLSON[5]. SMITH and CHURCH[6] have noted this difficulty arising out of internal excitation and have made use of a flowing-afterglow apparatus operated under collision-dominated plasma conditions to ensure that the reacting ions are first reduced to their ground state. Small cylindrical probes were used to measure electron, positive-ion, negative-ion concentrations, and relative ion masses in the high-density region of the plasma, and positive and negative-ion masses could be identified simultaneously at the downstream end of the tube. The positive and negative ions were produced by introducing controlled quantities of the reactants into a helium stream either upstream of a microwave cavity, when they would be subjected to an electrical discharge with the helium, or downstream of the cavity into the already established helium afterglow. The positive ions were thus produced by charge transfer with the He^+ and

[1] Mahan, B.H., Person, J.C. (1964): J. Chem. Phys. *40*, 392
[2] Eisner, P.N., Hirsh, M.A. (1971): Phys. Rev. Lett. *26*, 874
[3] Aberth, W.H., Peterson, J.R. (1970): Phys. Rev. A *1*, 158
[4] Peterson, J.R., Aberth, W.H., Moseley, J.T., Sheridan, J.R. (1971): Phys. Rev. A *3*, 1651
[5] Olson, R.E. (1972): J. Chem. Phys. 56, 2979
[6] Smith, D., Church, M.J. (1976): Int. J. Mass. Spectrom. Ion Phys. *19*, 185

He_2^+ ions and by Penning ionization by metastable helium atoms. The negative ions were formed by dissociative attachment, charge transfer, and electron attachment.

With these arrangements SMITH and CHURCH obtained values for the rate coefficients for the recombination of NO^+ with NO_2^- and NO_3^- at 300 K of $6.4 \cdot 10^{-14} m^3 s^{-1}$ and $5.7 \cdot 10^{-14} m^3 s^{-1}$, respectively, substantially smaller than those reported earlier but consistent with the theoretical predictions. SMITH et al.[7] have extended the measurements to the neutralisation of the water cluster ion $H^+ \cdot (H_2O)_4$ with a mixture of several negative ions by introducing water vapour in a subsidiary helium stream into the He/NO_2 afterglow. For a mixture of NO_3^-, NO_2^-, $NO_2^- \cdot H_2O$ and $NO_3^- \cdot H_2O$ they derive a rate coefficient of $5.5 \cdot 10^{-14} m^3 s^{-1}$. Such measurements for cluster ions clearly have direct application to theoretical models of the negative-ion chemistry of the D region and lower heights, as indicated in Sect. 28.

E. Experimental methods for measurement of neutral and ion composition

18. Direct and indirect approaches. The direct sampling of neutral and ionized constituents which has become possible during the past fifteen years has made a major contribution to our understanding of the neutral and ionized chemistry of the upper atmosphere. The identification of constituents and the measurement of their concentrations has both served to confirm existing ideas and to inspire new lines of thought. At the same time, less-direct approaches have also led to substantial progress. For instance, the interpretation by RATCLIFFE et al.[1] of ground-based measurements of electron concentrations with the conventional vertical-incidence method of radio-wave probing provided the basis for the present understanding of the ionization processes in the ionospheric F2 region, and complementary information was provided by the analysis of data on $O(^1D - ^3P)$ emissions at 630 nm in the nightglow by BARBIER[2]. More recently, observations of the ultra-violet dayglow emissions using rocket-borne Fastie-Ebert spectrometers by BARTH[3], MEIRA[4] and others have provided most data on the concentration of nitric oxide in the mesosphere and lower thermosphere, and photometer measurements of $O(^1S - ^1D)$ emissions at 557.7 nm on OGO 6 satellite have been interpreted by DONAHUE et al.[5] in terms of the atomic oxygen distribution at heights between

[7] Smith, D., Adams, N.G., Church, M.J. (1976): Planet. Space Sci. *24*, 697

[1] Ratcliffe, J.A., Schmerling, E.R., Setty, C.S.G.K., Thomas, J.O. (1956): Phil. Trans. R. Soc. London, Ser. A *248*, 621

[2] Barbier, D. (1957): Compt. Rend. *244*, 2077

[3] Barth, C.A. (1964): J. Geophys. Res. *69*, 3301

[4] Meira, L.G. (1971): J. Geophys. Res. *76*, 202

[5] Donahue, T.M., Guenther, B., Thomas, R.J. (1973): J. Geophys. Res. *78*, 6662

80 and 120 km. In another indirect approach, measurements of the atmospheric absorption of solar radiations have been employed to deduce information about neutral constituents. Rocket-borne spectrometer measurements of solar extreme ultra-violet radiations have been used by HINTEREGGER[6] and HALL et al.[7] to make deductions about the height distributions of atomic oxygen and molecular oxygen and nitrogen, and the technique has also been used for studies of ozone using ultra-violet radiations. NORTON and WARNOCK[8] have examined the occultation of solar Lyman-α radiation from measurements on Solrad 8 satellite and have shown that the molecular oxygen concentration near 100 km is greater in summer than in winter, and SCHMIDKE et al.[9] have also derived distributions of atomic oxygen between 200 and 500 km from observations on the AEROS satellite. In addition to these passive optical methods, measurements of resonance scattering of radiation emitted from rocket-borne ultra-violet lamps and from ground-based tuned lasers have been used to deduce the height distributions of atomic oxygen in the mesosphere and lower thermosphere[10], and sodium and potassium in the mesosphere[11,12], respectively. It is evident that these indirect approaches will continue to play an important role in the measurement of atmospheric constituents.

It seems appropriate in this brief description of experimental methods to pay particular attention to in-situ measurements with mass spectrometers which represent the major direct approach. As an indirect approach, the reception of radio waves incoherently scattered from electrons has proved to be a powerful technique for studying both neutral and ionized constituents, and therefore also deserves to be considered in some detail.

19. Mass-spectrometer experiments

α) *Ion-composition measurements.* Our present understanding of the chemical processes operating in the ionospheric regions is largely based on information derived from rocket and satellite-borne measurements with ion mass spectrometers, together with data on photoionization rates and on chemical reactions derived from laboratory measurements. The mass-spectrometer measurements have been confined to the past fifteen years but have ranged over all heights between 60 and 1,000 km [*19*]. See also RAWER's contribution in vol. 49/7, his Sect. 3.

At the greater heights the magnetic-sector field spectrometer (HOFFMAN[1]) and Bennett radio-frequency spectrometer (JOHNSON[2]) have been utilized on both rockets and satellites; the Bennett-type instrument has been most widely used down to about 140 km. Below about 90 km the mean free path of ions between collisions becomes comparable to the path length of the ions in the

[6] Hinteregger, H.E. (1962): J. Atmos. Sci. *19*, 351
[7] Hall, L.A., Schweizer, W., Hinteregger, H.E. (1965): J. Geophys. Res. *70*, 105
[8] Norton, R.B., Warnock, J.M. (1968): J. Geophys. Res. *73*, 5798
[9] Schmidke, G., Rawer, K., Fischer, T., Lotze, W. (1974): Space Res. *14*, 169
[10] Dickinson, P.H.G., Bolden, R.C., Young, R.A. (1974): Nature *252*, 289
[11] Bowman, M.R., Gibson, A.J., Sandford, M.C.W. (1969): Nature *221*, 456
[12] Megie, G., Bos, F., Blamont, J.E., Chanin, M.L. (1978): Planet. space Sci. *26*, 27
[1] Hoffman, J.H. (1969): Proc. IEEE *57*, 1063
[2] Johnson, C.Y. (1960) in: Encyclopedia of spectroscopy. Clark, G.L. (ed.), p. 587. New York: Reinhold

spectrometers, thus interfering with the operation of the instruments. Arrangements for pumping to pressures less than $0.1\,\mathrm{N\,m^{-2}}$ need to be incorporated to lengthen the mean free paths, and in almost all measurements at these heights quadrupole mass spectrometers[3] have been employed with cryogenically-cooled pumping systems or titanium getter pumps.

It is useful to consider in turn the general form of these three types of mass spectrometer. Since it is at D-region heights that the greatest number of species are encountered, and the greatest need for ion-composition data probably exists, it is appropriate to consider the quadrupole spectrometer in greater detail. Although emphasis is given to the determination of ion types, all three instruments have also been used for the measurement of neutral constituents by the incorporation of a suitable ionizing source, as described in Sect. 19β.

(i) *Magnetic-sector field mass spectrometer.* In this instrument, as used in Explorer 31 satellite[1], a high negative voltage, swept from about −4,000 to −150V, was used to draw ions from the ambient plasma into an analyser tube. Here the ions were separated into their constituent species by a magnetic analyser employing a field strength of about 0.2 T and were recorded by an electron multiplier and solid-state logarithmic electrometer amplifier. A particular feature of the instrument was its discrimination against ions of heavy mass. Such discrimination arose firstly from the use of the variable ion-accelerating voltage to scan the mass spectrum since the large mass ions were measured near the low voltage end where the transmission factor was less than at the high voltage/low mass end; secondly, the heavy ions being less mobile than the light ions were collected less efficiently from the ionospheric plasma; thirdly, the modulation due to satellite roll was more marked the heavier the ion. Laboratory calibrations were performed to assess mass-discrimination factors; in-flight calibration of the sampling efficiency was carried out by observing under conditions when the proportions of H^+ or O^+ ions in the F2 region were 90% or greater and comparing the total ion currents with electron or ion concentrations from other sensors. The roll-modulation effect was eliminated by measuring ion-abundance ratios only in the ram condition.

Coordinated measurements with such an instrument flown on a rocket and on Explorer 31 satellite by HOFFMAN et al.[4] during daytime in August 1966 showed that O^+ was the major ion from 200 to 1,000 km with H^+ contributing about 5% and He^+ less than 1%. From the chemical equilibrium situation between H^+ and O^+, a value of $2.4 \cdot 10^{10}\,\mathrm{m^{-3}}$ was deduced for the hydrogen atom concentration at 250 km. In addition, from the He^+ concentration near 400 km a rate coefficient for the loss of these ions with nitrogen molecules of $1.2 \cdot 10^{-15}\,\mathrm{m^3\,s^{-1}}$ was derived, in good agreement with subsequent laboratory measurements, Fig. 24. HOFFMAN[1] interpreted phase differences between the maxima of the satellite-roll modulations in the H^+ and O^+ signal currents from the magnetic-sector spectrometer on board Explorer 31 as evidence that H^+ ions were flowing upward with a velocity of 1.0 to $1.5 \cdot 10^3\,\mathrm{m\,s^{-1}}$ at 2,800 km altitude. This represented the first experimental observation of the polar wind (BANKS and HOLZER[5]).

(ii) *Bennett radio-frequency ion mass spectrometer.* Bennett-type mass spectrometers have been widely used for rocket-borne measurements of the E, F1 and F2 regions[6-10]. In addition, the instrument has been employed on the OGO series of satellites[11-13] and on Explorer 32[14]. In

---

[3] Lever, R.F. (1966): IBM J. Res. Dev. *10*, 26

[4] Hoffman, J.C., Johnson, C.Y., Holmes, J.C., Young, J.M. (1969): J. Geophys. Res. *74*, 6281

[5] Banks, P.M., Holzer, T.E. (1968): J. Geophys. Res. *73*, 6846

[6] Holmes, J.C., Johnson, C.Y., Young, J.M. (1965): Space Res. *5*, 756

[7] Brinton, H.C., Pharo, M.W., Mayr, H.G., Taylor, H.A. (1969): J. Geophys. Res. *74*, 2941

[8] Pharo, M.W., Scott, L.R., Mayr, H.G., Brace, L.H., Taylor, H.A. (1971): Planet. Space Sci. *19*, 15

[9] Giraud, A., Scialom, G., Pokhounkov, A., Poloskov, S., Tulinov, G. (1971): Space Res. *11*, 1057

[10] Johnson, C.Y. (1972): Radio Sci. *7*, 99

[11] Taylor, H.A., Brinton, H.C., Smith, C.R. (1965): J. Geophys. Res. *70*, 5769

[12] Taylor, H.A., Brinton, H.C., Pharo, M.W., Rahman, N.K. (1968): J. Geophys. Res. *73*, 5521

[13] Taylor, H.A., Mayr, H.G., Brinton, H.C. (1970): Space Res. *10*, 663

[14] Brinton, H.C., Pickett, R.A., Taylor, H.A. (1969): J. Geophys. Res. *74*, 4064

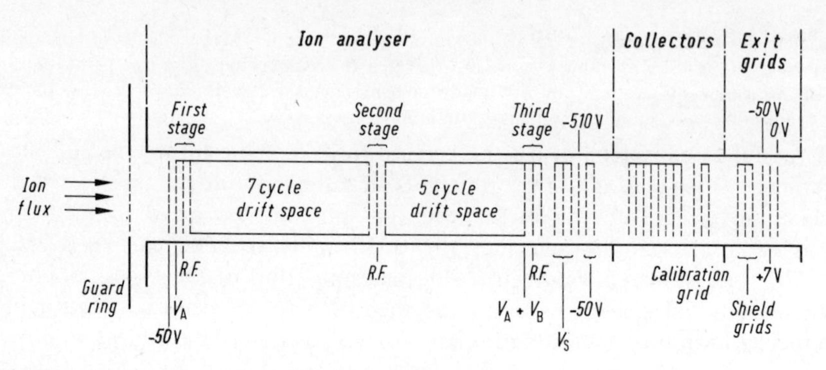

Fig. 31. A cross-sectional view of the ion analyser, collector and exit-grid assemblies in the radio-frequency ion mass spectrometer designed for Atmospheric Explorer C and E satellites, after BRINTON et al.[15]

order to understand its mode of operation, reference is made to the cross-sectional view in Fig. 31 of the instrument designed for the Atmospheric Explorer C and E missions (BRINTON et al.[15]). Ambient ions enter the instrument through the guard-ring grid and are accelerated by a slowly varying voltage sweeping between about -400 to -26 V. These ions pass through three series of grids to which a radio-frequency signal of frequency 2.5, 5 or 10 MHz is applied, according to the mass range being considered. For each ion there is a value of the accelerating voltage which causes the ion to move with the resonant velocity of the instrument. In this condition the ions gain energy from the r.f. fields in all three analyser stages, in the same manner as in a linear accelerator. The extra acceleration imparted to the ions is sufficient to allow them to pass through the retarding d.c. potential V_s and reach the collector plates; the value of this potential determines the balance between sensitivity and mass resolution of the instrument. The basic relationship between the mass of the ion, the tube geometry and the grid potentials in this condition is given by:

$$m_i = \frac{26.6(|V_A| + |V_B|)/\mathrm{V}}{(s/\mathrm{mm})^2 (f/\mathrm{MHz})^2} \qquad (19.1)$$

where m_i represents the ion mass in a.m.u., V_A is the sweep potential in volts, V_B is a constant bias potential in volts intended to hold the ion velocity constant between the first and second drift spaces, s is the spacing between the grids in the r.f. analyser stages in mm, and f is the radio frequency in MHz.

The three radio frequencies then provide overlapping mass ranges of 1 to 4, 2 to 18 and 8 to 72 a.m.u. for each sweep of the voltage V_A.

The Bennett mass spectrometer characteristically shows a reduced efficiency for ions which are resonant at small values of the sweep potential V_A. This mass discrimination is evaluated during the laboratory calibration of the instrument prior to flight.

This type of instrument operated on rockets has provided information on the day-to-night changes of ions at F1 and F2 region heights[6-9], the effects of increases in solar activity[9] and solar eclipses[16], and other variations. The height distributions of ions shown in Fig. 6 have been derived from r.f. ion mass spectrometer observations from rockets. Observations on the OGO 2

---

[15] Brinton, H.C., Scott, L.R., Pharo, M.W., Coulson, J.T. (1973): Radio Sci. *8*, 323

[16] Brace, L.H., Mayr, H.G., Pharo, M.W., Scott, L.R., Spencer, N.W., Carignan, G.R. (1972): J. Atmos. Terr. Phys. *34*, 673

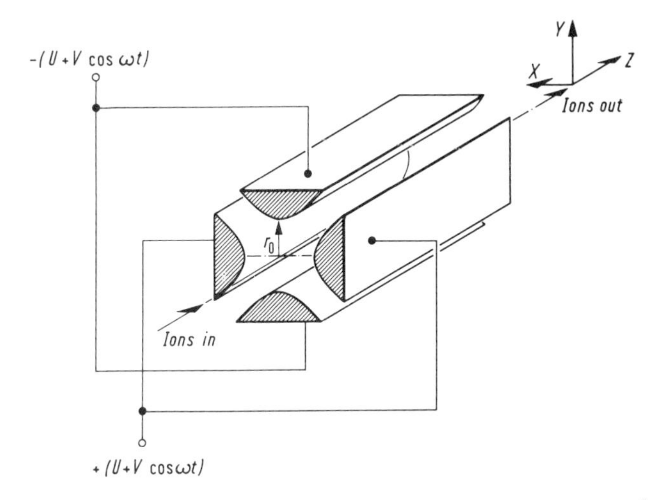

Fig. 32. A schematic view of the quadrupole mass filter, after LEVER[3]

satellite provided the first detailed measurements of the latitude variation of ion composition[12]. In a nearly polar dawn-dusk orbit, O^+ and H^+ were found to be the major ions between 415 and 1,525 km with N^+ and He^+ being minor ion constituents; at 1,000 km O^+ dominated in both northern and southern hemispheres but was superseded by H^+ at low latitudes. A persistent minimum or "trough" was observed in the concentrations of H^+ and He^+ near $\pm 60°$ geomagnetic latitude and this was confirmed with Bennett-type instruments on satellites OGO 4[13] and Explorer 32[14], the former also showing a trough in O^+ and N^+ concentrations near $\pm 70°$ geomagnetic latitude. The Explorer 32 observations also demonstrated the diurnal and seasonal variations of ion composition at greater heights, out to 2,700 km, and the OGO 4 data showed evidence that the geomagnetic field plays an important role in controlling the distribution of upper atmosphere ionization[17].

(iii) *Quadrupole mass-spectrometer experiment.* The light, compact construction of the quadrupole mass spectrometer combined with its insensitivity to ion velocity has rendered the instrument well-suited to rocket applications. It has been utilized in D-region studies at a number of sites and under a variety of conditions since 1963. The instrument is based on the mass-analyzing properties of linear quadrupole electric fields, the principle of which can be described by reference to the schematic view represented in Fig. 32, after Lever[3]. Potentials of $\pm(U+V\cos\omega t)$ are applied to opposite pairs of four electrodes of hyperbolic shape as shown. In this configuration the potential at a point is described by:

$$\varphi(x,y)=\left[\frac{x^2-y^2}{r_0^2}\right](U+V\cos\omega\, t). \tag{19.2}$$

This potential has the particular feature that the x and y components of the electric field are independent of y and x, respectively, so that the ions have independent oscillations in the x and y directions. The equations of motion of ions in the potential configuration $\varphi(x,y)$ are special cases of Mathieu's differential equations. It is known that two types of solution exist. For one the ion travels in direction z with constant velocity and exponentially increasing amplitudes of oscillation in directions x and y; these ions are lost to one of the four electrodes. The second solution again

[17] Taylor, H.A. (1971): Planet. Space Sci. *19*, 77

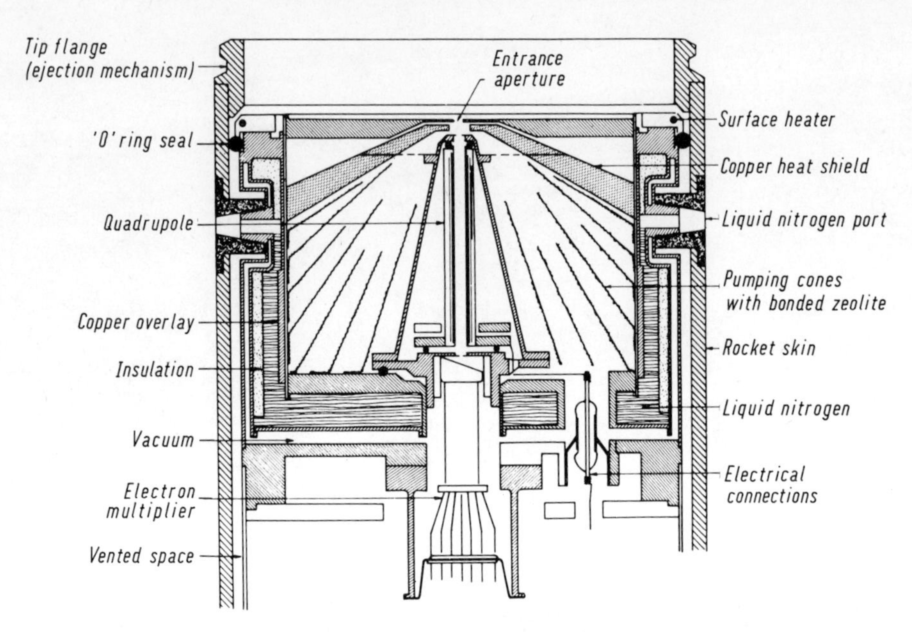

Fig. 33. A cross-sectional view of the quadupole mass spectrometer system with liquid-nitrogen cooled zeolite adsorption pump, after NARCISI and ROTH[19]

correponds to a constant velocity in direction z but with stable orbiting paths; these ions pass through the system to a collecting plate. It is found that these stable orbits are possible if the ratio U/V is equal to, or less than, 0.166. For operation near this critical value the mass-to-charge ratio is determined by the quotient $V/r_0^2\omega$. Mass analysis can then be affected by varying U and V with time while maintaining a constant ratio, the mass range being determined by r_0, the operating frequency ω and the range over which V is varied.

In the rocket experiment of NARCISI and BAILEY[18] the field-generating electrodes were made of stainless steel rods 7.6 cm long and 3.3 mm diameter, supported by a stainless steel conical structure. Ions were drawn into the entrance aperture by a potential of -8 V on the forward plate and were further accelerated to the quadrupole structure and rod system which were both biased at -128 V relative to the quadrupole vacuum envelope. The r.f. and d.c. voltages applied to the rods could then be adjusted so that only those ions with a specific mass-to-charge ratio would traverse the length of the rods and be recorded by an electron multiplier. Alternatively, by removing the d.c. voltage from the rods, the quadrupole could be operated in such a way that all ions above a specified mass could be focussed by adjusting the r.f. voltage. Calibration prior to flight was arranged using a standard electron source, and known gases and gas mixtures. A liquid-nitrogen cooled zeolite adsorption pump surrounded the conical support structure for the quadrupole rods as shown in Fig. 33, from NARCISI and ROTH[19]; this was capable of pumping speeds in excess of 100 litre s^{-1} at pressures near 0.013 N m^{-2}. An area of 0.2 m^2 of adsorptive zeolite was bonded onto seven concentric stainless steel cones contained in a liquid-nitrogen dewar. Before flight the system was pumped and baked for several hours to free the zeolite of most of its absorbed gas, and then sealed off after cooling.

In the operation of a quadrupole mass spectrometer by GOLDBERG and AIKIN[20] the instrument was housed in a chamber evacuated by a titanium getter pump and ion pump. Prior to launch a fresh coating of titanium was deposited on the inner walls of the vacuum chamber, and

---

[18] Narcisi, R.S., Bailey, A.D. (1965): J. Geophys. Res. *70*, 3687
[19] Narcisi, R.S., Roth, W. (1970): Adv. Electron. Electron Phys. *29*, 79
[20] Goldberg, R.A., Aikin, A.C. (1971): J. Geophys. Res. *76*, 8352

this deposit was maintained in an active pumping state until the front aperture to the housing and spectrometer was opened during flight.

As illustrated in Fig. 7 the general result of the observations carried out with quadrupole mass spectrometers under normal conditions[18–23] is that the middle and lower D region is dominated by water cluster ions, $H^+ \cdot (H_2O)_n$. The concentrations of these ions decrease rapidly with height above about 82 km in daytime and above about 86 km at twilight and at night. Above these heights NO^+ and O_2^+ ions dominate, and metal ions occur in a broad layer near 93 km, and also in more variable layers at greater heights; an observation by GOLDBERG et al.[24] during nighttime near the equator has shown the presence of light metallic ions, including Mg^+, Na^+ and possibly Si^+ up to 300 km and heavier ions Ca^+ and K^+ up to 200 km. The transition height between the water cluster ions and NO^+, O_2^+ has found to be lower during anomalous conditions in winter, both with quadrupole[25] and magnetic-sector-type[26] mass spectrometers.

The situation has also been found to change during high-latitude disturbed conditions. During a period of auroral activity NO^+ and O_2^+ were found to be the major ions down to about 75 km (KRANKOWSKY et al.[27]); during a Polar Cap Absorption (PCA) event NO^+ and O_2^+ appeared as the major ions down to 77 km at night, and down to 73 km in daytime and 78 km at sunset (NARCISI et al.[28]). These measurements during the PCA have helped to clarify the ion chemistry operating in the D region during disturbed conditions (NARCISI[29]).

The majority of mass-spectrometer observations have shown that $H^+ \cdot (H_2O)_2$ is the dominant water cluster ion in the quiescent D region, with significant concentrations of H_3O^+ and $H^+ \cdot (H_2O)_3$ also being present. There is, however, some doubt whether the ions $H^+ \cdot (H_2O)_2$ and H_3O^+ are representative of ambient conditions, since fragmentation of more complex weakly-bound ions could occur through thermodynamic break-up because of increased temperatures at the shock layer associated with the supersonic velocity of the rocket, or through collisions resulting from the draw-in electric field (NARCISI and ROTH[19]). In this connection, it has been found that larger cluster ions tend to be observed with lower-velocity rockets and also during the reduced shock

[21] Narcisi, R.S. (1967): Space Res. *7*, 186

[22] Narcisi, R.S., Bailey, A.D., Wlodyka, L.E., Philbrick, C.R. (1972): J. Atmos. Terr. Phys. *34*, 647

[23] Goldberg, R.A., Aikin, A.C. (1973): Science *180*, 294

[24] Goldberg, R.A., Aikin, A.C., Krishna Murthy, B.V. (1974): J. Geophys. Res. *79*, 2473

[25] Beynon, W.J.G., Williams, E.R., Krankowsky, D., Arnold, F., Bain, W.C., Dickinson, P.H.G. (1976): Nature *261*, 118

[26] Zbinden, P.A., Hidalgo, M.A., Eberhardt, P., Geiss, J. (1975): Planet. Space Sci. *23*, 1621

[27] Krankowsky, D., Arnold, F., Wieder, H., Kissel, J., Zahringer, J. (1972): Radio Sci. *7*, 93

[28] Narcisi, R.S., Philbrick, C.R., Thomas, D.M., Bailey, A.D., Wlodyka, L.E., Baker, D., Federico, G., Wlodyka, R.A., Gardner, M.E. (1972) in: Proc. COSPAR Symp. on "Solar Particle Event of November 1969", Air Force Cambridge Research Laboratories, Boston, Mass. Spec. Rep. 144, pp. 421–431

[29] Narcisi, R.S. (1972): In Proc. COSPAR Symp. on "Solar Particle Event of November 1969", Air Force Cambridge Research Laboratories, Boston, Mass. Spec. Rep. 144, pp. 557–569

conditions on the downleg of the trajectory[20, 27]. ARNOLD and KRANKOWSKY[30] have recently examined the effects of modifying the geometry of the sampling region and of reducing the draw-in potential. They found that the cluster ions were dominant below 85 km and, although all ions $H^+ \cdot (H_2O)_n$ with $n=1-7$ were observed, $H^+ \cdot (H_2O)_4$ dominated below 82 km with $H^+ \cdot (H_2O)_3$ and $H^+ \cdot (H_2O)_2$ being next in importance. In addition, new masses consistent with ions $NO^+ \cdot CO_2$, $NO^+ \cdot N_2$, $NO^+ \cdot H_2O \cdot CO_2$ and $NO^+ \cdot (H_2O)_2$ were observed, and also the corresponding complexes of O_2^+. It will be seen in Sect. 27 that the observation of such ions is of particular interest in connection with current ideas concerning the formation of water cluster ions.

As indicated in Fig. 8 the negative-ion measurements with cryogenically-pumped mass spectrometers by NARCISI et al.[31] and ARNOLD et al.[32] have shown conflicting results. NARCISI et al.[33] have also found that during a PCA event O_2^- was the dominant ion between 72 and 94 km in daytime and, although this ion was also present below 80 km at nighttime, O^- was the major ion between 74 and 94 km at that time. The observations of these ions above 80 km are difficult to understand in view of the rapid associative detachments by O atoms expected (e.g. reaction (12.35)). Satellite uses of this type of instrument are reported by RAWER in vol. 49/7, his Sects. 3 and 9.

β) Neutral-composition measurements. The three types of mass spectrometer described above can in principle be used for the measurement of neutral composition by the incorporation of suitable ionizing sources. See also RAWER's contribution in vol. 49/7, his Sect. 3.

Rocket-borne studies of thermospheric constituents have made use of magnetic-deflection instruments[34-36], radio-frequency instruments[37, 38] and quadrupole mass spectrometers[39, 40], and this latter technique has also been used with a liquid-nitrogen cooled cryopump system to extend the measurements down to about 80 km[41]. The application of these instruments to neutral-composition studies is accompanied by several difficulties, and the measurements have not shown systematic results. It is not clear whether some of the variations indicated are due to diurnal, seasonal and geographical effects, or arise from experimental errors. The incorporation of similar experiments in satellites has already helped to clarify some of the variations. A magnetic-deflection

---

[30] Arnold, F., Krankowsky, D. (1975) in: Proc. Int. Symp. on Solar Terrestrial Physics, Soc. Paulo, Brazil, May 1974, *3*, pp. 30–50

[31] Narcisi, R.S., Bailey, A.D., Wlodyka, L.E., Philbrick, C.R. (1972): J. Atmos. Terr. Phys. *34*, 647

[32] Arnold, F., Kissel, J., Krankowsky, D., Wieder, H., Zahringer, J. (1971): J. Atmos. Terr. Phys. *33*, 1169

[33] Narcisi, R.S., Sherman, S., Philbrick, C.R., Thomas, D.M., Bailey, A.D., Wlodyka, L.E., Wlodyka, R.A., Baker, D., Frederico, G. (1972) in: Proc. of COSPAR Symp. on "Solar Particle Event of November 1969", Air Force Cambridge Research Laboratories, Boston, Mass. Spec. Rep. 144, pp. 411–420

[34] Nier, A.O., Hoffman, J.H., Johnson, C.Y., Holmes, J.C. (1964): J. Geophys. Res. *69*, 979

[35] Nier, A.O., Hoffman, J.H., Johnson, C.Y., Holmes, J.C. (1964): J. Geophys. Res. *69*, 4629

[36] Kasprzak, W.T., Krankowsky, D., Nier, A.O. (1968): J. Geophys. Res. *73*, 6765

[37] Martynkevitch, G.M., Bjuro, E.D., Mal, G.V. (1968): Space Res. *13*, 947

[38] Pokhounkov, A. (1968): Space Res. *8*, 955

[39] Schaefer, E.J., Nichols, M.H. (1964): J. Geophys. Res. *69*, 4649

[40] Mauersberger, K., Muller, D., Offermann, D., von Zahn, U. (1968): J. Geophys. Res. *73*, 1071

[41] Philbrick, C.R., Faucher, G.A., Trzcinski, E. (1973): Space Res. *13*, 255

instrument was flown on Explorer 1 satellite[42] and also on the Atmosphere Explorer series of satellites[43], a Bennett-type of instrument on the Cosmos 274 satellite[44], and quadrupole instruments on the OV-36 satellite[45] and on the Atmospheric Explorer satellites[46]. In addition an r.f. monopole spectrometer has been operated on the ESRO 4 satellite[47].

The nature of the ion source is a major consideration in the use of mass spectrometers for measurements of neutral constituents; both closed and open sources have been adopted. In a closed source the ionizing region connects with the ambient atmosphere through a carefully defined small orifice and the incoming neutral particles are thermalised in this region. This type of source is well suited to the measurement of chemically inert constituents since it permits the ambient and measured concentrations to be related on the basis of kinetic theory considerations and a knowledge of the vehicle velocity. However, reactive species, such as atomic oxygen, can be lost by reactions at the ion-source walls and the measured concentrations could therefore be too small. In open sources the ambient gas particles are allowed to enter the ionizing region with the minimum of collisions with the structure. However, the measurements cannot easily be related to atmospheric concentrations because of the possibility of recombination of constituents prior to ionization and because of uncertainties in thermal accommodation.

Comparisons of rocket-borne measurements with closed and open sources have been made by Kasprzak et al.[36] and Hedin and Nier[48]. The former authors found that the closed and open-source instruments showed similar height distributions of atomic oxygen but the concentrations obtained with the closed source were very much smaller than those observed with the open source. They have suggested that the loss of atomic oxygen for each contact with a surface for both types of source was about 14%, which implied that the atomic oxygen concentrations derived from their open-source measurements needed to be increased by about 25%. However, Riley and Giese[49] proved from laboratory measurements that on surfaces such as those employed in mass spectrometers the loss per contact was about 50%. Since for an open-source instrument an atom may make several bounces before being measured, it appears that measurements may be in error by a factor of 2. In fact, Offermann and von Zahn[50] have suggested that the discrepancy might be still larger since measurements with a helium-cooled ion source have shown a ratio of atomic oxygen to molecular oxygen at 120 km of about 3.5 whereas other values reported for this height ranged from 0.4 to 1.6.

Both open-source and closed-source instruments have been included in the Atmospheric Explorer series of satellites[51,52]. The open-source version has been operated in the normal mode and in one which uses the velocity of the spacecraft to distinquish between incoming ambient particles and ambient particles which have struck instrument surfaces and become accommodated; the combined use of the two modes has made it possible to measure the concentrations of chemically active constituents, such as atomic oxygen, as well as inactive species[53]. Intercomparisons of concentrations measured by these two instruments, and by mass spectrometers carried on other satellites, at cross-over points of orbits in the 150 to 300 km height range have shown agreement generally within the experimental errors[53].

20. Incoherent Scatter. (See also Rawer's contribution in vol. 49/7, his Sect. 5.) In the conventional method of ionospheric sounding with radio waves,

---

[42] Reber, C.A., Nicolet, M. (1965): Planet. Space Sci. *13*, 617
[43] Nier, A.O., Potter, W.E., Hickman, D.R., Mauersberger, K. (1973): Radio Sci. *8*, 271
[44] Romanowsky, Y.A., Katyushina, K.V. (1974): Space Res. *14*, 163
[45] Philbrick, C.R. (1974): Space Res. *14*, 151
[46] Pelz, D.T., Reber, C.A., Hedin, A.E., Carignan, G.C. (1973): Radio Sci. *8*, 277
[47] Prölss, G.W., von Zahn, U. (1974): Space Res. *14*, 157
[48] Hedin, A.E., Nier, A.O. (1966): J. Geophys. Res. *71*, 4121
[49] Riley, J.A., Giese, C.F. (1970): J. Chem. Phys. *53*, 146
[50] Offermann, D., von Zahn, U. (1971): J. Geophys. Res. *76*, 2520
[51] Nier, A.O., Potter, W.E., Hickman, D.R., Mauersberger, K. (1973): Radio Sci. *8*, 271
[52] Pelz, D.T., Reber, C.A., Hedin, A.E., Carignan, G.R. (1973): Radio Sci. *8*, 277
[53] Trinks, H., von Zahn, U., Reber, C.A., Hedin, A.E., Spencer, N.W., Krankowsky, D., Lämmerzahl, P., Kayser, D.C., Nier, A.O. (1977): J. Geophys. Res. *82*, 1261

use is made of reflections which occur from levels where the sounding frequency is equal to the local plasma frequency. Such reflections will occur for frequencies up to the critical value corresponding to the peak electron density of the layer concerned. In addition, partially coherent scattering can occur from spread F or other ionospheric irregularities. No reflection or partially coherent scattering would be expected for frequencies substantially greater than the maximum plasma frequency along the path. However, individual electrons behave as independent weak scatterers, as first demonstrated by J.J. THOMSON. The relatively low value of the classical cross-section ($10^{-28}\,m^2$), together with the maximum electron density expected ($10^{12}\,m^{-3}$), imply that very powerful radar systems are required for the detection of incoherently-scattered radio waves. For selecting the height of the scattering volume use is made of radio-wave pulses, normally transmitted and received at the same site, or continuous waves transmitted and received with highly-directional antennae separated by 100 to 300 km. Following the first experimental test of the technique by BOWLES[1], a number of incoherent-scatter systems have been operated successfully to provide information on the neutral and ionised atmosphere at heights of the ionospheric E and F regions and indeed to several thousand kilometres on occasions.

α) Because of the randomness of position of electrons arising out of their thermal motion, incoherent scatter can occur even from regions which are homogeneous on the macroscopic scale. Theoretical studies have shown that true incoherent scatter only occurs when the sounding wavelength is much smaller than the Debye length, given by:

$$\lambda_D = \left(\frac{\varepsilon_0 k T_e}{N_e q^2}\right)^{\frac{1}{2}}, \tag{20.1}$$

where ε_0 is the permittivity of free space, k is Boltzmann's constant, T_e is the electron temperature, N_e the electron density and q the charge on an electron.

The wavelengths, λ, used for ionospheric studies are typically 0.2 to 5 m, considerably greater than the Debye length which is less than about 0.01 m at heights below 1,000 km, and in this situation the electrons cannot be displaced without movement of the ions. The scattering is then thought to arise from density fluctuations in the plasma introduced by the random thermal motion of the particles. More specifically, the radar is sensitive to those variations of plasma density which have a spacing such that for the wavelength and scattering angle involved the returned waves are in phase.

β) Doppler broadening of the scattered signal occurs owing to the random thermal motion of the irregularities. The power spectrum is narrowed to correspond to the characteristics of the ion gas even though the electrons are the effective scattering elements. The width of the spectrum is then given by $\frac{2}{\lambda}\left(\frac{8kT_i}{m_i}\right)^{\frac{1}{2}}$ for a Maxwellian distribution of ions having mass m_i and tem-

[1] Bowles, K.L. (1958): Phys. Rev. Lett. *1*, 454

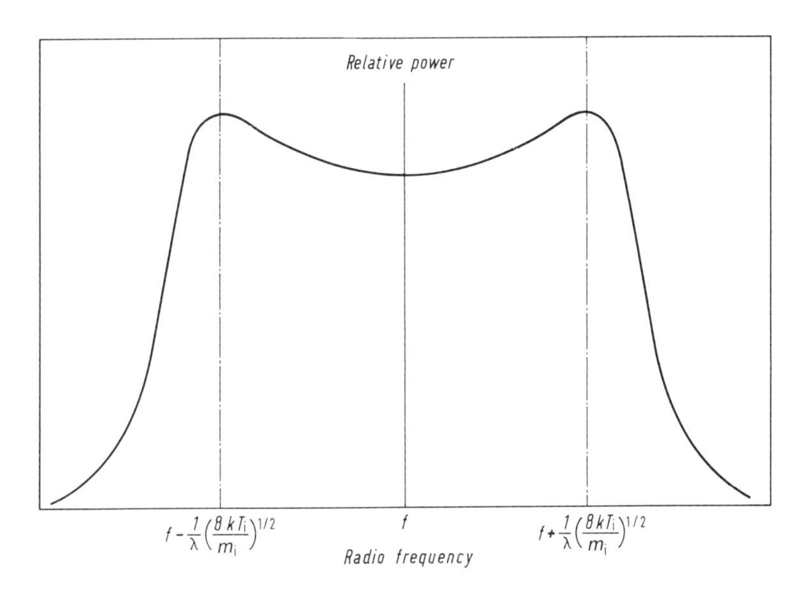

Fig. 34. The Doppler broadening of radio waves transmitted at frequency f and incoherently backscattered by the ionosphere for the case when electron and ion temperatures are equal

perature T_i equal to T_e, Fig. 34. Consequently, the scattered energy is concentrated in a much narrower frequency range than would be the case for electrons in the absence of electron-ion interactions. The narrowing of the effective spectrum width improves the signal-to-noise ratio.

Theoretical studies, reviewed by EVANS [*20*], have shown that the shape of the spectrum is modified when the electron temperature exceeds that of the ions. This effect was used to deduce the results shown in Fig. 5. The width of the spectra will also vary with m_i and, therefore, with ion composition. Fig. 35 shows the results derived for ionized gases containing single types of ions by MOORCROFT[2]; the spectra for atomic oxygen, helium and hydrogen ions are given for ratios of electron-to-ion temperature of 1.0, 2.0 and 3.0, only those parts of the spectra above the transmitted frequency being shown. These spectra are displayed in terms of a frequency scale normalised to the Doppler shift of an electron approaching at the mean thermal speed. With this presentation, the position of the peak in the spectrum for any particular ion remains in the same position for different ratios of electron-to-ion temperature. Furthermore, the width of the spectrum is nearly proportional to the square root of the ion mass. The corresponding curves for a mixture of ions are rather more complicated and the ratio of the power near the spectrum peak to that near the centre is then dependent on the relative ion concentration, in addition to the electron-to-ion temperature ratio. As a result, it can be difficult to distinguish between changes arising from variations in ion composition from those due to changes in T_i or T_e.

---

[2] Moorcroft, D.R. (1964): J. Geophys. Res. *69*, 955

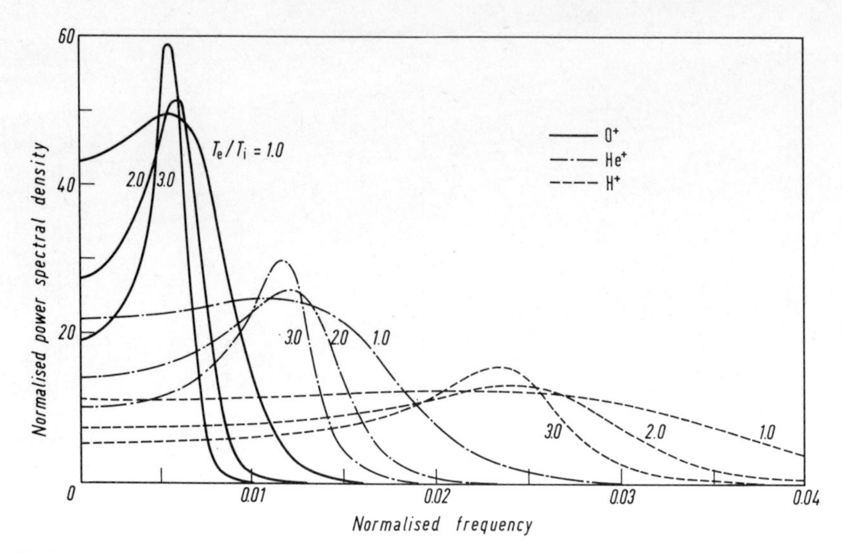

Fig. 35. The spectra deduced for O^+, He^+ and H^+ ions for ratios of electron-to-ion temperature of 1.0, 2.0 and 3.0, after MOORCROFT[2]. The ordinate is the power spectral density normalised to that for zero Doppler shift in the absence of electron-ion interactions. The frequency scale is normalised to the Doppler shift for an electron approaching at the mean thermal speed

In one approach adopted by PERKINS and WAND[3], it was assumed that only O_2^+ and NO^+ ions existed at altitudes below about 130 km and only O^+ ions above 250 km. The temperatures could then be deduced directly for these altitudes and a height variation of temperature assumed for the intervening heights. The transition region between the molecular ions and O^+ is extremely difficult to interpret. In the method employed by PETIT[4] a model for the ratio of O^+ concentration to the sum of O_2^+ and NO^+ concentrations as a function of height was adopted, based on rocket-borne measurements of ion composition. The observed spectra were then interpreted on the basis of this model and slightly modified models, until a temperature variation was found which was consistent with the reliable values established for greater and lower heights, where ions of one mass predominated. In an alternative approach, EVANS[5] assumed that the ion temperature was similar to that of the neutral gas, as given by the CIRA model atmosphere [4], and he was able by an iterative procedure to deduce both electron temperatures and ion composition from the measured spectra.

γ) Initially, the incoherent-scatter technique was applied to the determination of electron concentrations, electron and ion temperatures, and ion composition. However, the method is currently being exploited to provide additional information on ionospheric motions, the *properties of the neutral atmosphere* and the photoelectron flux. Of particular interest in the present study is the deductions made about neutral composition. Atomic oxygen concentrations above about 200 km can be derived from the solution of the heat balance equation for ions. Energy is transferred via collisions from electrons to ions, the rate of energy transfer being deduced from the measurements of ion and electron temperatures, and electron concentrations. The energy

[3] Perkins, F.W., Wand, R.H. (1967) in: Proc. Conf. on Thomson Scatter Studies of the Ionosphere, vol. 19. Evans, J.V. (ed.), p. 125. Aeronomy Laboratory, University of Illinois, Urbana

[4] Petit, M. (1968): Ann. Geophys. *24*, 1

[5] Evans, J.V. (1967): J. Geophys. Res. *72*, 3343

input to the ions is balanced by a cooling to atomic oxygen, the major neutral constituent, and measurements of the difference between ion and neutral temperature permit the determination of atomic oxygen concentrations. Using this approach, ALCAYDÉ et al.[6] have shown from data taken above Saint Santin, France, that the atomic oxygen concentration at 200 km is slightly larger in winter than summer.

ζ) The *molecular concentrations* at heights near 200 km can also be derived from a determination of the ratio of the molecular-ion concentration to electron concentration, p (COX and EVANS[7]). It can be shown from the steady-state form of the continuity equation for O^+ ions that

$$Q(O^+)[O]=\{k_{18}[N_2]+k_{19}[O_2]\}[O^+]=[kM][O^+], \qquad (20.2)$$

where $Q(O^+)$ represents the photoionization coefficient of oxygen atoms, and k_{18} and k_{19} represent the rate coefficients of reactions between O^+ ions and molecular nitrogen and oxygen (reactions (12.18) and (12.19)).

Then the ratio of atomic oxygen to weighted molecular sum is given by:

$$\frac{[O]}{[kM]}=\frac{(1-p)N_e}{Q(O^+)}, \qquad (20.3)$$

where N_e represents the electron concentration.

On the basis of this method ALCAYDÉ et al.[6] found a clear variation of this ratio from Saint Santin data. Estimates of molecular nitrogen concentrations could also be derived from the influence of ion-neutral collisions on the spectrum for heights below 120 km[8]. The results obtained and those derived simultaneously for atomic oxygen indicated relatively small seasonal changes, and the variation of the $[O]/[kM]$ ratio was, therefore, interpreted in terms of a summer-to-winter ratio of about 6 in molecular oxygen concentration.

F. Aeronomical models of neutral composition

21. Introduction. In the present study we are concerned with the heterosphere and homosphere at heights above 60 km in which the chemistry is not as complicated as at lower heights. Nevertheless, up to about 110 km chemical reactions still play a dominant part in determining the neutral composition. It is to be emphasised that it is not reactions involving the major constituents which characterise the chemistry but those which relate to minor constituents

[6] Alcayde, D., Bauer, P., Fontanari, J. (1974): J. Geophys. Res. *79*, 629
[7] Cox, L.P., Evans, J.V. (1970): J. Geophys. Res. *75*, 6271
[8] Waldteufel, P. (1969): Planet. Space Sci. *17*, 725

such as $O(^3P)$, $O(^1D)$, $O_2(^1\Delta_g)$, O_3, H_2O, OH, HO_2, NO and NO_2. It is found that these play central roles in the excitation of the airglow phenomena, the ion chemistry, the thermal balance and other aspects of aeronomy.

Relatively few measurements of neutral composition have been made in the region between 60 and 200 km and our present understanding is based largely on theoretical models. These involve for relevant constituents the solution of the continuity equations:

$$\frac{\partial N}{\partial t} = P - L - \operatorname{div}(NW), \tag{21.1}$$

where P and L represent the total production and loss rates, respectively, and include photodissociation processes and chemical reactions, and W is the velocity due to transport processes.

The mesosphere is a region of particular interest since it corresponds to the lowest heights at which the photodissociations of O_2 and H_2O are important, and the constituents resulting from these dissociations have chemical lifetimes which are significant but less than several hours. As a result major temporal variations, over time scales of a day or less, are shown. For constituents having long lifetimes, transport processes can seriously influence the spatial distributions and increasing attention has been paid during the past decade to incorporating such processes in theoretical models. To date, the greatest attention has been paid to the influence of vertical diffusion.

22. Vertical diffusion[0]. The study of the height distributions of atomic and molecular oxygen in the mesosphere and lower thermosphere by NICOLET and MANGE[1] showed that photochemical equilibrium could not be assumed, and this study and the early observations of the height distributions of gases above 110 km by BYRAM et al.[2] and MEADOWS and TOWNSEND[3] demonstrated the importance of vertical diffusion in the upper atmosphere.

The general case of the relative diffusion of constituents in a binary mixture of two gases was examined by CHAPMAN and COWLING [*21*]. On the basis of their formulation, and the assumption of the hydrostatic equation and the perfect gas law, it can be shown that the vertical velocity of constituent 1 diffusing in gas 2 is given by:

$$W_1 = -D_{12}\left\{\frac{1}{N_1}\frac{\partial N_1}{\partial h} + \frac{1}{H_1} + (1+\alpha_T)\frac{1}{T}\frac{\partial T}{\partial z}\right\}, \tag{22.1}$$

where D_{12} represents the molecular diffusion coefficient, N_1 and H_1 the concentration and scale height $(kT/m_1 g)$ of constituent 1, α_T the coefficient of thermal diffusion and T the atmospheric temperature. A detailed discussion of

[0] For more details see T. Yonezawa's contribution in this volume

[1] Nicolet, M., Mange, P. (1954): J. Geophys. Res. *59*, 16

[2] Byram, E.T., Chubb, T.A., Friedman, H. (1957) in: Threshold of space. Zelikoff, M. (ed.), p. 211. Oxford: Pergamon

[3] Meadows, E.B., Townsend, J.W. (1958): Ann. Geophys. *14*, 80

the subject is found in YONEZAWA's contribution in this volume, his Sects. 5...7.

Thermal diffusion arises from the presence of large temperature gradients and the coefficient is proportional to the difference in molecular/atomic mass between the diffusing species and the surrounding medium. It is found to be of major importance for H and He atoms in the regions of large temperature gradients in the thermosphere. It is probably legitimate to ignore the contribution of thermal diffusion in the mesosphere and lower thermosphere regions with which we are chiefly concerned.

Following COLEGROVE et al.[4], the diffusion velocity for a constituent i diffusing in an atmosphere of several constituents j can be written as:

$$(W_i)_\mathrm{M} = -D_i\left\{\frac{1}{N_i}\frac{\partial N_i}{\partial h} + \frac{1}{H_i} + \frac{1}{T}\frac{\partial T}{\partial z}\right\}, \tag{22.2}$$

where D_i represents an average diffusion coefficient and is given by

$$\frac{1}{D_i} = \sum_{j \neq i} \frac{N_j}{N D_{ij}}$$

and $N = \sum N_j$.

The individual coefficients D_{ij} can be estimated from gas kinetic theory.

For lower heights, below the turbopause, it is assumed that mixing is maintained by turbulence, the degree of turbulence being represented by an eddy diffusion coefficient K. This coefficient has been defined by LETTAU[5] in terms of an eddy transport velocity:

$$(W_i)_\mathrm{E} = -K\left[\frac{\partial\left(\frac{N_i}{\sum N_j}\right)\Big/\partial h}{N_i/\sum N_j}\right]. \tag{22.3}$$

This, combined with the differential form of the perfect gas law, gives:

$$(W_i)_\mathrm{E} = -K\left[\frac{1}{N_i}\frac{\partial N_i}{\partial h} + \frac{1}{H_\mathrm{ave}} + \frac{1}{T}\frac{\partial T}{\partial z}\right], \tag{22.4}$$

where H_ave represents the mean atmospheric gas scale height $(kT/\bar{m}g)$.

The total vertical transport velocity arising from diffusion processes, as represented by the sum of expressions (22.2) and (22.4), was first adopted by COLEGROVE et al.[4] in their examination of the oxygen case but has since been employed in the treatment of other constituents.

Although both measurements of turbulence and composition indicate that the values of molecular and eddy diffusion coefficients are similar in the 100 to 120 km region, the actual height variation of the eddy coefficient is the

[4] Colegrove, F.D., Hanson, W.B., Johnson, F.S. (1965): J. Geophys. Res. *70*, 4931

[5] Lettau, H. (1951) in: Compendium of meteorology. Malone, T.F. (ed.), p. 320. Boston: Amer. Met. Soc.

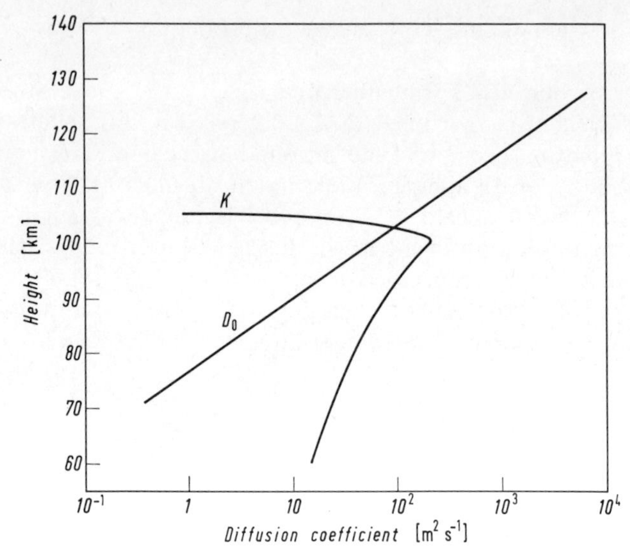

Fig. 36. The height variation of the eddy diffusion K and of the molecular diffusion coefficient for atomic oxygen D_0, after THOMAS and BOWMAN[6]

subject of considerable uncertainty. Figure 36 shows one variation adopted in model calculations for an oxygen-hydrogen atmosphere, together with that of the coefficient of molecular diffusion computed for oxygen atoms.

A major requirement in formulating a theoretical model of atmospheric composition is the incorporation of realistic upper and lower boundary conditions. For instance, in the case of an oxygen atmosphere, account needs to be taken of the downward flux of $O(^3P)$ from the thermosphere and the corresponding upward flux of O_2.

At great heights, in the exosphere, a critical level is reached at which collisions are so infrequent that a proportion of atoms have sufficient kinetic energy to escape from the earth's gravitational field.

23. Theoretical models of minor neutral species

α) Oxygen-hydrogen constituents. Early studies of the distributions of atomic oxygen and ozone were based on a simple reaction scheme involving the photodissociations of molecular oxygen and ozone, and the three-body recombination of oxygen atoms to yield oxygen molecules and of oxygen atoms with oxygen molecules to produce ozone; a two-body reaction of oxygen atoms with ozone represents a loss for each but a gain of molecular oxygen. Following the treatments of the photochemistry of the oxygen-hydrogen atmosphere by BATES and NICOLET[1], HUNT[2] and NICOLET[3], it became apparent that in con-

[6] Thomas, L., Bowman, M.R. (1972): J. Atmos. Terr. Phys. *34*, 1843
[1] Bates, D.R., Nicolet, M. (1950): J. Geophys. Res. *55*, 301
[2] Hunt, B.G. (1966): J. Geophys. Res. *71*, 1385
[3] Nicolet, M. (1970): Ann. Geophys. *26*, 531

sidering the mesosphere and lower thermosphere account needed to be taken of reactions with hydrogen species; at lower heights additional reactions involving nitrogen oxides become important (NICOLET[3]), and at greater heights the distribution of atomic oxygen and the thermal escape of hydrogen atoms are the subjects of primary interest.

The hydrogen constituents at mesospheric heights, H, OH, HO_2, H_2O_2 and H_2, result chiefly from the photodissociation of water vapour transported upwards from the Earth's surface. At still greater heights the photodissociation of H_2O is very rapid and the recombination of the products is negligible. Because of this and reactions with atomic oxygen, it is found that in the upper thermosphere hydrogen atoms represent the only significant form of hydrogen content.

The complexity of the chemistry controlling the oxygen-hydrogen constituent concentrations is illustrated by the photodissociation and chemical reaction scheme shown in Table 10, taken from THOMAS and BOWMAN[4]. It is now seen that reactions R14 and R30 are endothermic and will not proceed at mesospheric and low thermospheric heights; however, their omission from the scheme will have little significant effect. The realisation that predissociation in the Schumann-Runge bands (Sect. 4β (i)) leads to a considerable increase in the molecular oxygen dissociation rate in the mesosphere[5] represents one of the important recent developments in models incorporating such a scheme. Improvements in values of the rate coefficients involved represent a continuing process but perhaps the greatest need is to incorporate in the models transport of constituents by horizontal and vertical motions associated with the large-scale atmospheric circulation, in addition to the vertical molecular and eddy diffusion processes customarily treated.

A feature common to all theoretical models is a fairly rapid increase of $O(^3P)$ concentration with increasing height above 80 km in daytime, the height of maximum concentration depending on the assumed height variation of eddy diffusion coefficient. The noon and pre-sunrise distributions deduced for this constituent by THOMAS and BOWMAN[4] using the reaction scheme of Table 10 and the diffusion coefficients of Fig. 36 are presented in Fig. 37a. The $O(^3P)$ atoms are produced by the photodissociation of molecular oxygen and ozone, R1–R4 of Table 10, either directly or via $O(^1D)$ which is subsequently quenched. Reactions R9, 11 and 12 represent the major chemical loss processes for oxygen atoms, and the three-body recombinations of atoms, R8 a, b, make relatively little contribution.

Associated with the increase of $O(^3P)$ near 80 km in daytime is a more gradual increase of H concentration and rapid decreases of OH, HO_2 and H_2O_2 concentrations. The results for H and OH are shown in Fig. 37b.

The three-body reaction, R9, represents the major production mechanism for ozone over the height range of interest. Photodissociation by the Hartley band and continuum, processes R3a,b, are responsible for a major loss of ozone at all heights but are superseded by the reaction with hydrogen atoms,

---

[4] Thomas, L., Bowman, M.R. (1972): J. Atmos. Terr. Phys. *34*, 1843

[5] Hudson, R.D., Carter, V.C., Brieg, E.L. (1969): J. Geophys. Res. *74*, 4079

Table 10. Photodissociation processes and chemical reactions considered in model of oxygen-hydrogen atmosphere of THOMAS and BOWMAN[4]

	Process or reaction	Photodissociation or rate coefficient, s^{-1} or $m^3 s^{-1}$ or $m^6 s^{-1}$
R1	$O_2 + h\nu \rightarrow O(^1D) + O(^3P) - 7.08$ eV ($\lambda < 175$ nm)	$1.8 \cdot 10^{-6}$
R2	$O_2 + h\nu \rightarrow O(^3P) + O(^3P) - 5.11$ eV (175 nm $< \lambda <$ 243 nm)	$7.9 \cdot 10^{-8}$
R3a	$O_3 + h\nu \rightarrow O(^1D) + O_2 - 3.02$ eV ($\lambda < 310$ nm)	$1.1 \cdot 10^{-2}$
R3b	$O_3 + h\nu \rightarrow O(^1D) + O(^1\Delta_g) - 3.99$ eV ($\lambda < 310$ nm)	
R4	$O_3 + h\nu \rightarrow O(^3P) + O_2 - 1.05$ eV ($\lambda > 310$ nm)	$4.0 \cdot 10^{-4}$
R5	$H_2O + h\nu \rightarrow OH + H - 5.12$ eV (135 nm $< \lambda <$ 242 nm and Lyman-α)	$8.9 \cdot 10^{-6}$
R6	$H_2O_2 + h\nu \rightarrow 2OH - 2.15$ eV (187 nm $< \lambda <$ 577 nm)	$1.4 \cdot 10^{-4}$
R7	$HO_2 + h\nu \rightarrow OH + O(^3P) - 2.71$ eV (187 nm $< \lambda <$ 452 nm)	$1.4 \cdot 10^{-4}$
R8a	$O(^3P) + O(^3P) + M \rightarrow O_2 + M + 5.11$ eV	$2.7 \cdot 10^{-45}$
R8b	$O(^3P) + O(^3P) + M \rightarrow O_2(^1\Delta_g) + M + 4.13$ eV	$6.8 \cdot 10^{-46}$
R9	$O(^3P) + O_2 + M \rightarrow O_3 + M + 1.05$ eV	$8.2 \cdot 10^{-47} \exp\left(\frac{445}{T/K}\right)$
R10	$O(^3P) + O_3 \rightarrow 2O_2 + 4.06$ eV	$1.2 \cdot 10^{-17} \exp\left(\frac{-2{,}000}{T/K}\right)$
R11	$O(^3P) + OH \rightarrow H + O_2 + 0.72$ eV	$5.0 \cdot 10^{-17}$
R12	$O(^3P) + HO_2 \rightarrow OH + O_2 + 2.40$ eV	10^{-17}
R13	$O(^3P) + H_2O_2 \rightarrow OH + HO_2 + 0.56$ eV	10^{-21}
R14	$O(^3P) + H_2 \rightarrow OH + H - 0.08$ eV	$4.1 \cdot 10^{-17} \exp\left(\frac{-3{,}850}{T/K}\right)$
R15	$O(^3P) + OH + M \rightarrow HO_2 + M + 2.71$ eV	$1.4 \cdot 10^{-43}$
R16	$O(^3P) + H + M \rightarrow OH + M + 4.40$ eV	$8 \cdot 10^{-45}$
R17	$O(^1D) + M \rightarrow O(^3P) + M + 1.97$ eV	$2.2 \cdot 10^{-17}$
R18	$O(^1D) + O_3 \rightarrow 2O_2 + 6.03$ eV	$3 \cdot 10^{-16}$
R19	$O(^1D) + H_2 \rightarrow OH + H + 1.88$ eV	10^{-17}
R20	$O_2(^1\Delta_g) \rightarrow O_2 + h\nu + 0.98$ eV	$2.6 \cdot 10^{-4}$
R21	$O_2(^1\Delta_g) + M \rightarrow O_2 + 0.98$ eV	$2.5 \cdot 10^{-26} \sqrt{T/K}$
R22	$O_2(^1\Delta_g) + O_3 \rightarrow O_2 + O_3 + 0.98$ eV	$1.7 \cdot 10^{-20}$
R23	$O(^1D) + H_2O \rightarrow 2OH + 1.25$ eV	10^{-17}
R24	$H + O_3 \rightarrow OH + O_2 + 3.34$ eV	$2.6 \cdot 10^{-17}$
R25	$H + O_3 \rightarrow HO_2 + O(^3P) + 0.94$ eV	$2.0 \cdot 10^{-16} \exp\left(\frac{-2{,}000}{T/K}\right)$
R26	$H + HO_2 \rightarrow H_2O + O(^3P) + 2.40$ eV	$2.0 \cdot 10^{-16} \exp\left(\frac{-2{,}000}{T/K}\right)$
R27	$H + HO_2 \rightarrow H_2 + O_2 + 2.49$ eV	$2.0 \cdot 10^{-19}$
R28	$H + HO_2 \rightarrow 2OH + 1.68$ eV	10^{-17}
R29	$H + H_2O_2 \rightarrow H_2 + HO_2 + 0.65$ eV	10^{-19}
R30	$H + O_2 \rightarrow OH + O(^3P) - 0.72$ eV	$1.0 \cdot 10^{-15} \exp\left(\frac{-8{,}400}{T/K}\right)$
R31	$H + OH \rightarrow H_2 + O(^3P) + 0.08$ eV	$1.8 \cdot 10^{-18} \exp\left(\frac{-2{,}900}{T/K}\right)$
R32	$H + O_2 + M \rightarrow HO_2 + M + 1.99$ eV	$5.0 \cdot 10^{-44}$
R33	$H + H + M \rightarrow H_2 + M + 4.47$ eV	$2.6 \cdot 10^{-44}$
R34	$H + OH + M \rightarrow H_2O + M + 5.12$ eV	$2.5 \cdot 10^{-43}$
R35	$OH + HO_2 \rightarrow H_2O + O_2 + 3.12$ eV	10^{-17}
R36	$OH + H_2O_2 \rightarrow H_2O + HO_2 + 1.28$ eV	$4.0 \cdot 10^{-19}$
R37	$OH + OH \rightarrow H_2O + O(^3P) + 0.72$ eV	$2.0 \cdot 10^{-18}$
R38	$OH + O_3 \rightarrow HO_2 + O_2 + 1.66$ eV	$5.0 \cdot 10^{-19}$
R39	$OH + H_2 \rightarrow H_2O + H + 0.64$ eV	$1.0 \cdot 10^{-16} \exp\left(\frac{-2{,}950}{T/K}\right)$
R40	$HO_2 + HO_2 \rightarrow H_2O_2 + O_2 + 1.84$ eV	$1.5 \cdot 10^{-18}$
R41	$HO_2 + O_3 \rightarrow OH + 2O_2 + 1.35$ eV	10^{-20}

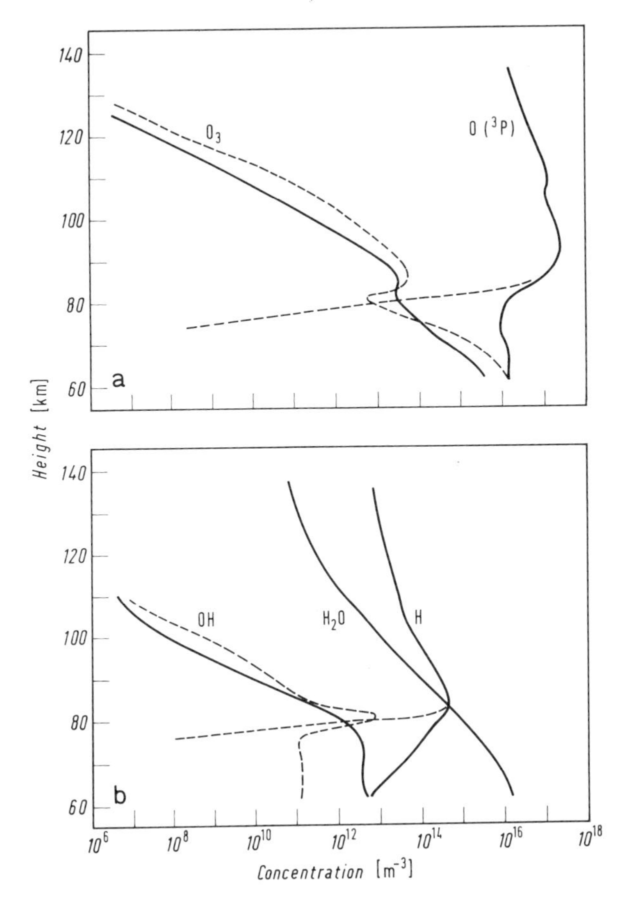

Fig. 37a, b. The height distributions of O(^{3}P), O_3, H_2O, OH and H computed for noon (——) and pre-sunrise (----) conditions using the chemical reaction scheme of Table 10 and the diffusion coefficients of Fig. 36, after THOMAS and BOWMAN[4]

R24, to produce vibrationally-excited hydroxyl radicals at heights between about 80 and 90 km. The height distributions of ozone presented in Fig. 37a show a general decrease with altitude with a small maximum near 85 km which is most pronounced at nighttime; calculations show that the distribution and the magnitude of this layer near 85 km are dependent on the assumed values of eddy diffusion coefficient.

The first electronically-excited states of atomic and molecular oxygen, O(^{1}D) and $O_2(^1\Delta_g)$, are each produced in the photodissociation of ozone by solar radiations of wavelengths less than 310 nm. For O(^{1}D) the photodissociation of molecular oxygen by the Schumann-Runge continuum is a more important source above the mesopause, and at heights above about 150 km the excited atoms are also produced by inelastic electron collisions with ground-state atoms and by dissociative recombination of molecular oxygen ions.

The radiative lifetime of the O(^{1}D) state is relatively long (110 s) and quenching by molecular oxygen or nitrogen represents the main loss process at heights below about 250 km since the quenching rate coefficients are in excess of $10^{-17}\,m^3\,s^{-1}$ for each molecule (Sect. 12ζ(i)). Such quenching by collisions with oxygen and nitrogen is slower than radiation down to about 70 km in the case of $O_2(^1\Delta_g)$, although the radiative lifetime is $2.7 \cdot 10^3$ s. The height distributions deduced for daytime for this electronically-excited molecule are reasonably consistent with what has been deduced from observations at 1.27 μm of the (0–0) band of the Infra-red Atmospheric system $O_2(a\,^1\Delta_g - X\,^3\Sigma_g)$ by EVANS et al.[6]. The concentration decreases from about $2 \cdot 10^{16}\,m^{-3}$ at 60 km to about $3 \cdot 10^{15}\,m^{-3}$ near 80 km with a sub-maximum near 85 km, similar in form to the distribution of O_3 in this height range.

For the second electronically-excited state of atomic oxygen, $O(^1S)$, excitation at mesopheric heights has been considered to occur directly from the reaction of three ground-state atoms, as suggested by CHAPMAN[7]. However, recent measurements have questioned this direct excitation mechanism[8], and it seems possible that an intermediate excited state of O_2 is involved[9]. At greater heights the dissociative recombination of molecular oxygen ions and photodissociation of molecular oxygen are also important sources. The excited state suffers collisional quenching only below about 95 km, since the rate coefficient for molecular oxygen is about $10^{-18}\,m^3\,s^{-1}$ and that for nitrogen is several orders of magnitude still smaller (Sect. 12ζ(i)).

The computed distribution of water vapour shown in Fig. 37b shows little evidence of the onset of photodissociation, which is the main loss process above about 60 km, and no noticeable diurnal variation. An upward flux compensates for this loss, the magnitude of this flux depending on the values of eddy diffusion coefficient adopted in the calculation. The production of H_2O arises from the interaction of HO_2 and OH, reaction R35, the rate coefficient of which is uncertain.

It is evident that the height distributions of $O(^3P)$, O_3 and H_2O in the middle mesosphere are dependent upon the concentrations of H, OH and HO_2. It is therefore useful to consider in some detail the photodissociation processes R5, 6 and the reactions R11, 12, 24, 32, 35, 38 and 40 which play the major part in controlling the concentrations of these reactants in the middle mesosphere.

These reactions are indicated for a height of 70 km in Fig. 38a, b which refer to conditions near noon and at the end of the nighttime period, just prior to sunrise, respectively. The line representing each photodissociation process or reaction is numbered in accordance with that given in Table 10 and the reactant involved is also shown. A distinction is drawn between production processes, represented by broken lines, and loss process represented by dotted lines. A process important simultaneously for production and loss is indicated by a chain line. This diagram illustrates that the photodissociation of H_2O represents the primary source of oxygen-hydrogen constituents, directly for H and OH, and by additional reactions for HO_2 and H_2O_2. The main loss processes of OH and HO_2 during the daytime are by recombination with $O(^3P)$ atoms, reactions R11 and R12, respectively: the corresponding loss process for H_2O_2 in daytime is about an order of magnitude slower than photodissociation and at nighttime the loss processes for this molecule are very slow. It is seen from Fig. 38a, b that the major difference between day and night is in the production and loss processes for HO_2. The large decrease in H concentrations below 80 km at nighttime illustrated in Fig. 37b renders production by reaction R32 unimportant compared with R38. The nighttime reduction in $O(^3P)$ below 80 km results in the major daytime loss of HO_2, reaction R12, being superseded by reactions R35 and R40. Although the production of OH from H atoms by reaction R24 is unimportant at nighttime, this reaction still represents a major loss for H atoms.

---

[6] Evans, W.F.J., Hunten, D.M., Llewellyn, E.J., Vallance Jones, A. (1968): J. Geophys. Res. *73*, 2885

[7] Chapman, S. (1931): Proc. R. Soc. London, Ser. A *132*, 353

[8] Slanger, T.G., Black, G. (1977): Planet. Space Sci. *25*, 79

[9] Barth, C.A. (1964): Ann. Geophys. *20*, 182

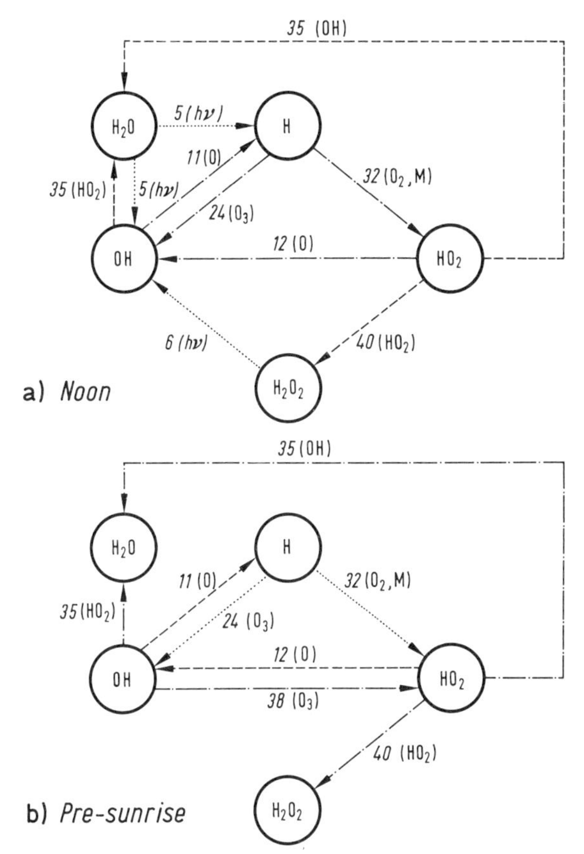

Fig. 38a, b. A schematic representation of processes controlling the concentrations of H, OH, HO_2, H_2O and H_2O_2 in the mesosphere at noon (**a**) and pre-sunrise (**b**). The individual processes of Table 10 which are important at a height of 70 km are shown together with the reactants involved. The significance of each process is also indicated: ---- production: ······ loss; –·–·– production and loss

It is to be noted that the three constituents, H, OH and HO_2 have chemical time constants in the mesosphere which are short in comparison with the mixing time constant, H_{ave}^2/K. These three constituents are, therefore, in chemical equilibrium with one another and in the temporal changes, as illustrated for OH in Fig. 37b, the total hydrogen content changes relatively slowly as determined by the reactions producing and destroying water, represented by processes R5 and R35 in Table 10.

For H_2, which is formed by the reaction of H and HO_2, R27, the main loss is by reaction with $O(^1D)$ atoms, reaction R19. This and the corresponding reaction for H_2O, R23, become increasingly important with decreasing height.

The results shown in Fig. 37a, b demonstrate the diurnal variations expected for the oxygen-hydrogen constituents. Analogous changes during eclipses have been predicted by Thomas and Bowman[10]. These show that the maximum effect occurs for $O(^3P)$ near 75 km, the concentration decreasing by about an order of magnitude.

[10] Thomas, L., Bowman, M.R. (1974): J. Atmos. Terr. Phys. *36*, 1421

β) Atomic nitrogen and nitrogen oxides. The importance of nitric oxide as a principal constituent in the formation of the ionospheric D region and as a charge-transfer agent in the E region has inspired a number of theoretical studies of nitrogen constituents in the past decade. Early photochemical studies based on the formation of the NO molecule from the reaction of ground-state atomic nitrogen and molecular oxygen yielded concentrations at least one order of magnitude smaller than those inferred from dayglow measurements of the nitric oxide bands by BARTH[11] and MEIRA[12]. Following a suggestion by NORTON and BARTH[13] and NICOLET[14], it is now realised that the corresponding reaction involving the first electronically-excited state of atomic nitrogen ($N(^2D)$) provides the major source of nitric oxide molecules (reaction (12.9)).

The rate coefficient of $1.4 \cdot 10^{-17}\,m^3\,s^{-1}$ reported by SLANGER et al.[15] is several orders of magnitude larger than that for the corresponding reaction involving $N(^4S)$ atoms. As already indicated in Sect. 4β the production of atomic nitrogen by direct photodissociation of molecular nitrogen is very slow, although predissociation in the N_2 absorption band between 80 and 100 nm is a potentially important source of $N(^2D)$. It is considered that the major sources of this excited state are associated with ionization processes, those examined in a recent study by STROBEL et al.[16] being shown in Table 11 which also shows the loss processes included.

ORAN et al.[17] suggested from measurements of the height profile of $N(^2D-^4S)$ emission at 520 nm by WALLACE and MCELROY[18] that a 50 to 100% limit could be estimated for the efficiency of $N(^2D)$ production by reaction R5 and an upper limit of $10^{-20}\,m^3\,s^{-1}$ for the quenching coefficient of the excited atom by oxygen atoms, reaction R11. With these limits, values of other parameters could be chosen to produce results consistent with the data available on NO concentrations, deduced from observations of fluorescence of the (1–0) band of the $\gamma(A\,^2\Sigma - X\,^2\Pi)$ system. However, an analysis by RUSCH et al.[19] of the high quality data for $N(^2D-^4S)$ emission at 520 nm provided by the Atmospheric Explorer C satellite indicated a much larger rate for the quenching of $N(^2D)$ by oxygen atoms than that deduced by ORAN et al.; this process was shown to be the dominant $N(^2D)$ loss process for heights below 240 km. Supporting evidence for the higher value has been provided by laboratory measurements of DAVENPORT et al.[20].

In their analysis of the Atmospheric Explorer data RUSCH et al. made use of simultaneous measurements of the NO fluorescence in the γ(1–0) band, neutral composition, and electron temperatures. STROBEL et al.[16] have applied the model developed by ORAN et al.[17] to these data and also to the nitric oxide emission data of FELDMANN and TAKACS[21]. In their analysis, reactions R20 and R21 were used to estimate the molecular oxygen densities. The height distributions of $N(^2D)$, NO and $N(^4S)$ derived by STROBEL et al.[16] are shown in Fig. 39 together with the NO concentrations deduced from the fluorescence experiment of BARTH et al.[22] on the Atmospheric Explorer C satellite. The production and loss rates for $N(^2D)$ deduced in this model are shown in Figs. 40 and 41, respectively, the corresponding rate coefficients being as shown in Table 11. On the basis of these results, STROBEL et al.[16] concluded that most of the NO^+ recombinations must produce $N(^2D)$ atoms, and that $N(^2D)$ is quenched by atomic oxygen with a rate coefficient of $1 \cdot 10^{-18}\,m^3\,s^{-1}$, about one-half the value reported by DAVENPORT et al.[20] from laboratory

[11] Barth, C.A. (1966): Ann. Geophys. *22*, 198
[12] Meira, L.G. (1971): J. Geophys. Res. *76*, 202
[13] Norton, R.B., Barth, C.A. (1970): J. Geophys. Res. *75*, 3903
[14] Nicolet, M. (1970): Planet. Space Sci. *18*, 1111
[15] Slanger, T.G., Wood, B.J., Black, G. (1971): J. Geophys. Res. *76*, 8430
[16] Strobel, D.F., Oran, E.S., Feldman, P.D. (1976): J. Geophys. Res. *81*, 3745
[17] Oran, E.S., Julienne, P.S., Strobel, D.F. (1975): J. Geophys. Res. *80*, 3068
[18] Wallace, L., McElroy, M.B. (1966): Planet. Space Sci. *14*, 677
[19] Rusch, D.W., Stewart, A.I., Hays, P.B., Hoffmann, J.H. (1975): J. Geophys. Res. *80*, 2300
[20] Davenport, J.E., Slanger, T.E., Black, G. (1976): J. Geophys. Res. *81*, 12
[21] Feldman, P.D., Takacs, P.Z. (1974): Geophys. Res. Lett. *1*, 169
[22] Barth, D.A., Rusch, D.W., Stewart, A.I. (1973): Radio Sci. *8*, 379

Table 11. Chemical reactions, photodissociation and photoionization processes considered in model of odd nitrogen in thermosphere by STROBEL et al.[16]

	Reaction or process	Rate coefficient and photo-dissociation or photoionization coefficient, $m^3 s^{-1}$ or s^{-1}
R1	$N^+ + O_2 \rightarrow O_2^+ + N(^2D) + 0.09$ eV	$3 \cdot 10^{-16}$
R2	$O^+ + N_2 \rightarrow NO^+ + N(^4S) + 1.10$ eV	$6 \cdot 10^{-13}$, $T_i \geqq 600$ K $6 \cdot 10^{-13}\left(\frac{600}{T_i/K}\right)$, $T_i < 600$ K
R3	$N_2^+ + e^- \rightarrow N(^4S) + N(^2D) + 3.44$ eV	$2.9 \cdot 10^{-13}\left(\frac{T_e/K}{300}\right)^{-0.33}$
R4	$N_2^+ + O \rightarrow NO^+ + N(^2D) + 0.68$ eV	$2.5 \cdot 10^{-16}\left(\frac{T_i/K}{300}\right)^{-0.44}$
R5	$NO^+ + e^- \rightarrow N(^2D) + O + 0.38$ eV	$3.5 \cdot 10^{-13}\left(\frac{T_e/K}{380}\right)^{-0.5}$ (R5, R6)
R6	$NO^+ + e^- \rightarrow N(^4S) + O + 2.76$ eV	
R7	$O_2^+ + N(^2D) \rightarrow N^+ + O_2 - 0.09$ eV	$2.5 \cdot 10^{-16}$
R8	$O_2^+ + N(^4S) \rightarrow NO^+ + O + 4.19$ eV	$1.8 \cdot 10^{-16}$
R9	$O_2^+ + NO \rightarrow NO^+ + O_2 + 2.80$ eV	$6.3 \cdot 10^{-16}$
R10	$N(^2D) + O_2 \rightarrow NO + O + 3.77$ eV	$7 \cdot 10^{-18}$
R11	$N(^2D) + O \rightarrow N(^4S) + O + 2.38$ eV	$1 \cdot 10^{-18}$
R12	$N(^2D) + NO \rightarrow N_2 + O + 5.63$ eV	$1.8 \cdot 10^{-16}$
R13	$N(^2D) + e^- \rightarrow N(^4S) + e^- + 2.38$ eV	$1.35 \cdot 10^{-16} (T_e/K - 220)^{0.5}$, $T_e/K \geqq 220$ K
R14	$N(^4S) + NO \rightarrow N_2 + O + 3.25$ eV	$2.1 \cdot 10^{-17}$
R15	$N(^4S) + O_2 \rightarrow NO + O + 1.39$ eV	$1.1 \cdot 10^{-20} T_n/K\, e^{-3150/T_n/K}$
R16	$NO + h\nu \rightarrow N(^4S) + O - 6.51$ eV	$J = 8.3 \cdot 10^{-6}$ at $\tau = 0$
R17	$NO + h\nu \rightarrow NO^+ + e^- - 9.27$ eV	$J = 6.0 \cdot 10^{-7}$ at $\tau = 0$
R18	$N_2 + e^-$ (fast) $\rightarrow N(^4S) + N(^2D) + e^- - 12.14$ eV	† (R18, R19)
R19	$N_2 + h\nu$(80–100 nm) $\rightarrow N(^4S) + N(^2D) - 12.14$ eV	
R20	$O_2^+ + e \rightarrow O + O(^1D) + 4.98$ eV	$2.2 \cdot 10^{-13}\left(\frac{T_e/K}{300}\right)^{-0.9}$
R21	$O^+ + O_2 \rightarrow O_2^+ + O + 1.55$ eV	$2.0 \cdot 10^{-17}\left(\frac{T_i/K}{300}\right)^{-0.4}$

† Outline of procedure adopted described in ORAN et al.[17]

measurements. From Fig. 40 it is seen that the dissociative recombination of NO^+ ions, and the reaction of molecular nitrogen ions and oxygen atoms represent the major sources of $N(^2D)$ atoms above 130 km; below this height photoelectron dissociation of molecular nitrogen is also important. Furthermore, Fig. 41 implies that it is only below about 140 km that the production of nitric oxide by the reaction of $N(^2D)$ atoms with molecular oxygen represents a dominant loss process for the excited atom.

A subsequent analysis of 520 nm emission and atmospheric composition data from Atmospheric Explorer C and D satellites by FREDERICK and RUSCH[23] confirmed the major production and loss processes for $N(^2D)$ atoms. However, the relative importance of quenching by atomic oxygen was reduced, the rate coefficient deduced being less than one-half that found by STROBEL et al.[16].

The major loss process for NO are the reaction with $N(^4S)$ atoms, R14 in Table 11, and the predissociation of the δ(0–0) and δ(1–0) bands (Sect. 6δ). At lower mesospheric heights the reaction with O_3 to form NO_2 is also important. Both STROBEL[24], and BRASSEUR and NICOLET[25] demon-

[23] Frederick, J.E., Rusch, D.W. (1977): J. Geophys. Res. *82*, 3509

[24] Strobel, D.F. (1971): J. Geophys. Res. *76*, 2441

[25] Brasseur, G., Nicolet, M. (1973): Planet. Space Sci. *21*, 939

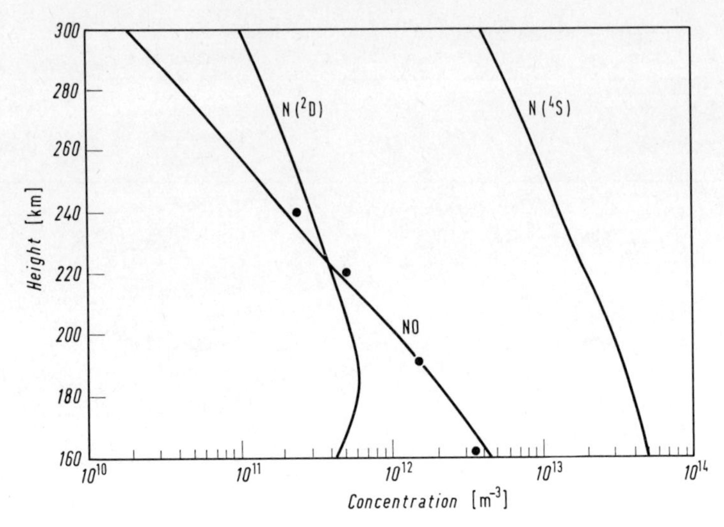

Fig. 39. The height distributions of $N(^2D)$, $N(^4S)$ and NO computed by STROBEL et al.[16] for conditions appropriate to Atmospheric Explorer satellite measurements[19]. The NO concentrations deduced from the fluorescence experiment of BARTH et al.[22] on Atmospheric Explorer C satellite are also shown by the solid circles

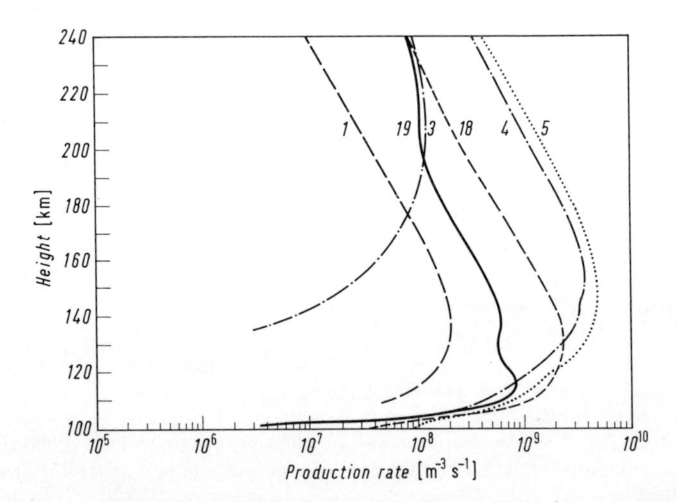

Fig. 40. The height variations of production rates of $N(^2D)$ computed by STROBEL et al.[16] for conditions appropriate to Atmospheric Explorer satellite observations[19]. The numbers associated with the curves correspond to those of reactions and processes in Table 11

strated the importance of eddy transport in controlling the mesospheric concentration of NO. Although some differences are shown between the results of different theoretical models of the height distribution of nitric oxide, arising in part from differences in the assumed eddy diffusion coefficients, general agreement is found in the sense of the diurnal variation of concentrations at different heights. Little day-to-night change is found in the 70 to 110 km region, whereas daytime increases are shown at greater heights. With suitable choice of input parameters, it has been possible to reproduce the height distributions deduced from observations of fluorescence of the

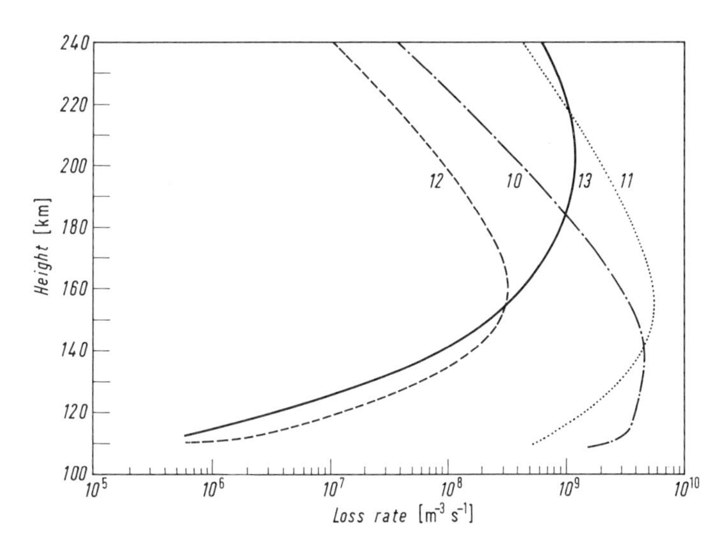

Fig. 41. The height variations of loss rates of $N(^2D)$ computed by STROBEL et al.[16] for conditions appropriate to Atmospheric Explorer satellite observations[19]. The numbers associated with the curves correspond to those of reactions in Table 11

γ(1–0) band in the dayglow. However, the major part of this emission originates at thermospheric heights and the values of NO concentrations deduced for mesospheric levels are not very reliable. Independent measurements of nitric oxide at these levels are an urgent requirement both for comparing with theoretically derived distributions and also for considerations of photoionization rates (Sect. 7γ) and negative-ion chemistry in the ionospheric D region (Sect. 28).

G. Aeronomical models of ion composition

I. Characteristics of ionospheric regions

24. Ionic Species. The designation of ionospheric regions given in Sect. 3 was based on the height distribution of electron concentration; the boundary between the D and E is generally taken to be at 95 km and that between E and F regions at 140 km. For our present purposes it is more appropriate to characterise the ionospheric regions in terms of the major ion chemistry processes operating [22]. These have been identified from the observed ion composition, laboratory reaction data, and the major spectral regions involved in the photoionization processes. Thus the D region can be considered as that part of the ionosphere in which O_2^+ and NO^+ ions formed during or following photoionization processes are converted to water cluster ions at heights below

86 km; the presence of negative ions at still lower heights represents a unique feature of this part of the ionosphere. In the E region, O_2^+ and NO^+ continue to be the major ions, the first of these being the main ion produced by photoionization, and layers of metal ions are also present. With increasing height the $NO^+:O_2^+$ concentration ratio increases until NO^+ becomes dominant at about 140 km. This coincides approximately with the ceiling of the E region, and at still greater heights the ion composition shows a predominance of O^+ ions which are characteristic of the F region. As shown in Sect. 7β, O^+ is the main product of photoionization in this part of the ionosphere and it will be found that the transition in ion composition from NO^+ to O^+ is associated with the change with height in relative concentrations of atomic oxygen and molecular oxygen or nitrogen. Before examining the processes controlling the ion composition in the different regions, it is useful to outline those which determine the height distribution of electron concentration [23].

25. The factors controlling height distribution of electron density

α) It has been seen in Sect. 7γ that the major *source of ionization in the lower ionisphere* during quiet conditions is the photoionization of nitric oxide by Lyman-α radiation. In addition, solar radiations of wavelength less than 111.8 nm are capable of ionizing $O_2(^1\Delta_g)$ molecules and, thereby, provide a significant contribution to the ionization rate at middle D-region heights. At lower heights the ionization of all constituents by cosmic rays represents the dominant source of ionization, and account also has to be taken of the photoionization of those constituents by hard X-rays during active solar conditions. The production of ionization by cosmic rays obviously continues throughout the night and other sources have also been invoked to contribute to the nighttime D region. Of primary interested is the photoionization of nitric oxide by Lyman-α radiation in the nightglow (OGAWA and TOHMATSU[1]) and the ionizing effects of precipitating high-energy particles (TULENOV et al.[2]). However, in view of the uncertainty in the NO concentrations and the variability in both scattered Lyman-α radiation and precipitating particle fluxes, it is not possible at present to establish the relative importance of these two sources of ionization at night. The possible influence of galactic X-ray sources on the nighttime D region has also been examined following reports of changes in radio-wave propagation characteristics associated with the transit of sources such as Scorpius XR-1. However, the ionization rate calculated for this source is only about 10^3 ion pairs $m^{-3} s^{-1}$, at least an order of magnitude smaller than those estimated from scattered Lyman-α or precipitating particle fluxes.

β) The *loss of electrons* in the *middle D region* arises from the dissociative recombination of molecular ions formed during the ionization processes or in subsequent ion-molecule reactions. At the lowest heights electron loss also occurs by three-body collisional attachment, predominantly to O_2 to form O_2^- (reaction (12.34)). Electrons can be released from this and negative ions

---

[1] Ogawa, T., Tohmatsu, T. (1966): Rep. Ionos. Space Res. Jpn *20*, 395

[2] Tulenov, V.F., Shibaeva, L.V., Jakovlev, S.G. (1969): Space Res. *9*, 231

formed by ionic reactions, either through photodetachment or associative detachment by atomic oxygen (e.g. reaction (12.35)). Alternatively, the negative ions can be destroyed by recombination with positive ions.

Based on these types of processes, the continuity equation for electrons at D-region heights can be written as:

$$\frac{\partial N_e}{\partial t}=\frac{\sum_i Q_i}{1+\lambda}-(\alpha_D+\lambda\,\alpha_I)\,N_e^2-\frac{N_e}{1+\lambda}\frac{d\lambda}{dt}, \tag{25.1}$$

where Q_i represents the ionization rate, λ is the ratio of negative-ion to electron concentrations, α_D is the rate coefficient for dissociative recombination for molecular ions and electrons, and α_I is the recombination coefficient for positive and negative ions.

The solution of this equation formed the basis of many early analyses of D-region changes, and the equilibrium form of the equation has often been used to define an effective loss coefficient given by $(1+\lambda)(\alpha_D+\lambda\,\alpha_I)$. However, since the identification from laboratory studies of the several ionic reactions involved, numerical studies based on positive and negative-ion reaction schemes [24] have been employed increasingly. It may be noted that no appreciable concentrations of negative ions are believed to exist at heights above the D region.

γ) For E and F-region heights the primary daytime sources of ionization are provided by the photoionization of constituents by extreme ultra-violet radiations and soft X-rays from the sun, as described in Sect. 7β. The lines corresponding to Lyman-β (102.6 nm), CIII (97.7 nm) and also HeI and II (58.4 and 30.4 nm) make important contributions. The principal ions produced are O_2^+ and N_2^+ up to about 160 km and O^+ at greater heights. The photoionization of nitric oxide by scattered Lyman-α contributes to the nighttime maintenance of the lower E region as well as the D region, but Lyman-β and HeII in the nightglow represent more important sources for the nighttime E region, with peak ionization rates occurring at about 115 km and 140 km, respectively (YOUNG et al.[3]).

The chemical loss of electrons at E-region heights arises from the dissociative recombination of NO^+ and O_2^+ ions. The electron concentrations during daytime are of the order of $10^{11}\,m^{-3}$ which with the values of dissociative recombination coefficients given in expressions (12.12) and (12.14), imply electron and ion lifetimes less than 100 s during which transport of ionization has no significant influence. However, the corresponding chemical lifetimes during nighttime are much longer, since the electron concentrations are smaller by about two orders of magnitude. Consequently, the transport of ionization by diffusion and neutral winds is an important feature of the nighttime E region[4-6].

[3] Young, J.M., Weller, C.E., Johnson, C.Y., Holmes, J.C. (1971): J. Geophys. Res. *76*, 3710

[4] Keneshea, T.J., Narcisi, R.S., Swider, W. (1970): J. Geophys. Res. *75*, 845

[5] Stubbe, P. (1972): J. Atmos. Terr. Phys. *34*, 519

[6] Strobel, D.F., Young, T.R., Meier, R.R., Coffey, T.P., Ali, A.W. (1974): J. Geophys. Res. *79*, 3171

δ) Transport processes are believed to play a particular role in the concentration of metal ions into layers of enhanced ionization corresponding to sporadic-E phenomena at middle latitudes[7]. These ions are thought to arise from the photoionization of metal atoms or charge transfer from O_2^+ or NO^+ ions, the atoms themselves being produced by meteor ablation (Sect. 30). The controlling influence of dynamical processes, such as shears in the East-West wind, are a consequence of the long chemical lifetime of the metal ion associated with their loss by radiative recombination. Although more rapidly combining metal oxide ions can be produced, it appears that their relative concentrations at E-region heights are probably insignificant.

ε) At *F-region heights*, above about 150 km, the principal ions are atomic, initially O^+ and at very great heights H^+, N^+ and He^+. It would be expected that the loss of O^+ ions by radiative recombination with electrons would be very slow (Sect. 12γ(i)), and BATES and MASSEY [*16*] proposed a sequence of reactions involving a preliminary conversion to molecular ions which then combine with electrons by dissociative recombination. Since N_2 and O_2 represent the main molecular constituents, it is believed that the ion-atom interchange and charge-transfer reactions (12.18) and (12.19) occur to produce NO^+ and O_2^+, respectively. Loss of electrons then occurs by recombination with these ions.

RATCLIFFE[8] has shown that the effective electron loss rate L, for the situation of an ion-molecule reaction of rate coefficient k followed by dissociative recombination of the molecular ion with rate coefficient α, is given for steady-state conditions by:

$$\frac{1}{L}=\frac{1}{k N_e[XY]}+\frac{1}{\alpha N_e^2}, \tag{25.2}$$

where [XY] represents the molecular concentration.

Two cases can then be distinguished, depending on the relative magnitudes of the two terms on the right hand side of (25.2):

(i) At low levels, $k[XY] \gg \alpha N_e$ and $L=\alpha N_e^2$; hence, for equilibrium conditions, $N_e=\sqrt{Q/\alpha}$ where Q is the ionization rate.

(ii) At higher levels, $k[XY] \ll \alpha N_e$ and $L=k[XY]N_e$ or βN_e; hence, for equilibrium conditions, $N_e=Q/\beta$.

It has been found by HIRSH[9] and others that the transition from an αN_e^2 to βN_e loss relation is responsible for the ledge in the electron concentration distribution characteristic of the F1 region.

It can be seen that for Case (ii), L can be expressed in terms of reactions (12.18) and (12.19):

$$L=\{k_{18}[N_2]+k_{19}[O_2]\}\,N_e=\beta N_e. \tag{25.3}$$

The variation of the ratio Q/β with height is then controlled by the relative variations of atomic oxygen and molecular oxygen and nitrogen. From Fig. 3 it

[7] Whitehead, J.D. (1971): J. Atmos. Terr. Phys. *20*, 49

[8] Ratcliffe, J.A. (1956): J. Atmos. Terr. Phys. *8*, 260

[9] Hirsh, A.J. (1959): J. Atmos. Terr. Phys. *17*, 86

is seen that the ratio of the atomic to molecular concentrations increases with increasing height which implies a corresponding continuous increase of N_e with height. However, the height distribution of ionization arising from production and loss can be influenced by diffusion. This diffusion will occur when the pressure-gradient force and gravity force do not balance each other, the resultant being opposed by the frictional drag arising from plasma motion. This latter force is smaller for electrons than for ions but the polarization field arising from any charge separation ensures that the electron and ion gases move together. The vertical velocity for the case of ion species i is given by an expression analogous to that for neutral constituents ((22.1)):

$$W_i = -D_a \left\{ \frac{1}{N_i(T_e+T_i)} \frac{\partial}{\partial h}(N_i T_i) + \frac{1}{N_e(T_e+T_i)} \frac{\partial}{\partial h}(N_e T_e) + \frac{m_i g}{k(T_e+T_i)} \right\}, \quad (25.4)$$

where D_a represents the ambipolar diffusion coefficient, N_i and m_i the concentration and mass of ion species i.

In the presence of the geomagnetic field of inclination I, the ambipolar diffusion coefficient is given by:

$$D_a = \frac{k(T_e+T_i)}{\nu_{in}\mu_{in}} \sin^2 I, \quad (25.5)$$

where ν_{in} represents the momentum-transfer collision frequency of ion species i with the neutral particles and μ_{in} is the ion-neutral reduced mass.

It has been shown that the downward movement of ionization caused by diffusion gives rise to the peak in the height variation of electron density in the F2 region[10].

In the presence of the geomagnetic field, additional vertical motion of the plasma can arise from a horizontal wind, except at the magnetic pole or equator, and from an electric field. It is found that neutral winds play an important role in controlling the behaviour of the F2 region, as first discussed by KING and KOHL[11] and reviewed by RISHBETH [25]. In addition, plasma drifts induced by electric fields are particularly important at low latitudes[12], giving rise to the equatorial anomaly, i.e. the variation of electron concentration with latitude which shows maxima in ionization either side of the magnetic dip equator; they are also important during magnetically disturbed conditions[13] and at high latitudes[14]. Detailed analyses of electron density changes are customarily based on solutions of the continuity equation:

$$\frac{\partial N_e}{\partial t} = P - L - \operatorname{div}(N_e W), \quad (25.6)$$

---

[10] Yonezawa, T. (1966): J. Radio Res. Lab. *3*, 1; see also Sect. 9 of his contribution in this volume, p. 129

[11] King, J.W., Kohl, H. (1965): Nature (London) *206*, 699

[12] Bramley, E.N., Peart, M. (1965): J. Atmos. Terr. Phys. *27*, 1201

[13] Stubbe, P., Chandra, S. (1970): J. Atmos. Terr. Phys. *32*, 1909

[14] Cole, K.D. (1971): Planet. Space Sci. *19*, 59

in which the plasma velocity W takes account of the contributions of diffusion, winds and electrodynamic drift.

The maintenance of a small ionization concentration at F2-region heights at *nighttime* has been the subject of considerable attention. It is still not established whether the persistence of ionization is due to an additional source of ionization, to a supply of ions by diffusion from greater heights, or to the effects of equatorward winds which serve to lift the existing ionization to greater heights at which the molecular densities, and hence the loss coefficient β, are smaller.

II. The ion chemistry of the D region

26. Ionic composition. The early theoretical model of the D region devised by NICOLET and AIKIN [*26*] predicted that NO^+ and O_2^+ would represent the dominant positive ions and that O_2^- ions would also be produced by three-body attachment of electrons to oxygen molecules at the lowest heights. Rocket-borne mass-spectrometer measurements of positive-ion composition carried out since 1963 over a range of latitudes and a variety of geophysical conditions (Sect. 19α(iii)) have shown that whilst O_2^+ and NO^+ do represent the major ions above 82 km in daytime, and above about 86 km at twilight or nighttime, water cluster ions, $H^+ \cdot (H_2O)_n$, dominate at the lower heights in each case, as illustrated for daytime in Fig. 7. These results and laboratory measurements of the relevant reactions have provided the basis for detailed studies of the positive-ion chemistry. Laboratory measurements have also shown that O_2^- ions enter into a sequence of reactions involving O_3, CO_2, NO and NO_2. It is on the basis of these reactions that our present understanding of the negative-ion chemistry has been developed since, as shown in Fig. 8, the two sets of mass-spectrometer measurements carried out to date have shown conflicting results.

In theoretical studies of the D region it is customary to assume that the chemical lifetimes of the ions and electrons are much shorter than the characteristic times for transport. For example, at altitudes near 80 km dissociative recombination imposes chemical lifetimes less than 10^3 s, since the recombination coefficients are in excess of $10^{-12}\,m^3\,s^{-1}$ (Sect. 12γ(ii)) and the molecular-ion and electron concentrations are about $10^9\,m^{-3}$ or greater. However, for the positive and negative ions at the lowest heights the coefficient for mutual neutralisation is one or two orders of magnitude smaller (Sect. 17) and this implies correspondingly longer chemical lifetimes. Movements might, therefore, have sufficient time to have some influence.

27. Positive-ion chemistry. As indicated in Sect. 7γ the ionization in the D region is produced largely through the ionization of nitric oxide molecules by solar Lyman-α radiation, of $O_2(^1\Delta_g)$ molecules by radiations of wavelength 102.7 to 111.8 nm, and of all constituents by short X-rays or cosmic rays. Consequently, the initial positive ions are expected to be NO^+, O_2^+ and N_2^+.

α) It has long been realised from laboratory measurements that *water cluster ions* can arise from the reaction:

$$H_2O^+ + H_2O \rightarrow H_3O^+ + OH + 1.21\,\mathrm{eV} \tag{27.1}$$

which has a rate coefficient of about $10^{-15}\,\mathrm{m^3\,s^{-1}}$ (TAL'ROZE and FRANKOVITCH[1]), followed by the successive three-body attachment of water molecules, as indicated in reaction (15.8).

However, in the atmosphere charge transfer to molecular oxygen can occur:

$$H_2O^+ + O_2 \rightarrow O_2^+ + H_2O + 0.56\,\mathrm{eV} \tag{27.2}$$

for which a rate coefficient of $2 \cdot 10^{-16}\,\mathrm{m^3\,s^{-1}}$ has been reported by FEHSENFELD et al.[2]. Because of the large $O_2 : H_2O$ concentration ratio (Fig. 2), this reaction supersedes (27.1).

The N_2^+ ions produced by photoionization are rapidly lost by reaction with O_2 to form O_2^+. Consequently O_2^+ and NO^+ ions, produced both by photoionization and also by ion-molecule reactions, are thought to be the precursors of the water cluster ions.

The reaction scheme involved in the formation of water cluster ions from O_2^+ was proposed by FEHSENFELD and FERGUSON[3] and GOOD et al.[4]. It was based on the three-body formation of O_4^+, subsequent two-body reactions to form H_3O^+ and $H^+ \cdot (H_2O)_2$, and successive three-body attachment of water molecules to produce higher-order cluster ions, as indicated in reactions (15.2) to (15.8).

Although there is some uncertainty about the branching ratio between reactions (15.5) and (15.6), the overall scheme, represented in Fig. 42, is capable of reproducing the general shape of the cluster ion height distributions. The rapid decrease in concentrations of water cluster ions above about 80 km is attributed to a reaction:

$$O_4^+ + O \rightarrow O_2^+ + O_3 + 0.64\,\mathrm{eV} \tag{27.3}$$

FERGUSON[5] has pointed out that the rapid increase in the rate of this reaction associated with the growth of atomic oxygen concentration with increasing height above 80 km, Figs. 2, 37a, could account for the decrease in cluster ions. However, detailed calculations based on the revised $O_2(^1\Delta_g)$ photoionization rates of HUFFMAN et al.[6] have shown that the supply of O_2^+ ions is too small to produce significant water cluster ion concentrations under normal conditions, particularly at the lower D-region heights[5,7]. On the other hand, NARCISI[8] has shown that the O_2^+ scheme is

---

[1] Tal'roze, V.L., Frankovitch, E.L. (1960): Z. Fiz. Him. *34*, 2709

[2] Fehsenfeld, F.C., Schmeltekopf, A.L., Ferguson, E.E. (1967): J. Chem. Phys. *46*, 2802

[3] Fehsenfeld, F.C., Ferguson, E.E. (1969): J. Geophys. Res. *74*, 2217

[4] Good, A., Durden, A., Kebarle, P. (1970): J. Chem. Phys. *52*, 222

[5] Ferguson, E.E. (1971) in: Mesospheric models and related experiments. Fiocco, G. (ed.), p. 188. Dordrecht: Reidel

[6] Huffman, R.E., Paulsen, D.E., Larrabee, J.C., Cairns, R.B. (1971): J. Geophys. Res. *76*, 1028

[7] Donahue, T.M. (1972): Radio Sci. *7*, 73

[8] Narcisi, R.S. (1972) in: Proc. COSPAR Symp. on "Solar Particle Event of November 1969", Air Force Cambridge Research Laboratories, Boston, Mass, Spec. Rep. 144, pp. 557–569

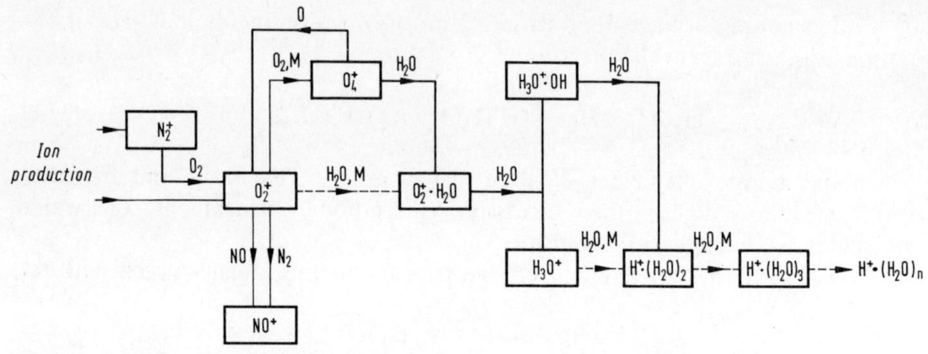

Fig. 42. A schematic representation of positive-ion reactions leading to the formation of water cluster ions from O_2^+, according to the model of FEHSENFELD and FERGUSON[3], GOOD et al.[4] and FERGUSON[5]. An indication of the relative speeds of the reactions at 80 km is given: changes corresponding to ion lifetimes of less than 2 s, 2 to 100 s, and greater than 100 s are shown by heavy lines, thin lines and broken lines, respectively

capable of explaining the positive-ion composition measurements during disturbed conditions at high latitudes (Sect. 19α(iii)) since the effective production rate of O_2^+ is sufficiently enhanced. It appears that the observed lowering of the transition height between O_2^+, NO^+ ions and water cluster ions probably arose from a lowering of the height at which the atomic oxygen concentration increased.

β) The establishment of the corresponding *reaction scheme* beginning with NO^+ has represented a major D-region problem in recent years. In addition to the need to reproduce the observed relative proportions of cluster ions and NO^+ ions below 82–86 km, such a scheme is necessary to reconcile the total ionization rates with the measured electron concentrations (DONAHUE[7]; REID[9]).

In considering suitable reaction schemes it is to be noted that reactions corresponding to (15.5) and (15.6) do not occur for NO^+. Instead, successive additions of water molecules are believed to take place, and the production of water cluster ions begins with the following reaction:

$$NO^+\cdot(H_2O)_3 + H_2O \rightarrow H^+\cdot(H_2O)_3 + HNO_2. \tag{27.4}$$

This difference in behaviour is due to the smaller ionization potential of NO compared with O_2.

The processes to be considered first then involve the successive hydration of NO^+ up to the third hydrate, as first proposed by FEHSENFELD and FERGUSON[3]. It is to be noted that the specific loss rate imposed on water cluster ions by dissociative recombination near 80 km implies a corresponding lower limit on the rate of production of these ions and consequently on the rate of hydration of NO^+ ions.

Attention was initially concentrated on the direct three-body processes:

$$NO^+\cdot(H_2O)_n + H_2O + M \rightarrow NO^+\cdot(H_2O)_{n+1} + M, \tag{27.5}$$

where $n=0$ to 2.

---

[9] Reid, G.C. (1971) in: Mesospheric models and related experiments. Fiocco, G. (ed.), p. 198. Dordrecht: Reidel

Laboratory measurements with flowing-afterglow[10,11], stationary-afterglow[12], and high-pressure mass spectrometer[13] equipment have revealed rate coefficients between about $1.1 \cdot 10^{-39}\,m^6\,s^{-1}$ and $1.6 \cdot 10^{-40}\,m^3\,s^{-1}$ for the three reactions. It has been found that with these values the production of water cluster ions by direct successive hydration of NO^+ followed by (27.4) is too slow to reproduce the observed ion composition of the D region. From the results of flowing-afterglow measurements, DUNKIN et al.[14] suggested that a more efficient mode of hydrating NO^+ would be provided by the three-body attachment of CO_2 to NO^+ (reaction (12.29)), followed by a switching reaction with H_2O, (12.33). These workers reported a rate coefficient of $2.5 \cdot 10^{-41}\,m^6\,s^{-1}$ for reaction (12.29) at 200 K with N_2 as the third body, and stationary-afterglow measurements by HEIMERL and VANDERHOFF[15] have yielded a value of $2.4 \cdot 10^{-41}\,m^6\,s^{-1}$ at 296 K with CO_2 as the third body. Although no measurements are available for reaction (12.33), the results of DUNKIN et al. indicate that it is very fast, a rate coefficient near $10^{-15}\,m^3\,s^{-1}$ being suggested. Depending on the concentration assumed for water vapour, it is found that the sequence of reactions (12.29) and (12.33) is about an order of magnitude more rapid than direct three-body hydration of NO^+.

HEIMERL and VANDERHOFF[16] suggested that the three-body association with molecular nitrogen, reaction (12.30), followed by the switching reaction with water, (12.32), might be more important than the sequence with carbon dioxide, because of the greater relative concentration of N_2. It was considered that the ion $NO^+ \cdot N_2$ might be too weakly bound to be significant for NO^+ loss, but NILES et al.[17] have proposed that a switching reaction with CO_2, (12.31), could yield the more stable ion $NO^+ \cdot CO_2$. Again, no value of the rate coefficient of this reaction is available but NILES et al. have stated that it is expected to be rapid.

It was pointed out in Sect. 19α(iii) that recent mass-spectrometer measurements from rockets[18] have shown ions of masses consistent with $NO^+ \cdot N_2$ and $NO^+ \cdot CO_2$, and this observation provides support for a scheme involving reactions (12.29) to (12.33).

γ) In view of the low bond energies expected for $NO^+ \cdot N_2$ and $NO^+ \cdot CO_2$ it might be expected that *collisional dissociation* could interfere with the formation of $NO^+ \cdot H_2O$ by these reactions. THOMAS[19] has demonstrated the importance of such collisional processes using rate coefficients based on the results of the laboratory measurements by DUNKIN et al.[14] and HEIMERL and VANDERHOFF[15]; because of the uncertainties in these results and absence of information on the bond energies no account could be taken of the temperature dependencies of these reactions.

JOHNSON et al.[20] have suggested from laboratory measurements using the drift-tube technique, that the forward reaction in (12.30) has a temperature variation of the form $2.0 \cdot 10^{-43} \left(\frac{300}{T/K}\right)^{4.4} m^6\,s^{-1}$; a similar value has been deduced by ARNOLD and KRANKOWSKY[21] from an analysis of D-

[10] Fehsenfeld, F.C., Mosesman, M., Ferguson, E.E. (1971): J. Chem. Phys. *55*, 2120
[11] Howard, C.J., Rundle, H.W., Kaufman, F. (1971): J. Chem. Phys. *55*, 4772
[12] Puckett, L.J., Teague, M.W. (1971): J. Chem. Phys. *54*, 2564
[13] French, M.A., Hills, L.P., Kebarle, P. (1973): Can. J. Chem. *51*, 456
[14] Dunkin, D.B., Fehsenfeld, F.C., Schmeltekopf, A.L., Ferguson, E.E. (1971): J. Chem. Phys. *54*, 3817
[15] Heimerl, J.M., Vanderhoff, J.V. (1974): J. Chem. Phys. *60*, 4362
[16] Heimerl, J.M., Vanderhoff, J.V. (1971): EOS Trans. AGU *52*, 870
[17] Niles, F.E., Heimerl, J.M., Keller, G.E. (1972): EOS Trans. AGU *53*, 456
[18] Arnold, F., Krankowsky, D. (1975) in: Proc. Int. Symp. on Solar Terrestrial Physics, Sao Paulo, Brazil, May 1974, *3*, pp. 30–50
[19] Thomas, L. (1976): J. Atmos. Terr. Phys. *38*, 61
[20] Johnsen, R., Huang, C.M., Biondi, M.A. (1975): J. Chem. Phys. *63*, 3374
[21] Arnold, F., Krankowsky, D. (1977): J. Atmos. Terr. Phys. *39*, 625

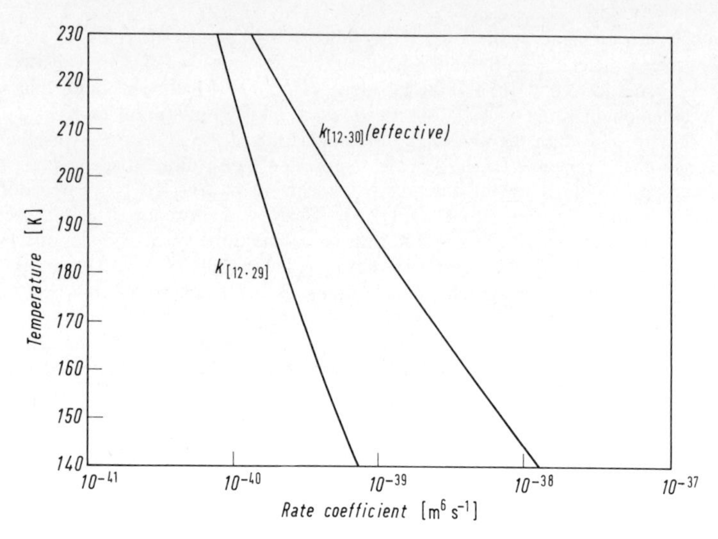

Fig. 43. The rate coefficients for the formation of $NO^+ \cdot CO_2$ by reaction (12.29) and by reaction (12.30) followed by (12.31) for different temperatures, based on temperature dependencies given in the text, after THOMAS[23]; the effective value for reaction (12.30) has been derived with allowance for competition between the reverse reaction and reaction (12.31)

region ion-composition measurements. From this expression of JOHNSON et al.[20] and the temperature dependence of the equilibrium constant also found by these workers it can be shown that the temperature variation of the reverse reaction in (12.30) is given by $1.1 \cdot 10^{-14} \left(\frac{300}{T/K}\right)^{4.4} \exp\left(-\frac{2.125}{T/K}\right) m^3 s^{-1}$. For the reverse reaction in (12.29), DUNKIN et al.[14] have suggested a rate coefficient within the range $1.2-5.0 \cdot 10^{-19} m^3 s^{-1}$ at 310 K, and a bond energy for $NO^+ \cdot CO_2$ of 0.48 eV, by analogy with measurements on the formation of $K^+ \cdot CO_2$ by KELLER and BEYER[22]. On the basis of these values it can be shown that the collisional dissociation of $NO^+ \cdot CO_2$ can be neglected in comparison with the loss of this ion by reaction (12.33) except for high mesospheric temperatures. However, the work of Heimerl and Vanderhoff[15] indicated that the binding energy of $NO^+ \cdot CO_2$ was less than that of $K^+ \cdot CO_2$, and dissociation could therefore be more significant.

With rate coefficients of $10^{-15} m^3 s^{-1}$ being assumed for reactions (12.31) and (12.32), the former represents the more rapid loss of $NO^+ \cdot N_2$ because of the greater concentration of CO_2 relative to H_2O, Fig. 2. The major production of $NO^+ \cdot H_2O$ is by the reaction of $NO^+ \cdot CO_2$ with water vapour, and the relative importance of reactions (12.29) and (12.30), (12.31) in the formation of $NO^+ \cdot CO_2$ for a range of temperatures representative of D-region conditions is shown in Fig. 43, after THOMAS[23].

These results have been derived for the height distributions of CO_2 and N_2 presented in Fig. 2, and a value of $2.5 \cdot 10^{-41} \left(\frac{300}{T/K}\right)^{4.4} m^6 s^{-1}$ for the forward reaction in (12.29), based on the value measured at 296 K by HEIMERL and VANDERHOFF[15] and a similar temperature dependence as that measured for the forward reaction in (12.30) by JOHNSON et al.[20]. In Fig. 43 the effective value for

[22] Keller, G.E., Beyer, R.A. (1971): Bull. Am. Phys. Soc. *16*, 214
[23] Thomas, L. (1976): J. Atmos. Terr. Phys. *38*, 1345

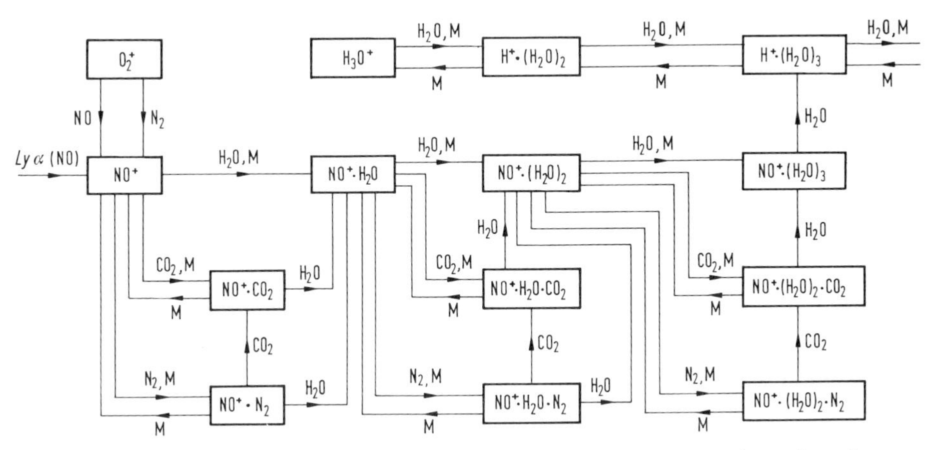

Fig. 44. A schematic representation of positive-ion reactions leading to the formation of water cluster ions from NO^+ ions, according to the model of FEHSENFELD and FERGUSON[3] extended by DUNKIN et al.[14], HEIMERL and VANDERHOFF[16], NILES et al.[17] and FERGUSON[25]

the forward reaction in (12.30) has been obtained allowing for the competition between the reverse reaction and reaction (12.31); this effective value is defined to provide comparison with $k_{(12.29)}$, the specific loss rate for NO^+ ions by each route being found by including the appropriate concentrations of CO_2 and the third body M. It is concluded that reactions (12.30) and (12.31) make the greater contribution to the formation of $NO^+\cdot CO_2$ and, hence, to the production of $NO^+\cdot H_2O$. This result is dependent only on the value and temperature dependence assumed for the forward reaction in (12.29), since the mixing ratio of CO_2 is reasonably well established over the height range of interest. Recent measurements of the forward reaction in (12.29) with N_2 as third body by SMITH et al.[24], when based on the measurement for 300 K and a T^{-6} temperature reaction (SMITH, private communication), indicate values similar to those for $k_{(12.29)}$ in Fig. 43.

δ) FERGUSON[25] has suggested that reactions analogous to (12.29) to (12.33) might be involved in forming the *second and third hydrates of* NO^+, e.g.:

$$NO^+\cdot H_2O + CO_2 + M \rightleftarrows NO^+\cdot H_2O\cdot CO_2 + M \tag{27.6}$$

$$NO^+\cdot H_2O + N_2 + M \rightleftarrows NO^+\cdot H_2O\cdot N_2 + M \tag{27.7}$$

$$NO^+\cdot H_2O\cdot N_2 + CO_2 \rightarrow NO^+\cdot H_2O\cdot CO_2 + N_2 \tag{27.8}$$

$$NO^+\cdot H_2O\cdot N_2 + H_2O \rightarrow NO^+\cdot(H_2O)_2) + N_2 \tag{27.9}$$

$$NO^+\cdot H_2O\cdot CO_2 + H_2O \rightarrow NO^+\cdot(H_2O)_2 + CO_2. \tag{27.10}$$

On the basis of these ideas, calculations have been carried out for the scheme represented in Fig. 44 by REID[26]. In the absence of information on reactions (27.6), (27.7) and the corresponding processes involving $NO^+\cdot(H_2O)_2$, Reid has assumed that the forward reactions have similar rate coefficients to (12.29) and (12.30), respectively. The corresponding reverse reaction coefficients have been deduced from the equilibrium coefficients which have been derived on the assumption that the enthalpy changes in the three reactions involving N_2 attachment, and in those involving CO_2 attachment, were in the same ratios as in the sequence of direct hydration processes (27.5).

[24] Smith, D., Adams, N.G., Grief, D. (1977): J. Atmos. Terr. Phys. *39*, 513

[25] Ferguson, E.E. (1971): Rev. Geophys. Space Phys. *9*, 997

[26] Reid, G.C. (1977): Planet. Space Sci. *25*, 275

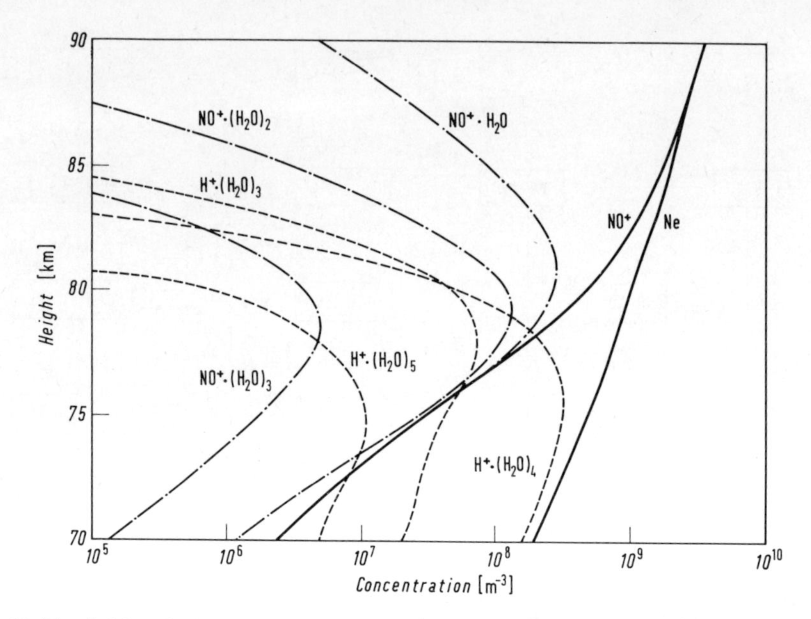

Fig. 45. The height distributions computed for water cluster ions, NO^+ and its hydrates, and electrons, using the reaction scheme represented in Fig. 44, after REID[26]

The results derived by Reid for a solar zenith angle of 56°, the nitric oxide distribution shown in Fig. 2, a water vapour distribution similar to that in Fig. 2, and temperatures representative of winter-time conditions at high latitudes are shown in Fig. 45.

It is evident that the scheme represented in Fig. 44 is capable of reproducing the general features of the positive-ion composition of the daytime D region, particularly the transition between the water cluster and NO^+ ions.

ε) It is to be noted that there is still some uncertainty about the *order of* the major ambient *water cluster ions*, as pointed out in Sect. 19α(iii). In order to reproduce the substantial concentrations of H_3O^+ observed in early observations, BURKE[27] invoked a reaction between $NO^+ \cdot H_2O$ and H atoms:

$$NO^+ \cdot H_2O + H \rightarrow H_3O^+ + NO + 1.97\,\text{eV}. \tag{27.11}$$

This reaction was not included in the model since FEHSENFELD et al.[28] have found that the rate constant is less than $7 \cdot 10^{-18}\,m^3\,s^{-1}$ and it is too slow to be important. These workers also showed that the corresponding reaction with OH proposed by HEIMERL et al.[29] is also too slow and the suggestion by these workers of an analogous reaction involving HO_2 has not yet been tested in the laboratory.

---

[27] Burke, R.R. (1970): J. Geophys. Res. *75*, 1345

[28] Fehsenfeld, F.C., Howard, C.J., Harrop, W.J., Ferguson, E.E. (1975): J. Geophys. Res. *80*, 2229

[29] Heimerl, J.M., Vanderhoff, J.A., Puckett, L.J., Niles, F.E. (1972): BRL Rep 1570, U.S. Army Aberdeen Res. and Develop. Centre, Aberdeen, Wash.

A notable feature of models based on reaction schemes similar to that of Fig. 44 is that large concentrations of $NO^+ \cdot H_2O$ and $NO^+ \cdot (H_2O)_2$ are predicted, as illustrated in Fig. 45. The mass-spectrometer measurements of JOHANNESSEN and KRANKOWSKY[30], ARNOLD and KRANKOWSKY[18], ZBINDEN et al.[31], and AIKIN et al.[32] all show evidence of $NO^+ \cdot H_2O$ ions but in relatively low concentrations. The inclusion of dissociation processes for $NO^+ \cdot H_2O$ in the theoretical models seems to be indicated. However, since the bond energy of this ion is relatively large, collisional dissociation should be negligible and recent laboratory measurements[33,34] have shown that photodissociation can probably be ignored also. REID[26] has suggested that the discrepancy between theory and observations arises because of dissociation of the NO^+ hydrates during mass-spectrometer sampling in the D region.

28. Negative-ion chemistry

α) In view of the conflict between the two sets of mass-spectrometer measurements of negative-ion composition in the D region (Fig. 8), most studies of the chemistry involved have been based on the results of *laboratory measurements* without serious reference to the rocket results. It has long been realised from laboratory measurements[1] that O_2^- can be formed by the rapid three-body attachment of electrons to O_2 molecules, reaction (12.34).

More recent measurements[2–12], chiefly with flowing-afterglow experiments, have shown that beginning with O_2^- ions a complicated sequence of reactions involving neutral constituents can occur and detailed schemes have been devised to incorporate these reactions. That due to FERGUSON[13] is shown in Fig. 46, the individual reactions being listed in Table 12. It is seen that two forms of NO_3^- are shown, the evidence for these being experimental[13]: NO_3^{-*} reacts with NO to form NO_2^- whereas NO_3^- does not. It has been suggested that NO_3^{-*} has a linear form O–N –O–O produced by switching reactions:

$$O_2^- \cdot X + NO \rightarrow O_2^- \cdot NO + X, \tag{28.1}$$

where $X = O_2$ or CO_2.

NO_3^- is the stable form and has equivalent O atoms each bonded to the N atom. This configuration is produced by the transfer of a single oxygen atom to NO_2^-:

$$NO_2^- + XO \rightarrow NO_3^- + X, \tag{28.2}$$

---

30 Johannessen, A., Krankowsky, D. (1972): J. Geophys. Res. *77*, 2888

31 Zbinden, P.A., Hidalgo, M.A., Eberhardt, P., Geiss, J. (1975): Planet. Space Sci. *23*, 1621

32 Aikin, A.C., Goldberg, R.A., Jones, W., Kane, J.A. (1977): J. Geophys. Res. *82*, 1869

33 Vanderhoff, J.A. (1977): J. Chem. Phys. *67*, 2332

34 Smith, G.P., Cosby, P.C., Moseley, J.T. (1977): J. Chem. Phys. *67*, 3818

1 Chanin, L.M., Phelps, A.V., Biondi, M.A. (1959): Phys. Rev. Lett. *2*, 344

2 Fehsenfeld, F.C., Schmeltekopf, A.L., Schiff, H.I., Ferguson, E.E. (1967): Planet. Space Sci. *15*, 373

3 Fehsenfeld, F.C., Ferguson, E.E., Bohme, D.K. (1969): Planet. Space Sci. *17*, 1759

4 Fehsenfeld, F.C., Albritton, D.L., Burt, J.A., Schiff, H.I. (1969): Can. J. Chem. *47*, 1793

5 Fehsenfeld, F.C., Howard, C.J., Ferguson, E.E. (1973): J. Chem. Phys. *58*, 5841

6 Fehsenfeld, F.C., Ferguson, E.E. (1968): Planet. Space Sci. *16*, 701

7 Fehsenfeld, F.C., Ferguson, E.E. (1974): J. Chem. Phys. *61*, 3181

8 Lelevier, R.E., Branscomb, L.M. (1968): J. Geophys. Res. *73*, 27

9 Parkes, D.A. (1972): J. Chem. Soc. Faraday Trans. 1 *68*, 627

10 Ferguson, E.E. (1969): Can. J. Chem. *47*, 1815

11 Fehsenfeld, F.C., Ferguson, E.E., Schmeltekopf (1966): J. Chem. Phys. *45*, 1844

12 Adams, N.G., Bohme, D.K., Dunkin, D.B., Fehsenfeld, F.C., Ferguson, E.E. (1970): J. Chem. Phys. *52*, 3133

13 Ferguson, E.E. (1974): Rev. Geophys. Space Phys. *12*, 703

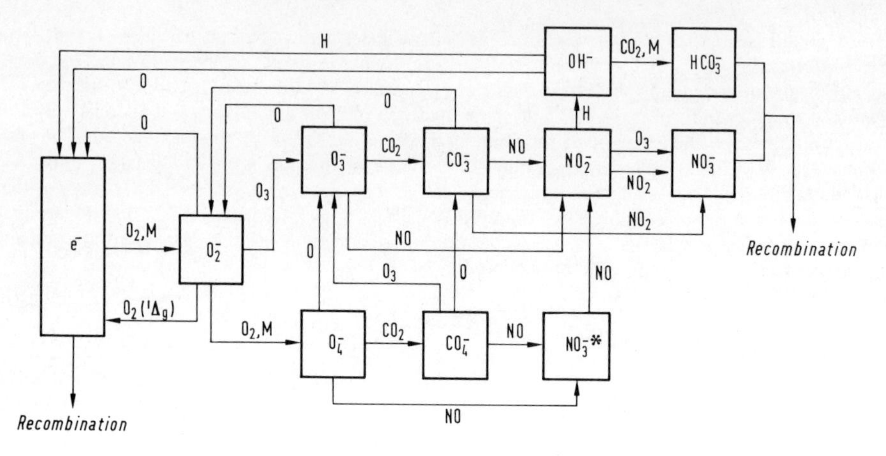

Fig. 46. A schematic representation of negative-ion reactions, after FERGUSION[13]

The scheme shown in Fig. 46 includes the ion HCO_3^- which is consistent with one of the ion masses recorded by ARNOLD et al.[14]. FERGUSON[13] has suggested that this can arise from the pair of reactions R12 and R17 in Table 12. However, he points out that because of the low concentration of H, and the rapid binary reactions of OH^- with H and O (R15 and R16), the concentration of HCO_3^- would be expected to be small.

β) It is believed that *hydration* of both positive and negative ions could occur. Since such hydration is a three-body process it will only be important for ions having long lifetimes. Recent measurements with flowing-afterglow systems have suggested that hydration might have little effect on the negative-ion neutral reactions[7] and on the negative-ion positive-ion recombination[15].

γ) Another consideration in the negative-ion chemistry is the *photodetachment* of electrons from negative ions. Measurements of the electron affinities of O^- (1.47 eV), O_2^- (0.43 eV), O_3^- (1.90 eV) and CO_4^- (1.22 eV) have been reviewed by PHELPS[16], and a value of 2.38 eV has been recently reported for NO_2^- by DUNKIN et al.[17] and of 3.9 eV for NO_3^- by FERGUSON et al.[18]. Experimental values of photodetachment coefficient are available for O^- (BRANSCOMB et al.[19]), O_2^- (WOO et al.[20]), and O_3^- (BYERLY and BEATY[21], BURT[22]), although the values for O_3^- differ by a factor of about six.

PETERSON[23] has recently drawn attention to the possible importance of *photodissociation* processes:

---

[14] Arnold, F., Kissel, J., Krankowsky, D., Wieder, H., Zahringer, J. (1971): J. Atmos. Terr. Phys. *33*, 1169

[15] Smith, D., Adams, N.G., Church, M.J. (1976): Planet. Space Sci. *24*, 697

[16] Phelps, A.V. (1969): Can. J. Chem. *47*, 1783

[17] Dunkin, D.B., Fehsenfeld, F.C., Ferguson, E.E. (1972): Chem. Phys. Lett. *15*, 257

[18] Ferguson, E.E., Dunkin, D.G., Fehsenfeld, F.C. (1972): J. Chem. Phys. *57*, 1459

[19] Branscomb, L.M., Burch, D.S., Smith, S.J., Geltman, S. (1958): Phys. Rev. *111*, 504

[20] Woo, D.B., Branscomb, L.M., Beaty, E.C. (1969): J. Geophys. Res. *74*, 2933

[21] Byerly, R., Beaty, E.C. (1971): J. Geophys. Res. *76*, 4596

[22] Burt, J.A. (1972): Ann. Geophys. *28*, 607

[23] Peterson, J.R. (1976): J. Geophys. Res. *81*, 1433

Table 12. Negative-ion reactions at 300 K, included in Fig. 46

	Reaction	Rate coefficient, $m^3 s^{-1}$ or $m^6 s^{-1}$	Reference
R1	$e^- + O_2 + O_2 \rightarrow O_2^- + O_2$	$1.6 \cdot 10^{-42}$	1
R2	$O_2^- + O \rightarrow O_3 + e^-$	$3.3 \cdot 10^{-16}$	2
R3	$O_2^- + O_2(^1\Delta_g) \rightarrow 2O_2 + e^-$	$2.0 \cdot 10^{-16}$	4
R4	$O_2^- + O_3 \rightarrow O_3^- + O_2$	$3.0 \cdot 10^{-16}$	2
R5	$O_2^- + O_2 + M \rightarrow O_4^- + O_2 + M$	$2.0 \cdot 10^{-41}$	3
R6	$O_3^- + O \rightarrow O_2^- + O_2$	$1.0 \cdot 10^{-16}$	8
R7	$O_3^- + CO_2 \rightarrow CO_3^- + O_2$	$5.5 \cdot 10^{-16}$	9
R8	$O_3^- + NO \rightarrow NO_2^- + O_2$	$1.0 \cdot 10^{-17}$	1
R9	$CO_3^- + O \rightarrow O_2^- + CO_2$	$8.0 \cdot 10^{-17}$	1
R10	$CO_3^- + NO \rightarrow NO_2^- + CO_2$	$9.0 \cdot 10^{-18}$	1
R11	$CO_3^- + NO_2 \rightarrow NO_3^- + CO_2$	$8.0 \cdot 10^{-17}$	7
R12	$NO_2^- + H \rightarrow OH^- + NO$	$4.0 \cdot 10^{-16}$	5
R13	$NO_2^- + O_3 \rightarrow NO_3^- + O_2$	$1.8 \cdot 10^{-17}$	6
R14	$NO_2^- + NO_2 \rightarrow NO_3^- + NO$	$\leqq 1.0 \cdot 10^{-19}$	
R15	$OH^- + H \rightarrow H_2O + e^-$	$1.8 \cdot 10^{-15}$	5
R16	$OH^- + O \rightarrow HO_2 + e^-$	$2.0 \cdot 10^{-16}$	11
R17	$OH^- + CO_2 + M \rightarrow HCO_3^- + M$	$7.6 \cdot 10^{-40}$	7
R18	$O_4^- + O \rightarrow O_3^- + O_2$	$4.0 \cdot 10^{-16}$	3
R19	$O_4^- + NO \rightarrow NO_3^{-*} + O_2$	$2.5 \cdot 10^{-16}$	12
R20	$O_4^- + CO_2 \rightarrow CO_4^- + O_2$	$4.3 \cdot 10^{-16}$	12
R21	$CO_4^- + O_3 \rightarrow O_3^- + O_2 + CO_2$	$1.3 \cdot 10^{-16}$	7
R22	$CO_4^- + O \rightarrow CO_3^- + O_2$	$1.5 \cdot 10^{-16}$	3
R23	$CO_4^- + NO \rightarrow NO_3^{-*} + CO_2$	$4.8 \cdot 10^{-17}$	3
R24	$NO_3^{-*} + NO \rightarrow NO_2^- + NO_2$	$1.5 \cdot 10^{-17}$	12

$$h\nu + CO_3^- \rightarrow CO_2 + O^- \tag{28.3}$$

$$h\nu + CO_3^- \cdot H_2O \rightarrow CO_3^- + H_2O. \tag{28.4}$$

He points out that such dissociations could be important in contributing to the marked changes in electron concentrations that occur in the D region during the pre-sunrise period[24, 25], before radiations capable of ionizing the gaseous constituents arrive. The dissociations occur for visible solar wavelengths, and since the O^- ions have an electron affinity of 1.47 eV, photodetachment can also be produced by visible and near infra-red radiations. From the cross-section measurements for the photodissociation processes, PETERSON has argued that all CO_3^- and $CO_3^- \cdot H_2O$ ions will be converted to O^- with an apparent rate of $0.2\,s^{-1}$, and that the effective zero optical depth photodetachment rate for the ions will be only slightly smaller than $0.2\,s^{-1}$, since photodetachment from O^- is very rapid. Clearly the importance of this sequence depends on the concentration of CO_3^- and $CO_3^- \cdot H_2O$ prior to sunrise. In addition, the importance of the sunrise changes of O atom concentration in converting CO_3^- ions to O_2^-, reaction R9 of Table 12, and then detaching

[24] Mechtly, E.A., Smith, L.G. (1968): J. Atmos. Terr. Phys. *30*, 363

[25] Thomas, L., Harrison, M.D. (1970): J. Atmos. Terr. Phys. *32*, 1

electrons from O_2^-, reaction R2, has been noted by TURCO and SECHRIST[26] and by BOWMAN and THOMAS[27].

No experimental data are available for the photodissociation coefficients of NO_2^- and NO_3^- or their hydrates. PETERSON[23] has argued that photodissociation is unlikely to be important for NO_3^- but that it could be for the isomer NO_3^{-*}. The importance of photodetachment for ions such as NO_3^-, which from model studies are believed to be the predominant negative ion in daytime, to the electron concentration in the lower D region has been examined by THOMAS et al.[28]. This and other theoretical studies have predicted a rapid decrease of negative-ion concentrations with increasing height above about 70 km in daytime, and some confirmation has been obtained from mass-spectrometer observations[29].

It is evident that no complete theoretical study of the D region will be possible until the ambient negative-ion composition and the relevant photodissociation and photodetachment processes have been established, and also the conversion of NO^+ ions to water cluster ions fully understood. However, considerable insight into D-region behaviour can be provided by simplified ion-chemistry models, such as that devised by MITRA and ROWE[30]. This was developed for the interpretation of D-region changes associated with solar flares but has since been applied to studies of the undisturbed region[31].

III. The ion chemistry of the E region

29. Processes controlling NO^+ and O_2^+ distributions. Several measurements of ion composition at E-region heights have shown that NO^+ and O_2^+ are the major ions in daytime, their concentrations being reasonably comparable. At nighttime NO^+ becomes dominant, its concentration being about an order of magnitude greater than that of O_2^+.

It is usually assumed that the daytime E region is the best understood part of the ionosphere, but detailed treatment is still limited by the uncertainties in the solar X-ray and extreme ultra-violet fluxes [*27*]. In addition, the height variations of temperature and density, and information on neutral composition, represent problem areas; the temperature dependence of chemical rate coefficients is an important factor (DONAHUE[1]). For nighttime conditions an added complication arises from the need to include transport processes. The electron density is smaller than that for daytime by about two orders of magnitude and this implies that the chemical time constants are correspondingly greater. It has been shown by WALKER and MCELROY[2] that transport processes then have time to be effective. Furthermore, improved information is

[26] Turco, R.P., Sechrist, C.F. (1973): Radio Sci. *7*, 725

[27] Bowman, M.R., Thomas, L. (1973): J. Atmos. Terr. Phys. *35*, 347

[28] Thomas, L., Gondhalekar, P.M., Bowman, M.R. (1973): J. Atmos. Terr. Phys. *35*, 397

[29] Arnold, F., Krankowsky, D. (1977) in: Dynamical and chemical coupling between the neutral and ionized atmosphere. Grandal, G., Holtet, J.A. (eds.), p. 93. Dordrecht: Reidel

[30] Mitra, A.P., Rowe, J.N. (1972): J. Atmos. Terr. Phys. *34*, 795

[31] Rowe, J.N., Mitra, A.P., Ferraro, A.J., Lee, H.S. (1974): J. Atmos. Terr. Phys. *36*, 755

[1] Donahue, T.M. (1966): Planet. Space Sci. *14*, 33

[2] Walker, J.C.G., McElroy, M.B. (1966): J. Geophys. Res. *71*, 3779

Table 13. Reactions controlling the ion chemistry of the E region

	Reaction	Rate coefficient, $m^3 s^{-1}$
R1	$O_2^+ + e^- \rightarrow O + O + 6.95\,eV$	$2.0 \cdot 10^{-13} \left(\frac{300}{T_e/K}\right)^{0.7}$
R2	$NO^+ + e^- \rightarrow N \rightarrow O + 2.76\,eV$	$4.4 \cdot 10^{-13} \left(\frac{300}{T_e/K}\right)^{1.2}$
R3	$N_2^+ + e^- \rightarrow N + N + 5.82\,eV$	$1.8 \cdot 10^{-13} \left(\frac{300}{T_e/K}\right)^{0.39}$
R4	$O^+ + O_2 \rightarrow O_2^+ + O + 1.55\,eV$	$2.0 \cdot 10^{-17}$
R5	$O^+ + N_2 \rightarrow NO^+ + N + 1.10\,eV$	$1.0 \cdot 10^{-18}$
R6	$O_2^+ + NO \rightarrow NO^+ + O_2 + 2.80\,eV$	$4.5 \cdot 10^{-16}$
R7	$O_2^+ + N \rightarrow NO^+ + O + 4.19\,eV$	$1.8 \cdot 10^{-16}$
R8	$N_2^+ + O \rightarrow NO^+ + N + 3.06\,eV$	$1.5 \cdot 10^{-16}$
R9	$N_2^+ + O_2 \rightarrow O_2^+ + N_2 + 3.52\,eV$	$4.0 \cdot 10^{-17}$

required on the scattered Lyman-α and Lyman-β intensities, to provide representative estimates of the nighttime ionization rates.

It is found that in order to describe the ion chemistry of the normal E region, relatively few reactions are required. These are shown in Table 13, the rate-coefficient data having been presented in Sect. 12γ and δ.

The contributions of R4 and R9 to O_2^+ formation, and of R8 to that of NO^+, can be neglected because of the small concentrations of O^+ and N_2^+ ions at E-region heights.

For daytime conditions, chemical equilibrium is expected and the concentrations (symbol [...]) of the ions O^+, O_2^+ and NO^+ can then be expressed as

$$[O^+] = \frac{Q(O^+)}{k_4[O_2] + k_5[N_2]} \tag{29.1}$$

$$[O_2^+] = \frac{Q(O_2^+)}{\alpha_1 N_e + k_6[NO] + k_7[N]} \tag{29.2}$$

$$[NO^+] = \frac{Q(N_2^+) + k_5[O^+][N_2] + k_6[O_2^+][NO] + k_7[O_2^+][N]}{\alpha_2 N_e} \tag{29.3}$$

in which the Q's represent the production rates of the appropriate ions arising from photoionization, and the k's and α's are the rate coefficients for the relevant ion-neutral and dissociative recombination reactions, respectively.

The appearance of $Q(N_2^+)$ in the expression for $[NO^+]$ arises from the rapid transfer from N_2^+ to NO^+ by reactions R8 and R9; a time constant of about 0.07 s applies for R8 at 120 km where the atomic oxygen concentration is near $10^{17}\,m^{-3}$, as seen from Fig. 3. In addition, reactions R6 and R7 provide the major loss processes for O_2^+ and, in fact, become more important than dissociative recombination when the ratio of nitric oxide or atomic nitrogen to electron concentration exceeds 10^3.

It is evident that the ion composition of the daytime E region is controlled by the nitric oxide and atomic nitrogen concentrations, in addition to the

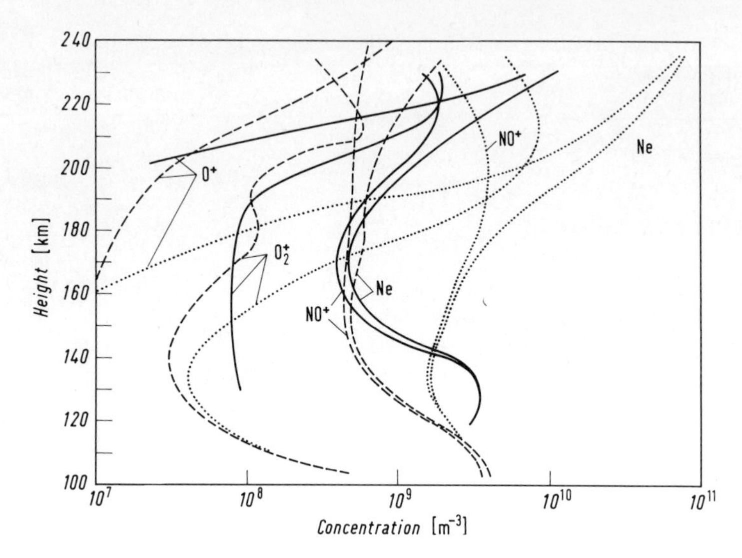

Fig. 47. A comparison of the ion composition in the E and lower F regions at nighttime derived from a theoretical model, including (– – –) and excluding (······) the effects of winds, with that measured with a mass spectrometer[5] (——), after STROBEL et al.[4]

intensities of solar X-ray and EUV radiations and the concentrations of the major constituents, O, O_2 and N_2. Information on all these input parameters are required to explain the observed changes in daytime ion compensation, particularly the observed increases in the ratio of nitric oxide ion to molecular oxygen ion concentrations between summer and winter and with enhanced solar activity.

A detailed time-dependent study of the E region, including the nighttime production of ionization by scattered Lyman-α and Lyman-β radiations and the relevant ion chemistry, was used by KENESHEA et al.[3] to provide a diurnal model of the layer. This model reproduced the gross features of the E region and confirmed the need for including nighttime ionization sources. With the increased number of observations of airglow emissions carried out in recent years, more representative models of the nighttime radiation field have been constructed for incorporation in quantitative studies of the nighttime ionosphere. Such a study for the E region and the lower F region has been carried out by STROBEL et al.[4] who also included the effects of ionization transport arising from diffusion and from diurnal winds deduced from pressure gradients given by model atmospheres. The results for O_2^+ and NO^+ ions are compared with measurements of ion composition carried out during solar minimum conditions[5] in Fig. 47, from STROBEL et al.[4]. They have concluded that above about 150 km the nighttime sources of ionization were sufficient to account for the maintenance of the nighttime ionization but between about 115 and

[3] Keneshea, T.J., Narcisi, R.S., Swider, S. (1970): J. Geophys. Res. 75, 845

[4] Strobel, D.F., Young, T.R., Meier, R.R., Coffey, T.P., Ali, A.W. (1974): J. Geophys. Res. 79, 3171

[5] Holmes, J.C., Johnson, C.Y., Young, J.M. (1965): Space Res. 5, 756

140 km local production is insufficient and vertical transport of ionization from other levels must take place. It seems certain that account will need to be taken of a more complex motion field, probably incorporating upward propagating tidal winds and gravity waves, to reproduce the nighttime E-region data.

In addition to this direct influence on the ion composition, the motion field will also modify the distributions of the neutral constituents NO, N, O and O_2. Attention has already been drawn in Sect. 23β to the importance of ionization processes in the production of atomic nitrogen and nitric oxide, and the importance of these constituents to the ion composition has been illustrated above. It is evident that the chemistry of neutral and ion constituents at E-region heights cannot realistically be separated.

30. The formation of metallic-ion layers. The occurrence of metal atoms and ions in the upper atmosphere, specifically K, Na and Ca^+, has long been recognised from observations of the twilight airglow. Mass-spectrometer observations by ISTOMIN[1] first demonstrated that metal ions Mg^+, Ca^+, Fe^+ and possibly Si^+, were present in stratified layers, and subsequent observations at nighttime by YOUNG et al.[2] showed that such layers could be associated with sporadic-E ionization. The layer actually studied by these latter workers had a horizontal extent of 60 km, a vertical thickness of 2 km, and a peak electron concentration of about $10^{11}\,m^{-3}$, which was one or two orders of magnitude greater than the ambient concentration in the vicinity of the layer. Additional mass-spectrometer studies, chiefly by NARCISI and colleagues[3], have confirmed that metal ions, particularly Mg^+, Fe^+ and Si^+, comprise the greater part of the positive-ion concentration of sporadic-E layers. The presence of the first two of these ions in such a layer has also been shown by dayglow observations of the Mg II doublet at 279.5 and 280.3 nm by ANDERSON and BARTH[4] and by incoherent-scatter measurements by BEHNKE and VICKREY[5].

α) The *general findings* of mass-spectrometer measurements are that the metal ions are located predominantly between 82 and 120 km; a layer of thickness between 5 and 10 km situated near 93 km is comprised chiefly of Mg^+ and Fe^+ with smaller amounts of Na^+, K^+, Al^+, Ca^+ and Ni^+; higher-altitude layers are made up of Si^+, Mg^+ or Fe^+. The metal-ion content is substantially increased during periods of meteoric activity[6,7], as well as in conditions of sporadic-E ionization, and at these times the height profiles of NO^+ and O_2^+ show decreases in association with the peaks of metal ions. This association between enhanced metal-ion concentrations and meteoric activity, and the agreement between the relative abundance of observed ions and atoms in chondrite meteorites[7] has provided the strongest evidence for a meteoric

[1] Istomin, V.G. (1963): Space Res. *3*, 209
[2] Young, J.M., Johnson, C.Y., Holmes, J.C. (1967): J. Geophys. Res. *72*, 1473
[3] Narcisi, R.S., Bailey, A.D., Della Lucca, L. (1967): Space Res. 7, 123
[4] Anderson, J.G., Barth, C.A. (1971): J. Geophys. Res. *76*, 3723
[5] Behnke, R.A., Vickery, J.F. (1975): Radio Sci. *10*, 325
[6] Narcisi, R.S. (1968): Space Res. *8*, 360
[7] Goldberg, R.A., Aikin, A.C. (1973): Science *180*, 294

origin for the corresponding metal atoms. It is generally assumed that meteor ablation is chiefly responsible, although a terrestrial origin has also been invoked, specifically the upward transfer of particulate matter from sea spray caused by stratospheric disturbances during the polar night (ALLEN[8]).

Considerations of the cross-sections for collisions at high energies have suggested that the production of metal ions during ablation is relatively unimportant. Instead, the larger proportion of ions is believed to arise from photoionization or charge-transfer processes. The ionization potentials of the metals of interest, e.g. Fe - 7.87 eV, Mg - 7.64 eV, correspond to wavelengths of solar radiations which do not penetrate below about 100 km, except within certain atmospheric windows. However, because of the relatively low values of these potentials, and the still lower values for other ions: Na - 5.14 eV, Ca - 6.11 eV, charge transfer can occur even with NO^+. Some examples of such transfer are shown in Table 14, the values of rate coefficients deduced from crossed-beam experiments as reviewed by FERGUSON[9] being shown.

Table 14. Charge transfer to metal atoms, after FERGUSON[9]

Reaction	Rate coefficient, $m^3 s^{-1}$
$O^+ + Mg \rightarrow O + Mg^+ + 5.97\,eV$	Small
$Fe \rightarrow O + Fe^+ + 5.75\,eV$	$2.9 \cdot 10^{-15}$
$Na \rightarrow O + Na^+ + 8.48\,eV$	Small
$Ca \rightarrow O + Ca^+ + 7.51\,eV$	$7.5 \cdot 10^{-16}$
$O_2^+ + Mg \rightarrow O_2 + Mg^+ + 4.42\,eV$	$1.2 \cdot 10^{-15}$
$Fe \rightarrow O_2 + Fe^+ + 4.19\,eV$	$1.2 \cdot 10^{-15}$
$Na \rightarrow O_2 + Na^+ + 6.93\,eV$	$1.4 \cdot 10^{-15}$
$Ca \rightarrow O_2 + Ca^+ + 5.95\,eV$	$4.1 \cdot 10^{-15}$
$N_2^+ + Fe \rightarrow N_2 + Fe^+ + 7.71\,eV$	$4.3 \cdot 10^{-16}$
$Na \rightarrow N_2 + Na^+ + 10.44\,eV$	$1.9 \cdot 10^{-15}$
$Ca \rightarrow N_2 + Ca^+ + 9.47\,eV$	$1.7 \cdot 10^{-15}$
$NO^+ + Mg \rightarrow NO + Mg^+ + 1.62\,eV$	$1.0 \cdot 10^{-15}$
$Fe \rightarrow NO + Fe^+ + 1.40\,eV$	$9.1 \cdot 10^{-16}$
$Ca \rightarrow NO + Ca^+ + 3.16\,eV$	$4.0 \cdot 10^{-15}$

β) Metal oxide ions MO^+ and MO_2^+ can also be formed from the corresponding oxides by photoionization or charge transfer:

$$MO + h\nu \rightarrow MO^+ + e^- \tag{30.1}$$

$$MO_2 + h\nu \rightarrow MO_2^+ + e^- \tag{30.2}$$

$$MO + O_2^+ \rightarrow MO^+ + O_2 \tag{30.3}$$

$$MO_2 + O_2^+ \rightarrow MO_2^+ + O_2. \tag{30.4}$$

It was considered that the dissociative recombination of these metal oxide ions might play an important part in the chemical loss of metal ions, since the direct loss of these by radiative recombination is so slow. However, as shown by FERGUSON and FEHSENFELD[10], the metal oxide ions are rapidly reduced by atomic oxygen to yield the metal ion:

$$MO^+ + O \rightarrow M^+ + O_2. \tag{30.5}$$

[8] Allen, F.R. (1970): J. Geophys. Res. *75*, 2947
[9] Ferguson, E.E. (1972): Radio Sci. *7*, 397
[10] Ferguson, E.E., Fehsenfeld, F.C. (1968): J. Geophys. Res. *73*, 6215

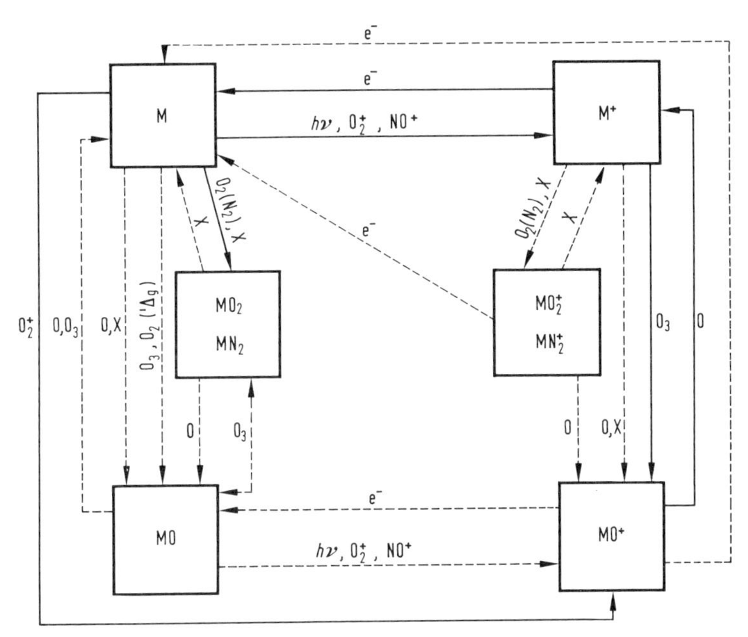

Fig. 48. A schematic representation of metal-atom and ion reactions, after BROWN[11]; reactions for which no rate-coefficient data are available in the case of magnesium are shown by broken lines. X signifies the third body in three-body processes and the second body in collisional dissociating reactions

There is still some uncertainty about the fate of MO_2^+ and particularly whether it is converted to MO^+ by the analogous reaction or lost by dissociative recombination. The significance of the loss processes for MgO_2^+ has been pointed out by ANDERSON and BARTH[4].

γ) It is evident that the neutral and ion chemistries of metals are intimately connected, as is illustrated in the *reaction scheme* devised by BROWN[11] shown in Fig. 48. In order to emphasise the need for laboratory measurements of reaction rate coefficients, those reactions for which no data are available in the case of magnesium are shown by broken lines; a similar lack of data is found for iron and other metal atoms. In view of the general slow loss rate of metal ions by two-body processes, it seems likely that transport into the lower ionosphere followed by three-body processes is chiefly responsible for the disappearance of these ions. In addition to the three-body reactions already indicated in Fig. 48, FERGUSON[9] has drawn attention to the hydration of metal ions, either directly or via the intermediate ion $M^+ \cdot CO_2$ and switching the CO_2 by H_2O. The formation of such intermediate ions by Na^+ ions has been examined in laboratory studies[12,13].

δ) Even when the ionization of metal atoms produced by meteor ablation, and the subsequent chemical reactions, are understood the question of their *layering and transport* is still to be

[11] Brown, T.L. (1973): Chem. Rev. *73*, 645
[12] Keller, G.E., Beyer, R.A. (1971): J. Geophys. Res. *76*, 289
[13] Niles, F.E., Heimerl, J.M., Keller, G.E., Puckett, L.J. (1972): Radio Sci. *7*, 117

answered. The present theoretical treatments are based on the convergence caused by wind shears[14, 15]. In the height region of interest the frequencies with which the ions gyrate around the geomagnetic field lines and the frequencies with which they collide with neutral particles are comparable. In these circumstances a horizontal East-West wind causes both horizontal transport of ions and also a Lorentz force given by $q_i(V \times B)$, where q_i is the ion charge and V is the component of the velocity normal to the geomagnetic field. In the case of a wind directed toward the East in the Northern hemisphere, an upward movement is imposed on the ions, whereas a wind toward the West imposes a downward movement; a convergence is then effected at a height of suitable wind shear. In each case, the electrons are constrained to follow the movement of the positive ions to maintain electrical neutrality. As an extension of these ideas it has been proposed by CHIMONAS and AXFORD[16] that the metal ions will remain at the nulls of the wind-pattern in the case of a periodic wind structure which has a downward phase velocity. This will continue until the collision frequencies are too high to permit the downward movement of the layer to continue and it is dispersed by diffusion, probably in the 80 to 90 km region.

It is evident that in order to understand the distributions of metal ions full account needs to be taken of these transport processes. However, the relatively small interest shown to date in ionospheric metal-ion chemistry, arising in part from the difficulty associated with carrying out the relevant laboratory measurements, has meant that the loss processes of both metal atoms and ions are still very poorly understood. It appears that developments in studies of relevant chemical processes and transport mechanisms, and information on relevant minor neutral constituents are essential requirements for understanding the data on metal ions provided so successfully by mass-spectrometer and less direct measurements.

IV. The ion chemistry of the F region

31. Molecular ions. From the examination of the loss processes for electrons at heights above about 150 km given in Sect. 25, it would be expected that the ion composition would be dominated by the molecular ions NO^+ and O_2^+ at heights up to the F1 ledge of ionization and by O^+ at greater heights. Mass-spectrometer observations have confirmed this prediction, as illustrated in Fig. 6.

The ion chemistry at F1-region heights represents an extension of that of the E region as summarised in Table 13, although there is a still greater need to take account of the effect of temperature on chemical rate coefficients. In addition to the information relating to the coefficients of dissociative recombination given in Sects. 12γ and 16, data on the temperature dependence of reactions R4–6 and R9 of Table 13 have been derived from flowing-afterglow measurements, as illustrated in Fig. 24 after LINDINGER et al.[1], and the energy dependence of reaction R8 has been examined in a crossed-beam experiment by RUTHERFORD and VROOM[2] and in drift-tube measurements by MCFARLAND et al.[3]. Deductions from Atmospheric Explorer satellite data [*15*] have shown

[14] Whithead, J.D. (1961): J. Atmos. Terr. Phys. *20*, 49

[15] Keneshea, T.J., Macleod, M.A. (1970): J. Atmos. Sci. *27*, 981

[16] Chimonas, G., Axford, W.I. (1968): J. Geophys. Res. *73*, 111

[1] Lindinger, W., Fehsenfeld, F.C., Schmeltekopf, A.L., Ferguson, E.E. (1974): J. Geophys. Res. *79*, 4753

[2] Rutherford, J.A., Vroom, D.A. (1974): J. Chem. Phys. *61*, 2514

[3] McFarland, M., Albritton, D.L., Fehsenfeld, F.C., Ferguson, E.E., Schmeltekopf, A.L. (1974): J. Geophys. Res. *79*, 2925

good agreement with laboratory measurements of reactions R5 and R8, although the increase in the rate coefficient of R8 with ion temperature indicated over the range 600–900 K was in contrast with the decrease found in the laboratory by McFarland et al.[3].

Although account has been taken of the variation with temperature of the dissociative recombination coefficients of NO^+, O_2^+ and N_2^+ in theoretical models, no detailed treatment has, to date, included the temperature dependence of the ion-molecule reactions. Nevertheless, good agreement has been obtained between theoretical models of ion composition and observations (e.g. Pharo et al.[4]). A major limitation in the model calculations is the uncertainty in the neutral composition, and specifically in the height distributions of atomic oxygen and nitric oxide.

Attention has already been paid in Sect. 23β to those aspects of F-region ion chemistry involved in the production of $N(^2D)$ atoms. Reactions leading to the formation of NO^+ featured strongly in that analysis of the odd nitrogen constituents.

One such reaction involved charge exchange of N_2^+ ions with O atoms, this representing one of the principal loss processes for the molecular ion. It is realised that the ionization of N_2 molecules results from the absorption of solar radiations of wavelengths below 79.6 nm and by charge exchange with metastable oxygen ions, reaction (12.40). This additional source of N_2^+ has been invoked by a number of workers in order to explain observations of the First Negative band system of $N_2^+(B\,^2\Sigma_u^+ - X\,^2\Sigma_g^+)$ in the dayglow and twilight airglow. Feldman[5] has derived the production rate of $O^+(^2D)$ ions from the photoionization of atomic oxygen by solar radiations of wavelength less than 73.2 nm and has shown that the charge-exchange reaction dominates over direct photoionization of N_2 as a source of N_2^+ at heights above about 240 km. Furthermore, Oppenheimer et al.[6] have shown from a detailed study based on daytime composition measurements from the Atmospheric Explorer C satellite that below 300 km most of the $O^+(^2D)$ ions undergo charge exchange with N_2 molecules but at greater heights quenching by electrons, atomic oxygen, and molecular nitrogen become more important.

The formation of N_2^+ ions by this charge-exchange reaction has also been included by Rishbeth et al.[7] in a chemical scheme used to account qualitatively for rocket-borne ion-composition measurements. Furthermore, these workers attributed an anti-correlation between O^+ and NO^+ concentrations observed at heights of 400 to 500 km by the OGO 6 satellite to the balance between the loss by dissociative recombination of NO^+ and its formation by the reaction of N_2^+ ions and O atoms at nighttime; in this balance the O^+ concentration closely represented the electron concentration.

A further complication in the molecular-ion chemistry of the F2 region has been identified by Torr et al.[8]. They have attempted to account for the height distributions of NO^+ and O_2^+ ions observed at nighttime with the magnetic ion mass spectrometer on Atmospheric Explorer C satellite in terms of production by reactions of O^+ with N_2 and O_2 ((12.18) and (12.19)) and loss by dissociative recombination. Marked discrepancies were found between the calculated and observed distributions for heights below about 240 km, the deduced NO^+ concentrations being too small and the deduced O_2^+ concentrations too large. In order to account for these discrepancies, Torr et al.[8] invoked loss of O_2^+ and the production of NO^+ ions by reaction (12.23). On the basis

[4] Pharo, M.W., Scott, L.R., Mayr, H.G., Brace, L.G., Taylor, H.A. (1971): Planet. Space Sci. *19*, 15

[5] Feldman, P.D. (1973): J. Geophys. Res. *78*, 2010

[6] Oppenheimer, M., Dalgarno, A., Brinton, H.C. (1976): J. Geophys. Res. *81*, 3762

[7] Rishbeth, H., Bauer, P., Hanson, W.B. (1972): Planet. Space Sci. *20*, 1287

[8] Torr, M.R., Torr, D.G., Walker, J.C.G., Hays, P.B., Hanson, W.B., Hoffman, H.H., Kayser, D.C. (1975): Geophys. Res. Lett. *2*, 385

of the rate coefficient value of $1.8 \cdot 10^{-16}\ m^3\ s^{-1}$, they concluded that the N concentrations required were about $7 \cdot 10^{12}\ m^{-3}$ at 220 km.

For F2-region heights transport processes, notably diffusion and neutral air winds, become increasingly important for the major charged particles: electrons and O^+ ions. However RISHBETH et al.[7] have shown that for molecular ions the height distributions are determined by chemical processes even up to 400 to 500 km. This is because the chemical lifetimes imposed by dissociative recombination are too short for diffusion to effect any marked redistribution.

32. Atomic ions

α) N^+ is the second most abundant ion in the F region but the photochemistry of this constituent is still not established. The most detailed treatment carried out to date is probably that of BAILEY and MOFFETT[1] who considered the following sources:

$$N_2 + h\nu(<51\ \text{nm}) \rightarrow N^+ + N + e^- - 24.29\ \text{eV} \tag{32.1}$$

$$N(^4S) + h\nu(\lambda<86\ \text{nm}) \rightarrow N^+ + e^- - 14.53\ \text{eV} \tag{32.2}$$

$$N(^2D) + h\nu(\lambda<102\ \text{nm}) \rightarrow N^+ + e^- - 12.15\ \text{eV} \tag{32.3}$$

$$N(^2D) + O_2^+ \rightarrow N^+ + O_2 - 0.09\ \text{eV} \tag{32.4}$$

$$N_2^+ + N \rightarrow N_2 + N^+ + 1.05\ \text{eV} \tag{32.5}$$

$$He^+ + N_2 \rightarrow He + N + N^+ + 0.28\ \text{eV}. \tag{32.6}$$

Reaction (32.4) is slightly endothermic but is likely to proceed at F-region temperatures; BAILEY and MOFFETT have adopted a value of $4 \cdot 10^{-16}\ m^3\ s^{-1}$ for the rate coefficient, following a suggestion by DALGARNO[2]; for reaction (32.5) an upper value of $10^{-17}\ m^3\ s^{-1}$ proposed in a private communication by FERGUSON was adopted; for reaction (32.6) a rate coefficient of $1.5 \cdot 10^{-15}\ m^3\ s^{-1}$ was assumed and this is seen to be consistent with the data shown in Fig. 24. The height variation of the production rate corresponding to each of these sources is shown in Fig. 49.

The major loss processes for N^+ ions are believed to be ion-atom interchange and charge transfer with O_2, reactions (12.27) and (12.28).

BAILEY and MOFFETT deduced the height distribution of N^+, with account being taken of diffusion, for conditions similar to that prevailing during observations carried out by HOFFMAN et al.[3] using a magnetic-type mass spectrometer. A comparison of the calculated and observed distributions revealed that the predicted N^+ concentrations were too small by a factor of about 2 near 300 km, and BAILEY and MOFFETT suggested that an unidentified source of the atomic ion was operating near this altitude.

β) For H^+ ions photoionization of atomic hydrogen, transported upwards by molecular diffusion from the turbopause region (Sect. 22), is likely to be important at the greatest altitudes but the principal source in the F2 region is believed to be the charge-exchange reaction:

$$O^+(^4S) + H(^2S) \rightleftarrows H^+ + O(^3P). \tag{32.7}$$

[1] Bailey, G.J., Moffett, R.J. (1972): Planet. Space Sci. *20*, 616
[2] Dalgarno, A. (1970): Ann. Geophys. *26*, 601
[3] Hoffman, J.H., Johnson, C.Y., Holmes, J.C., Young, J.M. (1969): J. Geophys. Res. *74*, 6281

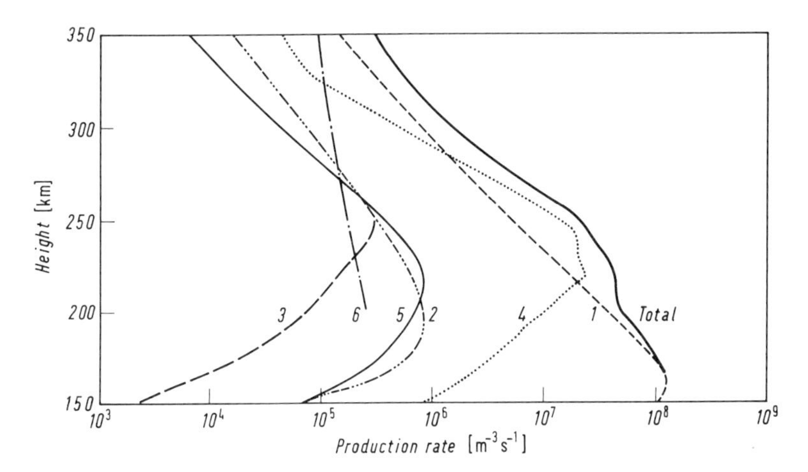

Fig. 49. The height variations of production rates of N^+ ions computed by BAILEY and MOFFETT[1]. The numbers associated with the curves correspond to the reactions (32.1) to (32.6)

It has been assumed that owing to the near resonance condition arising from the similar values of O and H ionization potentials, the forward and reverse reactions would be in equilibrium.

This assumption of equilibrium has been justified by measurements of FEHSENFELD and FERGUSON[4] who obtained the following values of the equilibrium constant for different temperatures: $T=300$ K, $K=0.55$; $T=600$ K, $K=0.71$; $T=1{,}000$ K, $K=0.79$; $T=2{,}000$ K, $K=0.84$.

The forward reaction provides a source of protons in the upper ionosphere in daytime. Furthermore, the long-lived protons constitute a reservoir of ionization from which the nighttime F region can be replenished through the reverse reaction (HANSON and ORTENBURGER[5]).

From mass-spectrometer measurements, BRINTON et al.[6] and HOFFMAN et al.[3] suggested, respectively, that equilibrium prevailed below about 350 km and 450 km. On the basis of this chemistry, deductions have been made about the concentrations of atomic hydrogen. From observations of O^+ and H^+ carried out with the r.f. ion mass spectrometer and of atomic oxygen by the open-source neutral mass spectrometer on Atmospheric Explorer C satellite, BRINTON et al.[7] have been able to examine the seasonal and diurnal variations of atomic hydrogen concentration at 250 km altitude and to examine the relationship between the concentration and atmospheric temperature.

γ) The helium ions observed in the upper atmosphere (Fig. 6) are produced by the photoionization by solar radiations of wavelength less than 50.4 nm of neutral atoms arising from radioactive decay of thorium and uranium at the ground and also introduced into the upper atmosphere during solar flares. The major loss processes are believed to be charge transfer and dissociative charge transfer with N_2, reactions (12.21) and (12.20), and the analogous processes

---

4 Fehsenfeld, F.C., Ferguson, E.E. (1972): J. Chem. Phys. *56*, 3066
5 Hanson, W.B., Ortenburger, I.B. (1961): J. Geophys. Res. *66*, 1425
6 Brinton, H.C., Pharo, M.W., Mayr, H.G., Taylor, H.A. (1969): J. Geophys. Res. *74*, 2941
7 Brinton, H.C., Mayr, H.G., Potter, W.E. (1975): Geophys. Res. Lett. *2*, 389

involving O_2. The rate coefficients of these reactions have been measured using a number of different techniques, as reported by FERGUSON[8], and are well established. The branching ratios in each case are less well known but it is believed that the main products are the atomic ions. Since the concentrations of N_2 exceed those of O_2 above about 300 km, the principal loss for He^+ ions is thought to be by the N_2 reactions. As mentioned in Sect. 12δ(ii), the total rate coefficient of these are independent of temperature below 900 K, and also do not vary with the N_2 vibrational temperature although the branching ratio does. Comparisons of the He^+ distributions computed on the basis of the photoionization rates and the chemical loss rates with observations have shown reasonable agreement up to about 400 km[3,6]. At greater heights, diffusion processes determine the height distribution.

δ) The mass-spectrometer measurements of metallic ions at F-region heights in the equatorial ionosphere (Sect. 19α(iii)) confirmed the early inference of Fe^+ ions from observations with a retarding potential analyser on OGO 6 satellite by HANSON and SANATANI[9] and of Mg^+ ions from observations of solar ultra-violet scattering on satellite TD1 (BOKSENBERG and GÉRARD[10]) and more recently on OGO 4 (GÉRARD[11]). HANSON et al.[12] attributed such metallic ions to vertical transport from a source region below 100 km where the ions have been produced by charge transfer between O_2^+ or NO^+ and metallic atoms. They suggested that the vertical polarisation electric field associated with the equatorial electrojet current would allow ions to move upwards through the region of relatively high collisional frequency. Once the ions attain the height of 160 km they would undergo the normal electromagnetic drift arising from the East-West electric field observed with the incoherent-scatter technique (WOODMAN[13]). The subsequent redistribution of these ions carried by diffusion along geomagnetic field lines has also been considered by HANSON et al.[12].

H. The production of excited species and their roles in atmospheric and ionospheric processes

33. Production of excited species. It has been seen in Chaps. B and C that excited species can result from several processes in the upper atmosphere: fluorescence and resonance scattering of solar radiation, photodissociation, photoionization, neutral-neutral and ion-neutral chemical reac-

[8] Ferguson, E.E. (1973): At. Data Nucl. Data Tables *12*, 159
[9] Hanson, W.B., Sanatani, S. (1970): J. Geophys. Res. *75*, 5503
[10] Boksenberg, A., Gérard, J.C. (1973): J. Geophys. Res. *78*, 4641
[11] Gérard, J.C. (1976): J. Geophys. Res. *81*, 83
[12] Hanson, W.B., Sterling, D.L., Woodman, R.F. (1972): J. Geophys. Res. *77*, 5530
[13] Woodman, R.F. (1970): J. Geophys. Res. *75*, 6249

tions, and dissociative recombination of molecular ions. Observations of emissions from airglow[1-6], as illustrated for daytime in Fig. 50, and auroral features have demonstrated that appreciable concentrations of excited atoms and molecules can be present in the atmosphere [*28*]. It has long been realised that the internal energy of reactants can strongly influence the rate coefficients and even the reaction mechanisms of chemical reactions, and this is illustrated by the roles of excited species in the neutral and ion chemistry of the upper atmosphere. Before considering these roles it is useful to discuss first the formation of particular excited species and also the concentrations of these in the height range of interest [*29*].

α) Atomic oxygen O. In daytime the main production mechanism of $O(^1D)$ atoms up to about 80 km is the photolysis of ozone and this is superseded at greater heights by the photodissociation of molecular oxygen by the Schumann-Runge continuum and Lyman-α radiation, up to about 150 km. Above this height, photoelectron excitation and dissociative recombination of O_2^+ ions represent the main sources of the excited atoms, and such recombination is also responsible for the major production at nighttime. Because of the large values of rate coefficients referred to in Sect. 12ζ(i), quenching by molecular oxygen and nitrogen imposes a lower height limit on the occurrence of $O(^1D)$ atoms, in spite of their large production rate, as shown in the height variation of $O(^1D-^3P)$ emission presented in Fig. 50.

Observations of the $O(^1S-^1D)$ emission at 557.7 nm reveal that the main peak in the height distribution of $O(^1S)$ is located near 100 km, as indicated in Fig. 50. It was formerly believed that laboratory measurements by FELDER and YOUNG[7] showed that the three-body recombination of $O(^3P)$ atoms is responsible for this peak, as proposed by CHAPMAN[8], but this conclusion has been questioned following more recent measurements[9]. Furthermore, the photodissociation of molecular oxygen by Lyman-α radiation might also make a contribution. A second, broader peak is found near 160 km, for which dissociative recombination of O_2^+ ions and excitation by photoelectron impact are thought to be largely responsible. As already mentioned in Sect. 12ζ(i), quenching of the excited atoms arises from collisions with both atomic and molecular oxygen at mesospheric heights.

β) Molecular oxygen O_2. Emissions from the Atmospheric Infra-red bands $O_2(^1\Delta_g-^3\Sigma_g^-)$ at 1.27 μm and 1.58 μm represent the strongest features in the dayglow. Excitation of the $O_2(^1\Delta_g)$ in daytime arises chiefly from the photol-

[1] Nagata, T., Tohmatsu, T., Ogawa, T. (1968): J. Geomagn. Geoelectr. *20*, 315

[2] Rusch, D.W., Stewart, A.I., Hays, P.B., Hoffman, J.R. (1975): J. Geophys. Res. *80*, 2300

[3] Wallace, L., McElroy, M.B. (1966): Planet. Space Sci. *14*, 677

[4] Llewellyn, E.J., Evans, W.J.F. (1971) in: The radiating atmosphere. McCormac, B.M. (ed.), p. 17. Dordrecht: Reidel

[5] Evans, W.F.J., Hunten, D.M., Lewellyn, E.J., Vallance Jones, A. (1968): J. Geophys. Res. *73*, 2885

[6] Wallace, L., Hunten, D.M. (1968): J. Geophys. Res. *73*, 4813

[7] Felder, W., Young, R.A. (1972): J. Chem. Phys. *56*, 6028

[8] Chapman, S. (1931): Proc. R. Soc. London, Ser. A *132*, 353

[9] Slanger, T.G., Black, G. (1977): Planet. Space Sci. *25*, 79

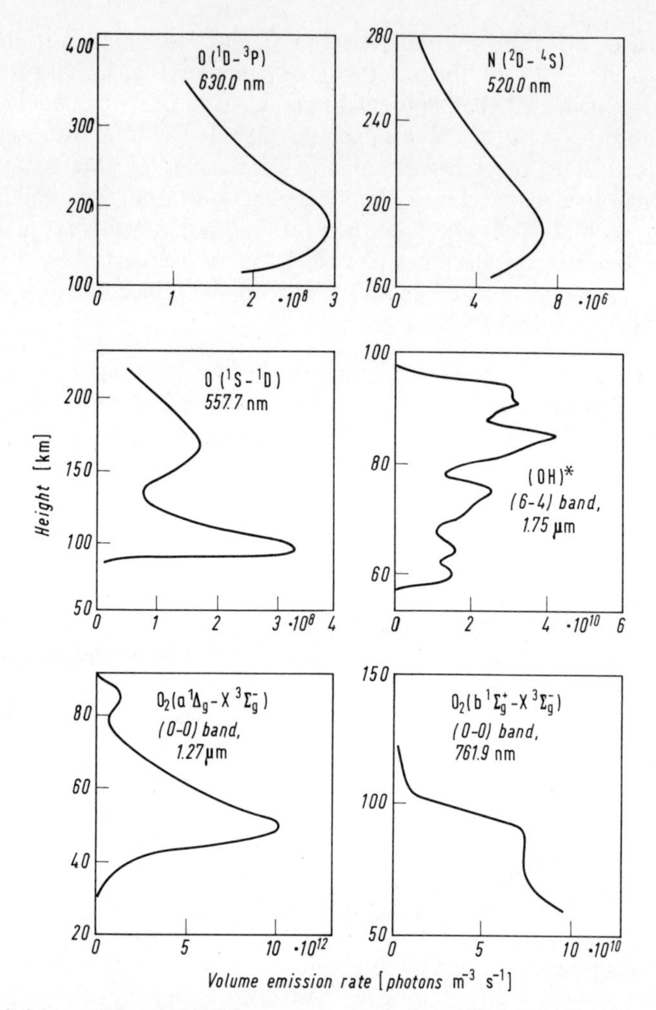

Fig. 50. The height variations of volume emission rates in the dayglow of 630.0 nm[1], 520.0 nm[2], 557.7 nm[3], 1.75 µm[4], 1.27 µm[5], and 761.9 nm[6]. Compare Figs. 43–47 (pp. 64/65) in this Encyclopedia, vol. 49/5 (A.T. and E. VASSY)

ysis of ozone in the Hartley bands, and the relatively slow rate of quenching by molecular oxygen and nitrogen, together with the large radiative lifetime, results in daytime concentrations in excess of $10^{15}\,m^3\,s^{-1}$ at mesospheric heights, and still greater in the stratosphere. Very few estimates of the height distribution of $O_2(^1\Delta_g)$ are available for nighttime and some doubt exists about the excitation mechanism.

Possible mechanisms must provide at least 0.98 eV, the excitation energy of the state. At the present time the three-body association of ground state atoms, in which an energy of 5.11 eV is released, is considered as a probable candidate since the height distribution of nightglow from $O_2(^1\Delta_g)$ is similar to that of other molecular oxygen emissions. The Chapman process[8], in which the third body in the association is also an oxygen atom, also represents a possible source of $O_2(^1\Delta_g)$ molecules at nighttime.

For $O_2(b\,^1\Sigma_g^+)$ the production mechanisms are believed to be resonance absorption of solar radiations between 60 and 100 km, and at greater heights the reaction between $O(^1D)$ atoms and ground-state molecules (12.39). Quenching of the excited molecules occurs at low mesospheric heights and arises chiefly from collisions with molecular nitrogen and oxygen. The interpretation of the rocket measurement shown in Fig. 50 of the Atmospheric system $O_2(b\,^1\Sigma_g^+ - X\,^3\Sigma_g^-)$ at 761.9 nm by WALLACE and HUNTEN[6] has yielded a height distribution for the excited molecule with a concentration constant at about $10^{12}\,m^{-3}$ in the height range 50–80 km and decreasing to about $10^{10}\,cm^{-3}$ near 140 km. Excitation at nighttime is again thought to arise, at least in part, from the three-body recombination of atomic oxygen.

The Herzberg band, originating from the $O_2(A\,^3\Sigma_u^+ - X\,^3\Sigma_g^-)$ transition, is the third of the band systems of molecular oxygen originating at heights below 100 km in the nightglow. As with the $b\,^1\Sigma_g^+$ and $^1\Delta_g$ states, it is believed that at nighttime excitation might originate from the three-body association of oxygen atoms but the relative yield of this and the other two states has not been established.

γ) Atomic nitrogen N. The main production and loss processes for $N(^2D)$ atoms in the daytime have been examined in Sect. 23β. It was shown that these processes have been studied in relation to the observations of dayglow emission at 520 nm of $N(^2D - ^4S)$ made on rockets and also on the Atmospheric Explorer C satellite, the results of one analysis being presented in Figs. 40 and 41. Both the recent atmospheric results and laboratory measurements have indicated the importance of quenching by $O(^3P)$ atoms, especially above about 140 km. At nighttime the situation is more simple since dissociative recombination of NO^+ ions is believed to be the only important source whilst collisions with atomic and molecular oxygen are probably the principal deactivation mechanisms.

δ) Vibrationally-excited molecular nitrogen N_2 and oxygen O_2. The collisional deactivation of $O(^1D)$ by molecular nitrogen results in vibrational excitation of the molecule (Sect. 12ζ(i)). On the basis of experimental data and theoretical calculations it appears that about 30% of the $O(^1D)$ electronic energy is converted into vibrational energy of the molecule. DALGARNO[10] proposed that in addition to this mechanism, vibrationally-excited nitrogen could be produced by the atom-interchange reaction between atomic nitrogen and nitric oxide, and also by the impact of photoelectrons with molecular nitrogen.

The excited nitrogen molecules cannot radiate and are not readily quenched in collision with other nitrogen or oxygen molecules. However, the following processes might play important deactivation roles:

$$(N_2)^*(v=1) + CO_2 \rightarrow (N_2)(v=0) + (CO_2)^*(001) \tag{33.1}$$

$$(N_2)^* + O \rightarrow N_2 + O + \text{translational energy.} \tag{33.2}$$

---

[10] Dalgarno, A. (1963): Planet. Space Sci. *10*, 19

TAYLOR and BITTERMAN[11] have shown that owing to a natural resonance the transfer of energy into the 001 vibrational mode of CO_2 is rapid, a rate coefficient of $5 \cdot 10^{-19}\,m^3\,s^{-1}$ being found for a temperature of 400 K. For quenching by atomic oxygen, recent measurements by MCNEAL et al.[12] have indicated a rather larger value than previously accepted, and BREIG et al.[13] have shown that as a result quenching by atomic oxygen is a dominant loss process for vibrationally-excited nitrogen in the lower thermosphere. Its inclusion results in a significant reduction in the theoretical vibrational temperature which has important aeronomic implications, as in the energy transfer to the electron gas and in the ion-atom interchange reaction with O^+ (reaction (12.18)) which is known to be very dependent on the vibrational temperature of the molecule[14].

Laboratory measurements by GILPIN et al.[15] have shown that the reaction of atomic oxygen with ozone is capable of producing oxygen molecules vibrationally excited to high levels:

$$O + O_3 \rightarrow O_2(^1\Delta_g) + (O_2)^* + 3.09\,\text{eV}. \tag{33.3}$$

This reaction and the predissociation of molecular oxygen in the Schumann-Runge band system (Sect. 4β(i)) could be important sources of the vibrationally-excited molecule at mesospheric heights[16] but for greater heights energy transfer from $O(^1D)$ atoms and photoelectron excitation represent the main production mechanisms. Quenching is believed to occur by collision with $O(^3P)$ atoms, a rate coefficient of $6.9 \cdot 10^{-15} \exp\left(-\frac{76.75}{(T/K)^{\frac{1}{3}}}\right) m^3\,s^{-1}$ being recently reported by TAYLOR[17] for the $v=1$ state.

34. Roles of excited species. Examples of possible roles of excited species have already been encountered and it seems likely that these species might have a much greater influence than is presently appreciated. The most direct result of such species would be the enhancement of chemical rate coefficients, as already noted for the reactions of $N(^2D)$ atoms with molecular oxygen (reaction (12.9)) and of ground-state oxygen ions with vibrationally-excited nitrogen (reaction (12.18)). Alternatively, the reaction products might be altered. Although considerable progress has already been made in studies of reactions involving relevant electronically-excited states, as in the quenching of $O(^1D)$, $O(^1S)$, $N(^2D)$ and $O_2(^1\Delta_g)$, the temperature dependences of these reactions have not been measured and in many cases the products have not been identified. The kinetics of vibrationally and rotationally-excited molecules is still not well developed.

In order to illustrate some aspects of our present understanding of the impact of excited species their importance in three areas will be described.

α) Energy balance. The photodissociation of molecular oxygen in the Schumann-Runge continuum is considered to be the principal source of $O(^1D)$ atoms

[11] Taylor, R.L., Bitterman, S.A. (1969): Rev. Mod. Phys. *41*, 26
[12] McNeal, R.J., Whitson, M.E., Cook, G.R. (1972): Chem. Phys. Lett. *16*, 507
[13] Breig, E.L., Brennan, M.E., McNeal, R.J. (1973): J. Geophys. Res. *78*, 1225
[14] Schmeltekopf, A.L., Ferguson, E.E., Fehsenfeld, F.C. (1968): J. Chem. Phys. *48*, 2968
[15] Gilpin, R., Schiff, H.I., Welge, K.H. (1971): J. Chem. Phys. *55*, 1087
[16] Hudson, R.D., Mahle, S.H. (1972): J. Geophys. Res. *77*, 2902
[17] Taylor, R.L. (1974): Can. J. Chem. *52*, 1436

below about 150 km. For heights near and above this level radiation is an important deactivation process, and the emission profiles give an indication of the photodissociation rate, and hence of the level of solar intensity in the Schumann-Runge continuum spectral region. An interpretation of rocket-borne measurements by SCHAEFFER et al.[1] indicated that the solar fluxes were considerably below those of DETWILER et al.[2]. Agreement between theory and experiment required a reduction by about one-third, as revealed by the solar-flux measurements of PARKINSON and REEVES[3] (Sect. 5β).

At heights below 150 km the excitation energy of $O(^1D)$ atoms is rapidly quenched by collisions and contributes to the peak in heating rate near 100 km caused by the photodissociation of molecular oxygen. In the quenching resulting from collisions with molecular nitrogen, the molecules are vibrationally excited and then represent a source of energy for heating of the ambient electron gas and for infra-red emission of the ν_3 band of CO_2 resulting from the near-resonance vibrational-energy transfer reaction (33.1). An additional transfer to thermal electrons of energy derived initially from solar radiations occurs from $N(^2D)$, both directly (Table 11) and through the vibrationally-excited molecular nitrogen resulting from the reaction of nitric oxide and ground-state nitrogen atoms.

β) Photoionization. As mentioned in Sect. 7γ the photoionization of $O_2(^1\Delta_g)$ by solar radiations of wavelength 102.7 to 111.8 nm is believed to be an important source of ionization in the D region[4,5]. In view of the large intensity of solar Lyman-α radiation, the photoionization of highly excited O_2 molecules could represent a potentially more importance source.

An excitation energy of 1.9 eV would be required for ionization by the Lyman-α photon energy, 10.2 eV, corresponding to a vibrational quantum number of 11 or greater for vibrationally-excited molecules. This possibility was first considered by INN[6,7] and more recently by NORTON and REID[8]. These latter workers also considered, as alternatives, the possible ionizations of $O_2(^1\Sigma_u^-)$, $O_2(^3\Delta_u)$, $O_2(^3\Sigma_u^+)$, $O_2(^3\Sigma_u^-)$ states. However, as they point out, it is very doubtful whether sufficient concentrations of these excited states will exist to contribute appreciably to the production of O_2^+ ions in the D region.

For greater heights information on photoelectron fluxes could, in principle, be derived from twilight observations of $O(^1D-{}^3P)$ emission at 630 nm, $O(^1S-{}^1D)$ emission at 557.7 nm, the First Negative band system of $N_2^+(B\,^2\Sigma_u^+-X\,^2\Sigma_g^+)$, and also of 1,083 nm radiation resonantly scattered from $He\,2(^3S)$ metastable atoms.

γ) Ion chemistry. Laboratory measurements by FEHSENFELD et al.[9] have shown that the $O_2(^1\Delta_g)$ molecules are sufficiently energetic to detach electrons

1 Schaeffer, R.C., Feldman, P.D., Fastie, W.G. (1971): J. Geophys. Res. *76*, 3168
2 Detwiler, C.R., Garrett, D.L., Purcell, J.D., Tousey, R. (1961): Ann. Geophys. *17*, 263
3 Parkinson, W.H., Reeves, E.M. (1969): Sol. Phys. *10*, 342
4 Hunten, D.M., McElroy, M.B. (1968): J. Geophys. Res. *73*, 2421
5 Huffman, R.E., Paulsen, D.E., Larrabee, J.C., Cairns, R.B. (1971): J. Geophys. Res. *76*, 1028
6 Inn, E.C.Y. (1961): Planet. Space Sci. *5*, 76
7 Inn, E.C.Y. (1961): Planet. Space Sci. *8*, 200
8 Norton, R.B., Reid, G.C. (1972): J. Geophys. Res. *77*, 6287
9 Fehsenfeld, F.C., Albritton, D.L., Burt, J.A., Schiff, H.I. (1969): Can. J. Chem. *47*, 1793

from O_2^-, the negative ion formed by electron attachment in the D region, as shown in reaction R3 of Table 12. This and the corresponding reaction for $O(^3P)$ are important in the early stages of the negative-ion scheme. The excited molecules, and also $O(^3P)$ atoms, are also capable of interfering with the formation of positive water cluster ions from the primary O_2^+ ions, because of the destruction of the intermediate O_4^+ ion:

$$O_4^+ + O_2(^1\Delta_g) \rightarrow O_2^+ + 2O_2 + 0.56\,\text{eV}. \quad (34.1)$$

At greater heights the formation of N_2^+ ions from reactions with N_2 of $O^+(^2D)$, $O^+(^2P)$, and $O_2^+(a\,^4\Pi_u)$ ions produced during photoionization (reactions (12.40), (12.41) and (12.42)) could affect the ion composition since the corresponding reactions for the ground-state ions yield NO^+ ions and also proceed much more slowly. In addition, the possibility of producing metallic ions by charge transfer from the excited molecular oxygen ion has been demonstrated in the case of calcium (reaction (15.1)).

In view of these types of observations, and the marked effect of vibrational excitation in the reaction of O^+ with N_2 mentioned previously, it seems evident that in a quantitative treatment of the ionization balance of the E and F regions, attention will need to be given to ions which are excited electronically, and perhaps vibrationally, and to the possible vibrational excitation of reactants such as molecular nitrogen and oxygen.

General references

[*1*] Whitten, R.C., Poppoff, I.G. (1971): Fundamentals of aeronomy. New York: Wiley
[*2*] Nicolet, M. (1960): The properties and constitution of the upper atmosphere. In: Physics of the upper atmosphere. Ratcliffe, J.A. (ed.), chapt. 2. New York: Academic Press
[*3*] Cole, A.E., Court, A., Kantor, A.J. (1965): Model atmospheres. In: Handbook of geophysics and space environments. Valley, S.L. (ed.), chapt. 2. New York: McGraw-Hill
[*4*] CIRA (1972): COSPAR International Reference Atmosphere. Berlin: Akademie-Verlag
[*5*] Ratcliffe, J.A., Weekes, K. (1960): The ionosphere. In: Physics of the upper atmosphere. Ratcliffe, J.A. (ed.), chapt. 9. New York: Academic Press
[*6*] Rawer, K., Bilitza, D., Ramakrishnan, S. (1978): Goals and status of the international reference ionosphere. Rev. Geophys. Space Phys. *16*, 177
[*7*] Hudson, R.D. (1971): Critical review of ultra-violet photoabsorption cross-sections for molecules of astrophysical and aeronomic interest. Rev. Geophys. Space Phys. *9*, 305
[*8*] Banks, P.M., Kockarts, G. (1973): Aeronomy, parts A, B. New York: Academic Press
[*9*] Delaboudinière, J.P., Donnelly, R.F., Hinteregger, H.E., Schmidtke, G., Simon, P.C. (1978): Intercomparison/compilation of relevant solar flux data related to aeronomy. COSPAR Technique Manual No. 7
[*10*] Marr, C.V. (1967): Photoionization processes in gases. New York: Academic Press
[*11*] Chapman, S. (1931): The absorption and dissociative or ionizing effect of monochromatic radiation in an atmosphere on a rotating earth. I and II, Proc. Phys. Soc. London, *43*, 26 and 484
[*12*] Allen, C.W. (1965): The interpretation of the XUV solar spectrum. Space Sci. Rev. *4*, 91
[*13*] Hasted, J.B. (1964): Physics of atomic collisions. London: Butterworth
[*14*] Danilov, A.D. (1970): Chemistry of the ionosphere. New York: Plenum Press
[*15*] Torr, D.G., Torr, M.R. (1978): Review of rate coefficients of ionic reactions determined from measurements made by the atmospheric explore satellites. Rev. Geophys. Space Phys. *16*, 327
[*16*] Bates, D.R., Massey, H.S.W. (1948): The basic reactions in the upper atmosphere II. The theory of recombination in the ionized layers. Proc. R. Soc. London, Ser. A *192*, 1

[*17*] McDaniel, C.W., Cermak, V., Dalgarno, A., Ferguson, E.E. (1970): Ion-molecule reactions. New York: Wiley-Interscience
[*18*] Ferguson, E.E. (1975): Ion-molecule reactions. Ann. Rev. Phys. Chem. *26*, 17
[*19*] Narcisi, R.S. (1973): Mass-spectrometer measurements in the ionosphere. In: Physics and chemistry of upper atmospheres. McCormac, B.M. (ed.), p. 171. Dordrecht: Reidel
[*20*] Evans, J.V. (1969): Theory and practice of ionospheric study by Thomson scatter radar. Proc. IEEE *57*, 496
[*21*] Chapman, S., Cowling, T.C. (1970): The mathematical theory of non-uniform gases. London: Cambridge University Press
[*22*] Bates, D.R. (1970): Reactions in the ionosphere. Contemp. Phys. *11*, 105
[*23*] Rishbeth, H., Garriott, O.K. (1969): Introduction to ionospheric physics. New York: Academic Press
[*24*] Thomas, L. (1974): Recent developments and outstanding problems in the theory of the D region. Radio Sci. *9*, 121
[*25*] Rishbeth, H. (1972): Thermospheric winds and the F region. A review. J. Atmos. Terr. Phys. *34*, 1
[*26*] Nicolet, M., Aikin, A.C. (1960): The formation of the D region of the ionosphere. J. Geophys. Res. *68*, 1469
[*27*] Strobel, D.F. (1974): Physics and chemistry of the E region. A review. Radio Sci. *9*, 159
[*28*] Chamberlain, J.W. (1961): Physics of the aurora and airglow. New York: Academic Press
[*29*] Vlasov, M.N. (1976): Photochemistry of excited species. J. Atmos. Terr. Phys. *38*, 807

Note added in proof. For recent collections of cross-section etc. data consult:

Torr, D.G. (1979): Ionospheric Chemistry. Rev. Geophys. Space Res. *17*, 510

Torr, M.R., Torr, D.G. (1982): The role of metastable species in the thermosphere. Rev. Geophys. Space Res. *20*, 91

Diffusion in the High Atmosphere

By

T. YONEZAWA

With 27 Figures

I. Introduction and the general theory of diffusion

1. General remarks. When a system of different kinds of particles, such as the atmosphere, is not in thermal equilibrium, changes will be brought about in the distribution of particle concentrations and the system will approach a state of equilibrium. This process is called diffusion. If particle concentrations in a mixture of fluids at rest are not uniform, a change will occur which makes them tend towards uniformity. If temperature distribution is not uniform, thermal diffusion will take place. If an external force acts on the particles, their distribution will undergo a change so that it tends to a particular (non-uniform) one corresponding to the external force. These are processes which make some aspects of diffusion in general.

The Earth's atmosphere is subjected to time-varying heating due to solar electromagnetic and corpuscular radiation and is constantly disturbed by large-scale atmospheric circulation and small-scale winds; hence it is not in the thermal equilibrium condition. On the other hand, processes such as diffusion and thermal conduction which make the atmosphere tend towards thermal equilibrium are constantly going on so that, in the majority of cases, it may be considered to be in a steady-state condition brought about by the balance between the above-mentioned processes. In particular, it is readily conceivable that diffusion plays an important role in determining the concentration distribution of atmospheric constituents in a steady state. Since atmospheric particles, including charged ones, are directly or indirectly influenced by electric and magnetic fields present in the high atmosphere and by gravity, they are often in a steady state which is determined by the balance between these field forces and the action of diffusion, and hence their concentration distribution is also determined by this balance. In the following description consideration is given mainly as to how diffusion processes influence the distribution of the concentration of atmospheric particles, including charged ones, in the high atmosphere.

Only the high atmosphere above about 60 km is considered here. For details on the atmosphere below this height the reader is requested to refer to articles on meteorological phenomena. While in meteorology diffusion processes come into question mainly in relation to turbulence (eddy diffusion), molecular diffusion is more important in upper atmospheric physics. However, since mixing of different atmospheric constituents due to turbulence takes place in the high atmosphere (i.e. at heights less than about one hundred and a few tens of km), the effect of eddy diffusion must also be taken into consideration. In particular, eddy diffusion plays an important role in the redistribution of molecular and atomic oxygen, accompanying the dissociation of molecular oxygen, and it also affects the distribution of some minor constituents at levels of about 100 km and somewhat higher. This has a further influence on their distribution at much higher altitudes. It should also be noted that although the effect of thermal diffusion may be usually neglected, it can have an appreciable influence in some special cases, e.g. in the distribution of light atoms such as hydrogen and helium.

The scope of the present article is the following. In Chap. I fundamental materials necessary for the consideration of the diffusion processes in the high atmosphere, i.e. molecular diffusion, thermal diffusion, eddy diffusion, and ambipolar diffusion are explained together with the description of the fundamentals of the methods of gas-kinetic-theoretical and hydrodynamical treatment of these processes. In Chap. II separation of respective constituents of the high atmosphere due to molecular diffusion and their mixing due to eddy diffusion are discussed and consideration is given to chemical release experiments relevant to the above phenomena. In Chap. III the problem of the vertical distribution of molecular and atomic oxygen is considered; this is the best example of manifestation of molecular and eddy diffusion. In Chap. IV the formation of the ionospheric F2-layer[1] is treated in so far as ambipolar diffusion plays an important role; the behavior of meteor trails and artificial ion clouds released from rockets are also typical examples of ambipolar diffusion. In Chap. V consideration is given to the vertical distribution of atomic hydrogen, helium, their ions, and some other minor constituents and, in particular, the influence of the polar wind on the distribution of hydrogen and helium ions.

2. Fundamental expressions. The normal method of discussing the effect of diffusion in the atmosphere is to use formulas which CHAPMAN and COWLING [*1*] worked out by approximately solving Boltzmann's equation. However, if this method is applied rigorously, expressions for gas mixtures composed of three or more components become very complicated, and therefore very troublesome to deal with. Moreover, their method, as it is, cannot be applied for example to a plasma with different electron and ion temperatures because they only treated cases in which different kinds of gas have a common temperature. In such cases it is more convenient to use the hydrodynamical approximation which is, for example, adopted in SPITZER's book [*2*]. The relevant equations can also be derived from Boltzmann's equation, and as equations can be obtained for respective kinds of gas separately, different

[1] As for the general problem see the contribution of THOMAS in this volume, p. 7

temperatures can be assigned to them and no very great difficulty arises in dealing with a multicomponent system with different temperatures. However, in order to simplify the analysis, it is usual to express the net momentum gain which gas molecules acquire per unit volume by collision ($\boldsymbol{P}_i$ in Eq. (4.2) below) in terms of the rather ambiguous quantity of effective collision frequency (cf. Eq. (4.6) below); this may make the degree of approximation worse. Moreover, this method has the weak point that thermal diffusion cannot be worked out. In the following, either of the above two methods are used according to the circumstances. Readers should bear in mind their weak points. For example, in Sect. 36, gas-kinetic-theoretical expressions with different temperatures are applied to the electron and ion populations in a plasma, but it is not clear to what degree this approximation is justified in the Earth's environment.

Before entering into the main discussion, it is useful to record some fundamental expressions obtained from the gas-kinetic theory. Each component of a gas mixture is distinguished by the subscript i, and the velocity with respect to an arbitrary reference frame and the number density of the i-th molecule (or atom - this will not be repeated in the following) are denoted by $\boldsymbol{v}_i$ and n_i, respectively; then the velocity of the gas as a whole, i.e. the average of the velocities of all gas molecules, $\bar{\boldsymbol{v}}$, is given by

$$n\,\bar{\boldsymbol{v}}=\sum_i n_i\,\bar{\boldsymbol{v}}_i, \tag{2.1}$$

where a horizontal bar denotes the average and n is the molecular number density of the gas as a whole:

$$n=\sum_i n_i. \tag{2.2}$$

The mass velocity of the gas as a whole, $\boldsymbol{v}_0$, is the mean of the average velocities of the respective kinds of molecules with the mass densities of the respective components attached as weight; this generally differs from $\bar{\boldsymbol{v}}$ and is defined by the following expression:

$$\rho\,\boldsymbol{v}_0=\sum_i \rho_i\,\bar{\boldsymbol{v}}_i=\sum_i n_i\,m_i\,\bar{\boldsymbol{v}}_i, \tag{2.3}$$

$$\rho=\sum_i \rho_i=\sum_i n_i\,m_i, \tag{2.4}$$

where m_i is the mass of the i-th molecule, ρ_i the mass density of the i-th component, and ρ is the mass density of the gas as a whole. By the law of conservation of momentum, $\boldsymbol{v}_0$ does not change with time if no external force acts on the gas. The peculiar velocity of the i-th molecule, $\boldsymbol{V}_i$, is defined as its velocity relative to $\boldsymbol{v}_0$:

$$\boldsymbol{V}_i=\boldsymbol{v}_i-\boldsymbol{v}_0. \tag{2.5}$$

The average of the peculiar velocities $\boldsymbol{V}_i$ over the i-th molecules, $\bar{\boldsymbol{V}}_i$, is the diffusion velocity of the i-th component. When two kinds of molecules with $i=1$ and 2 are considered, the difference of their velocities and that of their peculiar velocities are equal:

$$\boldsymbol{V}_1-\boldsymbol{V}_2=\boldsymbol{v}_1-\boldsymbol{v}_2 \tag{2.6}$$

and these differences do not depend on the velocity of the reference frame. Therefore the same is true with the relative diffusion velocity of two kinds of component gases.

3. Gas-kinetic-theoretical formulas for a multicomponent system

α) When a *rarefied gas mixture* composed of ν kinds of components is considered, it can be shown that the following relations hold with a sufficient degree of approximation for the diffusion velocity of each component gas and among those of different kinds [3a, b]:

$$m_i \boldsymbol{j}_i = n_i m_i \bar{\boldsymbol{V}}_i = \frac{n^2}{\rho}\sum_{j=1}^{\nu} m_i m_j D_{ij}\boldsymbol{d}_j - D_i^T \nabla \ln T, \qquad i=1,2,3,\ldots,\nu \tag{3.1}$$

with

$$\boldsymbol{d}_j = \nabla\left(\frac{n_j}{n}\right)+\left(\frac{n_j}{n}-\frac{n_j m_j}{\rho}\right)\nabla \ln p - \frac{n_j m_j}{p\rho}\left(\frac{\rho}{m_j}\boldsymbol{X}_j-\sum_{k=1}^{\nu} n_k \boldsymbol{X}_k\right), \tag{3.2}$$

and

$$\sum_{j=1}^{\nu}\frac{n_i n_j}{n^2 \mathscr{D}_{ij}^{(1)}}(\bar{\boldsymbol{V}}_j-\bar{\boldsymbol{V}}_i)=\boldsymbol{d}_i-\nabla \ln T\sum_{j=1}^{\nu}\frac{n_i n_j}{n^2 \mathscr{D}_{ij}^{(1)}}\left(\frac{D_j^T}{n_j m_j}-\frac{D_i^T}{n_i m_i}\right), \tag{3.3}$$

where $\boldsymbol{j}_i$ is the diffusion flux of the i-th molecules, p the hydrostatic total pressure of the gas, T the absolute temperature of the gas, $\boldsymbol{X}_i$ the external force acting on the i-th molecule, D_{ij} and D_i^T the multicomponent diffusion coefficients and multicomponent thermal diffusion coefficients, respectively, and $\mathscr{D}_{ij}$ is the diffusion coefficient of a binary gas mixture composed of only the i-th and j-th components, the superscript (1) indicating the first-order approximate value of the variable to which it is attached. Here the first-order approximate value is the one obtained by solving Boltzmann's equation using the approximate method of expanding the velocity distribution function in a series of Sonine polynomials and taking only the first term. This value of $\mathscr{D}_{ij}$ is given by the following expression [3c]:

$$\mathscr{D}_{ij}^{(1)}=\frac{3(m_i+m_j)}{16\, n\, m_i m_j}\frac{kT}{\Omega_{ij}^{(1,1)}}, \tag{3.4}$$

where k is Boltzmann's constant and

$$\Omega_{ij}^{(l,s)}=\sqrt{\frac{(m_i+m_j)kT}{2\pi m_i m_j}}\int_0^{\infty} e^{-\gamma^2}\gamma^{2s+3} Q_{ij}^{(l)}(g)\,d\gamma, \tag{3.5}$$

$$\begin{aligned} Q_{ij}^{(l)}(g) &= 2\pi\int_0^{\pi}\sigma_{ij}(\chi,g)(1-\cos^l\chi)\sin\chi\, d\chi \\ &= 2\pi\int_0^{\infty}(1-\cos^l\chi)\, b\, db, \end{aligned} \tag{3.6}$$

with

$$\gamma=\sqrt{\frac{m_i m_j}{2(m_i+m_j)kT}}\, g;$$

here χ is the angle by which the i-th and j-th molecules are deflected in the center of gravity coordinate system when they collide with each other, g the relative speed of the colliding molecules, $\sigma_{ij}(\chi, g)$ the differential collision cross-section of i- and j-molecules at the relative speed of g, i.e. the probability that the relative velocity is deflected by χ during the collision and falls into the solid angle of $2\pi \sin\chi\, d\chi$ thereafter is given by $2\pi\sigma_{ij}(\chi, g)\sin\chi\, d\chi$, and b is the impact parameter. From these expressions it is readily seen that

$$D_{ij}^{(1)} = D_{ji}^{(1)}. \tag{3.7}$$

D_{ij} and D_i^T are given by [3d]

$$D_{ij} = \frac{\rho n_i}{2n m_j}\sqrt{\frac{2kT}{m_i}}\, c_{i0}^{ji}, \tag{3.8}$$

$$D_i^T = \frac{n_i m_i}{2}\sqrt{\frac{2kT}{m_i}}\, a_{i0}, \tag{3.9}$$

and a_{i0} and c_{i0}^{ji} are determined by the following expressions [3e, 4]:

$$\sum_{j=1}^{\nu}\sum_{p=0}^{M} q_{ij}^{mp}\, a_{jp} = -\frac{15\sqrt{\pi}\, n_i}{2}\,\delta_{m1}, \tag{3.10}$$

$$\sum_{j=1}^{\nu}\sum_{p=0}^{M} q_{ij}^{mp}\, c_{jp}^{hk} = 3\sqrt{\pi}\,(\delta_{ik}-\delta_{ih})\,\delta_{m0}, \tag{3.11}$$

$$(i=1,2,\ldots,\nu;\ m=0,1,\ldots,M)$$

where M is less than the degree of approximation or the number of terms retained in the Sonine polynomial expansion by one.

β) *Approximations.* Let $\boldsymbol{q}_{ij}^{mp}$ be the following matrix

$$\boldsymbol{q}_{ij}^{mp} = \begin{pmatrix} q_{11}^{mp} & q_{12}^{mp} & \cdots & q_{1\nu}^{mp} \\ q_{21}^{mp} & q_{22}^{mp} & \cdots & q_{2\nu}^{mp} \\ \vdots & \vdots & \ddots & \vdots \\ q_{\nu 1}^{mp} & q_{\nu 2}^{mp} & \cdots & q_{\nu\nu}^{mp} \end{pmatrix}; \tag{3.12}$$

then the fourth approximations to D_{ij} and D_i^T are given by

$$D_{ij}^{(4)} = \frac{3\rho n_i}{2n m_j |\boldsymbol{q}|}\sqrt{\frac{2\pi kT}{m_i}} \begin{vmatrix} \boldsymbol{q}_{hk}^{00} & \boldsymbol{q}_{hk}^{01} & \boldsymbol{q}_{hk}^{02} & \boldsymbol{q}_{hk}^{03} & \delta_{hj}-\delta_{hi} \\ \boldsymbol{q}_{hk}^{10} & \boldsymbol{q}_{hk}^{11} & \boldsymbol{q}_{hk}^{12} & \boldsymbol{q}_{hk}^{13} & 0 \\ \boldsymbol{q}_{hk}^{20} & \boldsymbol{q}_{hk}^{21} & \boldsymbol{q}_{hk}^{22} & \boldsymbol{q}_{hk}^{23} & 0 \\ \boldsymbol{q}_{hk}^{30} & \boldsymbol{q}_{hk}^{31} & \boldsymbol{q}_{hk}^{32} & \boldsymbol{q}_{hk}^{33} & 0 \\ \delta_{ki} & 0 & 0 & 0 & 0 \end{vmatrix}, \tag{3.13}$$

$$D_i^{T(4)} = \frac{15 n_i \sqrt{2\pi m_i kT}}{4|\boldsymbol{q}|} \begin{vmatrix} q_{hk}^{00} & q_{hk}^{01} & q_{hk}^{02} & q_{hk}^{03} & 0 \\ q_{hk}^{10} & q_{hk}^{11} & q_{hk}^{12} & q_{hk}^{13} & n_h \\ q_{hk}^{20} & q_{hk}^{21} & q_{hk}^{22} & q_{hk}^{23} & 0 \\ q_{hk}^{30} & q_{hk}^{31} & q_{hk}^{32} & q_{hk}^{33} & 0 \\ \delta_{ki} & 0 & 0 & 0 & 0 \end{vmatrix}, \tag{3.14}$$

where $|\boldsymbol{q}|$ is the determinant formed from the numerator by deleting the last row and last column. To obtain the third, second, or first approximations to these coefficients, the blocks with m or $p=3$, m or $p \geqq 2$, or m or $p \geqq 1$, respectively, are deleted. It is seen from these expressions that D_{ii} and $D_i^{T(1)}$ vanish identically. The elements q_{ij}^{mp} can be expressed in terms of

$$\bar{Q}_{ij}^{(l,s)}(T) = \frac{4(l+1)}{(s+1)!\,[2l+1-(-1)^l]} \int_0^\infty e^{-\gamma^2} \gamma^{2s+3} Q_{ij}^{(l)}(g)\,d\gamma. \tag{3.15}$$

For example [3f, 4],

$$q_{ij}^{00} = 8 \sum_l n_l \sqrt{\frac{m_i}{m_i+m_l}}\, \bar{Q}_{il}^{(1,1)} \left[n_i \sqrt{\frac{m_l}{m_j}} (\delta_{ij} - \delta_{jl}) - n_j \frac{\sqrt{m_l m_j}}{m_i} (1-\delta_{il}) \right], \tag{3.16}$$

$$q_{ij}^{01} = 8 n_i \left(\frac{m_i}{m_j}\right)^{\frac{3}{2}} \sum_l n_l \left(\frac{m_l}{m_i+m_l}\right)^{\frac{3}{2}} \left(\tfrac{5}{2} \bar{Q}_{il}^{(1,1)} - 3 \bar{Q}_{il}^{(1,2)}\right) (\delta_{ij} - \delta_{jl}), \tag{3.17}$$

$$q_{ij}^{10} = \frac{m_j}{m_i} q_{ij}^{01}, \tag{3.18}$$

$$q_{ij}^{11} = 8 n_i \left(\frac{m_i}{m_j}\right)^{\frac{3}{2}} \sum_l n_l \frac{m_l^{\frac{1}{2}}}{(m_i+m_l)^{5/2}} \{ (\delta_{ij} - \delta_{jl}) [\tfrac{5}{4}(6 m_j^2 + 5 m_l^2) \bar{Q}_{il}^{(1,1)} - 15 m_l^2 \bar{Q}_{il}^{(1,2)} + 12 m_l^2 \bar{Q}_{il}^{(1,3)}] + (\delta_{ij} + \delta_{jl}) 4 m_j m_l \bar{Q}_{il}^{(2,2)} \}. \tag{3.19}$$

Expressions for other q_{ij}^{mp}'s with m and p equal to integral values not more than 3 are given by DEVOTO [4]. The tables at the end of the book by HIRSCHFELDER et al. [3] are very useful for calculating $\bar{\Omega}_{ij}^{ls}$. Actual calculations of diffusion coefficients can be performed with the use of this literature. SCHUNK and WALKER [5] also summarize the expressions obtained in the case of collisions among electrons, atomic oxygen, and its ion which are very closely related to upper atmospheric phenomena.

γ) In the special case of a *ternary mixture* whose three components are indicated by subscripts i, j, and k, the multicomponent diffusion coefficient can be expressed in terms of binary ones $\mathcal{D}_{ij}$ in the following way [3g]:

$$D_{ij} = \mathcal{D}_{ij} \left[1 + \frac{n_k \left(\frac{m_k}{m_j} \mathcal{D}_{ik} - \mathcal{D}_{ij} \right)}{n_i \mathcal{D}_{jk} + n_j \mathcal{D}_{ki} + n_k \mathcal{D}_{ij}} \right], \quad (i \neq j \neq k \neq i). \tag{3.20}$$

As D_{ij} can be expressed in a simple form to the first approximation, as shown in Eq. (3.4), use of Eq. (3.20) makes it easier to deal with the multicomponent diffusion coefficient of a ternary mixture. Using Eq. (3.20), the following relation can also be proved:

$$D_{ji} = D_{ij} + m_k \left(\frac{D_{kj}}{m_i} - \frac{D_{ki}}{m_j} \right). \tag{3.20a}$$

This relation was deduced by SCHUNK and WALKER [5].

δ) For a *binary mixture* consisting of two components with $i=1$ and 2, Eq. (3.3) can be simplified to a considerable degree, giving

$$\bar{V}_1 - \bar{V}_2 = -\frac{n^2}{n_1 n_2} D_{12} \left[\nabla \left(\frac{n_1}{n} \right) + \frac{n_1 n_2 (m_2 - m_1)}{n \rho} \nabla \ln p \right.$$
$$\left. + \frac{\rho_1 \rho_2}{p \rho} (\boldsymbol{F}_2 - \boldsymbol{F}_1) + k_T \nabla \ln T \right], \tag{3.21}$$

if Eq. (3.2) is taken into account; here $\boldsymbol{F}_1$ and $\boldsymbol{F}_2$ are the external forces acting on the components 1 and 2, respectively, per unit mass and k_T is the thermal diffusion ratio defined by

$$k_T = \frac{\rho}{n^2 m_1 m_2} \frac{D_1^T}{D_{12}}. \tag{3.22}$$

In deriving Eq. (3.21), it is necessary to use the relations $D_{12} = D_{21}$ and $D_1^T = -D_2^T$. Eq. (3.7) has shown that $D_{12} = D_{21}$ is valid at least to the first approximation; it will be seen in Sect. 6 that this is also true of $D_1^T = -D_2^T$.

4. Hydrodynamical formulas. The equation of continuity

$$\frac{\partial n_i}{\partial t} + \nabla \cdot (n_i \bar{\boldsymbol{v}}_i) = 0 \tag{4.1}$$

and the equation of momentum transfer (or the equation of motion)

$$n_i m_i \left(\frac{\partial \bar{\boldsymbol{v}}_i}{\partial t} + (\bar{\boldsymbol{v}}_i \cdot \nabla) \bar{\boldsymbol{v}}_i \right) = n_i \bar{\boldsymbol{X}}_i - \nabla \cdot \tilde{\psi}_i + \boldsymbol{P}_i \tag{4.2}$$

can be deduced from Boltzmann's equation for each component separately [2a]; here $\tilde{\psi}_i$ is the stress tensor

$$\tilde{\psi}_i = n_i m_i \overline{(\boldsymbol{v}_i - \bar{\boldsymbol{v}}_i)(\boldsymbol{v}_i - \bar{\boldsymbol{v}}_i)}, \tag{4.3}$$

$\bar{\boldsymbol{X}}_i$ the mean force acting on the i-particles (it should be noted that, if the force acting on a particle depends on its velocity, different forces will act on the same kind of molecules), and $\boldsymbol{P}_i$ is the net momentum gain per unit volume and time which the i-particle acquires through collisions with other kinds of particles. In deducing Eqs. (4.1) and (4.2) it is assumed that the x-, y-, and z-components of the force $\boldsymbol{X}_i$ acting on the i-particle do not depend on the x-, y-, and z-components of $\boldsymbol{v}_i$, respectively; the gravitational and electromagnetic forces satisfy this condition.

As $\bar{\boldsymbol{v}}_i$'s are different for different kinds of gases, the form of the expression for $\tilde{\psi}_i$ given by Eq. (4.3) is not adequate when a mixture of different kinds of gases is considered as a whole. It is more reasonable to refer to the mass-velocity of the gas mixture, i.e. to define the stress tensor by

$$\tilde{\psi}_i' = n_i m_i \overline{(\boldsymbol{v}_i - \boldsymbol{v}_0)(\boldsymbol{v}_i - \boldsymbol{v}_0)}. \tag{4.4}$$

This definition is adopted by CHAPMAN-COWLING [1] and by HIRSCHFELDER et al. [3]. However,

$$\begin{aligned}\tilde{\psi}_i' - \tilde{\psi}_i &= n_i m_i \overline{[(\boldsymbol{v}_i - \boldsymbol{v}_0)(\boldsymbol{v}_i - \boldsymbol{v}_0) - (\boldsymbol{v}_i - \bar{\boldsymbol{v}}_i)(\boldsymbol{v}_i - \bar{\boldsymbol{v}}_i)]}\\ &= n_i m_i (\boldsymbol{v}_0 - \bar{\boldsymbol{v}}_i)(\boldsymbol{v}_0 - \bar{\boldsymbol{v}}_i)\end{aligned}$$

and each element of this tensor is generally small compared with $|\boldsymbol{v}_i|^2$ (in the upper atmospheric problems $|\boldsymbol{v}_i|$ is usually of the order of 10^3 m sec^{-1} while $|\boldsymbol{v}_0|$ and $|\bar{\boldsymbol{v}}_i|$ are of the order of 10^2 m s^{-1} at the most). Furthermore, if $n_i m_i (\boldsymbol{v}_0 - \bar{\boldsymbol{v}}_i)(\boldsymbol{v}_0 - \bar{\boldsymbol{v}}_i)$ does not markedly change with the position, $\tilde{\psi}_i$ in Eq. (4.2) can be replaced by $\tilde{\psi}_i'$. We shall only consider such cases in the following and neglect the difference between $\tilde{\psi}_i$ and $\tilde{\psi}_i'$.

If the gas is isotropic, the non-diagonal elements of the stress tensor given by Eq. (4.3) vanish and its diagonal elements become equal to p_i, the partial pressure of the i-particles; then, on the right-hand side of Eq. (4.2)

$$\nabla \cdot \tilde{\psi}_i = \nabla p_i. \tag{4.5}$$

In the following we shall consider only the cases in which this equation holds.

It is usual for $\boldsymbol{P}_i$ on the right-hand side of Eq. (4.2) to be expressed in the following way[1]:

$$\boldsymbol{P}_i = n_i \sum_j \frac{m_i m_j}{m_i + m_j} (\bar{\boldsymbol{v}}_j - \bar{\boldsymbol{v}}_i) \nu_{ij}, \tag{4.6}$$

where ν_{ij} is the effective frequency per unit time of collisions which an i-molecule suffers with j-molecules of a different kind[2] because, when an i-molecule and a j-molecule with velocities $\boldsymbol{v}_i$ and $\boldsymbol{v}_j$, respectively, collide with each other, the net momentum gain acquired by the i-molecule is on average $[m_i m_j/(m_i + m_j)](\bar{\boldsymbol{v}}_j - \bar{\boldsymbol{v}}_i)$.

5. Molecular diffusion

α) For a *non-uniform gas mixture* composed of two components the relative diffusion velocity between them is given by the following expression according

[1] Shabansky, V.P. (1971): Space Sci. Rev. *12*, 299

[2] As for a more detailed discussion see SUCHY's contribution in vol. 49/7, p. 1

to Eqs. (3.21) and (2.6):

$$\bar{\boldsymbol{v}}_1 - \bar{\boldsymbol{v}}_2 = \boldsymbol{v}_d + \boldsymbol{v}_p + \boldsymbol{v}_F + \boldsymbol{v}_T, \tag{5.1}$$

where

$$\boldsymbol{v}_d = -\frac{n^2}{n_1 n_2} D_{12} \nabla \left(\frac{n_1}{n}\right), \tag{5.2}$$

$$\boldsymbol{v}_p = -\frac{n(m_2 - m_1)}{\rho} D_{12} \nabla \ln p, \tag{5.3}$$

$$\boldsymbol{v}_F = \frac{m_1 m_2}{m} D_{12} \frac{\boldsymbol{F}_1 - \boldsymbol{F}_2}{kT}, \tag{5.4}$$

$$\boldsymbol{v}_T = -D_{12} \alpha_T \nabla \ln T, \tag{5.5}$$

m being the average molecular mass of the gas mixture and α_T defined by

$$\alpha_T = \frac{n^2}{n_1 n_2} k_T \tag{5.6}$$

and called the thermal diffusion factor. As described in the next section, this quantity is almost independent of the concentration ratio of the two components so that it is in some cases more convenient to use than k_T. In deriving Eq. (5.4) the equation of state

$$p = nkT \tag{5.7}$$

has been used. The velocities given by Eqs. (5.2)–(5.5) are, respectively, ordinary diffusion velocity, pressure diffusion velocity, forced diffusion velocity, and thermal diffusion velocity. They are diffusion velocities of the component of species 1 relative to those of the component of species 2, manifesting themselves when there are gradients in relative concentrations of the components, a hydrostatic pressure gradient of the mixture as a whole, inequality of external forces acting on the components per unit mass, and a temperature gradient, respectively.

According to DALTON's law, hydrostatic pressure is the sum of the partial pressures p_i of the component gases:

$$p = p_1 + p_2; \tag{5.8}$$

p_i satisfies the equation of state:

$$p_i = n_i kT. \tag{5.9}$$

Although gravitational force acts in the atmosphere, the force per unit mass is equal for any component gas so that it does not directly bring about relative motion between any two components. However, there is a large pressure gradient in the vertical direction due to gravity and, according to Eq. (5.3), this induces relative motion among the components, indirectly manifesting the effect of gravity. The most important example of forced diffusion is the effect of electric and magnetic fields on ionized gases. Thermal diffusion is dealt with in the next section.

β) Theoretically the *diffusion coefficient* for a binary gas mixture D_{12} can be obtained to the first approximation by substituting $i=1$ and $j=2$ in Eqs. (3.4) -(3.6). $Q_{12}^{(1)}$ given by Eq. (3.6) is a kind of collision cross section and its values are different according to the type of intermolecular force acting between the molecules of species 1 and 2. If each molecule behaves like a rigid elastic sphere, $Q_{12}^{(1)}$ is given by

$$Q_{12}^{(1)}=\frac{\pi}{4}(\sigma_1+\sigma_2)^2, \tag{5.10}$$

where σ_1 and σ_2 are the diameters of the spheres. When $i=1$ and $j=2$ Eq. (3.4) becomes

$$D_{12}^{(1)}=\frac{3}{4n(\sigma_1+\sigma_2)^2}\sqrt{\frac{2kT(m_1+m_2)}{\pi m_1 m_2}}. \tag{5.11}$$

Eqs. (3.4) or (5.11) shows that the diffusion coefficient D_{12} does not to the first approximation depend on the proportions in which the components are contained in the mixture. In a higher degree of approximation D_{12} varies slightly with the mixing ratios of the components. As a matter of fact, however, this change is of the order of magnitude of experimental errors so that it is disregarded in the following. If pressure is reckoned in units of atmosphere and σ_1 and σ_2 in nm, Eq. (5.11) is transformed into the following by inserting numerical values into the constants:

$$D_{12}^{(1)}=2.629\cdot 10^{-9}\,\mathrm{m^2\,s^{-1}}\cdot\left(\frac{T}{\mathrm{K}}\right)^{\frac{3}{2}}\left(\frac{M_1+M_2}{2M_1 M_2}\right)^{\frac{1}{2}}\bigg/\left(\frac{p}{\mathrm{P}}\left(\tfrac{1}{2}(\sigma_1+\sigma_2)/\mathrm{nm}\right)^2\right), \tag{5.11a}$$

where M_1 and M_2 are the molecular weights of the molecules of species 1 and 2. Since actual molecules are not rigid elastic spheres, but they exert an *intermolecular force* on each other, the question arises as to what value is to be taken for $\sigma_1+\sigma_2$. An answer to this question can be obtained from HIRSCHFELDER et al. [3], in which the detailed procedure of calculating Eqs. (3.5) and (3.6) are described, intermolecular forces being taken into account, and the results are tabulated at the end of the book. Using these tables, one can readily find theoretical values of $\sigma_1+\sigma_2$ for pairs of common gases. For further details, the reader is requested to refer to Chap. 8 of HIRSCHFELDER's book.

γ) Even in the case of binary mixtures, *experimental measurement* of diffusion coefficients is difficult and liable to experimental errors. However Table 1 shows experimental values of D_{12} for some pairs of gases which may be relevant to the upper atmospheric problems [*1*a], under conditions when the total pressure is 1 at and the temperature is 273 K (=0 °C). Using these values and Eq. (5.11a), the values of $(1/2)(\sigma_1+\sigma_2)$ can be obtained; these are reasonable. For instance, $3.01\cdot 10^{-10}$ m, $3.45\cdot 10^{-10}$ m, and $3.74\cdot 10^{-10}$ m are obtained

Table 1. Values of the diffusion coefficient D_{12} (in $10^{-5}\,m^2\,s^{-1}$)

Pair of gases	H_2-O_2	H_2-N_2	H_2-CO_2	H_2-air	O_2-N_2	O_2-CO_2	O_2-air	CO_2-air
D_{12}	6.97	6.74	5.50	6.11	1.81	1.39	1.78	1.38

Pair of gases	He-A	H_2-D_2	H_2-CO	H_2-N_2O	O_2-CO	CO-N_2	CO-CO_2	CO_2-N_2	CO_2-N_2O
D_{12}	6.41	12.0	6.51	5.35	1.85	1.92	1.37	1.44	0.96

for the pairs of H_2-N_2, O_2-N_2, and CO_2-N_2 and these values are in agreement with our knowledge on the dimension of molecules. Corresponding theoretically deduced values are respectively $3.325 \cdot 10^{-10}$ m, $3.557 \cdot 10^{-10}$ m, and $3.839 \cdot 10^{-10}$ m [*3*i] which are in good agreement with the values obtained above.

It should be noted that, if one of the gases forming a pair is air, the use of Eqs. (5.11) or (5.11a) is not strictly justified because air itself is a mixture. However, expressions for a multicomponent system are so complicated that they cannot be readily utilized. Fortunately, air is largely composed of nitrogen (at very high altitudes the predominant constituent is O, He or H for different ranges of height and temperature) so that other components may be considered as minor constituents. Then, if the interactions between any pairs of minor constituents are neglected, the expressions and laws for a binary mixture can be separately applied to a system composed of the main and a minor constituent. It is very difficult to treat a diffusion problem of a constituent in the atmosphere strictly without resorting to the above-mentioned approximation.

δ) As regards the *temperature dependence* of D_{12}, Eq. (5.11) seems to indicate that it is proportional to the square root of absolute temperature. However, if intermolecular force is taken into account, the value adopted for $\sigma_1+\sigma_2$ changes with temperature, making the temperature dependence of D_{12} more complicated. If the intermolecular force acting between molecules of species 1 and 2 is a repulsive force inversely proportional to the ν_{12}-th power of intermolecular distance, it can be shown that [*1*b]

$$D_{12} \propto \frac{T^{\mu}}{n}, \tag{5.12}$$

where

$$\mu = \frac{1}{2} + \frac{2}{\nu_{12}-1}, \tag{5.13}$$

and, if pressure is constant,

$$D_{12} \propto T^{1+\mu}. \tag{5.14}$$

According to experiments, μ takes the values of 0.755 and 0.792 for the pairs of gases H_2-O_2 and O_2-N_2, respectively [*1*c], and corresponding to these μ's ν_{12} is found to be 8.8 and 7.9 from Eq. (5.13). These values of μ mean that D_{12} depends on temperature to a greater extent than is apparent from Eq. (5.11). On the basis of these experimental values the following relations can be seen to be valid:

$$D_{O_2, N_2} = 1.81 \cdot 10^{-5}\,\text{m}^2\,\text{s}^{-1}\,(T/273\,\text{K})^{0.79}(n_0/n), \tag{5.15}$$

$$D_{H_2, \text{air}} = 6.11 \cdot 10^{-5}\,\text{m}^2\,\text{s}^{-1}\,(T/273\,\text{K})^{0.76}(n_0/n), \tag{5.16}$$

where $n_0 = 2.69 \cdot 10^{25}\,\text{m}^{-3}$. For other pairs of gases expressions of similar form may be employed, but the numerical values to be adopted are not always known.

6. Thermal diffusion. As an account is given in Sect. 3, the thermal diffusion coefficient or the thermal diffusion ratio k_T for a binary mixture defined by Eq. (3.22) vanishes to the first approximation. Its second approximation is given by [3j]

$$k_T = \frac{n_1 n_2}{6n^2 \lambda_{12}} \frac{S^{(1)} n_1 - S^{(2)} n_2}{n(X_\lambda + Y_\lambda)} (6 C_{12}^* - 5). \tag{6.1}$$

(In references [1] and [3] this degree of approximation is called the first approximation.) Here

$$S^{(1)} = \frac{M_1 + M_2}{2M_2} \frac{\lambda_{12}}{\lambda_1} - \frac{15}{4A_{12}^*} \left(\frac{M_2 - M_1}{2M_1}\right) - 1, \tag{6.2}$$

$$S^{(2)} = \frac{M_2 + M_1}{2M_1} \frac{\lambda_{12}}{\lambda_2} - \frac{15}{4A_{12}^*} \left(\frac{M_1 - M_2}{2M_2}\right) - 1, \tag{6.3}$$

$$X_\lambda = \frac{1}{n^2} \left(\frac{n_1^2}{\lambda_1} + \frac{2n_1 n_2}{\lambda_{12}} + \frac{n_2^2}{\lambda_2}\right), \tag{6.4}$$

$$Y_\lambda = \frac{1}{n^2} \left(\frac{n_1^2}{\lambda_1} U^{(1)} + \frac{2n_1 n_2}{\lambda_{12}} U^{(12)} + \frac{n_2^2}{\lambda_2} U^{(2)}\right), \tag{6.5}$$

$$U^{(1)} = \frac{4}{15} A_{12}^* - \frac{1}{12} \left(\frac{12}{5} B_{12}^* + 1\right) \frac{M_1}{M_2} + \frac{1}{2} \frac{(M_1 - M_2)^2}{M_1 M_2}, \tag{6.6}$$

$$U^{(2)} = \frac{4}{15} A_{12}^* - \frac{1}{12} \left(\frac{12}{5} B_{12}^* + 1\right) \frac{M_2}{M_1} + \frac{1}{2} \frac{(M_2 - M_1)^2}{M_1 M_2}, \tag{6.7}$$

$$U^{(12)} = \frac{4}{15} A_{12}^* \frac{(M_1 + M_2)^2}{4M_1 M_2} \frac{\lambda_{12}^2}{\lambda_1 \lambda_2} - \frac{1}{12} \left(\frac{12}{5} B_{12}^* + 1\right) - \frac{5}{32 A_{12}^*} \left(\frac{12}{5} B_{12}^* - 5\right) \frac{(M_1 - M_2)^2}{M_1 M_2}, \tag{6.8}$$

where A_{12}^*, B_{12}^* and C_{12}^* are quantities expressed in terms of $\Omega_{ij}^{(l,s)}$ given by Eq. (3.5) in the following fashion:

$$A_{12}^* = \Omega_{12}^{(2,2)} / (2\Omega_{12}^{(1,1)}), \tag{6.9}$$

$$B_{12}^* = (5\Omega_{12}^{(1,2)} - \Omega_{12}^{(1,3)}) / (3\Omega_{12}^{(1,1)}), \tag{6.10}$$

$$C_{12}^* = \Omega_{12}^{(1,2)} / (3\Omega_{12}^{(1,1)}); \tag{6.11}$$

λ_i is the coefficient of thermal conduction for the gas of species i ($i=1, 2$) and has the following relation to the coefficient of self-diffusion D_i:

$$\lambda_i = \frac{25}{8} \frac{p D_i}{T A_i^*}, \tag{6.12}$$

where A_i^* is a quantity corresponding to A_{12}^* given by Eq. (6.9) when the same kind of molecules of species i collide with each other and D_i is obtained by setting $m_1 = m_2 = m_i$ and $\sigma_1 = \sigma_2 = \sigma_i$ in Eq. (5.11):

$$D_i = \frac{3}{8 n \sigma_i^2} \sqrt{\frac{kT}{\pi m_i}}. \tag{6.13}$$

The thermal conduction coefficient λ_{12} is given by the following expression corresponding to Eq. (6.12):

$$\lambda_{12} = \frac{25}{8} \frac{p D_{12}}{T A_{12}^*}. \tag{6.14}$$

Lastly, A_{12}^*, B_{12}^*, C_{12}^*, and A_i^* ($i=1, 2$) are quantities expressed in terms of $\Omega_{ij}^{(l,s)}$ defined in Eq. (3.5) by Eqs. (6.9)–(6.11) and by an expression corresponding to Eq. (6.9) when colliding molecules are of the same species i, respectively, and can be calculated if the law of intermolecular force is given. They are slowly varying functions of temperature and take the value of unity if each molecule is considered to be a rigid sphere; even in more general cases their values are almost unity. The details of the method of calculation of these quantities are described in Chap. 8 of HIRSCHFELDER et al. [3].

According to experiments the values of k_T are usually of the order of magnitude of 0.1–0.01 and this decreases with the degree of inequality between n_1 and n_2. Therefore, the effect of thermal diffusion can generally be neglected in the atmosphere unless $|\nabla \ln T|$ is very great. However, this is not necessarily allowed in the discussion of the distribution of hydrogen and helium in the upper atmosphere (cf. Chap. V).

As regards the relation $D_1^T = -D_2^T$ to which readers' attention is drawn in Sect. 3, it is clear that this holds at least to the second approximation if one notes Eq. (3.22) and considers that the sign of k_T is reversed when variables with the subscript 1 and 2 are interchanged in Eq. (6.1). Also actual calculation shows that $S^{(1)}$ and $S^{(2)}$ given by Eqs. (6.2) and (6.3) have opposite signs and absolute values of about the same magnitude so that, as may be supposed from the expression of k_T given by Eq. (6.1), $(n^2/n_1 n_2) k_T$ is almost independent of the concentration ratio of the two components. Hence the same is true of α_T given by Eq. (5.6), as stated in the previous section.

As may be seen from Eq. (6.1), k_T or α_T can be positive or negative according to the value of C_{12}^*. As C_{12}^* is a function of temperature, k_T may be positive or negative for different temperatures. Positive values of k_T or α_T signifies that the component of species 1 tends to move into a region of lower temperature and the component of species 2 tends to move into a region of higher temperature.

7. Thermal diffusion factors of H and He in N_2. For the sake of later necessity, an attempt is made here to obtain the value of the thermal diffusion factor of helium in N_2-gas, as an example; the number density of helium, n_1, is assumed to be sufficiently low compared with that of N_2, n_2 ($n_1 \ll n_2$). In this case from Eqs. (5.6), (6.1), (6.4), and (6.5) we obtain

$$\alpha_T = -\frac{1}{6}\frac{\lambda_2}{\lambda_{12}}\frac{S^{(2)}}{1+U^{(2)}}(6C_{12}^* - 5). \tag{7.1}$$

Assuming at first that each molecule behaves like a rigid elastic sphere, we can show that A_{12}^*, B_{12}^* and C_{12}^* given by Eqs. (6.9)–(6.11) all become unity by calculating the integrals in Eq. (3.5) and (3.6) and the factor $(6C_{12}^* - 5)$ in Eq. (7.1) also becomes unity. $U^{(2)}$ can be calculated by Eq. (6.7) and $S^{(2)}$ by Eq. (6.3) if only the value of the thermal conductivity ratio λ_{12}/λ_2 is known. For the values of σ_1 and σ_2 in Eqs. (5.11), (5.11a), and (6.13), we adopt the values tabulated in Table I–A at the end of the book by HIRSCHFELDER et al. [3]:

$$\sigma_1 = 0.2576 \text{ nm for He}, \qquad \sigma_2 = 0.3749 \text{ nm for } N_2.$$

Using these values and Eqs. (5.11) or (5.11a), (6.12), (6.13), and (6.14), λ_{12}/λ_2 can be obtained and hence also α_T. The result is $\alpha_T = -0.38$ which is in agreement with KOCKARTS[1]. However, if the intermolecular force is taken into consideration, the value of α_T becomes somewhat different, mainly because the factor $(6C_{12}^* - 5)$ in Eq. (7.1) deviates from unity. If LENNARD-JONES' potential is taken for that of intermolecular force $\varphi(r)$ as a function of intermolecular distance r,

$$\varphi(r) = 4\varepsilon_{12}[(\sigma_{12}/r)^{12} - (\sigma_{12}/r)^6], \tag{7.2}$$

calculation according to the method given in HIRSCHFELDER et al. [3k] gives $\varepsilon_{12}/k = 28.56$ K and

$$\sigma_{12} = \tfrac{1}{2}(\sigma_1 + \sigma_2) \tag{7.3}$$

and the values of A_{12}^*, B_{12}^*, C_{12}^* and A_2^* can be obtained by interpolation of those tabulated in Table I–N in HIRSCHFELDER et al. [3]. Of these values those parameters other than C_{12}^* practically do not affect the final result at all but C_{12}^* takes a constant value of 0.948 in the temperature range of 500 K...2,000 K, and so $\alpha_T = -0.27$. Although the quantum mechanical effect becomes important for light gases such as hydrogen and helium, particularly at low temperatures [6a], it is unnecessary to discuss this effect under such rather high temperatures as found in upper atmospheric problems.

In the next place, let us consider the thermal diffusion of atomic hydrogen in nitrogen gas. In this case, however, we cannot take the same method as described above for He in N_2, because the values of σ and other parameters have not been obtained for atomic hydrogen from experiments concerning transport phenomena and are not tabulated in HIRSCHFELDER et al. [3]. We shall therefore estimate their rough values using the methods explained below.

[1] Kockarts, G. (1972): J. Atmos. Terr. Phys. *34*, 1729

According to London, the second term on the right-hand side of Eq. (7.2), that is, the potential for the induced-dipole-induced-dipole interaction or Van der Waals force can be expressed approximately in the following form for spherical molecules [31]

$$\psi_{12} = -\frac{3}{4}\frac{\mathscr{E}_1 \mathscr{E}_2}{\mathscr{E}_1 + \mathscr{E}_2}\frac{\alpha_1 \alpha_2}{r^6}, \tag{7.4}$$

where $\mathscr{E}_1$ and $\mathscr{E}_2$ are the characteristic energies for the molecules of species 1 and 2, respectively, approximately equal to their ionization potentials and α_1 and α_2 are their polarizabilities.

Assuming that this formula can be applied to the binary systems He–N_2 and H–N_2 and noting that the ionization potentials of H, He, and N_2 are 13.595 eV, 24.581 eV, and 15.576 eV, respectively, and that the experimental values of polarizability of the former two are 4.5 a_0^3 and 1.37a_0^3 (a_0 is the BOHR radius [3 pm]), we can see that the coefficient of the second term in Eq. (7.2), i.e. $4\varepsilon_{12}\sigma_{12}^6$, is 2.5 times larger for the system of H–N_2 than for that of He–N_2; hence the value of $\varepsilon_{12}\sigma_{12}^6$ can be found. As regards the effective diameter σ_1 of a H-atom, we have estimated it in the following two ways. First, considering that the minimum of the Lennard-Jones potential occurs at an intermolecular distance $r=2^{\frac{1}{6}}\sigma$ and that this may correspond to the internuclear distance of an H_2 molecule of 0.74 Å (74 pm), σ_1 is assumed to be equal to $0.74/2^{\frac{1}{6}}$ or 0.66 Å (66 pm). Secondly, we have estimated the internuclear distance at which the energy of interaction of two H-atoms becomes zero, using its experimental curve [3n] and this has been taken as σ_1, giving $\sigma_1=0.40$ Å (40 pm). When the values of σ_1 are taken in these ways, σ_{12} can be obtained from Eq. (7.3) and when this is combined with the above-obtained value of $\varepsilon_{12}\sigma_{12}^6$, ε_{12} can also be found. In this fashion calculation of α_T has been made for a binary mixture composed mainly of N_2 with a small amount of H. The results are as shown in Table 2.

Table 2. Calculated tentative values of α_T/cm^{3}g^{-1} for the H–N_2 gas

σ_H / pm	ε_{12}/k / K	α_T			
		$T=500$ K	1,000 K	1,500 K	2,000 K
40	896	0.03	−0.03	−0.10	−0.15
66	622	0.02	−0.09	−0.16	−0.20

From this table it is seen that α_T changes sign from plus to minus with an increase in temperature between 500 K and 1,000 K and its absolute values are small in this neighborhood, but it tends to increase in its absolute value with rising temperature after the change of sign; for higher temperatures it seems to approach a value which is of the same order of magnitude as in the case of the He–N_2 system. However, these results have been obtained on the basis of a rough estimate of the intermolecular force between H and N_2 and may be considerably different for other ways of evaluating the values of the parameters so that, regarding the quantitative aspect, they are to be taken with sufficient reserve.

In the high atmosphere the thermal diffusion of H and He in atomic oxygen gas is also important. However, the intermolecular interaction in these cases is all the more poorly understood than the cases described above, making the treatment of this problem very difficult. Moreover the thermal diffusion in the O atmosphere is somewhat less important than that in the N_2 atmosphere because N_2 is the main constituent of the atmosphere in the height range where the height gradient of temperature is the greatest, i.e. where thermal diffusion has the greatest influence in the high atmosphere. Therefore we do not intend to pursue this problem any further here. It may be worth mentioning that BRINKMANN[2] has given some consideration to the interaction between an O atom and an H or He atom.

8. Eddy diffusion. If the atmosphere is in a turbulent condition due to the action of randomly changing winds and other agents and if there is non-uniformity in the distribution of heat, momentum, and the concentration of a specific kind of molecules, etc., the transport of these quantities will occur mainly by turbulence and this acts to make them uniform. In particular, the dispersion of mass of matter during this process is called eddy diffusion and its transport takes place irrespective of any difference in average molecular velocity among the various constituents. Let s be a certain physical quantity per unit volume having a conservative property and it is assumed that there is no external supply and loss of this quantity. Let us consider what kind of law governs the variation of the quantity s with the space coordinates and time when it is carried along by a flow under the influence of turbulence [7, *8*].

Let $\boldsymbol{v}$ be the velocity of the flow of a fluid and this is considered to be composed of two components, i.e. the average velocity $\bar{\boldsymbol{v}}$ and the random deviation therefrom $\boldsymbol{v}'$:

$$\boldsymbol{v} = \bar{\boldsymbol{v}} + \boldsymbol{v}'. \tag{8.1}$$

Similarly s is considered to be composed of the corresponding two components:

$$s = \bar{s} + s'. \tag{8.2}$$

Then, since the flux of s moving across a unit area is $s\boldsymbol{v}$, its average value $\bar{\boldsymbol{F}}$ is given by

$$\begin{aligned}\bar{\boldsymbol{F}} &= \overline{(\bar{s} + s')(\bar{\boldsymbol{v}} + \boldsymbol{v}')} \\ &= \bar{s}\bar{\boldsymbol{v}} + \overline{s'\,\boldsymbol{v}'}.\end{aligned} \tag{8.3}$$

The first term on the right-hand side expresses the flux carried by the average flow and the second term represents the one brought about by turbulence. In order to transform the latter, let us suppose that: the physical quantity carried by turbulence moves in any direction accompanying the random motion of a portion of the fluid, or the eddy; this keeps its characteristics while it moves from its starting point, whose radius vector is $\boldsymbol{r}$, to another covering a vectorial distance of $\boldsymbol{l}$; when it reaches the point $\boldsymbol{r} + \boldsymbol{l}$, it loses its characteristics by being mixed with the surrounding portion of the fluid (hence $|\boldsymbol{l}|$ is called the mixing length). Thus, the physical quantity at the point $\boldsymbol{r}, \bar{s}(\boldsymbol{r})$, is carried to the

[2] Brinkmann, R.T. (1970): Planet. Space Sci. *18*, 449

point $\boldsymbol{r}+\boldsymbol{l}$, and the difference between this and the original s at the latter point, i.e. $\bar{s}(\boldsymbol{r})-\bar{s}(\boldsymbol{r}+\boldsymbol{l})$, is considered to be the random component of s, i.e. s', at the latter point. Formulated mathematically

$$s'(\boldsymbol{r})=\bar{s}(\boldsymbol{r}-\boldsymbol{l})-\bar{s}(\boldsymbol{r})=-(\boldsymbol{l}\cdot\nabla)\bar{s}(\boldsymbol{r}). \tag{8.4}$$

Hence

$$\overline{s'\boldsymbol{v}'(\boldsymbol{r})}=-\overline{\boldsymbol{v}'(\boldsymbol{l}\cdot\nabla)\bar{s}(\boldsymbol{r})}=-\overline{\boldsymbol{v}'(\boldsymbol{l}}\cdot\nabla)\bar{s}(\boldsymbol{r})=-(\bar{\tilde{K}}\cdot\nabla)\bar{s}(\boldsymbol{r}), \tag{8.5}$$

where $\bar{\tilde{K}}$ is the average of the tensor $\boldsymbol{v}'\boldsymbol{l}$:

$$\bar{\tilde{K}}=\overline{\boldsymbol{v}'\boldsymbol{l}}. \tag{8.6}$$

If an appropriate set of coordinate axes are selected, this tensor can be brought into a diagonal form, whose principal elements are denoted by K_{xx}, K_{yy}, and K_{zz}. Then Eq. (8.5) takes the following simple form:

$$(\overline{s'v'})_x=-K_{xx}\frac{\partial}{\partial x}\bar{s}, \quad (\overline{s'v'})_y=-K_{yy}\frac{\partial}{\partial y}\bar{s}, \quad (\overline{s'v'})_z=-K_{zz}\frac{\partial}{\partial z}\bar{s}. \tag{8.7}$$

Usually atmospheric turbulence can be considered as isotropic and K_{xx}, K_{yy}, and K_{zz} become equal to one another. In the following these are set equal to K:

$$K=K_{xx}=K_{yy}=K_{zz}. \tag{8.8}$$

K is called the coefficient of eddy diffusion. As $\boldsymbol{l}$ and $\boldsymbol{v}'$ are determined by atmospheric turbulence, they should be independent of the physical quantity transported. Hence, the same can be said of K_{xx}, K_{yy}, K_{zz}, and K from Eqs. (8.6) and (8.8).

In the above description we have deduced Eq. (8.7) on the basis of some plausible assumptions and showed that the coefficient of eddy diffusion and some other parameters do not depend on the physical quantity transported. However, this method of reasoning may not be convincing enough because it uses a rather ambiguous physical quantity called mixing length. It may be better to consider the above-obtained law of eddy diffusion and the properties of the parameters as assumptions to be proved empirically [*8*a].

9. Ambipolar diffusion

Formulas given in Sect. 3 can be applied to ionized gases, but it is well known that the integral in Eq. (3.6) becomes logarithmically infinite and does not give a definite value for collisions between charged particles. The procedure usually employed to avoid this difficulty is to cut off the integration with respect to b at either the interelectron distance or, more commonly, the Debye length. We do not intend to enter into details about the problems associated with this procedure, but only wish to mention the interesting studies by KIHARA et al.[1,2], to which readers interested in this sort of problem are requested to refer.

---

[1] Kihara, T., Aono, O. (1963): J. Phys. Soc. Jpn. *18*, 837
[2] Kihara, T., Aono, O., Itikawa, Y. (1963): J. Phys. Soc. Jpn. *18*, 1043

α) *Simplifications.* The ionized gas has some characteristics which will be utilized to simplify some of the formulas obtained hitherto. One is that the mass of an electron is relatively very small, and the other is that neutrality of charge as a whole is maintained. Denoting electrons by the subscript 1, the different kinds of ions and neutral molecules by those from 2 to ζ and from $\zeta+1$ to ν, respectively, and the charge on a particle by qZ_i, we get

$$\sum_{i=1}^{\zeta} n_i Z_i = 0. \tag{9.1}$$

If we set $\varepsilon_i \equiv (m_1/m_i)^{\frac{1}{2}}$, $1 < i \leqq \zeta$, ε_i is less than about 1/100 for the ions present in the upper atmosphere, except for hydrogen ions, and therefore in the following discussion we shall neglect ε_i against 1, unless stated otherwise.

In the presence of a concentration, pressure or temperature gradient, electrons will tend to diffuse much more rapidly than the ions because of their relatively small mass. The natural tendency is to upset condition (9.1) and to build up a net charge distribution. However, as the electrons diffuse away from the ions, an electric field will be set up which will reduce the diffusion rate of the electrons and augment that of the ions. A steady state is reached in which the net current vanishes at any point, giving

$$\sum_{i=1}^{\zeta} Z_i \boldsymbol{j}_i = 0, \tag{9.2}$$

where $\boldsymbol{j}_i$ is the flow of the particles of species i given by Eq. (3.1).

In the special case in which the ionized gas is a mixture of atoms, their ions, and electrons (hereafter called the ideal partially ionized gas), the ion and electron number fluxes are equal and the diffusion is called ambipolar.

Let the electric field be denoted by $\boldsymbol{E}$; then, noting Eq. (9.1), we get from Eq. (3.2) the following expression for charged particles:

$$\boldsymbol{d}_i = \nabla\left(\frac{n_i}{n}\right) + \left(\frac{n_i}{n} - \frac{n_i m_i}{\rho}\right)\nabla \ln p - \frac{n_i}{p} Z_i q \boldsymbol{E}, \qquad i = 1, \ldots, \zeta; \tag{9.3}$$

it should be noted that, even under the action of gravity, this does not enter explicitly into the above expression. In a similar fashion we get for neutral particles

$$\boldsymbol{d}_i = \nabla\left(\frac{n_i}{n}\right) + \left(\frac{n_i}{n} - \frac{n_i m_i}{\rho}\right)\nabla \ln p, \qquad i = \zeta+1, \ldots, \nu. \tag{9.4}$$

Inserting these expressions of $\boldsymbol{d}_i$ into Eq. (3.1) and using Eq. (5.7), we obtain

$$\boldsymbol{j}_i = \frac{n^2}{\rho} \sum_{j=1}^{\nu} m_j D_{ij} \left[\nabla\left(\frac{n_j}{n}\right) + \left(\frac{n_j}{n} - \frac{\rho_j}{\rho}\right)\nabla \ln p - \frac{n_j Z_j q \boldsymbol{E}}{nkT}\right] - \frac{D_i^T}{m_i}\nabla \ln T, \tag{9.5}$$

where $Z_i = 0$ for $i > \zeta$. For ambipolar diffusion Eq. (9.2) holds; if $\boldsymbol{j}_i$ given by Eq. (9.5) is inserted into this equation and the resulting equation is solved for

$\boldsymbol{E}$, it gives

$$-\frac{nq}{\rho kT}\boldsymbol{E}=\frac{1}{\beta}\left\{\frac{n^2}{\rho}\sum_{i=1}^{\zeta}Z_i\sum_{j=1}^{\nu}m_jD_{ij}\left[\nabla\left(\frac{n_j}{n}\right)\right.\right.$$
$$\left.\left.+\left(\frac{n_j}{n}-\frac{\rho_j}{\rho}\right)\nabla\ln p\right]-\sum_{i=1}^{\zeta}\frac{Z_iD_i^T}{m_i}\nabla\ln T\right\} \tag{9.6}$$

with

$$\beta\equiv-\sum_{i=1}^{\zeta}\sum_{j=1}^{\zeta}Z_iZ_jn_jm_jD_{ij}\simeq\sum_{j=2}^{\zeta}Z_jn_jm_jD_{1j}; \tag{9.7}$$

in deriving the last approximate equation we have used the following inequalities as well as $D_{ii}=0$ referred to in Sect. 3:

$$D_{1j}\gg D_{ij}, \qquad 1<i,j\leqq\zeta, \tag{9.8}$$

$$m_iD_{1i}\gg m_1D_{i1}, \qquad 1<i\leqq\zeta. \tag{9.9}$$

The validity of these inequalities may be seen in the following way. On the basis of Eq. (3.8), we can readily obtain

$$\frac{D_{ij}}{D_{1j}}=\frac{n_i}{n_1}\left(\frac{m_1}{m_i}\right)^{\frac{1}{2}}\frac{c_{i0}^{ji}}{c_{10}^{j1}}, \qquad 1<i,j\leqq\zeta,$$

and

$$\frac{m_1D_{i1}}{m_iD_{1i}}=\frac{n_i}{n_1}\left(\frac{m_1}{m_i}\right)^{\frac{1}{2}}\frac{c_{i0}^{1i}}{c_{10}^{i1}}, \qquad 1<i\leqq\zeta.$$

Now, considering that $n_i/n_1\leqq 1$ (the number of negative ions being assumed to be small compared with that of electrons) and $\varepsilon_i\equiv(m_1/m_i)^{\frac{1}{2}}\ll 1$, it can be understood that inequalities (9.8) and (9.9) hold good if $|c_{i0}^{ji}/c_{10}^{j1}|$ and $|c_{i0}^{1i}/c_{10}^{i1}|$ are of the order of unity or less.

That this is indeed the case can be confirmed by solving the linear simultaneous equations (3.11) for c_{jp}^{hk}, some plausible values being assumed for the cross sections given by Eq. (3.15) and certain values pertinent to the actual ionospheric conditions being taken for electron, ion, and neutral densities.

If Eq. (9.6) is inserted into Eq. (9.5) to eliminate the electric field $\boldsymbol{E}$, we get the flux of particles of species i:

$$j_i=\frac{n^2}{\rho}\sum_{j=1}^{\nu}m_j\left(D_{ij}+\frac{1}{\beta}\sum_{k=1}^{\zeta}n_km_kZ_kD_{ik}\sum_{l=1}^{\zeta}Z_lD_{lj}\right)$$
$$\cdot\left[\nabla\left(\frac{n_j}{n}\right)+\left(\frac{n_j}{n}-\frac{\rho_j}{\rho}\right)\nabla\ln p\right]$$
$$-\left(\frac{D_i^T}{m_i}+\frac{1}{\beta}\sum_{k=1}^{\zeta}n_km_kZ_kD_{ik}\sum_{l=1}^{\zeta}\frac{Z_lD_l^T}{m_l}\right)\nabla\ln T. \tag{9.10}$$

β) For an *ideal weakly ionized ternary gas* ($\nu=3$, $\zeta=2$, and $n_1=n_2\ll n_3$) it is possible to work out some approximate relationships among the diffusion coefficients D_{ij}. $Q_{ij}^{(l)}$ given by Eq. (3.6) is a certain sort of collision cross-section and this is much larger for an electron-ion collision than for one between an electron or ion and a neutral atom; thus,

$$Q_{12}^{(l)} \gg Q_{13}^{(l)} \approx Q_{23}^{(l)}. \tag{9.11}$$

Therefore, we obtain from Eqs. (3.4) and (3.5)

$$\frac{D_{13}^{(1)}}{D_{12}^{(1)}} \approx \frac{2n_1\,\Omega_{12}^{(1,1)}}{n_3\,\Omega_{13}^{(1,1)}}, \qquad D_{23}^{(1)}/D_{13}^{(1)} \approx \sqrt{2}(m_1/m_2)^{\frac{1}{2}} = O(\varepsilon_2). \tag{9.12}$$

and $D_{13}^{(1)}/D_{12}^{(1)}$ is large or small compared with unity according to whether $n_1/n_3 \gg$ or $\ll \Omega_{13}^{(1,1)}/\Omega_{12}^{(1,1)}$. It is to be noted that, as $D_{ij}^{(1)}$ is invariant for the interchange of subscripts i and j, similar relations as those given by (9.12) can be deduced by interchange of subscripts. An example of approximate relationships can be obtained by expressing D_{12} and D_{21} in terms of D_{ij} using Eq. (3.20) and by noting the relations $n_1=n_2\ll n_3$ as well as the second one of (9.12):

$$\frac{D_{12}}{D_{21}} = \frac{n_1 D_{23}+(n_2+n_3)\,D_{13}}{\left(n_1+n_3\dfrac{m_3}{m_1}\right)D_{23}+n_2 D_{13}} = \frac{n_1\,O(\varepsilon_2)+(n_2+n_3)}{(n_1+n_3/\varepsilon_2^2)\,O(\varepsilon_2)+n_2} = O(\varepsilon_2). \tag{9.13}$$

Similar arguments will enable us to estimate the relative magnitude of various D_{ij}'s but this also depends on how small a quantity $n_1/n_3(=n_2/n_3)$ is, so that it may not always be possible to state uniquely that a certain D_{ij} is negligible compared with another D_{hk}.

At any rate, the following approximate relationships can be derived by using Eqs. (3.20) and (9.12) if ionization is sufficiently weak:

$$D_{23} \approx D_{32} \approx D_{23}, \tag{9.14a}$$

$$D_{12} \approx D_{13}. \tag{9.14b}$$

Utilizing some of these and other similar relationships we can get from Eq. (9.10) the following expressions for the flux of electrons, positive ions and neutrals of an ideal weakly ionized ternary gas to the approximation in which ε_2 is neglected beside 1:

$$j_1 = j_2 = -\frac{n^2 m_3}{\rho} D_{\mathrm{A}}\left[\nabla\left(\frac{n_1}{n}\right)+\left(\frac{n_1}{n}-\frac{\rho_2}{2\rho}\right)\nabla\ln p\right]-\frac{D_{\mathrm{A}}^T}{m_2}\nabla\ln T, \tag{9.15}$$

$$\begin{aligned} j_3 = {} & \frac{n^2 m_3}{\rho} 2D_{32}\left[\left(1-\frac{m_1 D_{31}}{m_2 D_{32}}\right)\frac{D_{13}-D_{23}}{D_{12}}+\frac{m_1 D_{31}}{m_2 D_{32}}\right] \\ & \cdot\left[\nabla\left(\frac{n_1}{n}\right)+\left(\frac{n_1}{n}-\frac{\rho_2}{2\rho}\right)\nabla\ln p\right] \\ & -\frac{D_3^T}{m_3}\left[1-\frac{D_1^T}{D_{12}D_3^T}\left(\frac{m_2}{m_1}D_{32}-D_{31}\right)+\frac{D_{32}D_2^T}{D_{12}D_3^T}\left(1-\frac{m_1 D_{31}}{m_2 D_{32}}\right)\right]\nabla\ln T, \end{aligned} \tag{9.16}$$

where the ambipolar diffusion coefficient D_A and the thermal ambipolar diffusion coefficient D_A^T are defined by the following expression:

$$D_A = 2D_{23}\left[1 - \frac{m_1 D_{21}}{m_2 D_{23}}\left(1 - \frac{D_{13} - D_{23}}{D_{12}}\right)\right], \tag{9.17}$$

$$D_A^T = D_2^T\left\{1 + \frac{D_{21}}{D_{12}}\left[\frac{D_1^T}{D_2^T} - \frac{m_1}{m_2}\left(1 + \frac{D_{21} D_1^T}{D_{12} D_2^T}\right)\right]\right\}. \tag{9.18}$$

Using Eqs. (3.7) and (3.20), it can be shown that approximately

$$\left|\frac{D_{12} - D_{13}}{D_{23}}\right| = \left|\frac{n_1(D_{12} - D_{13})}{n_2 D_{13} + n_3 D_{12}}\right| \leq 1, \tag{9.19}$$

from which, together with Eqs. (9.13) and (9.14a), the second term between the brackets of Eq. (9.17) is seen to be of the order of $O(\varepsilon_2)$ and therefore

$$D_A \approx 2D_{23}. \tag{9.14c}$$

This is a well-known relation in the case of usual ambipolar diffusion; it is, however, demonstrated here even in the case where thermal diffusion is taken into account.

γ) Lastly, it should be noticed that the present treatment of the problem refers to the *reference frame* with respect to which the mass-velocity of the gas as a whole, $\boldsymbol{v}_0$, vanishes, i.e. the expression on the extreme right-hand side in Eq. (2.3) is zero, giving

$$\sum_{i=1}^{3} m_i \boldsymbol{j}_i = 0. \tag{9.20}$$

If Eq. (9.5) is substituted for $\boldsymbol{j}_i$ in the left-hand side of this equation, the resulting expression must always vanish irrespective of the values of the independent variables. Now, on the right-hand side of Eq. (9.5), T is contained in the term of the extreme right-hand side and in the third term (involving $\boldsymbol{E}$) between the brackets, but it can readily be shown by using Eq. (3.20a) that the latter term, when inserted into Eq. (9.20), does not make any contribution to the left-hand side of Eq. (9.20). Therefore it is concluded that the following equation should hold true:

$$\sum_{i=1}^{3} D_i^T = 0. \tag{9.21}$$

Now, it can be shown by solving simultaneous linear equations (3.10) that $|a_{10}| \lesssim |a_{20}|$ for the atmospheric conditions in the F2-layer of the ionosphere and for plausible values of collision cross sections, so that, by using Eq. (3.9), we get in the first approximation

$$|D_1^T| \lesssim |D_2^T|\, O(\varepsilon_2). \tag{9.22}$$

From this relation and Eq. (9.21) we get further

$$D_2^T \approx -D_3^T \tag{9.23}$$

in the approximation that small quantities of the order of $O(\varepsilon_2)$ are neglected.

δ) *Static atmosphere.* It can be readily seen that the expression for the flux of electrons or ions given by Eq. (9.15) can be brought into the following form if it is considered that

$$\nabla p = \rho \boldsymbol{g}, \tag{9.24}$$

where $\boldsymbol{g}$ is the acceleration due to gravity, supposing the Earth's atmosphere to be static (note that $n_1 = n_2$ and the equation of state (5.7)):

$$n_1 \boldsymbol{v}_1 = n_2 \boldsymbol{v}_2 = -\frac{n m_3}{\rho} D_{\mathrm{A}} \left(\nabla n_2 + \frac{n_2}{T} \nabla T - n_2 \frac{m_2 \boldsymbol{g}}{2kT} \right) - \frac{D_{\mathrm{A}}^T}{m_2} \nabla \ln T. \tag{9.25}$$

Assuming also the ionospheric layers to be stratified uniformly in the horizontal direction, only change in the vertical direction need be considered and, inserting Eq. (9.25) into the equation of continuity (4.1) with ion production and loss terms added and using the approximate relation $\rho \simeq n m_3$, we get

$$\frac{\partial n_2}{\partial t} = Q_2 - L_2 + \frac{\partial}{\partial z} \left\{ D_{\mathrm{A}} \left[\frac{\partial n_2}{\partial z} + \frac{n_2}{T} \frac{\partial T}{\partial z} (1 + \alpha_{\mathrm{a}T}) + \frac{n_2}{2H_2} \right] \right\}, \tag{9.26}$$

where z is height, Q_2 the rate of ion production, L_2 the rate of ion loss, H_2 the scale height for the ion given by

$$H_2 = \frac{kT}{m_2 g} \tag{9.27}$$

and $\alpha_{\mathrm{a}T}$ is the ambipolar thermal diffusion factor defined by

$$\alpha_{\mathrm{a}T} = \frac{D_{\mathrm{A}}^T}{n_2 m_2 D_{\mathrm{A}}}. \tag{9.28}$$

This has a form similar to that of the thermal diffusion factor for a neutral binary system (cf. Eqs. (3.22) and (5.6)). It is needless to say that the same expression as Eq. (9.26) also holds for electrons, since n_1 is equal to n_2.

ε) *Deviation from thermal equilibrium.* In the above argument it is assumed that electrons, ions, and neutral atoms have a common temperature because it is based on gas-kinetic formulas for a multicomponent system. In order to deal with a case in which temperatures are different for electrons, ions, and neutrals, a hydrodynamical method must be employed. The results of such treatments show that the continuity equation of electrons corresponding to Eq. (9.26) is given by [3,4]

$$\frac{\partial N}{\partial t}=Q-L+\frac{\partial}{\partial z}\left\{D_{\mathrm{a}}\left[\frac{\partial N}{\partial z}+(1+\alpha_{\mathrm{a}T})\frac{N}{T_{\mathrm{p}}}\frac{\partial}{\partial z}T_{\mathrm{p}}+\frac{N}{H_{\mathrm{p}}}\right]\right\}, \tag{9.29}$$

where m_e, T_e, and m_i, T_i are the mass and temperature of electrons and ions, Q and L the rates of production of electrons and their loss, D_a, T_p, and H_p are given by the following equations:

$$D_{\mathrm{a}}=\frac{k(T_{\mathrm{i}}+T_{\mathrm{e}})}{m_{\mathrm{i}}\,\nu_{\mathrm{in}}+m_{\mathrm{e}}\,\nu_{\mathrm{en}}}, \tag{9.30}$$

$$T_{\mathrm{p}}=\tfrac{1}{2}(T_{\mathrm{e}}+T_{\mathrm{i}}), \tag{9.31}$$

and

$$H_{\mathrm{p}}=\frac{k(T_{\mathrm{i}}+T_{\mathrm{e}})}{(m_{\mathrm{i}}+m_{\mathrm{e}})\,g}, \tag{9.32}$$

in which ν_{in} and ν_{en} are the gas-kinetic collision frequencies of an ion or electron with neutral particles [5]; in deducing Eq. (9.29) it has been assumed that the temperatures of electrons, ions, and neutrals maintain constant ratios independent of height throughout the atmosphere. D_a is the ambipolar diffusion coefficient deduced hydrodynamically and T_p and H_p are called the plasma temperature and the plasma scale height. When Eqs. (9.29) and (9.32) are compared with Eqs. (9.26) and (9.27) respectively, it will be seen that each term corresponds with each other but T is replaced by plasma temperature T_p. It must be remarked that the thermal diffusion term in Eq. (9.29) cannot be deduced by the hydrodynamical method, however it is included in Eq. (9.29) assuming the same form as appears in Eq. (9.26).

Lastly, the electric field $\boldsymbol{E}$ given by Eq. (9.6) can be simplified for the ideal weakly ionized gas as follows (note Eq. (9.13)):

$$q\boldsymbol{E}=\left(1-2\frac{D_{13}-D_{23}}{D_{12}}\right)\frac{1}{n_1}\nabla p_1-\left(1-\frac{D_{13}-D_{23}}{D_{12}}\right)\frac{\rho_2}{\rho}\frac{1}{n_1}\nabla p$$
$$-\frac{\rho\, k}{n\, n_1\, m_2 D_{12}}\left(\frac{D_1^T}{m_1}-\frac{D_2^T}{m_2}\right)\nabla T. \tag{9.33}$$

According to SCHUNK and WALKER [5], however, $(D_{13}-D_{23})/D_{12}$ is roughly equal to unity; thus, if this quantity is replaced by 1, Eq. (9.33) is further simplified to give

$$q\boldsymbol{E}=-\frac{1}{n_1}\nabla p_1-\alpha\, k\nabla T \tag{9.33a}$$

---

[3] Kendall, P.C., Pickering, W.M. (1967): Planet. Space Sci. *15*, 825

[4] Yonezawa, T. (1972): J. Radio Res. Lab. (Jpn) *19*, 109

[5] These are different from the averaged transport collision frequencies used when radio wave propagation in a plasma is considered. For a thorough treatment see SUCHY's contribution in vol. 49/7

with

$$\alpha = \frac{\rho}{n n_1 m_2 D_{12}} \left(\frac{D_1^T}{m_1} - \frac{D_2^T}{m_2} \right). \tag{9.34}$$

The same authors also note that D_2^T in Eq. (9.34) is substituted by D_A^T given by Eq. (9.18) if the approximation is raised by one step, i.e.

$$\alpha = \frac{\rho}{n n_1 m_2 D_{12}} \left(\frac{D_1^T}{m_1} - \frac{D_A^T}{m_2} \right). \tag{9.34a}$$

It may be worthy of notice that the first term on the *right*-hand side of Eq. (9.33a) corresponds to the *left*-hand side of Eq. (35.2) in Chapter V; that is, Eq. (9.33a) includes the effect of thermal diffusion but ignores the gravity acting on electrons while Eq. (35.2) includes the latter but ignores the former.

ζ) In the above discussion we have not taken into account the existence of the *magnetic field*. However, in the F2-region and above where the atmosphere is tenuous enough to make electron-ion diffusion a subject for discussion, the collision frequency of a charged particle is much less than its gyrofrequency around geomagnetic field lines so that charged particles are constrained around field lines and can move freely only along them. Therefore, in order to take the geomagnetic field into consideration, it is sufficient to take the charged particles to be able to move in the direction of geomagnetic field lines. As a result of this the vertical component of the velocity of charged particles is reduced by a factor of $\sin^2 I$, where I is the magnetic dip[6]. For example, we must multiply the terms expressing the vertical electron flux in Eq. (9.29) by $\sin^2 I$, which makes the continuity equation of electrons as follows:

$$\frac{\partial N}{\partial t} = Q - L + \frac{\partial}{\partial z} \left\{ \sin^2 I \cdot D_a \left[\frac{\partial N}{\partial z} + (1 + \alpha_{aT}) \frac{N}{T_p} \frac{\partial}{\partial z} T_p + \frac{N}{H_p} \right] \right\}. \tag{9.29a}$$

In the E-region (at about 100 km height), however, the motion of meteor trails and artificial ion clouds ejected from rockets also comes under discussion in connection with ambipolar diffusion. In these cases the collision frequency of charged particles is much greater and the effect of the geomagnetic field is more complicated than in the F2-region. Hence these kinds of problems are treated separately in Sects. 26–30.

II. Diffusive equilibrium, mixing, and turbulence in the high atmosphere

10. Diffusion and mixing. Let us consider how the height distributions of the constituent molecules, atoms, ions, and electrons in the high atmosphere

[6] See Sect. 38 in: Rawer, K., Suchy, K., this Encyclopedia, vol. 49/2, pp. 1–546

are determined. It is a matter of course that the mechanism of production and loss of each constituent is of primary importance for its distribution except for the permanent constituents. If the mean lifetime of a constituent produced by certain chemical reactions is very short, it will disappear before it is carried away by diffusion and atmospheric motion or mixed with other constituents by turbulence so that its distribution will depend exclusively on the chemical reactions which take part in its production and removal. In this case a discussion should be made separately for each constituent. In this article such cases will not be considered.

On the other hand, if the mean lifetime of a constituent is very long, it will be carried a sufficiently long distance by diffusion and atmospheric motion or mixed with other constituents by turbulence before it disappears by chemical reactions so that its distribution becomes very different from that at the time of its production and it is determined by the action of other factors than chemical reactions.

α) Of these factors, *molecular diffusion* acts, as is explained in the next section, in such a way as to make lighter constituents concentrate relatively more at higher altitudes and heavier ones concentrate relatively more at lower altitudes, when the atmosphere in the gravitational field has reached the equilibrium condition. In other words, molecular diffusion tends to make atmospheric constituents separate from one another according to their weight. The time necessary for making the molecules move by a distance of L, on average, as a result of molecular diffusion is of the order of L^2/D (cf. later Eq. (13.5); D is the molecular diffusion coefficient). If L is taken to be the mean scale height of the atmosphere $\bar{H}$ given by Eq. (11.11) below, the above-mentioned time becomes the one in which molecular diffusion exerts an appreciable influence on the vertical distribution of atmospheric molecules. In the case of charged particles the Earth's magnetic field and the electric field in the high atmosphere exert force on them and affect their distribution. When there is a temperature gradient, thermal diffusion occurs, but the values of its coefficient is small for common atmospheric constituents (an account is given in Sect. 6) and its effect can be neglected. However, this is not the case with light atoms, such as hydrogen and helium, and with charged particles. These particles are discussed in other chapters and here the effect of thermal diffusion is not taken into consideration.

β) *Mixing* is effected by turbulence which appears when there are atmospheric movements brought about by local winds and heating and atmospheric wave motion such as internal gravity waves. This tends to make the concentration proportion of each constituent of the atmosphere uniform and independent of height. As is well known in meteorology, this action has a great effect in a region where temperature decreases with height, such as near the ground and it can be formulated in terms of the eddy diffusion coefficient K introduced in Sect. 8. Like the case of molecular diffusion, it takes a time of the order of L^2/K for the particles to be moved by a distance of L, on average, by the action of eddy diffusion. As a matter of fact, however, it is not possible to specify the value of K definitely by theory alone and the utmost we can do is to regard the eddy diffusion coefficient as an adjustable parameter and to determine its value so as to give the best fit with observations.

Atmospheric motion includes various phenomena from large-scale ones such as atmospheric circulation, atmospheric tides, and planetary waves to small-scale ones such as local winds and internal gravity waves. The small-scale motions produce mixing due to turbulence, as mentioned above. The large-scale motions set each constituent into motion directly and influence their distribution, but the velocity of the motion is generally low compared with that of molecular diffusion and can usually exert only a secondary effect.

γ) From the above considerations it will be seen that the distribution of a constituent with a very long mean lifetime, including permanent ones, is determined by the two *opposing actions of molecular and eddy diffusion.* In reality, the mean lifetime of a constituent molecule is often comparable to the time $\bar{H}^2/D$ or $\bar{H}^2/K$ in which molecular or eddy diffusion comes to manifest its effect on the vertical distribution of the molecules, so that, in discussing the distribution of a constituent in a number of cases, consideration must also be given to chemical reactions (cf. for instance Chapter III).

11. Diffusive equilibrium distribution and the model atmosphere

α) *Molecular diffusion.* When thermal diffusion is neglected, we get from Eqs. (3.2) and (3.3) the following equation for the equilibrium condition under the action of gravity:

$$\frac{\mathrm{d}}{\mathrm{d}z}\left(\frac{n_i}{n}\right)+\left(\frac{n_i}{n}-\frac{n_i m_i}{\rho}\right)\frac{1}{p}\frac{\mathrm{d}p}{\mathrm{d}z}=0, \tag{11.1}$$

where z is height and the conditions are assumed to be uniform in the horizontal direction. Using the equation of state (5.7) and noting that the variation in total pressure p with height is given by Eq. (9.24):

$$\frac{\mathrm{d}p}{\mathrm{d}z}=-\rho g, \tag{11.2}$$

we can transform Eq. (11.1) into the following:

$$\frac{\mathrm{d}}{\mathrm{d}z}\ln(n_i T)=-\frac{m_i g}{kT}. \tag{11.3}$$

When integrated, this gives

$$n_i=n_{i0}\frac{T_0}{T}\exp\left(-\int_{z_0}^{z}\frac{m_i g}{kT}\,\mathrm{d}z\right), \tag{11.4}$$

where variables with subscript 0 denote their values at the datum level. This equation expresses the vertical distribution of the number density of the constituent of species i, showing that the respective constituents are distributed independently of one another according to Eq. (11.4) under the action of molecular diffusion, when the equilibrium condition is reached. This is called the condition of diffusive equilibrium and the distribution given by Eq. (11.4) is called the diffusive equilibrium distribution. It will readily be seen from it that the lighter a constituent is, the greater the proportion of the atmosphere it will occupy at high altitudes. Thus, molecular diffusion tends to separate the constituents from one another according to their weight.

β) If *eddy diffusion* is also taken into account, the approximation given later in Sect. 19 leads to the following equation corresponding to Eq. (11.1) (cf. Eqs. (3.2), (3.3), (19.4), and (19.7, 8)):

$$\left(1+\frac{K}{D_i}\right)\frac{\mathrm{d}}{\mathrm{d}z}\left(\frac{n_i}{n}\right)+\left(\frac{n_i}{n}-\frac{n_i m_i}{\rho}\right)\frac{1}{p}\frac{\mathrm{d}p}{\mathrm{d}z}=0, \qquad (11.1\,\mathrm{a})$$

where K is the eddy diffusion coefficient and D_i is given by Eq. (19.8) below. Simple calculations show that the equation corresponding to Eq. (11.4) is

$$\frac{n_i}{n}=\frac{n_{i0}}{n_0}\,\frac{\exp\left(-\int\limits_{z_0}^{z}\frac{D_i}{D_i+K}\,\frac{m_i g}{kT}\,\mathrm{d}z\right)}{\exp\left(-\int\limits_{z_0}^{z}\frac{D_i}{D_i+K}\,\frac{\bar{m} g}{kT}\,\mathrm{d}z\right)}, \qquad (11.4\,\mathrm{a})$$

where $\bar{m}$ is the mean mass of the atmospheric molecules. It will be seen from Eq. (11.4a) that, if $K \gg D_i$, $n_i/n \approx n_{i0}/n_0$. This means that, if molecular diffusion can be neglected compared with eddy diffusion, the concentration proportion of the constituent of species i becomes the same irrespective of height, indicating that eddy diffusion tends to mix the atmospheric constituents and to make each one form a constant proportion of the atmosphere independent of height.

γ) The *scale height* for the constituent molecule of species i is given by

$$H_i=\frac{kT}{m_i g} \qquad (11.5)$$

and if the change of gravity with height is disregarded, the ratio of H_i to T can be regarded as a constant independent of height and Eq. (11.4) becomes

$$n_i=n_{i0}\frac{H_{i0}}{H_i}\exp\left(-\int\limits_{z_0}^{z}\frac{\mathrm{d}z}{H_i}\right). \qquad (11.6)$$

In particular, when H_i increases linearly with height, i.e.

$$H_i=H_{i0}+\Gamma_i(z-z_0), \qquad (11.7)$$

where Γ_i is the height gradient of H_i, Eq. (11.6) becomes

$$n_i=n_{i0}\left(\frac{H_{i0}}{H_i}\right)^{1+1/\Gamma_i}. \qquad (11.8)$$

In this case H_i is a variable corresponding to height through Eq. (11.7).

On the other hand, if the atmosphere is well mixed, it will readily be seen, by noting the equation

$$\rho = n\bar{m} \tag{11.9}$$

and Eqs. (11.2) and (5.7) and the equations which are derived from Eqs. (11.3) and (11.4) by replacing n_i and m_i with n and $\bar{m}$, respectively, that:

$$n = n_0 \frac{T_0}{T} \exp\left(-\int_{z_0}^{z} \frac{\bar{m}g}{kT} \mathrm{d}z\right). \tag{11.10}$$

Similarly when n_i, H_i, and Γ_i in Eqs. (11.6)...(11.8) are replaced with n, $\bar{H}$, and $\bar{\Gamma}$, respectively, where $\bar{H}$ is the mean scale height given by

$$\bar{H} = \frac{kT}{\bar{m}g} \tag{11.11}$$

and $\bar{\Gamma}$ is the height gradient of $\bar{H}$, the resulting equations are also true. In particular

$$n = n_0 \left(\frac{\bar{H}_0}{\bar{H}}\right)^{1+1/\bar{\Gamma}}. \tag{11.12}$$

From Eqs. (11.5) and (11.11) it follows that the relation between H_i and $\bar{H}$ is:

$$H_i = \frac{\bar{m}}{m_i} \bar{H}, \tag{11.13}$$

and the same relation holds between their height gradients:

$$\Gamma_i = \frac{\bar{m}}{m_i} \bar{\Gamma}. \tag{11.14}$$

Hence from Eqs. (11.8) and (11.13) we get

$$\frac{n_i}{n_{i0}} = \left(\frac{\bar{H}_0}{\bar{H}}\right)^{1+m_i/(\bar{m}\bar{\Gamma})}. \tag{11.15}$$

Let the vertical distribution of the number density of the constituent of species i be set as follows:

$$\frac{n_i}{n_{i0}} = \left(\frac{\bar{H}_0}{\bar{H}}\right)^{1+x/\bar{\Gamma}}; \tag{11.16}$$

then Eq. (11.15) or Eq. (11.12) shows that Eq. (11.16) gives the distribution of the constituent in the state of diffusive equilibrium or complete mixing if x is taken as $m_i/\bar{m}$ or 1, respectively. Further, the total number of particles contained in a semi-infinite vertical column with unit cross section above the level

$z=z_0$ is found to be

$$\int_{z_0}^{\infty} n_i \, dz = n_{i0} \int_{\bar{H}_0}^{\infty} \left(\frac{\bar{H}_0}{\bar{H}}\right)^{1+x/\bar{\Gamma}} \frac{d\bar{H}}{\bar{\Gamma}} = \frac{n_{i0}\bar{H}_0}{x}. \tag{11.17}$$

δ) If the atmospheric temperature or *scale height varies non-linearly* with height, let us take it that the values of H_i for z equal to z_j ($j=0, 1, 2, \ldots, n$) are given. Denoting these values of H_i by $H_{i,j}$, we may be able to regard H_i approximately as a linear function of z in the narrow height range (z_j, z_{j+1}); then we can readily express the value of n_i at $z=z_k$, $n_{i,k}$, in the following way by using Eq. (11.8):

$$n_{i,k} = n_{i0} \prod_{j=0}^{k-1} \left(\frac{H_{i,j}}{H_{i,j+1}}\right)^{1+1/\Gamma_{i,j}} \quad (k=1, 2, 3, \ldots, n), \tag{11.18}$$

where $\prod_{j=0}^{k-1}$ is the product of the factors on its right side over j from 0 to $k-1$ and

$$\Gamma_{i,j} = \frac{H_{i,j+1} - H_{i,j}}{z_{j+1} - z_j} \quad (j=0, 1, 2, \ldots, n-1). \tag{11.19}$$

ε) In JACCHIA's 1964 *model atmosphere*[1] the atmospheric temperature T is given by

$$T = T_\infty - (T_\infty - T_{120}) \exp[-s(z - 120 \text{ km})], \tag{11.20}$$

where T_{120} is the atmospheric temperature at the 120 km level and T_∞ is the one which is asymptotically approached at sufficiently high altitudes, i.e. the exospheric temperature; s is a constant whose value is different for different T_∞'s and given by

$$s = 0.0291 \text{ km}^{-1} \exp(-x^2/2), \tag{11.21}$$

$$x = \frac{T_\infty - 800 \text{ K}}{750 \text{ K} + 1.722 \times 10^{-4} (T_\infty - 800 \text{ K})^2}. \tag{11.22}$$

As shown later, each kind of atmospheric constituent molecules can be regarded as in diffusive equilibrium above 120 km; hence the following analytic expression is obtained for the number density of each kind of molecules above 120 km by inserting Eq. (11.20) into Eq. (11.4):

$$n_i(z) = n_{i,120} \frac{y^{\gamma_i}}{[y + a(1-y)]^{1+\gamma_i}} \quad (z \geqq 120 \text{ km}), \tag{11.23}$$

where

$$y = \exp[-s(z - 120 \text{ km})], \tag{11.24}$$

$$\gamma_i = \frac{m_i g_{120}}{k T_\infty s}, \tag{11.25}$$

$$a = T_\infty / T_{120}; \tag{11.26}$$

[1] Jacchia, L.G. (1964): Smithson. Astrophys. Observat. Spec. Rep. *170*

the subscript 120 means the value at the height of 120 km. The variation of g with height has been neglected in deriving the above expressions.

It seems that Eq. (11.20) is not a sufficient approximation to the actual upper atmospheric temperature distribution and the following expression is given instead of Eq. (11.20) in JACCHIA's 1971 model atmosphere[2] for the height range above 125 km:

$$T = T_{125} + A \tan^{-1}\left\{\left[\left(\frac{\mathrm{d}T}{\mathrm{d}z}\right)_{125} \Big/ A\right] \cdot (z - 125\,\mathrm{km})\left[1 + 4.5 \cdot 10^{-6}\,\mathrm{km}^{-2.5}(z - 125\,\mathrm{km})^{2.5}\right]\right\}, \tag{11.27}$$

$$\begin{aligned} T_{125}/\mathrm{K} &= 371.6678 + 0.0518806\, T_\infty/\mathrm{K} \\ &\quad - 294.3505\, \mathrm{e}^{-0.00216222 T_\infty/\mathrm{K}}, \\ A &= (2/\pi)(T_\infty - T_{125}), \\ \left(\frac{\mathrm{d}T}{\mathrm{d}z}\right)_{125} &= 1.90\,(T_{125} - 183\ \mathrm{K})/35\ \mathrm{km}. \end{aligned} \tag{11.28}$$

Unlike the case of Eq. (11.20), we cannot obtain here an analytic expression which simply expresses the number density of molecules. Incidentally, the temperature T between heights of 90...125 km is expressed as a quartic of z whose five coefficients are determined by the five conditions that $T_{90} = 183$ K, $\left(\frac{\mathrm{d}T}{\mathrm{d}z}\right)_{90} = 0$, T and $\mathrm{d}T/\mathrm{d}z$ are continuous with the values given by equations in (11.28) at 125 km, and $(\mathrm{d}^2T/\mathrm{d}z^2)_{125} = 0$. These values of the coefficients are different for different T_∞'s and must be determined separately for each value of T_∞.

12. Diffusive separation

α) It will be seen from Eq. (11.4a) that the *relative importance of molecular diffusion and mixing* depends on the ratio of the coefficients of molecular and eddy diffusion, D_i and K. D_i is a weighted harmonic mean of $D_{ij}^{(1)}$ over the constituents of species j, as will be seen from Eq. (19.8) below, and can be regarded as approximately inversely proportional to the total number density if Eq. (5.11) is noted. K is defined by Eqs. (8.6) and (8.8) in terms of the random velocity and the mixing length (or the scale of the eddy); the latter being a physically ambiguous quantity, the dependence of K on height is not certain. However, it seems unlikely that K increases rapidly and without restriction with height [7]. If this is true, the effect of molecular diffusion will predominate over that of mixing at sufficiently high altitudes. On the other hand, the atmosphere is well mixed near the ground level and the proportion of each constituent is constant and independent of height and location. Therefore, the vertical distribution of the atmospheric constituents will be determined mainly by molecular diffusion above a certain level and the constituents will be distributed according to Eq. (11.4). Below this level the atmosphere will be in a

[2] Jacchia, L.G. (1971): Smithson. Astrophys. Observat. Spec. Rep. *332*

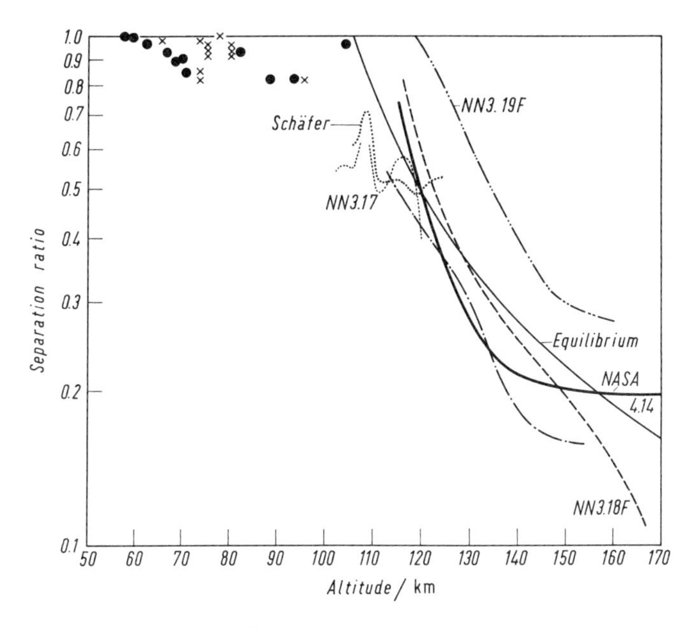

Fig. 1*. Summary of data on the relative abundance of argon and nitrogen at various altitudes. ● New Mexico, WENZEL et al., × USSR, MIRTOV, —— Virginia, MEADOWS-REED and SMITH, —·—, ----, —··—Canada, MEADOWS and TOWNSEND, —— Equilibrium, ... Virginia, SCHAEFER and NICHOLS

well-mixed condition. As regards its height, theoretical estimation is rather difficult and it is advisable to determine it on the basis of observations.

The phenomenon that each constituent of the atmosphere is distributed according to Eq. (11.4) independently of other constituents, i.e. lighter ones concentrate relatively more at higher altitudes and heavier ones at lower altitudes, is called diffusive separation. Rocket observations have been repeatedly attempted since their early stage to determine at what altitude diffusive separation sets in. One method is to collect samples of the upper atmosphere and to examine them as to whether there is any difference in composition in comparison with the atmosphere near the ground. The other method is to study the composition of the upper atmosphere directly by a mass spectrometer on board a rocket. In reality, N_2 and Ar, which are not subject to chemical reactions are usually studied.

β) MEADOWS-REED and SMITH[1] summarize the *results of experiments* performed not later than about the end of 1960. Figure 1 shows the results in which the separation ratio r of Ar vs. N_2 is plotted against height. r is defined by

$$r = \frac{(\rho_{\mathrm{Ar}}/\rho_{\mathrm{N_2}})_{\mathrm{alt}}}{(\rho_{\mathrm{Ar}}/\rho_{\mathrm{N_2}})_{\mathrm{ground}}}, \tag{12.1}$$

* From[1], p. 3206, Fig. 8

[1] Meadows-Reed, E., Smith, C.R. (1964): J. Geophys. Res. *69*, 3199

where ρ is the mass density of Ar or N_2 shown by the subscript and "alt" and "ground" mean the values at the altitude under consideration and on the ground, respectively. The location of observation and the names of observer or analyzer of the data are indicated in the figure. Circle-and-dots and crosses show the results obtained by the former method mentioned above. The American experiments were performed between 1950 and 1956 at White Sands, New Mexico (32.4 °N, geomag. lat. 41.2 °N) while the Soviet ones were performed between 1951 and 1956. Curves indicate the results obtained by mass spectrometers as well as a calculated result based on a model atmosphere.

The observations at Fort Churchill (58.7 °N, geomag. lat. 68.7 °N) in Canada were performed three times before and during the International Geophysical Year by MEADOWS and TOWNSEND[2] using mass spectrometers. The first (NN3.17) and the second (NN3.18F) experiments were performed during the night, rockets being launched at 2321 CST on November 20, 1956 and at 2002 CST on February 21, 1958, respectively. These results may be considered as typical of the nocturnal atmosphere in high latitudes. The third one (NN3.19F) was observed by a rocket which was launched in the daytime at 1207 CST on March 23, 1958 during a polar blackout; this may account for the result which deviates from the other two. At Wallops Island, Virginia (37.8 °N, geomag. lat. 49.2 °N) observations were made at 1141 EST on November 15, 1960 by MEADOWS-REED and SMITH[1] and at 1302 EST on May 18, 1962 by SCHAEFER and NICHOLS[3] using mass spectrometers. The former is shown in Fig. 1 by a thick solid line and may be regarded as typical of midlatitude daytime. However, values are not quite certain for heights above 150 km. It is noteworthy that this result is in good agreement with the observational ones typical of the nocturnal atmosphere in *high* latitudes. SCHAEFER and NICHOLS' measurements indicate irregular fluctuations, but they do not deviate widely from other results of observation. The calculated curve based on a model atmosphere assuming equilibrium is also in good agreement with the observations. Besides these measurements, Soviet scientists made similar rocket observations using mass spectrometers at midnight on September 23, 1960, in the morning on July 14 and 22, 1959 and late in the afternoon on November 15, 1961[4]. Although the results of these measurements are not depicted in the figure, they are generally in good agreement with those shown in Fig. 1.

Summarizing these results of rocket observations, MEADOWS-REED and SMITH[3] concluded that the height of the level where diffusive separation of Ar and N_2 becomes appreciable varies between about 100 km and 119 km according to circumstances and that its typical values lie between 105 km and 112 km.

13. Chemical release

α) *Turbopause*. If a chemical substance is released from a rocket into the atmosphere during morning or evening twilight hours, it will receive sunlight which is still shining in the upper atmosphere and its resonance emission may be observed on the ground as a cloud of the substance. Alkali vapor such as lithium, sodium, or cesium is mainly employed for this purpose but trimethyl aluminium (TMA) is also available and can emit light by itself by chemical reaction. Figure 2 is a photograph of a cloud of lithium vapor which was ejected from a rocket[1], taken 470 s after the ejection, at 0440 CST on Novem-

[2] Meadows, E.B., Townsend, J.W. (1960): Space Res. *1*, 175
[3] Schaefer, E.J., Nichols, M.H. (1964): Space Res. *4*, 205
[4] Pokhunkov, A.A. (1963): Space Res. *3*, 132
[1] Lloyd, K.H., Low, C.H., Vincent, R.A. (1973): Planet. Space Sci. *21*, 653

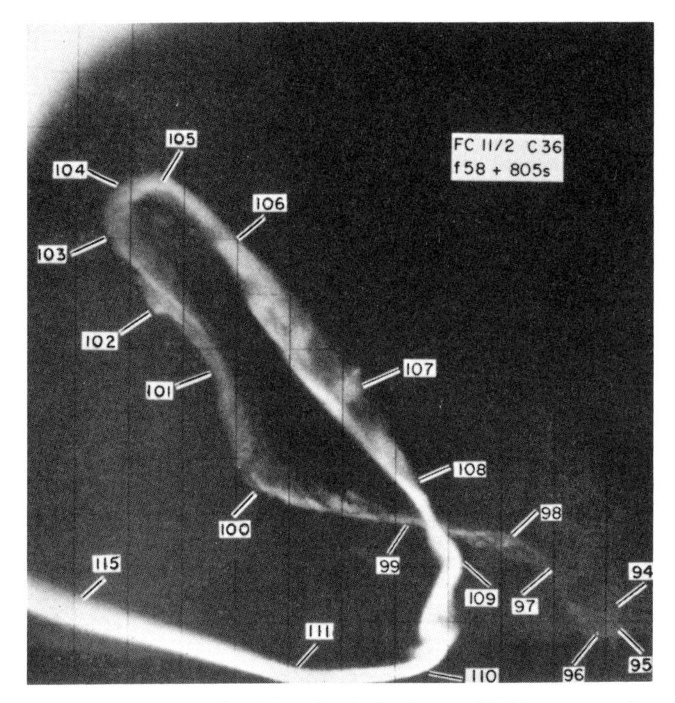

Fig. 2*. The turbulent structure obtained by chemical release of lithium vapor from a rocket at Woomera, Australia, at 0440 CST, 6 November 1970, 470 s after release. Numbers indicate the altitude in km

ber 6, 1970 at Woomera (31.0 °S, geomag. lat. 41.2 °S), Australia. The numbers beside the cloud indicate the height in km. It will be seen that while the vapor trail has a smooth structure above 110 km, it shows a turbulent one below this height. The upper boundary level of this turbulent region is called the turbopause. It appears from the results of many rocket experiments that this boundary level is usually situated at an altitude of 110...130 km.

β) *General diffusion equation.* In considering the behavior of a chemical substance after its ejection into the atmosphere, let us use the approximation that the substance and the main atmosphere are regarded as the two components of a binary mixture, denoted by subscripts 1 and 2, respectively, and apply the formulas for a binary mixture to them. Before a sufficient length of time has elapsed, the gradient of number density of molecules or atoms ejected from a rocket is so large that the pressure gradient and the number density gradient of the atmosphere can be neglected compared therewith; thus the problem can be regarded as spherically or axially symmetric. If it is assumed that the atmospheric molecules are at rest as a whole ($\bar{\mathfrak{v}}_2=0$), that the number density of atoms of the chemical substance is small compared with that of atmospheric molecules ($n_1 \ll n_2$), and that there is no temperature gradient, the mass velocity

* From Lloyd, Low, Vincent (1973): Planet. Space Sci. *21* (No. 4), opposite to p. 654, Fig. 2

of this binary system can be regarded as zero ($\boldsymbol{v}_0=0$). Then we get from Eqs. (2.5) and (3.21)

$$n_1 \bar{\boldsymbol{v}}_1 = -D_{12} \nabla n_1 .$$

Inserting this into the equation of continuity (4.1), we have

$$\frac{\partial n_1}{\partial t} = \nabla \cdot D_{12} \nabla n_1 . \tag{13.1}$$

In the case of spherical or axial symmetry, the following formula holds for any vector $\boldsymbol{A}$:

$$\nabla \cdot \boldsymbol{A} = \frac{1}{r^m} \frac{\partial}{\partial r} (r^m A_r),$$

where r is the coordinate in the radial direction and m is taken as 2 or 1 for the spherically or axially symmetric case. Then Eq. (13.1) is transformed into

$$\frac{\partial n}{\partial t} = \frac{1}{r^m} \frac{\partial}{\partial r} \left(D r^m \frac{\partial n}{\partial r} \right), \tag{13.2}$$

where the subscript 1 of n_1 is omitted and D_{12} is simply written as D; r may be considered as the radial distance from the point of ejection of the chemical substance. Although D is a function of the position in space, its variation may be regarded as small compared with that of n with space coordinates and time. If this is admitted, Eq. (13.2) becomes

$$\frac{\partial n}{\partial t} = D \left(\frac{\partial^2 n}{\partial r^2} + \frac{m}{r} \frac{\partial n}{\partial r} \right). \tag{13.2a}$$

γ) *Cloud formation.* In the three dimensional case ($m=2$) one of the solutions of the present problem is

$$n(r,t) = \frac{n_t}{[\pi(4Dt+a^2)]^{3/2}} \exp\left(-\frac{r^2}{4Dt+a^2} \right); \tag{13.3}$$

this can be confirmed by directly substituting n from Eq. (13.3) for n in Eq. (13.2a). n_t is the total number of atoms ejected. By putting $t=0$ in Eq. (13.3), this is seen to be the solution corresponding to the initial distribution of n of

$$n(r,0) = \frac{n_t}{(\sqrt{\pi}\, a)^3} \exp\left(-\frac{r^2}{a^2} \right). \tag{13.4}$$

In particular, for a point source, by letting a tend to zero, we have the solution

$$n(r,t) = \frac{n_t}{(4\pi D t)^{3/2}} \exp\left(-\frac{r^2}{4Dt} \right). \tag{13.3a}$$

This equation shows that the probability of an ejected atom being found at a distance r from the point of ejection after the lapse of time t is given by the right-hand side of Eq. (13.3a) excluding the factor n_t. Therefore, for an arbitrary initial distribution of n as a function of r the solution can be obtained in the form of an integral by multiplying the function by the above probability and integrating the product. We do not intend to enter into actual examples here, however.

It can readily be seen by using this probability that the mean of the distances of displacement of the atoms at time t is given by

$$\bar{r}=\frac{4}{\sqrt{\pi}}\sqrt{Dt}. \tag{13.5}$$

In the two-dimensional case ($m=1$), the solution corresponding to Eq. (13.3) is readily seen to be given by

$$n(r,t)=\frac{n_t}{\pi(4Dt+a^2)}\exp\left(-\frac{r^2}{4Dt+a^2}\right). \tag{13.6}$$

If the number density of atoms at the radial distance of r_e from the center or the central axis of their distribution is smaller than that at the center or on the central axis by a factor of $1/e$, this r_e may be regarded as a measure of the dimensions of the atomic cloud. We shall call it the effective radius of the atomic cloud. Then, in both two and three dimensional cases we have

$$r_e^2=4Dt+a^2. \tag{13.7}$$

Hence the value of D should be determinable by observing the temporal variation in r_e.

δ) Chemical release experiments have been performed many times. We may mention as one of the most typical of these the one carried out in the morning (06.09 CST) and in the evening (18.19 CST) on 31 May, 1968 at Woomera, Australia[2]. In this experiment trimethyl aluminium (TMA) was used. According to the results, the trail appeared laminar and diffused with cylindrical symmetry above the turbopause, but it was observed to become evidently turbulent after a certain time below this level. Figure 3 shows the temporal increase in r_e^2 at a height of 105 km. The growth in the first 8 s took place faster than that in later hours by about one order of magnitude and may be regarded as the release phase independent of the environmental circumstances. At $t=8$ s, the effective radius reached about 130 m; up to $t=32$ s the process may be considered to be due to molecular diffusion. Turbulence began to appear at $t=33$ s and r_e^2 increased with time in proportion to t^3, but at $t=54$ s the trail became very irregular and the estimation of the radius of the cloud on the basis of the Gauss distribution (as given by Eq. (13.6)) proved to be meaningless. It was difficult to obtain the value of D, the coefficient of molecular diffusion, within a factor of 3 from the variation of r_e^2 in the time interval $8\,\mathrm{s}\leqq t\leqq 32\,\mathrm{s}$ and they did not attempt it. In the time interval $33\,\mathrm{s}\leqq t\leqq 54\,\mathrm{s}$ turbulence plays a predominant role in the process of growth, but this will be considered separately in the next section.

In the experiment by Rees et al.[2] the behavior of the trail later than about 60 s after its formation could not be known clearly, but the experiment by Lloyd et al.[1], referred to earlier, supplements this. According to this latter, the expansion of the trail of lithium vapor during the time interval from about 100 to 900 s after its formation at a height of 107 km, where a strong wind shear exists, was as shown in Fig. 4 and the trail expanded with its radius in proportion to

---

[2] Rees, D., Roper, R.G., Lloyd, K.H., Low, C.H. (1972): Philos. Trans. R. Soc. London Ser. A *271*, 631

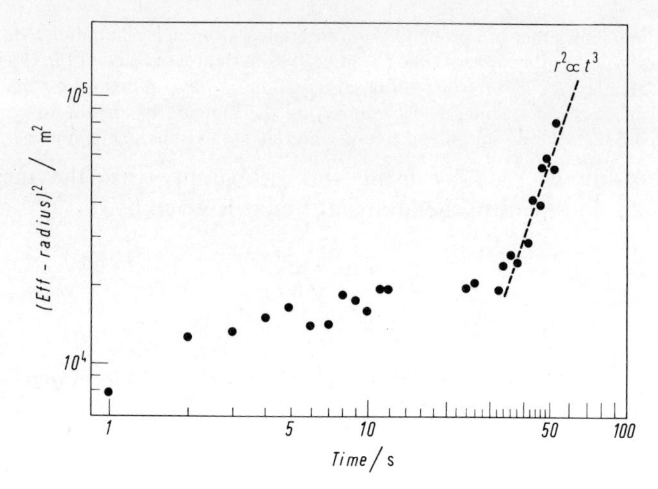

Fig. 3*. Variation of the effective radius r_e (squared) of a TMA cloud with time after release at 105 km

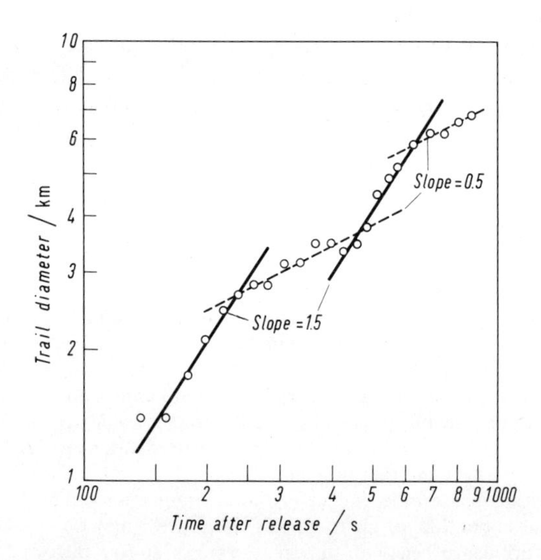

Fig. 4.** Expansion of the trail of lithium vapor from about 100 to 900 s after release in the high shear region at 107 km

the 3/2-th power of t during one interval and to the 1/2-th power of t during another, to the 2/3-th power on the average. This indicates that, after the lapse of time of this order of magnitude, eddies of different types and dimensions come to take part in the diffusion process and makes it a very complicated phenomenon. Some aspects of this phenomenon will be considered in the next Section.

---

* From Rees, Roper, Lloyd, Low (1972): Philos. Trans. R. Soc. London A*271* (No. 1218), 637, Fig. 6

** From Lloyd, Low, Vincent (1973): Planet. Space Sci. *21* (No. 4), 657, Fig. 5

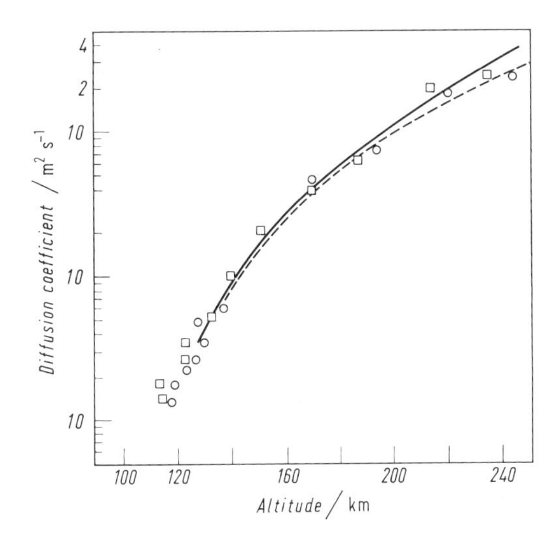

Fig. 5*. Diffusion coefficients measured on the morning and evening of 31 May 1968 at Woomera, Australia against altitude. ○, morning measurement; □, evening measurement. Solid and broken lines show curves fitted to evening and morning data points, respectively

As stated earlier, the growth of a trail is smooth at altitudes higher than about 110 km and the values of the coefficient of diffusion can be determined by measuring the temporal variation of the dimension of the trail and applying Eq. (13.7) to it. REES et al.[2] obtained the results shown in Fig. 5.

ε) On the other hand, the *coefficient of molecular diffusion* can be calculated for the present instance. TMA, after its ejection into the atmosphere, reacts with atomic oxygen and produces AlO which is considered to be most important in the diffusion process. Let us consider the molecular diffusion of this substance through the atmosphere which is assumed for simplicity to be composed of only nitrogen molecules.

As the values of some parameters such as the molecular radius of AlO cannot be found in the tables of HIRSCHFELDER et al. [*3*], they are inferred from those of other molecules or atoms on the assumption, after REES et al.[2], that they are smooth functions of molecular or atomic weight. These authors estimated that the exospheric temperatures at the times of the experiments shown in Fig. 5 were 980 K in the morning and 1,150 K in the evening; hence we have adopted the model atmosphere which is the average of JACCHIA's 1971 model atmospheres[3] for the exospheric temperatures of 1,000 K and 1,100 K. In this way, using Eqs. (5.11) and (5.11a), the values of the coefficient of molecular diffusion expected at the heights of 110, 120, 130, and 140 km were calculated as 0.15, 0.77, 2.6, and $6.6 \cdot 10^3\,\mathrm{m^2\,s^{-1}}$, respectively.

When such calculated values are compared with those depicted in Fig. 5, it will be seen that, while the latter are appreciably larger than the calculated values of the coefficient of molecular diffusion at the height of 120 km, the differences between both sets diminish above this level and become almost insignificant at 140 km. Thus, the values of the coefficient of diffusion de-

* From Rees, Roper, Lloyd, Low (1972): Philos. Trans. R. Soc. London A*271* (No. 1218), 658, Fig. 22

[3] Jacchia, L.G. (1971): Smithson. Astrophys. Observat. Spec. Rep. *332*

termined by chemical release experiments are still somewhat larger than those of molecular diffusion at and below 130 km, and this is in agreement with other workers' results.[4] REES et al.[2] interpreted this fact as showing that considerable turbulence still remains even in the region above the turbopause. It is interesting that the same conclusion is reached in the consideration of the distribution of atomic and molecular oxygen which is dealt with in the next Chapter.

14. Turbulence in release experiments. We are not in a position to have any theory which can describe the turbulence in the high atmosphere in sufficient detail. Hence, we wish here to outline BOOKER's theory of turbulence[1] as an example only, and to apply it with some success to chemical release experiments described in the previous section.

α) *Eddy transport.* First, let us assume that large-scale eddies with the scale of L_1 are produced by some causes such as wind shear or internal gravity waves and that energy is used in this process at the rate of ε per unit time and per unit mass of the atmosphere. Since large-scale eddies have a large Reynolds number, they will lose energy by viscosity only to a small extent so that, as they split successively into smaller and smaller eddies, energy is transferred to the small-scale eddies at the same rate. When they become small enough with the scale L_2 and with a Reynolds number of the order of unity, they will rapidly disappear, their energy being lost by viscosity and converted into the heat energy of the atmosphere. Thus, it will be seen that a spectrum of scales of eddies ranging from L_1 down to L_2 exists and that energy is transferred at first to large scale eddies at the rate of ε per unit time and mass and then successively to smaller and smaller eddies until at last it is converted into heat at practically the same rate.

Let the time constant and turbulence-velocity of large eddies be denoted by t_1 and v_1, respectively. Then, since the energy per unit mass, $\frac{1}{2}v_1^2$, has been acquired during the time interval t_1 at the rate ε, we have approximately

$$v_1^2 \simeq \varepsilon t_1. \tag{14.1}$$

The scale of an eddy may be considered to be given by

$$L_1 \simeq v_1 t_1. \tag{14.2}$$

It will readily be seen that the same expressions hold also for small eddies indicated by the subscript 2:

$$v_2^2 \simeq \varepsilon t_2, \tag{14.3}$$

$$L_2 \simeq v_2 t_2. \tag{14.4}$$

As the Reynolds number for small eddies $L_2 v_2/\nu$ (ν here is kinematic viscosity) is of the order of unity, the following approximate relation is derived if

[4] Zimmerman, S.P., Champion, K.S.W. (1963): J. Geophys. Res. *68*, 3049
[1] Booker, H.G. (1956): J. Geophys. Res. *61*, 673

Eq. (14.4) is noted:

$$L_2^2 \simeq v t_2. \tag{14.5}$$

β) *Cloud radius.* BOOKER[1] assumed that molecular diffusion plays the predominant role in the growth of the trail for $t<t_2$ while eddy diffusion takes over this role for $t>t_2$. He has further assumed that, as the trail expands with time, larger and larger eddies play the predominant role and eventually the large eddies with the scale of L_1 take part exclusively in the diffusion of the trail for $t>t_1$. If it is assumed that, at time t, the diffusion of the trail is mainly effected through the action of the eddies with the time constant t, this is in agreement with the above assumption. Thus, the diffusion coefficient at time t between t_2 and t_1 ($t_2<t<t_1$) is seen to be given by εt^2 if one notes the relation

$$v L = \varepsilon t^2, \tag{14.6}$$

which is obtained from Eqs. (14.3) and (14.4) with the subscript 2 taken away, and Eqs. (8.6) and (8.8) (here v is the turbulence-velocity which corresponds to v' in Sect. 8). εt^2 is linked continuously with $v_1 L_1$ at t equal to t_1. In this way the effective diffusion coefficient of the trail is seen to become approximately as follows:

$$\begin{array}{lll} \text{first:} & D \text{ (coefficient of molecular diffusion)} & \text{for } 0<t<t_2, \\ \text{then:} & \varepsilon t^2 & \text{for } t_2<t<t_1, \\ \text{finally:} & v_1 L_1 \text{ (independent of } t\text{)}, & \text{for } t_1<t. \end{array} \tag{14.7}$$

If the coefficient of diffusion is given by εt^2 as a function of time, Eq. (13.2a) becomes

$$\frac{\partial n}{\partial (t^3)} = \frac{\varepsilon}{3}\left(\frac{\partial^2 n}{\partial r^2} + \frac{m}{r}\frac{\partial n}{\partial r}\right),$$

whose solution is obtained by replacing t with t^3 and D with $\varepsilon/3$ in Eq. (13.3), (13.3a) or (13.6). Hence the following relations are obtained instead of Eq. (13.7):

$$r_e^2 = \begin{cases} 4Dt + a^2 & \text{for } t<t_2, \quad (14.8) \\ \frac{4}{3}\varepsilon(t^3 - t_2^3) + r_{e2}^2 & \text{for } t_2<t<t_1, \quad (14.9) \\ 4(v_1 L_1)(t-t_1) + r_{e1}^2 & \text{for } t>t_1, \quad (14.10) \end{cases}$$

where r_{e1} and r_{e2} are the values of r_e at the times $t=t_1$ and $t=t_2$[2].

In considering the reflection of radio waves by meteor trails, BOOKER[1] took t_2 and t_1 to be equal to 0.4 and 50 s, respectively. Judging from the experimental results shown in Fig. 3, this value of t_2 seems to be too small for the consideration of the chemical release experiments concerned here.

---

[2] Booker, H.G., Cohen, R. (1956): J. Geophys. Res. *61*, 707

γ) REES et al.[3] applied Eq. (14.9) to their *experimental results* (Fig. 3) and determined the value of ε:

$$\varepsilon = 0.37\ \mathrm{W\,kg^{-1}}. \tag{14.11}$$

This is much smaller than that given by BOOKER[1] (25 W kg^{-1}), but various workers[4-7] have obtained values in the range 0.5...0.005 W kg^{-1} so that the value given above does not seem too small; possibly it may be to a certain extent too large.

In the chemical release experiment of lithium vapor by LLOYD et al.[4], data have been obtained between about 100 and 900 s after the release which supplement the observed results of REES et al.[3] In a high shear region at a height of 107 km the expansion of the trail was as shown in Fig. 4 and the trail increased in its diameter in proportion to the 3/2-th or 1/2-th power of time t. This may be taken to suggest that the trail is influenced by various types of turbulence according to circumstances, e.g. in the presence of strong or weak wind-shear (in this case r_e^2 should vary in proportion to t^2 or t, respectively[8]) besides isotropic turbulence already explained. If it is assumed that BOOKER's above-mentioned theory can be applied to the first part of the expansion of the trail, the value of ε can be obtained with the result that $\varepsilon = 0.04$ W kg^{-1}. Further, one will notice from Fig. 4 that the slope of the line changes abruptly from 1.5 to 0.5 or *vice versa* at about 250 and 450 sec after the release; this means that the relation between r_e^2 and t changes abruptly at these times from the 3/2-th power law to the 1/2-th power law or *vice versa*.

It may be doubtful that such times can be regarded as corresponding to BOOKER's time t_1, but if the former time is assumed to represent the time t_1, the coefficient of eddy diffusion by large-scale eddies can be calculated using Eq. (14.7) and the above-obtained value of ε of 0.04 W kg^{-1}:

$$v_1 L_1 \simeq 2.5 \cdot 10^3\ \mathrm{m^2\,s^{-1}}.$$

From Eq. (14.1) the turbulence-velocity of large eddies is derived:

$$v_1 \simeq 3.2\ \mathrm{m\,s^{-1}},$$

and from Eq. (14.2) the scale of large eddies:

$$L_1 \simeq 800\ \mathrm{m}.$$

These values agree in their order of magnitude with those estimated by BLAMONT[6], i.e. $t_1 \simeq 500$ s, $v_1 \simeq 2$ m s^{-1}, and $L_1 \simeq 1$ km. It may be almost meaningless to give these values a significance more than their orders of magnitude.

δ) As regards *small eddies*, if we take t_2 to be equal to 33 s as explained in the previous section and use the value of ε given by Eq. (14.11), we get the

[3] Rees, D., Roper, R.G., Lloyd, K.H., Low, C.H. (1972): Philos. Trans. R. Soc. London Ser. A *271*, 631

[4] Lloyd, K.H., Low, C.H., Vincent, R.A. (1973): Planet. Space Sci. *21*, 653

[5] Zimmerman, S.P., Champion, K.S.W. (1963): J. Geophys. Res. *68*, 3049

[6] Blamont, J.E. (1963): Planet. Space Sci. *10*, 89

[7] Greenhow, J.S., Neufeld, E.L. (1959): J. Geophys. Res. *64*, 2129
Greenhow, J.S. (1959): J. Geophys. Res. *64*, 2208; Blamont, J.E., de Jager, C. (1961): Ann. Geophys. *17*, 134

[8] Tchen, C.M. (1959): Adv. Geophys. *6*, 65

following from Eqs. (14.3) and (14.4):

$$v_2 \simeq 3.5 \text{ m s}^{-1}, \quad L_2 \simeq 120 \text{ m}.$$

These values are to be compared with those[6] of BLAMONT: $t_2 \simeq 30$ s, $v_2 \simeq 1$ m s^{-1}, and $L_2 \simeq 30$ m. Here again, only the order of magnitude is significant.

REES et al.[3] estimated the value of ε by a different method. Using Eqs. (14.3...5), ε is expressed as follows:

$$\varepsilon \simeq \nu / t_2^2. \tag{14.12}$$

If an appropriate model atmosphere is assumed, the kinematic viscosity at the height of z km is represented by

$$\nu/\text{m}^2 \text{ s}^{-1} = \exp[0.17(z/\text{km} - 80.0)]. \tag{14.13}$$

Substituting ν from Eq. (14.13) for that in Eq. (14.12), ε is expressed as a function of height. This gives 0.064 W kg^{-1} as the value of ε at the height of 105 km which is considerably smaller than given by Eq. (14.11) but fairly close to that derived from the experiment by LLOYD et al.[4] If this value of ε is to be used, those of v_2 and L_2 given at the beginning of this Subsect. should be reduced by a factor of 2.4, making the agreement with BLAMONT's values[6] closer. Lastly it may be worth mentioning that REES et al.[3] noted a steep increase in the time constant of small eddies with height at turbopause, inferred from their experimental results and theoretical calculations.

KELLOGG[9] evaluated the coefficient of eddy diffusion by his own special method based on observed results of internal gravity waves and vapor trail experiments. He gave values of $4 \cdot 10^3$ and $2 \cdot 10^3$ m^{2}s^{-1} at 100 and 80 km, respectively. These are roughly in agreement with the value obtained above.

15. Evaluation of the time required for attaining diffusive separation. As stated in Sect. 12, it has been confirmed by rocket observations that the diffusive separation of each constituent of the upper atmosphere takes place above a level of about 110 km. Let us consider the problem of the height where this level is expected to exist on theoretical grounds. For this purpose, we assume that each constituent of the atmosphere is at first in a completely mixed condition at every height; if it is then left free, with no restraint from outside except for gravity, it will tend toward the condition of diffusive equilibrium. We evaluate the time necessary for each constituent to reach diffusive equilibrium as a function of height. If, at a certain height, this time is short compared with the characteristic times of the processes taking part in mixing the atmosphere, for instance the time which is needed for winds to exert an appreciable influence on the distribution of the constituent, it can be concluded that diffusive separation of the constituent should prevail at this height.

[9] Kellogg, W.W. (1964): Space Sci. Rev. *3*, 275

α) However, the Earth's atmosphere, at least below about 200 km, is composed of two or three main constituents N_2, O_2, and O in comparable proportions. In this case we must consider the problem of *major constituent diffusion* which is very difficult to deal with. Let us assume that the atmosphere is composed of two main permanent constituents distinguished by subscripts 1 and 2 and that the mean scale height given by Eq. (11.11), $\bar{H}$, is a linear function of height[1]. As the total number of each kind of particle is constant, from Eq. (11.17) we get:

$$\int_{z_0}^{\infty} n_{iD}\,\mathrm{d}z = n_{iD_0}\bar{H}_0\frac{\bar{m}}{m_i} = \int_{z_0}^{\infty} n_{iM}\,\mathrm{d}z = n_{iM_0}\bar{H}_0, \qquad i=1,2, \tag{15.1}$$

where subscripts D and M indicate the condition of diffusive equilibrium and complete mixing, respectively, and z_0 is the value of z at the ground level. Hence we have

$$n_{iD_0} = \frac{m_i}{\bar{m}}\,n_{iM_0}. \tag{15.2}$$

The total number densities of particles at a level are derived with Eqs. (11.15, 16) as follows:

$$n_{1M} + n_{2M} = (n_{1M_0} + n_{2M_0})\left(\frac{\bar{H}_0}{\bar{H}}\right)^{1+1/\Gamma}, \tag{15.3}$$

$$n_{1D} + n_{2D} = \frac{m_1}{\bar{m}}\,n_{1M_0}\left(\frac{\bar{H}_0}{\bar{H}}\right)^{1+m_1/(\bar{m}\Gamma)} + \frac{m_2}{\bar{m}}\,n_{2M_0}\left(\frac{\bar{H}_0}{\bar{H}}\right)^{1+m_2/(\bar{m}\Gamma)} \tag{15.4}$$

for the states of complete mixing and diffusive equilibrium, respectively. The right-hand sides of these two equations are generally not equal to each other so that the total local concentration at any level changes during the process of transition from the state of complete mixing to that of diffusive equilibrium; thus, even at constant temperature, the total pressure at a level is not maintained constant during the process.

In this phenomenon the number densities n_1 and n_2 of the respective species of molecules and their mean velocities $\bar{v}_1$ and $\bar{v}_2$ vary in accordance with the continuity equation (4.1) and $\bar{v}_1 - \bar{v}_2$ should be equal to the right-hand side of Eq. (3.21) owing to Eq. (2.6). However, as the total local concentration at any level is not maintained constant, another simple conditional equation which must be satisfied by $\bar{v}_1, \bar{v}_2, n_1$, and n_2 cannot be found and it is not possible to determine these variables as functions of time and height.

In the following, therefore, we do not intend to deal with such a major constituent diffusion problem but confine ourselves only to the treatment of a minor constituent (this is the same in the next Chapter where the vertical distributions of O_2 and O are considered). Fortunately, the atmosphere is for the greater part composed of molecular nitrogen in a well-mixed state and it

[1] Mange, P. (1957): J. Geophys. Res. *62*, 279

will not lead to serious errors to treat other constituents as minor ones. At higher levels, atomic oxygen, atomic helium, atomic hydrogen or even protons are the predominant constituent and in such regions minor constituent diffusion is sufficient to describe the behavior of other constituents.

β) *Minor constituent diffusion.* Let the constituent of species 1 be the minor one and that of species 2 be the predominant one and assume species 1 diffuses in the vertical direction through species 2 which is stable so that its distribution is not influenced by the behavior of the minor constituent. In this case we can set $\bar{v}_2=0$ and this makes the fourth conditional equation which cannot be obtained in the case of major constituent diffusion. If the effect of thermal diffusion is neglected, we get from Eqs. (2.6), (3.21), (4.1), (5.7), (9.24), and (11.8) and the relations $n_1 \ll n_2 \simeq n$ and $v_2=0$ the following equation of diffusion:

$$\frac{\partial n_1}{\partial t}=\frac{\partial}{\partial z}\left\{D_{12}\left[\frac{\partial n_1}{\partial z}+\left(\frac{m_1}{m_2}+\Gamma_2\right)\frac{n_1}{H_2}\right]\right\}. \tag{15.5}$$

Since D_{12} is inversely proportional to $n_1+n_2 \simeq n_2$ as seen from Eq. (5.11) and proportional to the square root of temperature T (apart from the T-dependence of $\sigma_1+\sigma_2$ in the denominator of the right-hand side of Eq. (5.11)), we have the following expression for D_{12} if Eqs. (11.5) and (11.8) are noted:

$$D_{12}=D_0\left(\frac{H}{H_0}\right)^{\frac{3}{2}+1/\Gamma}, \tag{15.6}$$

where we have omitted subscripts 2 and 12 on the right-hand side (this is the same for the following expressions). Introducing a new variable y:

$$y=\left(\frac{H}{H_0}\right)^{(\Gamma-2)/4\Gamma} \tag{15.7}$$

and noting the relation (11.7) or

$$H=H_0+\Gamma(z-z_0), \tag{15.8}$$

we get

$$\frac{\partial}{\partial z}=-\frac{2-\Gamma}{4H_0}y^{(2+3\Gamma)/(2-\Gamma)}\frac{\partial}{\partial y}. \tag{15.9}$$

By Eqs. (15.6) and (15.7)

$$D=D_0\,y^{-2(2+3\Gamma)/(2-\Gamma)} \tag{15.10}$$

and

$$H=H_0\,y^{-4\Gamma/(2-\Gamma)}, \tag{15.11}$$

where we have simply written D instead of D_{12}. Using Eqs. (15.10, 11), Eq. (15.5) is transformed into

$$\delta^{-2}\frac{\partial u}{\partial t}=\frac{\partial^2 u}{\partial y^2}+\frac{2\gamma+1}{y}\frac{\partial u}{\partial y}, \tag{15.12}$$

where

$$u = n_1\, y^{4(m_1/m + \Gamma)/(\Gamma - 2)}, \tag{15.13}$$

$$\gamma \equiv \frac{m_1/m}{1 - \Gamma/2} - 1, \tag{15.14}$$

and

$$\delta \equiv \frac{(2-\Gamma)\, D_0^{1/2}}{4 H_0}. \tag{15.15}$$

u given by Eq. (15.13) can be expressed in terms of γ as follows:

$$u = n_1\, y^{-2(\gamma + (2+\Gamma)/(2-\Gamma))}. \tag{15.16}$$

The *flux of electrons or ions* in the vertical *upward* direction, F, is given by

$$F \equiv -D\left[\frac{\partial n_1}{\partial z} + \left(\frac{m_1}{m} + \Gamma\right)\frac{n_1}{H}\right] = \frac{2-\Gamma}{4H_0} D_0\, y^{2\gamma+1} \frac{\partial u}{\partial y}. \tag{15.17}$$

γ) A *solution* of Eq. (15.12) can be obtained by the conventional method of separation of variables. If ω is an arbitrary constant, its particular solution is given by

$$u_{\text{par}} = \mathrm{e}^{-\delta^2 \omega^2 t}\, y^{-\gamma}\, \mathrm{J}_{\pm\gamma}(\pm\omega y), \tag{15.18}$$

where $\mathrm{J}_n(x)$ is the Bessel function of the first kind of order n and argument x.

At sufficiently high altitudes H tends to infinity, or, as $0 < \Gamma < 2$ in the actual atmosphere, y tends to zero there by Eq. (15.7). Thus, asymptotically we may write for Eq. (15.18):

$$u_{\text{par}} = \mathrm{e}^{-\delta^2 \omega^2 t} \times O(y^{\pm\gamma - \gamma}). \tag{15.19}$$

So long as no sink or source exists at infinity, the vertical flux F should tend to zero at sufficiently high altitudes. Eq. (15.17) shows that this condition is not satisfied if the plus-minus sign in the exponent of y in Eq. (15.19) is negative. Hence the double sign before γ in Eq. (15.18) should be taken as positive.

The general solution for the present problem is given by

$$u(t, y) = \int_{-\infty}^{\infty} \mathrm{C}(\omega)\, \mathrm{e}^{-\delta^2 \omega^2 t}\, y^{-\gamma}\, \mathrm{J}_\gamma(\omega y)\, \mathrm{d}\omega, \tag{15.20}$$

where $\mathrm{C}(\omega)$ is an arbitrary function of ω. It should be noted that, as the limits of integration are $\pm\infty$, the double sign which should be inserted before ωy in $\mathrm{J}_\gamma(\omega y)$ can be omitted. Setting $t = 0$ in Eq. (15.20), we get

$$y^\gamma u(0, y) = \int_{-\infty}^{\infty} \mathrm{C}(\omega)\, \mathrm{J}_\gamma(\omega y)\, \mathrm{d}\omega. \tag{15.21}$$

If $u(0, y)$ is given as an initial condition, we express this as a Fourier-Bessel integral[2] in the following way (noting that $0<y<\infty$)

$$y^{\gamma} u(0, y)=\int_0^\infty \int_0^\infty \alpha^{\gamma} u(0, \alpha) \mathrm{J}_{\gamma}(\omega y) \mathrm{J}_{\gamma}(\omega \alpha) \alpha \omega \,\mathrm{d}\alpha \,\mathrm{d}\omega, \qquad \gamma>-\tfrac{1}{2}. \tag{15.22}$$

Comparing Eqs. (15.21) and (15.22), we see that, if $\mathrm{C}(\omega)$ is taken as

$$\mathrm{C}(\omega)=\begin{cases}\int_0^\infty \alpha^{\gamma+1} u(0, \alpha) \mathrm{J}_{\gamma}(\omega \alpha) \omega \,\mathrm{d}\alpha & \omega>0, \\ 0 & \omega<0,\end{cases} \tag{15.23}$$

$u(t, y)$ given by Eq. (15.20) is the solution satisfying the initial condition. Substituting $\mathrm{C}(\omega)$ from Eq. (15.23) for that in Eq. (15.20) and changing the order of integration by the use of the following formula[3]

$$\int_0^\infty \exp(-p^2 t^2) \cdot \mathrm{J}_{v}(a t) \mathrm{J}_{v}(b t) t \,\mathrm{d}t=\frac{1}{2p^2} \exp\left(-\frac{a^2+b^2}{4p^2}\right) \cdot \mathrm{I}_{v}\left(\frac{a b}{2p^2}\right),$$

$$R(v)>-1, \qquad |\arg p|<\frac{\pi}{4}, \tag{15.24}$$

we get the following final solution in which $\mathrm{I}_{\gamma}(x)$ is the Bessel function of purely imaginary argument:

$$u(t, y)=\frac{y^{-\gamma}}{2\delta^2 t} \exp\left(-\frac{y^2}{4\delta^2 t}\right) \int_0^\infty u(0, \alpha) \alpha^{\gamma+1} \exp\left(-\frac{\alpha^2}{4\delta^2 t}\right) \cdot \mathrm{I}_{\gamma}\left(\frac{\alpha y}{2\delta^2 t}\right) \mathrm{d}\alpha. \tag{15.25}$$

This solution has been obtained for the present under the condition $\gamma>-\frac{1}{2}$, or by Eq. (15.14)

$$\frac{m_1}{m}>\frac{1}{4}(2-\Gamma), \tag{15.26}$$

however, owing to the principle of analytic continuation, it is still the solution of the problem for γ's beyond the above limitation, so long as it is analytic. If u is expressed in terms of n_1 using Eq. (15.16), Eq. (15.25) becomes[1]

$$n_1(t, y)=\frac{y^{\gamma+\frac{2(2+\Gamma)}{2-\Gamma}}}{2\delta^2 t} \exp\left(-\frac{y^2}{4\delta^2 t}\right)$$

$$\cdot \int_0^\infty n_1(0, \alpha) \alpha^{-\left(\gamma+\frac{2+3\Gamma}{2-\Gamma}\right)} \exp\left(-\frac{\alpha^2}{4\delta^2 t}\right) \cdot \mathrm{I}_{\gamma}\left(\frac{\alpha y}{2\delta^2 t}\right) \mathrm{d}\alpha. \tag{15.27}$$

---

[2] Watson, G.N. (1952): A treatise on the theory of Bessel functions, 2nd edn., p. 453. Cambridge: University Press

[3] Watson, G.N. (1952): A treatise on the theory of Bessel functions, 2nd edn., p. 395. Cambridge: University Press

The solution of this form was first obtained by EPSTEIN[4] in 1932 and by SUTTON[5].

δ) *Diffusive separation.* Let us consider the case in which the minor constituent is initially in a state completely mixed with the main constituent. Then, from Eqs. (11.8) and (15.11) we get

$$n_1(0, y) = n_{10}\, y^{4(1+\Gamma)/(2-\Gamma)}. \tag{15.28}$$

When this is inserted, the integral in Eq. (15.27) can be calculated using the power series of $\mathrm{I}_\gamma(x)$ with the result that

$$\begin{aligned}&\int_0^\infty n_1(0,\alpha)\,\alpha^{-\gamma-(2+3\Gamma)/(2-\Gamma)} \exp\left(-\frac{\alpha^2}{4\delta^2 t}\right)\cdot \mathrm{I}_\gamma\left(\frac{\alpha y}{2\delta^2 t}\right) \mathrm{d}\alpha \\ &= \tfrac{1}{2} n_{10} \sum_{s=0}^{\infty} \frac{\Gamma\left(s+\frac{2}{2-\Gamma}\right)}{s!\,\Gamma(\gamma+s+1)}\, y^{\gamma+2s} (4\delta^2 t)^{-\gamma-s+2/(2-\Gamma)},\end{aligned} \tag{15.29}$$

where $\Gamma(x)$ is the gamma function. Thus $n_1(t, y)$ given by Eq. (15.27) becomes

$$\begin{aligned} n_1(t,y) &= n_{10} (4\delta^2 t)^{-\gamma+\frac{\Gamma}{2-\Gamma}}\, y^{2\gamma+\frac{2(2+\Gamma)}{2-\Gamma}} \\ &\cdot \frac{\Gamma\left(\frac{2}{2-\Gamma}\right)}{\Gamma(\gamma+1)} \exp\left(-\frac{y^2}{4\delta^2 t}\right)\cdot {}_1\mathrm{F}_1\left(\frac{2}{2-\Gamma}; \gamma+1; \frac{y^2}{4\delta^2 t}\right),\end{aligned} \tag{15.30}$$

where ${}_1\mathrm{F}_1(\alpha; \rho; z)$ is the confluent hypergeometric series given by

$${}_1\mathrm{F}_1(\alpha;\rho;z) = 1 + \frac{\alpha}{1!\,\rho} z + \frac{\alpha(\alpha+1)}{2!\,\rho(\rho+1)} z^2 + \dots. \tag{15.31}$$

If $t\to\infty$, $\exp\left(-\frac{y^2}{4\delta^2 t}\right)\cdot {}_1\mathrm{F}_1\left(\frac{2}{2-\Gamma}; \gamma+1; \frac{y^2}{4\delta^2 t}\right) \to 1$ and $n_1(t, y)$ approaches a distribution proportional to $y^{2\gamma+\frac{2(2+\Gamma)}{2-\Gamma}}$ asymptotically. It is readily seen by the use of Eqs. (11.15), (15.8), (15.11), and (15.14) that this is the diffusive equilibrium distribution for the constituent of species 1. Therefore, the following function appearing in Eq. (15.30) as a function of time,

$$\mathrm{f}\left(\frac{y^2}{4\delta^2 t}; \Gamma; \gamma\right) \equiv \exp\left(-\frac{y^2}{4\delta^2 t}\right)\cdot {}_1\mathrm{F}_1\left(\frac{2}{2-\Gamma}; \gamma+1; \frac{y^2}{4\delta^2 t}\right), \tag{15.32}$$

---

[4] Epstein, P.S. (1932): Gerlands Beitr. Geophys. *35*, 153

[5] Sutton, W.G.L. (1943): Proc. R. Soc. London A *182*, 48

may be considered as a measure showing to what extent the initial distribution of complete mixing approaches that of diffusive equilibrium with the lapse of time.

Actual calculations show that $f(z;\Gamma;\gamma)$ $(z\equiv y^2/(4\delta^2 t))$ is a monotonously decreasing function of z for the values of Γ and γ adopted in the following. Hence, if for example $f(z;\Gamma;\gamma)$ takes the value of 0.90 for certain given values of time t and height y, the values of f are nearer to unity than 0.90 at that time and at higher levels (note that smaller values of y correspond to higher levels by Eq. (15.7) in which $\Gamma-2<0$). In other words, the distribution of the minor constituent is in agreement with that of diffusive equilibrium within a 10% error at that time and at higher levels.

Using the following asymptotic expansions of the confluent hypergeometric series[6]:

$$
\begin{aligned}
{}_1F_1(\alpha;\rho;z) &= W_1(\alpha;\rho;z)+W_2(\alpha;\rho;z),\\
W_1(\alpha;\rho;z) &\sim \frac{\Gamma(\rho)}{\Gamma(\rho-\alpha)}(-z)^{-\alpha}\,G(\alpha,\alpha-\rho+1;\,-z),\\
W_2(\alpha;\rho;z) &\sim \frac{\Gamma(\rho)}{\Gamma(\alpha)}e^{z}z^{\alpha-\rho}\,G(1-\alpha,\rho-\alpha;\,z),\\
G(\alpha;\beta;z) &= 1+\frac{\alpha\beta}{1!}\frac{1}{z}+\frac{\alpha(\alpha+1)\beta(\beta+1)}{2!}\left(\frac{1}{z}\right)^2+\dots,
\end{aligned}
$$

$n_1(t,y)$ given by Eq. (15.30) can be transformed into the following form for large values of $y^2/(4\delta^2 t)$:

$$
n_1(t,y)=n_{10}\,y^{4(1+\Gamma)/(2-\Gamma)}\,G\left(-\frac{\Gamma}{2-\Gamma},\gamma-\frac{\Gamma}{2-\Gamma};\frac{y^2}{4\delta^2 t}\right). \tag{15.33}
$$

In this form it is evident that the minor constituent of species 1 is certainly in a state of complete mixing with the main constituent for $t\to 0$ (cf. Eq. (15.28)) and the factor $G\left(-\frac{\Gamma}{2-\Gamma},\gamma-\frac{\Gamma}{2-\Gamma};\frac{y^2}{4\delta^2 t}\right)$ may be considered as a measure indicating to what extent the actual distribution at any time approximates to the initial distribution of complete mixing.

ε) *An example.* Assuming that argon is at first completely mixed with the atmosphere, let us evaluate the time which is necessary for it to approach closely the state of diffusive equilibrium by molecular diffusion when it is left free. For simplicity, the atmosphere is assumed to be composed of only nitrogen molecules and their number density n_0, the scale height H_0 and its height gradient Γ_0 at the 100 km level have been taken as $1.198\cdot 10^{19}\,\mathrm{m}^{-3}$, 6.083 km, and 0.3818, respectively. If the height gradient of scale height is independent of height, $[N_2]$, the number density of N_2, varies between 100 and 200 km as shown in Fig. 6. In the same figure the total number density of the atmosphere is also depicted according to JACCHIA's 1971 model atmosphere for the exospheric temperature of 1,100 K. (Above 110 km this model is identi-

[6] Morse, P.M., Feshbach, H. (1953): Methods of theoretical physics, part I, p. 607. New York: McGraw-Hill

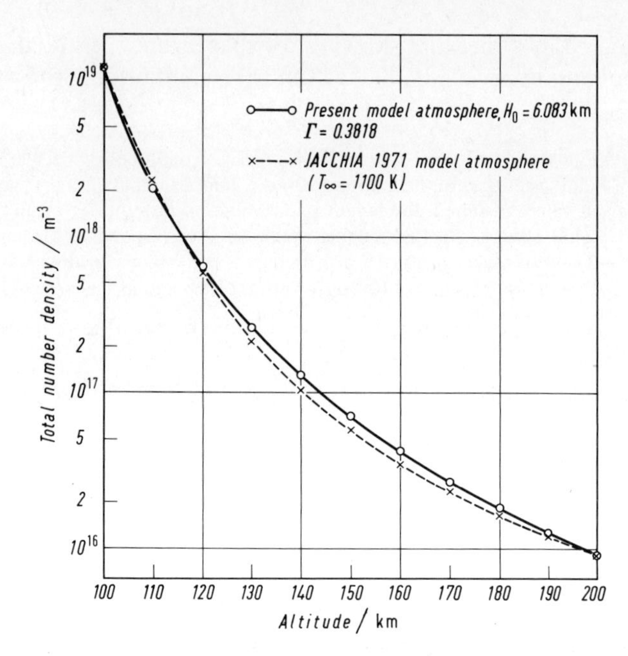

Fig. 6. Total number density of atmospheric molecules adopted for the present calculation against altitude and that of JACCHIA's 1971 Model Atmosphere ($T_\infty = 1{,}100$ K)

cal with the official CIRA 1972.) It is seen that both curves agree fairly well with each other, the former being in excess of the latter by 26% at the most. Hence, the values of n_0, H_0, and Γ_0 adopted above may be regarded as reproducing the actual atmosphere fairly well between 100 and 200 km, as far as the total number density is concerned.

The coefficient of molecular diffusion D for the binary gas mixture composed of N_2 and Ar can be calculated according to the method explained by HIRSCHFELDER et al. [3] (cf. Sect. 5). If after JACCHIA[7] the atmospheric temperature at the 100 km level is taken as 194.8 K, the value of D at the same level is 28.02 m^2 s^{-1}. Between 100 km and 200 km, we have calculated, for each level at 10 km intervals, the time necessary for argon atoms, starting from the initial distribution of complete mixing, to reach the diffusive equilibrium distribution within 5, 10 or 20%. The results are shown in Fig. 7. It is seen that it takes rather a long time for argon atoms to reach the diffusive equilibrium distribution by molecular diffusion. If the upper atmosphere is disturbed by heating and cooling, tidal motion, winds, gravity waves, etc. with a period of one day, half a day or shorter, with consequent mixing of the constituents, these agents will probably manifest their effects in several days. Figure 7 shows that below about 160 km an atmospheric constituent will not have enough time to settle by diffusive separation. Considering that, as described in Sect. 12, the diffusive separation between N_2 and Ar sets in at a height of about 110 km, we are obliged to conclude that the above-mentioned agents are not so effective

[7] Jacchia, L.G. (1971): Smithson. Astrophys. Observat. Spec. Rep. *332*

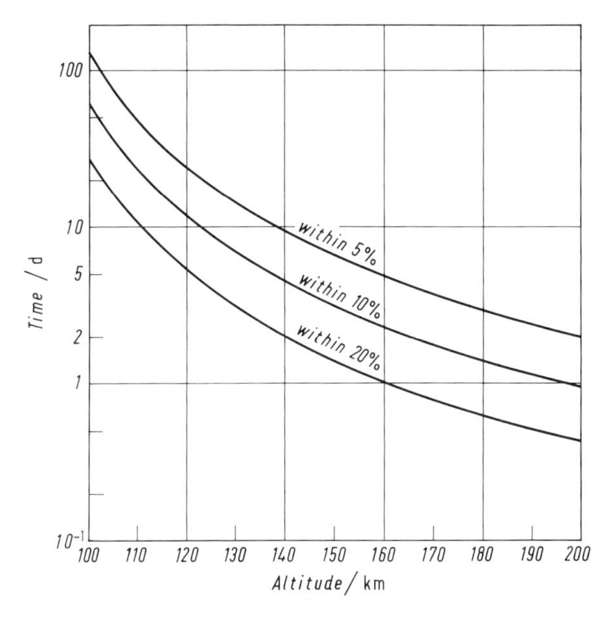

Fig. 7. Time necessary for the Ar atoms completely mixed with an N_2 atmosphere to attain the diffusive equilibrium condition within the error of 5, 10, or 20%

as supposed above in mixing the atmosphere. The rapid termination of turbulence above about 110 km and the steep rise in atmospheric temperature at and above the same level may act unfavorably for atmospheric mixing.

III. Vertical distributions of molecular and atomic oxygen

16. Dissociation of oxygen molecules. Since the problem of the dissociation of O_2 in the high atmosphere was considered by CHAPMAN in 1930[1] and by WULF and DEMING in 1936[2] and 1938[3], it has been attacked by a large number of workers. As the typical one in the early stage of this kind of work PENNDORF's detailed study[4] may be worth mentioning. In this paper, tables are given which show the number densities of oxygen atoms and molecules as a function of height for nine cases where different assumptions have been made with regard to the effective temperature of the Sun, the three-body recombination coefficient of atomic oxygen, the solar zenith angle, etc. However these number densities are widely different from our present knowledge; even taking the case nearest to the present model atmosphere, we see that, while the number density of O_2, $[O_2]$, is $1.5 \cdot 10^{18}\,m^{-3}$ at 110 km, it falls to $2 \cdot 10^{14}\,m^{-3}$ at

[1] Chapman, S. (1930): Philos. Mag. *10*, 369
[2] Wulf, O.R., Deming, L.S. (1936): Terr. Magn. Atmos. Electr. *41*, 299
[3] Wulf, O.R., Deming, L.S. (1938): Terr. Magn. Atmos. Electr. *43*, 283
[4] Penndorf, R. (1949): J. Geophys. Res. *54*, 7

130 km, showing a very steep rate of decrease with height. As PENNDORF only took photochemical and chemical processes into account and did not consider the effect of diffusion at all, these results are not surprising. With the advancement in the rocket measurement of $[O_2]$, it became clear that his calculation was not satisfactory. It was NICOLET[5] who first pointed out the importance of the diffusion process in interpreting the discrepancy between calculated and observed distributions of $[O_2]$.

Although molecular diffusion was first taken into consideration in the process of diffusion, this is not a simple case at all. Since the height range where dissociation of O_2 takes place overlaps with that of atmospheric mixing due to turbulence (see Chap. II), eddy diffusion must also be taken into account. On the other hand, the vertical distribution of $[O_2]$ has been made fairly clear by rocket observations, either using a mass spectrometer or measuring the attenuation of solar ultraviolet radiation due to absorption by O_2. The distribution of $[O]$ is also being revealed gradually, though with less accuracy, so that it has become possible to make a quantitative inference, in particular, on eddy diffusion by comparing the results of theoretical estimates and observations. In this respect the vertical distributions of O_2 and O has an important significance for the disclosure of diffusion problems.

O_2 molecules in the upper atmosphere are mainly dissociated by absorption of solar ultraviolet radiation in the wavelength range 125...175 nm corresponding to the Schumann-Runge continuum. The absorption coefficients of O_2 molecules in this wavelength region have been repeatedly measured for the past few decades and are fairly well known but not to such an extent as is sufficiently satisfactory. In a review paper by HUDSON[6], results of many measurements[7-13] of O_2 absorption cross-sections in the Schumann-Runge continuum have been compared. These are shown in Fig. 8(a)...(d) proving that the experimental results by different workers are in fair agreement with one another. HUDSON concludes that absorption cross sections in the 137...180 nm range are probably known within 5% (excluding systematic errors) but the shape of this continuum is not so well known. In the past there was a constant discrepancy between measured values using photographic and photoelectronic techniques which has partly been ascribed to the different means of correcting for scattered light. However, the photographic measurements by GOLDSTEIN and MASTRUP shown in Fig. 8(c) were made using an improved experimental method so that they are in better agreement with the results of photoelectronic technique also shown in the same figure. However, there still seem to be some systematic errors remaining.

---

[5] Nicolet, M. (1954) in: The earth as a planet. G. Kuiper (ed.), p. 644. Chicago: University of Chicago Press

[6] Hudson, R.D. (1971): Rev. Geophys. Space Phys. *9*, 305

[7] Watanabe, K., Inn, E.C.Y., Zelikoff, M. (1953): J. Chem. Phys. *21*, 1026

[8] Watanabe, K., Zelikoff, M. (1953): J. Opt. Soc. Am. *43*, 753

[9] Metzger, P.H., Cook, G.R. (1964): J. Quant. Spectrosc. Radiat. Transfer *4*, 107

[10] Huffman, R.E., Tanaka, Y., Larrabee, J.C. (1964): Discuss. Faraday Soc. *37*, 159

[11] Blake, A.J., Carver, J.H., Haddad, G.N. (1966): J. Quant. Spectros. Radiat. Transfer *6*, 451

[12] Goldstein, R., Mastrup, F.N. (1966): J. Opt. Soc. Am. *56*, 765

[13] Hudson, R.D., Carter, V.L., Stein, J.A. (1966): J. Geophys. Res. *71*, 2295

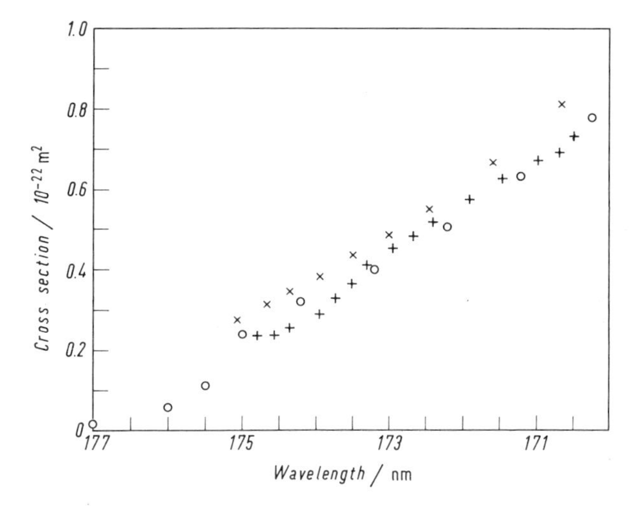

Fig. 8a*. Comparison of measured total absorption cross sections for O_2 between 170 and 177 nm; +, WATANABE et al. (1953); ×, BLAKE et al. (1966); ○, HUDSON et al. (1966)

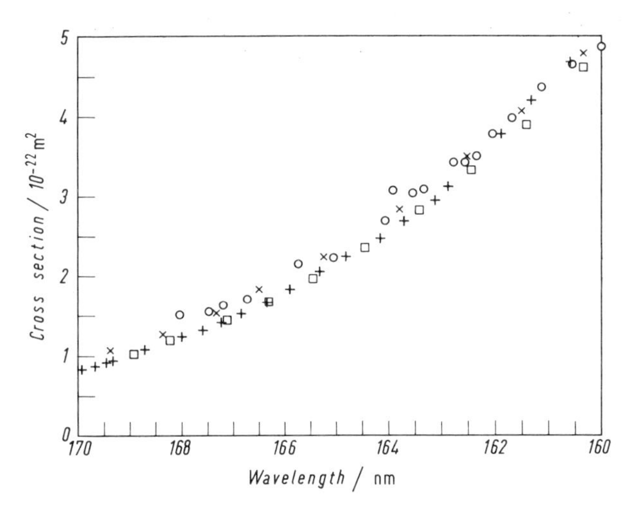

Fig. 8b**. Comparison of measured total absorption cross sections for O_2 between 160 and 170 nm; +, WATANABE et al. (1953); ×, BLAKE et al. (1966); ○, METZGER and COOK (1964); □, HUDSON et al. (1966)

The Schumann-Runge bands exist in the wavelength region 175...195 nm adjacent to the Schumann-Runge continuum on the longer wavelength side. Absorption of radiation in this wavelength range does not cause the dissociation of O_2 in principle, but there is evidence that dissociation occurs to some extent owing to predissociation[6]. Therefore, it is necessary to take into account

* From Hudson (1971): Rev. Geophys. Space Phys. *9* (No. 2), 339, Fig. 14
** From Hudson (1971): Rev. Geophys. Space Phys. *9* (No. 2), 340, Fig. 15

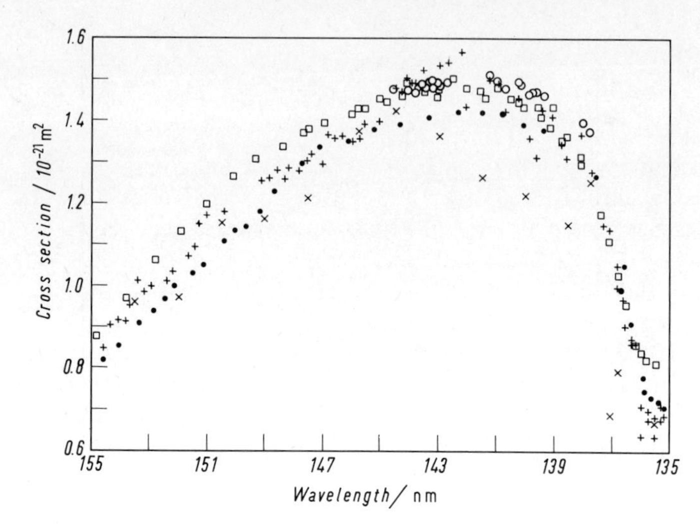

Fig. 8c*. Comparison of measured total absorption cross sections for O_2 between 135 and 155 nm; ·, WATANABE et al. (1953); □, BLAKE et al. (1966); +, METZGER and COOK (1964); ○, HUFFMAN et al. (1964); ×, GOLDSTEIN and MASTRUP (1966)

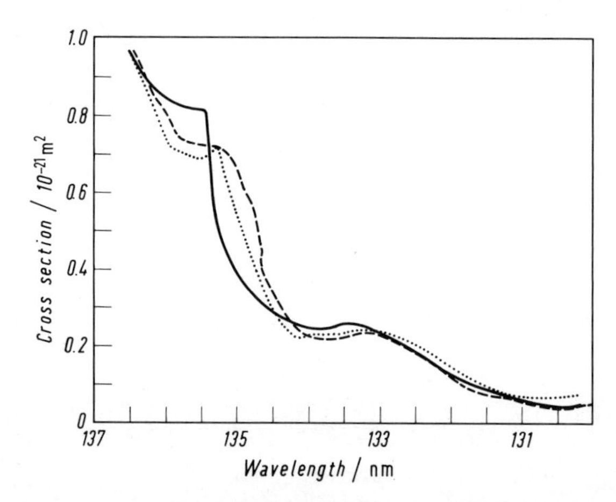

Fig. 8d.** Comparison of measured total absorption cross section data for O_2 between 130 and 137 nm; solid line, BLAKE et al. (1966); dashed line, WATANABE et al. (1953); dotted line, METZGER and COOK (1964)

solar radiation to a wavelength of at least as long as 180 nm in order to give detailed consideration to the problem of O_2-dissociation at levels above about 100 km.

17. Loss processes of atomic oxygen. The most important of the loss processes of atomic oxygen is its recombination through the three-body collision

* From Hudson (1971): Rev. Geophys. Space Phys. 9 (No. 2), 338, Fig. 13
** From Hudson (1971): Rev. Geophys. Space Phys. 9 (No. 2), 341, Fig. 16

among two oxygen atoms and a third body:

$$O + O + M \rightarrow O_2 + M^*, \tag{17.1}$$

where M is the third body and the symbol * indicates that the atom or molecule to which it is attached is in an excited state. Concerning this reaction we have a considerable amount of experimental measurements using discharge-flow systems and shock tubes which have been summarized by SCHOFIELD[1]. The agreement of the values of its rate coefficient k_1 among different measurements is not always good, but it is $28 \cdot 10^{-46}\,m^6\,s^{-1}$ for $M = N_2$ and a temperature of 294 K and $(4.4 \ldots 25) \cdot 10^{-46}\,m^6\,s^{-1}$ for $M = O_2$ and high temperatures of 2,000...3,500 K; in the case of M=O it is $1.5 \cdot 10^{-46}\,m^6\,s^{-1}$ for a temperature of 300 K and $(17 \ldots 38) \cdot 10^{-46}\,m^6\,s^{-1}$ for high temperatures of 2,000...3,500 K. As regards temperature dependence, this is not very clear but above 2,500 K forms of the dependence of $T^{-1/2}$ to T^{-2} have been indicated. KIEFER and LUTZ[2] obtained the following result when the third body is O_2:

$$k_1 = 2.2 \cdot 10^{-41}\,m^6\,s^{-1}(T/K)^{-1.22} \qquad \text{(for 1,500...2,800 K)}.$$

CAMPBELL and GRAY[3] give

$$k_1 = 9.9 \cdot 10^{-46}\,m^6\,s^{-1} \exp(470\,K/T)$$

for reaction (17.1). However, our present knowledge about this reaction is by no means satisfactory.

Atomic oxygen disappears also through the following other reactions:

$$O + O \rightarrow O_2 + h\nu, \tag{17.2}$$

$$O + O_2 + M \rightarrow O_3 + M^*, \tag{17.3}$$

$$O + O_3 \rightarrow O_2 + O_2. \tag{17.4}$$

α) *Radiative recombination.* Among the reactions of type (17.2) most important is the process in which two ground-electronic-state oxygen atoms $O(^3P)$ approach each other along the potential curve leading to the $A\,^3\Sigma_u^+$-state of O_2 with subsequent radiative transition to the ground state $(X\,^3\Sigma_g^-)$ of O_2 when they come sufficiently near to each other. This is the inverse to the Herzberg continuum absorption process:

$$O(^3P) + O(^3P)\,|\,(A\,^3\Sigma_u^+) \rightarrow O_2(X\,^3\Sigma_g^-) + h\nu(\lambda < 242\,\text{nm}). \tag{17.5}$$

Since the Herzberg continuum absorption is due to a forbidden transition, the above reaction is expected to proceed fairly slowly; indeed, according to

[1] Schofield, K. (1967): Planet. Space Sci. *15*, 643
[2] Kiefer, J.H., Lutz, R.W. (1965): J. Chem. Phys. *42*, 1709
[3] Campbell, I.M., Gray, C.N. (1973): Chem. Phys. Lett. *18*, 607

YOUNG and SHARPLESS' measurement[4], its reaction rate coefficient is $3.6 \cdot 10^{-27}\,m^3\,s^{-1}$ at ordinary temperatures; this may be in error by a factor of 2 which is fairly small. We have other reactions belonging to the category of reaction (17.2) such as a reaction like (17.5) in which the state $A\,^3\Sigma_u^+$ is replaced by $c\,^1\Sigma_u^-$ or $C\,^3\Delta_u$, etc., or by replacing $X\,^3\Sigma_g^-$ by $b\,^1\Sigma_g^+$, but these correspond to forbidden transitions also and the probability of their occurrence is small.

β) *Three body and radiative removal.* Of course, for three body removal through reaction (17.1) the rate coefficient is proportional to the number density of the third body and this decreases rapidly with height. On the other hand, reaction (17.2) is a two-body collision process and its reaction rate is independent of height. Therefore a certain level exists above which this latter reaction is more effective in removing atomic oxygen than reaction (17.1). It is readily seen that at this level

$$[M]=k_2/k_1$$

where [M] is the number density of the third body and k_1 and k_2 are the rate coefficients of reactions (17.1) and (17.2). From the foregoing considerations we may take k_1 and k_2 to be equal to $3 \cdot 10^{-45}\,m^6\,s^{-1}$ and $5 \cdot 10^{-27}\,m^3\,s^{-1}$; then [M] given by the above equation becomes $1.7 \cdot 10^{18}\,m^{-3}$, which, according to JACCHIA's 1971 model atmosphere ($T_\infty = 1{,}100$ K), corresponds to a height of 112 km (cf. Fig. 6). Above about this level, reaction (17.2) cannot be neglected compared with reaction (17.1). However, as transport of particles by diffusion becomes increasingly important above the same level, reaction (17.2) is often disregarded for simplicity. For a detailed discussion it may be desirable to take into account its effect. It must, however, be noted that, if reaction (17.5) takes place at its full rate, the Herzberg continuum should be emitted very strongly in the airglow. Taking [O] and the local scale height for O at the 100-km level to be equal to $4.8 \cdot 10^{17}\,m^{-3}$ and 11.0 km (CIRA 1972), respectively, the number of reactions (17.5) which occur above 100 km per unit time and column of unit cross section is calculated to be roughly $6.3 \cdot 10^{12}\,m^{-2}\,s^{-1}$ corresponding to an intensity of the Herzberg continuum of 630 rayleighs. Such strong emission, however, does not seem to have ever been observed in the airglow; hence it is possible that the value of k_2 adopted above is too large.

γ) *Formation of ozone.* Let us consider reaction (17.3) next. According to SCHOFIELD's review[1], there are some fluctuations among the measured values of rate coefficient, but $6 \cdot 10^{-46}\,m^6\,s^{-1}$ at 300 K may be a reasonable value; this is in agreement with the results of later measurements[5,6]. Thus reaction (17.3) may be considered to be slower than reaction (17.1) by a factor of about five so that it needs not to be discussed except at atmospheric levels below about 100 km where $[O_2]$ is greater than [O] by a factor of not less than about 5. Furthermore ozone, O_3, produced by this reaction is not only removed by reaction (17.4) but dissociated into O and O_2 by solar radiation during the

[4] Young, R.A., Sharpless, R.L. (1963): J. Chem. Phys. *39*, 1071

[5] Slanger, T.G., Black, G. (1970): J. Chem. Phys. *53*, 3717

[6] Huie, R.E., Herron, J.T., Davis, D.D. (1972): J. Phys. Chem. *76*, 2653

day. In the latter case reaction (17.3) has almost no effect for the disappearance of O atoms. On the other hand, if O_3 is removed by reaction (17.4), the combined effect of reactions (17.3) and (17.4) is equivalent to that of reaction (17.1), once a balance has been established between the removal of O_3 by reaction (17.4) and its production by reaction (17.3).

δ) *Ozone removal.* Now, let us compare the importance of the O_3 removal by reaction (17.4) and its photodissociation by solar radiation. While the mean lifetime of O_3 against photodissociation[7] is of the order of 100 s, that against reaction (17.4) is $1/(k_4[O])$, where k_4 is its rate coefficient, and this lifetime is variable with height. According to SCHOFIELD[1] there is discrepancy of about one order of magnitude in the measured values of k_4. Let us take a value about in the middle, $1 \cdot 10^{-20}\,m^3\,s^{-1}$ at 300 K, for k_4. (This is in agreement within about 10% with the rate coefficient $1.9 \cdot 10^{-17}\,m^3\,s^{-1}\,\exp(-2{,}300\,K/T)$ given by HAMPSON et al.[8].) Since the number density of oxygen atoms is at most about $5 \cdot 10^{17}\,m^{-3}$ occurring near 100...110 km, the mean lifetime of O_3 against reaction (17.4) is of the order of 200 s or more. At these levels, this reaction is of about the same importance as the photodissociation of O_3; below and above this height range, its importance decreases, but it still remains important during the night.

ε) *Removal of O.* We now consider reactions (17.4), (17.1) and (17.2) in view of their importance in removing atomic oxygen. Taking the same values as adopted above for the rate coefficients k_1, k_2, and k_4, the following relations hold at the critical levels where the reaction rates for (17.1) and (17.4) or (17.2) and (17.4) become equal:

$$[O_3]/[O]\,[M]=k_1/k_4=3 \cdot 10^{-25}\,m^3 \tag{17.6}$$

$$[O_3]/[O]=k_2/k_4=5 \cdot 10^{-7}. \tag{17.7}$$

Therefore, referring to CIRA 1972 model atmosphere, it is seen that, if $[O_3]$ is considerably smaller than $3.5 \cdot 10^{12}$, $1.4 \cdot 10^{12}$, $2.1 \cdot 10^{11}$, and $7.1 \cdot 10^{10}\,m^{-3}$ at 90, 100, 110, and 120 km, respectively, reaction (17.4) can be neglected compared with at least one of the reactions (17.1) and (17.2) and is not so important at these heights. The number density of O_3 is not very well known, but it seems that it is of the order of $3 \cdot 10^{11}\,m^{-3}$ during the day and $1 \cdot 10^{12}\,m^{-3}$ during the night at 100 km and it falls rapidly with increasing height above this level (by more than one order of magnitude per 10 km)[7]. Hence we may be able to consider that reaction (17.4) is not so important as a mechanism of removing atomic oxygen above about 100 km during the day and above about 110 km during the night. Summarizing the above-obtained results, we may say that at levels lower than about 100 km, particularly during the night, the simultaneous occurrence of reactions (17.3) and (17.4) is equivalent to make the apparent

---

[7] Ogawa, T., Shimazaki, T. (1975): J. Geophys. Res. *80*, 3945

[8] Hampson, R.F., Braun, W., Brown, R.L., Garvin, D., Herron, J.T., Huie, R.E., Kurylo, M.J., Laufer, A.H., McKinley, J.D., Okabe, H., Scheer, M.D., Tsang, W., Stedman, D.H. (1973): J. Phys. Chem. Ref. Data *2*, 267

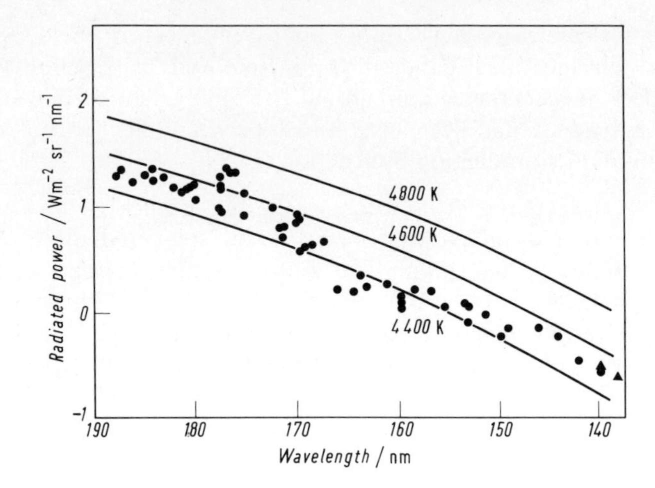

Fig. 9*. Intensity of solar radiation in the wavelength range 140…187.5 nm measured by rocket technique at White Sands, New Mexico at 16 31 UT on 24 September 1968. The ordinate is the power (W) radiated from unit area (m^2) on the Sun into unit solid angle (sr) per unit wavelength (nm), in logarithmic scale. ($1\ Wm^{-2}\,sr^{-1}\,nm^{-1} = 10^{10}\,erg\,cm^{-2}\,s^{-1}\,sr^{-1}\,cm^{-1}$.) Triangles show the results obtained by the satellite OSO 4. The curves represent values for equivalent black body temperatures

rate coefficient of reaction (17.1) augment by a factor of roughly 2 or more and that, up to the heights of 100…110 km, the process (17.4) contributes to some extent to the disappearance of oxygen atoms.

18. Solar radiation. As stated in Sect. 16, the sole agent which contributes to the dissociation of O_2 molecules near the level of 100 km height is the solar radiation in the wavelength range 125…180 nm. The intensity of this solar radiation given by DETWILER et al.[1] has been used frequently by various workers in the past. They gave average values of the intensity over successive 1 nm intervals based on rocket measurements; for the wavelength range of 85…155 nm measurements were performed on 19 April, 1960 using photographic-photometric techniques and similarly for 170…200 nm performed on 13 March, 1959. They found that the intensity at 180 nm corresponds to that of the black body at 4,900 K and, as the wavelength decreases, the equivalent black body temperature gradually falls to about 4,750 K at 150 nm. However, according to PARKINSON and REEVES's rocket observation performed on 24 September, 1968 using a photomultiplier[2], the intensity of solar radiation in this wavelength range was found to be weaker by a factor of about 3, the greater part corresponding to equivalent black body temperatures of 4,400–4,600 K. Figure 9 shows these results together with two values indicated by triangles obtained by spectrometer-spectroheliometer measurements on board

* From Parkinson, Reeves (1969): Solar Physics *10* (No. 2), 346, Fig. 2

[1] Detwiler, C.R., Garrett, D.L., Purcell, J.D., Tousey, R. (1961): Ann. Geophys. *17*, 263

[2] Parkinson, W.H., Reeves, E.M. (1969): Sol. Phys. *10*, 342

the satellite OSO 4. The results are in good agreement with those obtained by photomultiplier techniques. The ordinate of Fig. 9 shows the energy received from the sun. The difference between the two rocket measurements[1,2] seems to be too large to be attributed to variations in solar activity. HINTEREGGER[3], noting that the experiment in 1968 was conducted using photoelectric spectrophotometry while the older experiments employed photographic photometry technique, stated that the newer data are probably more reliable[4].

19. Fundamental equations governing the steady-state vertical distributions of oxygen molecules and atoms. As stated in Sect. 16, the problem of the vertical distribution of oxygen molecules and atoms has been discussed by a large number of workers for many years. However, COLEGROVE et al.[1,2] in 1965 and 1966 only considered this problem in a more satisfactory way taking into account not only molecular diffusion but also the effect of mixing due to eddy diffusion. As a result of this study, they arrived at the conclusion that eddy diffusion takes place to some extent even at levels above the turbopause; it is interesting to note that this is in agreement with the conclusion drawn from chemical release experiments described in Sect. 13. In the following we shall deal with the problem of the vertical distribution of O_2 and O after the method of these authors.

α) *Simplifications.* We assume a ternary gas mixture composed of N_2, O_2, and O with only one major constituent which is N_2 whose distribution is given and remains unchanged while O_2 and O are taken as minor constituents. These latter may move about in the N_2 atmosphere without causing change in total pressure and temperature. Subscripts 1, 2, and 3 are used to indicate quantities referring to O, O_2, and N_2, respectively. If the effect of thermal diffusion is disregarded, the following equations are derived from Eqs. (3.2) and (3.3) with the aid of Eqs. (5.7), (9.24), and (11.5):

$$\sum_{j=1}^{3} n_i n_j(\bar{V}_j - \bar{V}_i)/nD_{ij}^{(1)} = \frac{d}{dz} n_i + \left(\frac{1}{T}\frac{dT}{dz} + \frac{1}{H_i}\right) n_i, \qquad i=1,2,3. \tag{19.1}$$

Since Eq. (3.3) is not independent for all i's, the same is true for Eq. (19.1).

As regards eddy diffusion, if the quantity s in Eq. (8.7) is set equal to n_i/n, the following expression for the z-component of the flux of n_i/n due to turbulence is obtained:

$$\overline{\left(\frac{n_i}{n}\right)' v'} = -K \frac{d}{dz}\left(\frac{n_i}{n}\right), \tag{19.2}$$

where $(n_i/n)'$ is the fluctuation in n_i/n (Subscripts z to be attached to v' and K are omitted because only variations in the z direction are considered here).

---

[3] Hinteregger, H.E. (1970): Ann. Geophys. *26*, 547

[4] See NIKOL'SKIJ's contribution in this volume, p. 309

[1] Colegrove, F.D., Hanson, W.B., Johnson, F.S. (1965): J. Geophys. Res. *70*, 4931

[2] Colegrove, F.D., Johnson, F.S., Hanson, W.B. (1966): J. Geophys. Res. *71*, 2227

Considering that

$$\left(\frac{n_i}{n}\right)' = \frac{n_i'}{n} - \frac{n_i}{n^2} n' \tag{19.3}$$

and that on average $n_i'/n' = O((n_i/n)^{1/2})$, the second term on the right-hand side of Eq. (19.3) can be neglected for minor constituents O and O_2 ($n_1, n_2 \ll n$) and we have for the flux of n_i due to turbulence, $f_i (= n_i' v_i')$ $(i=1,2)$, the following expression:

$$f_i = -K n \frac{\mathrm{d}}{\mathrm{d}z}\left(\frac{n_i}{n}\right) \qquad (i=1,2) \tag{19.4}$$

$$= -K\left[\frac{\mathrm{d}n_i}{\mathrm{d}z} + \left(\frac{1}{\bar{H}} + \frac{1}{T}\frac{\mathrm{d}T}{\mathrm{d}z}\right) n_i\right], \tag{19.5}$$

where $\bar{H}$ is the mean scale height given by Eq. (11.11). In deriving Eq. (19.5) the following equation was used:

$$\frac{1}{n}\frac{\mathrm{d}n}{\mathrm{d}z} = -\frac{1}{T}\frac{\mathrm{d}T}{\mathrm{d}z} - \frac{1}{\bar{H}} \tag{19.6}$$

which can be deduced by differentiating Eq. (11.10).

β) If the flux of the molecules or atoms of species i due to both *molecular and eddy diffusion* is denoted by ϕ_i, we have from Eqs. (19.1) and (19.4)

$$\begin{aligned} \phi_i &\equiv n_i \bar{V}_i + f_i \qquad (i=1,2) \\ &= -D_i\left\{\frac{\mathrm{d}n_i}{\mathrm{d}z} + \left(\frac{1}{H_i} + \frac{1}{T}\frac{\mathrm{d}T}{\mathrm{d}z}\right) n_i - \frac{n_i}{n}\sum_{j=1}^{2}\frac{1}{D_{ij}^{(1)}}\left[\phi_j + K n \frac{\mathrm{d}}{\mathrm{d}z}\left(\frac{n_j}{n}\right)\right]\right\} \\ &\quad - K n \frac{\mathrm{d}}{\mathrm{d}z}\left(\frac{n_i}{n}\right), \end{aligned} \tag{19.7}$$

where

$$1/D_i \equiv \sum_{j=1}^{3}\left(\frac{n_j}{n D_{ij}^{(1)}}\right) \tag{19.8}$$

and we have used $V_3 = 0$ resulting from the assumption made at the beginning of this section. In the right-hand side of Eq. (19.7), the second term in the square brackets can be neglected compared with the last term (as it is multiplied by the small quantity n_i/n). So we get the following two equations by writing down Eq. (19.7) for $i=1$ and 2 and using Eq. (19.5):

$$\begin{aligned} &\left(1 - \frac{n_1}{n}\frac{D_1}{D_{11}^{(1)}}\right)\phi_1 - \frac{n_1 D_1}{n D_{12}^{(1)}}\phi_2 \\ &\qquad = -(D_1 + K)\frac{\mathrm{d}n_1}{\mathrm{d}z} - \left[\left(\frac{D_1}{H_1} + \frac{K}{\bar{H}}\right) + (D_1 + K)\frac{1}{T}\frac{\mathrm{d}T}{\mathrm{d}z}\right] n_1, \end{aligned} \tag{19.9}$$

$$\begin{aligned} &-\frac{n_2 D_2}{n D_{21}^{(1)}}\phi_1 + \left(1 - \frac{n_2 D_2}{n D_{22}^{(1)}}\right)\phi_2 \\ &\qquad = -(D_2 + K)\frac{\mathrm{d}n_2}{\mathrm{d}z} - \left[\left(\frac{D_2}{H_2} + \frac{K}{\bar{H}}\right) + (D_2 + K)\frac{1}{T}\frac{\mathrm{d}T}{\mathrm{d}z}\right] n_2. \end{aligned} \tag{19.10}$$

As regards the equation of continuity for O atoms and for O_2 molecules, COLEGROVE et al. in [1] considered that, for most of the calculations, only reaction (17.1) led to the removal of atomic oxygen, though in [2] they also considered reaction (17.3). In the first paper they additionally examined the case where reactions (17.3) and (17.4) are taken into account at the same time and showed that the difference in the final result is a decrease of only 5% in the value obtained for the eddy diffusion coefficient. The *continuity equations* which they employed in their second paper are:

$$d\phi_1/dz = 2Jn_2 - 2k_1 n_1^2 n - k_3 n_1 n_2 n, \tag{19.11}$$

$$d\phi_2/dz = -Jn_2 + k_1 n_1^2 n, \tag{19.12}$$

where J is the photodissociation rate of an oxygen molecule. Eqs. (19.9...12) make the four simultaneous differential equations for the four unknown functions ϕ_1, ϕ_2, n_1, and n_2 and if these are solved under appropriate boundary conditions, the vertical distributions of O_2 and O can be deduced. By comparing the former distribution with the observed results we can determine the most adequate values of the coefficient of eddy diffusion in this height range.

20. Parameter values adopted

From the intensity of solar radiation responsible for the photodissociation of O_2 (Fig. 9) and the corresponding values of the absorption coefficient of O_2 (Fig. 8) the photodissociation rate of O_2 at the top of the atmosphere is found to be of the order of $10^{-6}\,s^{-1}$, i.e. the mean lifetime of O_2 against photodissociation is of the order of 10 days at that level but becomes longer at lower heights. If the values of DETWILER et al.[1] for the intensity of solar radiation are used, this time is reduced by a factor of about 1/3, but is still considerably longer than one day. On the other hand, the mean lifetime of an O atom against its removal by reaction (17.1) is of the order of several tens of days at 90 km and this becomes longer at higher levels. Therefore, the concentrations of O and O_2 do not closely follow the day and night variation in the rate of production of O and undergo no appreciable diurnal variation. The characteristic time for diffusion is given by the square of the local scale height divided by the coefficient of diffusion as explained in Sect. 10. This takes the values of 160, 29, 7, and 3 days at 90, 100, 110, and 120 km for the molecular diffusion of O atoms; the corresponding times are 2, 3, 4, and 8 days for the eddy diffusion of O atoms if K is taken as $5 \cdot 10^2\,m^2\,s^{-1}$. In the case of O_2 molecules the above figures are reduced by a factor of about 1/4. At any rate, diffusion of O and O_2 does not take place so rapidly at these heights. Therefore in the following we shall assume that the number densities of O and O_2 undergo no appreciable diurnal variation and concentrate our attention only on the diurnal average values of all quantities.

In the calculations of COLEGROVE et al.[2] the intensity of solar radiation is taken to be half the actual one as a diurnal average and the solar zenith angle is taken to be 45°. Standard temperatures of the atmosphere have been taken for heights at and below 100 km from the U.S. Standard Atmosphere, 1962 and from F.S. JOHNSON[3] above 100 km. Calculations have also been made where temperatures are higher or lower by 20 K than standard ones or when their rate of increase with height is steeper or gentler than standard ones. The values of the intensity of solar radiation obtained by DETWILER et al.[1] were adopted and for the absorption coefficient of O_2 those of

---

[1] Detwiler, C.R., Garrett, D.L., Purcell, J.D., Tousey, R. (1961): Ann. Geophys. *17*, 263

[2] Colegrove, F.D., Hanson, W.B., Johnson, F.S. (1965): J. Geophys. Res. *70*, 4931; Colegrove, F.D., Johnson, F.S., Hanson, W.B. (1966): J. Geophys. Res. *71*, 2227

[3] Johnson, F.S. (ed.) (1965): Satellite environment handbook, 2nd edn. Stanford (Cal.): Stanford University Press

WATANABE et al.[4,5] and METZGER and COOK[6] were used. Actually the photodissociation rate of O_2 was calculated for each successive interval of 5 nm width in the wavelength range 117.5...177.5 nm and by adding them the photodissociation rate of O_2 was obtained as a function of the column density of O_2. Then the values of J to be used in Eqs. (19.11) and (19.12) can be obtained for each step of integration as the photodissociation rate of O_2 calculated above for the O_2 column density corresponding to that step of integration. As regards the rate coefficients of reactions (17.1) and (17.3), it was assumed in the first paper of [2] that $k_1=2.6\cdot 10^{-46}(T/\mathrm{K})^{\frac{1}{2}}\,\mathrm{m}^6\,\mathrm{s}^{-1}$ and $k_3=8.7\cdot 10^{-47}(T/\mathrm{K})^{\frac{1}{2}}\,\mathrm{m}^6\,\mathrm{s}^{-1}$ and that O_3 molecules produced by reaction (17.3) recombine immediately with O atoms through reaction (17.4) and are transformed into two O_2 molecules. However, in the second paper it was simply assumed that $k_1=4.5\cdot 10^{-45}\,\mathrm{m}^6\,\mathrm{s}^{-1}$ and $k_3=2\cdot 10^{-45}\,\mathrm{m}^6\,\mathrm{s}^{-1}$; reaction (17.4) was not taken into account.

The value of the coefficient of molecular diffusion is not certain if atomic oxygen takes part in the process. However, COLEGROVE et al.[2] adopted the following values:

$$D_{23}^{(1)}=1.81\cdot 10^{-5}(T/T_0)^{1.75}(P_0/P)\,\mathrm{m}^2\,\mathrm{s}^{-1}, \tag{20.1}$$

$$D_{12}^{(1)}=D_{13}^{(1)}=2.6\cdot 10^{-5}(T/T_0)^{1.75}(P_0/P)\,\mathrm{m}^2\,\mathrm{s}^{-1}, \tag{20.2}$$

where T_0 and P_0 are the standard temperature and pressure. Eq. (20.1) should be compared with Eq. (5.15). Eq. (20.2) is a guess based on some experimental results[7] and theoretical inferences[8]. The eddy diffusion coefficient K is discussed in the next section.

21. The solution of the fundamental equations and the vertical distributions of O_2 and O

α) *Solution method.* In their first paper COLEGROVE et al.[1] took the atmospheric temperatures and the distribution of N_2 as fixed and gave them standard values; the relation between the adopted value of K and the distribution of O_2 and O was then examined. In this treatment the third term on the right-hand side in Eq. (19.11) is neglected so that the relation $\phi_1=-2\phi_2$ is obtained from Eqs. (19.11, 12). If the starting values of n_1, n_2, ϕ_1, and ϕ_2 at 120 km and the value of K (independent of height) are taken and integration is performed in both the upward and downward directions, the above starting values and the value of K are found to be subject to stringent limitation in order that n_1 and ϕ_2 remain positive, that O_2 occupies about 1/5 of the atmosphere at sufficiently low heights, and that n_1, n_2, ϕ_1, and ϕ_2 take reasonable values at sufficiently high levels. Consistent distributions and fluxes of O_2 and O and the corresponding value of K are obtained by "trial and error". In particular, it is interesting to note that the product of the O to O_2 ratio at 120 km by K takes a constant value of $4\cdot 10^2\,\mathrm{m}^2\,\mathrm{s}^{-1}$. Thus we have values of K of 8, 4, 2, and $0.8\cdot 10^2\,\mathrm{m}^2\,\mathrm{s}^{-1}$ for values of the density ratio $n(\mathrm{O})/n(\mathrm{O}_2)$ of 0.5, 1, 2, and 5, respectively. This is a reasonable range of values, such that a value of K of about $4\cdot 10^2\,\mathrm{m}^2\,\mathrm{s}^{-1}$ seems to be most probable. It was examined to what extent the value of K is changed if the intensity of solar radiation, the rate coefficient of reaction (17.1) or the coefficient of diffusion in Eq. (20.2) are

[4] Watanabe, K., Inn, E.C.Y., Zelikoff, M. (1953): J. Chem. Phys. *21*, 1026

[5] Watanabe, K., Zelikoff, M. (1953): J. Opt. Soc. Am. *43*, 753

[6] Metzger, P.H., Cook, G.R. (1964): J. Quant. Spectrosc. Radiat. Transfer *4*, 107

[7] Walker, R.E. (1961): J. Chem. Phys. *34*, 2196; Morgan, J.E., Schiff, H.I. (1964): Can. J. Chem. *42*, 2300

[8] Yun, K.S., Weissman, S., Mason, E.A. (1962): Phys. Fluids *5*, 672

[1] Colegrove, F.D., Hanson, W.B., Johnson, F.S. (1965): J. Geophys. Res. *70*, 4931

increased or decreased by a factor of 1.5...2, or if reactions (17.3) and (17.4) are taken into consideration. It was found that such changes remain within a few tens of percent, except for the effect of solar radiation, variation of which may bring about a decrease of about 40% or an increase of about 60% in K if the intensity is multiplied by 2/3 or 1.5, respectively. Since, as stated in Sect. 18, there is a possibility that the intensity of solar radiation is considerably weaker than COLEGROVE et al.[1] assumed, the above-mentioned point may deserve further examination.[2] These authors considered also the case where K varies with height and found that this does not affect the distribution of $[O_2]$ to any great extent but has a considerable influence on that of $[O]$.

The values of K obtained above are about one order of magnitude smaller than those inferred from rocket vapor trail experiments described in Sects. 13 and 14. This discrepancy is probably explicable by the fact that the values from vapor trail experiments represent the coefficient of diffusion in the horizontal direction while the values obtained above represent the coefficient of diffusion in the vertical direction, and that temperature increases with height in the height region concerned here; this tends to prevent atmospheric mixing in the vertical direction, as is well known in meteorology.

In the second paper of COLEGROVE et al.[3] the atmospheric composition including the concentrations of Ar, He, and H was calculated for seven cases, in some of which the temperature distribution differs from the standard one. The distribution of He and H is discussed in Chap. V.

β) In Fig. 10 the vertical distribution of O_2 calculated by COLEGROVE et al.[3] is compared with the *results of rocket observations* made by JURSA et al.[4] using the technique of photographic photometry of absorption spectrograms and by SMITH and WEEKS[5] analyzing Lyman-α absorption profiles. Day, local hour, place of observation, and the solar zenith angle are indicated in the Fig. 10. $[O_2]$-distributions from JACCHIA's 1971 model atmosphere ($T_\infty = 1{,}100$ K) which is identical with the 'COSPAR International Reference Atmosphere' CIRA 1972, as well as the elder CIRA 1965 are also depicted in the figure. The calculated results of COLEGROVE et al.[3] are in good agreement with the observations and CIRA 1965, the agreement with the CIRA 1965 being almost complete in the 90 to 115 km range. CIRA 1972 deviates somewhat from these but not to any great degree.

A similar comparison of atomic oxygen is made in Fig. 11. The precision of rocket observation of (O) is probably considerably lower than that of (O_2). We show results of observations with silver film sensor[6], from the intensity of the 557.7 nm airglow emission (green line of O)[7] and with mass spectrometers[8,9]. Further, results obtained from measurements of hydroxyl airglow[10]

[2] See NIKOL'SKIJ's contribution in this volume, p. 309

[3] Colegrove, F.D., Johnson, F.S., Hanson, W.B. (1966): J. Geophys. Res. *71*, 2227

[4] Jursa, A.S., Nakamura, M., Tanaka, Y. (1965): J. Geophys. Res. *70*, 2699

[5] Smith, L.G., Weeks, L. (1966): NASA Contract. Rep. CR-392

[6] Henderson, W.R. (1971): J. Geophys. Res. *76*, 3166

[7] Dandekar, B.S., Turtle, J.P. (1971): Planet. Space Sci. *19*, 949

[8] Hedin, A.E., Avery, C.P., Tschetter, C.D. (1964): J. Geophys. Res. *69*, 4637

[9] Schaefer, E.J., Nichols, M.H. (1964): J. Geophys. Res. *69*, 4649

[10] Good, R.E. (1976): Planet. Space Sci. *24*, 389

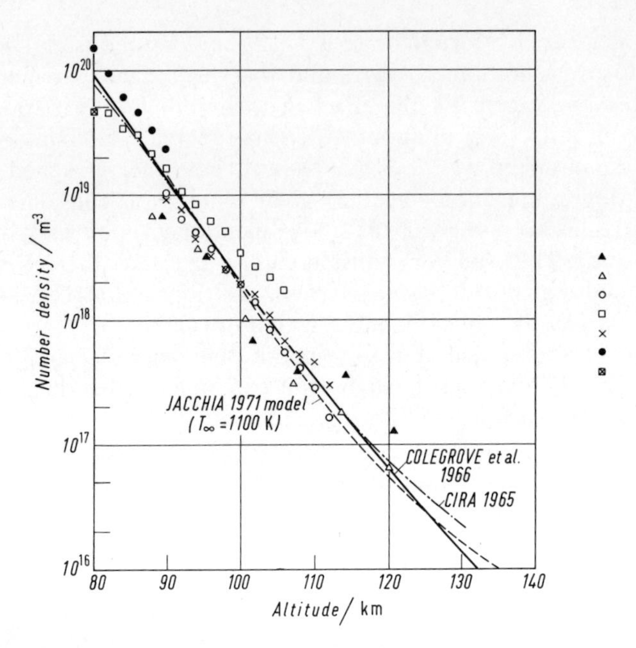

Fig. 10. Comparison of calculated and observed number densities of O_2. Those tabulated in JACCHIA's 1971 ($T_\infty = 1{,}100$ K) and CIRA 1965 Model Atmospheres are also indicated

△ 19 Mar. 1963 $\chi = 37°$, ▲ 26 Sep. 1963 $\chi = 39°$ — White Sands[4]
○ 19 Nov. 1964 $\chi = 95°$, □ 15 Jul. 1964 $\chi = 84°$, × 15 Jul. 1964 $\chi = 95°$ — Wallops Is.[5]
● 20 Jul. 1963 $\chi = 48\text{–}56°$, ⊠ 27 Feb. 1963 $\chi = 56°$ — Fort Churchill[5]
(χ = Solar zenith angle)

and of the OI green line nightglow emission[11], respectively, and by TRINKS et al.[12] using a rocket-borne mass spectrometer are included in the figure. The times and places of observation are indicated in the figure legend. As for theoretical results, apart from those of COLEGROVE et al.[3], we show for noon conditions the more recent results of THOMAS and BOWMAN[13], and, for comparison, the distribution of [O] according to CIRA 1972 (JACCHIA).

As can be seen, the curve of COLEGROVE et al.[3] has a very broad peak in the height range 90...100 km and its shape does not agree with the observed results. Also the calculated peak value is considerably lower than observed values; considering that the intensity of solar radiation has probably been assumed too strong, the true discrepancy is even larger than apparent.

THOMAS and BOWMAN[13] use more recent values of the parameters, in particular, a smaller intensity of solar radiation based on the measurements of PARKINSON and REEVES[14]; consideration is given to predissociation in the

---

[11] Wasser, B., Donahue, T.M. (1979): J. Geophys. Res. *84*, 1297
[12] Trinks, H., Offermann, D., von Zahn, U., Steinhauer, C. (1978): J. Geophys. Res. *83*, 2169
[13] Thomas, L., Bowman, M.R. (1972): J. Atmos. Terr. Phys. *34*, 1843
[14] Parkinson, W.H., Reeves, E.M. (1969): Sol. Phys. *10*, 342

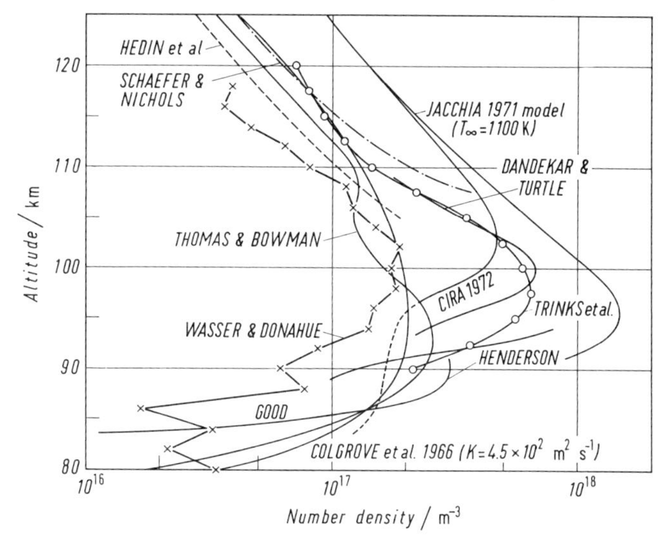

Fig. 11. Comparison of calculated and observed number densities of O. Those tabulated in JACCHIA's 1971 Model Atmosphere. ($T_\infty=1{,}100$ K) and in CIRA 1972 are also indicated. Times and places of observation are the following:

HENDERSON:	1141 EST, 14 Oct. 1970, Wallops Is., Virginia
DANDEKAR and TURTLE:	0300 HST, 22 May 1969, Kauai, Hawaii
HEDIN et al.:	0730 MST, 6 June 1963, White Sands, New Mexico
SCHAEFER and NICHOLS:	1302 EST, 18 May 1962, Wallops Is., Virginia
GOOD:	1104 UT, 16 Apr. 1973, White Sands, New Mexico
WASSER and DONAHUE:	2120 LT, 25 Nov. 1969, OGO 6, 48.8 °N
TRINKS et al.:	1514 LST, 29 June 1974, Wallops Is., Virginia

Schumann-Runge bands which may augment the dissociation of O_2; a variation of the coefficient of eddy diffusion with height is adopted which, above about 100 km, is assumed to tend to zero rapidly. However, the calculated results show only a slight improvement in the near peak shape of the O-distribution so that considerable disagreement between calculated and observed results still remains. It may be conceivable that the effect of molecular and eddy diffusion are of comparable order of magnitude in the neighborhood of the peak and both act to transport O atoms in the downward direction there, making the shape of the peak broader. On the other hand, CIRA 1972 seems to represent the characteristics of the shape of O-distribution reasonably well, but not the absolute values. We conclude that, although calculations according to the method explained above are of theoretical interest, they could not yet give full satisfaction[15].

22. Other theoretical studies on the distribution of O_2 and O. A number of workers such as SHIMAZAKI[1], SHIMAZAKI and LAIRD[2], HESSTVEDT[3], and KE-

[15] See also THOMAS' contribution to this volume, p. 7

[1] Shimazaki, T. (1967): J. Atmos. Terr. Phys. *29*, 723

[2] Shimazaki, T., Laird, A.R. (1970): J. Geophys. Res. *75*, 3221

[3] Hesstvedt, E. (1968): Geofys. Publ. *27*, *4*, 1

NESHEA and ZIMMERMAN[4] have dealt in detail with the problem of the vertical distribution of atomic and molecular oxygen. Except for HESSTVEDT's[3], these mark an improvement insofar as an attempt is made to assess the temporal variation in the vertical distribution of $[O_2]$ and $[O]$ by preserving the terms containing time derivatives in the fundamental equations and integrating them with respect to time and height.

Unfortunately, SHIMAZAKI[1] adopted an inappropriate boundary condition that at the lower boundary the vertical flux of O_2 or O should have an extremum value (usually maximum or minimum)[5]. He took the 70 km level as the lower boundary, but according to the calculations of COLEGROVE et al.[6], the vertical flux of O atoms reaches a maximum at a height of about 90 km and it will probably be far from being maximum or minimum at 70 km. Hence SHIMAZAKI's results calculated under the boundary condition that the flux takes a maximum or minimum value at 70 km cannot be regarded as representing physical reality. In SHIMAZAKI and LAIRD's paper[2] a step forward has been made in taking more numerous chemical reactions into account than earlier studies and consideration has also been given to various minor constituents, but the above-mentioned defect of the method of calculation regarding the lower boundary condition (at 40 km now) still remains. However, for some minor constituents including O_3 another lower boundary condition is imposed so that the calculated results are free from this defect. Thus, as far as the distribution of O_2 and O is concerned, it is not clear to what extent the results of [1,2] represent the actual conditions of the upper atmosphere.

In KENESHEA and ZIMMERMAN's calculations[4] it is assumed that eddy diffusion occurs only at levels below a certain height and in some cases this assumption seems to be favorable for the interpretation of observed facts. However, as they adopt SHIMAZAKI's boundary condition[1], their study is subject to the same weakness as SHIMAZAKI's. Indeed we can see from their curves showing the vertical flux of oxygen atoms that these reach extremum values at the lower boundary of 70 km height.

A detailed calculation has been made in HESSTVEDT's study[3] of the concentrations of constituents from the mesosphere to the lower part of the thermosphere, i.e. between 45 and 115 km, for an oxygen-hydrogen atmosphere. However importance is ascribed to rather lower heights and the author does not intend to make a detailed analysis of composition in the height range 100...115 km; for the calculation of the densities of O_2 and O, use is made of the observed value of the $[O]$ to $[O_2]$ ratio.

It appears from the foregoing discussion that on the problem of the vertical redistribution of O_2 and O accompanying the photodissociation of O_2 by solar radiation a clear understanding has not yet been reached even though it is an old problem. For a better understanding it is desirable that our knowledge concerning the intensity of solar radiation, the eddy diffusion coefficient as a function of height, and molecular diffusion processes involving O atoms be-

[4] Keneshea, T.J., Zimmerman, S.P. (1970): J. Atmos. Sci. *27*, 831
[5] Shimazaki, T. (1972): Radio Sci. *7*, 695
[6] Colegrove, F.D., Hanson, W.B., Johnson, F.S. (1965): J. Geophys. Res. *70*, 4931

comes more abundant. This may be particularly important to remove the apparent disagreement between theory and observation with regard to the shape of the distribution of O near its peak.

IV. Ambipolar diffusion including the formation of the F2-layer

23. Mechanism of formation of the F2-layer. Among the ionized layers existing in the ionosphere the formation of the F2-layer is intimately associated with the diffusion of an ionized gas in the Earth's gravitational field. If diffusion did not take place at all, an ionized layer would not be formed at heights where the F2-layer is actually observed. It is well known that the F2-layer is much more liable to geomagnetic control than other ionized layers; this fact can be interpreted appropriately only if it is considered that vertical transport takes part in the mechanism of its formation.

Elias[1] and Chapman[2-4] were the first to theoretically consider the ionization of the Earth's upper atmosphere, on the assumption of an isothermal atmosphere.

α) Nicolet[5,6] extended the results to the case where there is a *constant temperature* (or scale height) *gradient* in the atmosphere (see[7] for other assumptions). At the height of the F2-layer the atmosphere is mainly composed of atomic oxygen. We consider as a first approximation that only atomic oxygen is ionized and denote its scale height by H and the height gradient of H by Γ. Then, as in Eq. (11.7) we have

$$H = H_0 + \Gamma(z - z_0) \tag{23.1}$$

and

$$\Gamma = \frac{\mathrm{d}H}{\mathrm{d}z}. \tag{23.2}$$

By Eq. (11.8) the number density of O atoms, $n(\mathrm{O})$, is given by

$$n(\mathrm{O}) = n_0(\mathrm{O}) \cdot (H_0/H)^{1+1/\Gamma}. \tag{23.3}$$

The expression for the rate of production of electrons deduced by Nicolet[5] if applied to the present case of ionization of O atoms gives

$$Q = Q_{\mathrm{m}}\, \mathrm{e}^{1+\Gamma} (H_{\mathrm{m}}/H)^{1+1/\Gamma} \exp\left[-(1+\Gamma)(H_{\mathrm{m}}/H)^{1/\Gamma} \sec\chi\right]. \tag{23.4}$$

[1] Elias, G.J. (1923): Tiydschr. Ned. Radio Gen. *2*, 1
[2] Chapman, S. (1931): Proc. Phys. Soc. Lond. *43*, 26, 483
[3] Chapman, S. (1939): Proc. Phys. Soc. London *51*, 93
[4] Chapman, S. (1953): Proc. Phys. Soc. London Sect. B*66*, 710
[5] Nicolet, M. (1951): J. Atmos. Terr. Phys. *1*, 141
[6] Nicolet, M., Bossy, L. (1949): Ann. Geophys. *5*, 275
[7] Argence, E., Mayot, M., Rawer, K. (1950): Ann. Geophys. *6*, 242

where Q is the rate of electron production due to ionization of O atoms by solar radiation, χ the solar zenith angle, Q_m the peak value of Q as a function of height when χ is equal to zero, and H_m is the value of H at the level where Q becomes equal to Q_m for χ equal to zero.

β) *Dissociative recombination.* The disappearance of electrons and O^+-ions in the F2-region is effected mainly through the following two-stage reactions[8]:

$$O^+ + N_2 \rightarrow NO^+ + N, \tag{23.5}$$

$$NO^+ + e^- \rightarrow N + O; \tag{23.6}$$

or

$$O^+ + O_2 \rightarrow O_2^+ + O, \tag{23.7}$$

$$O_2^+ + e^- \rightarrow O + O. \tag{23.8}$$

Reactions (23.5) and (23.7) are ion-atom interchange reactions while reactions (23.6) and (23.8) describe the dissociative recombination between electrons and molecular ions. Of these two kinds of reactions, the latter proceeds more quickly than the former in the F2-region so that the rates of the above two-stage reactions as a whole are controlled by the slower rates of reactions (23.5) and (23.7). If these rate coefficients are denoted by k_5 and k_7 the rate of removal of electrons L is given by

$$L = (k_5\, n(N_2) + k_7\, n(O_2)) \cdot n(O^+) \tag{23.9}$$

or

$$L = (k_5\, n(N_2) + k_7\, n(O_2)) \cdot N_e. \tag{23.10}$$

If Eqs. (11.5, 7, 8) are applied, the variations with height of $n(N_2)$ and $n(O_2)$ are seen to be given by

$$n(N_2) = n_0(N_2) \cdot (H_0/H)^{1+7/4\Gamma}, \tag{23.11}$$

$$n(O_2) = n_0(O_2) \cdot (H_0/H)^{1+2/\Gamma}. \tag{23.12}$$

γ) *In the steady state* in which electron production by Eq. (23.4) and its disappearance by Eqs. (23.9) or (23.10) balance with each other the electron density is given by the following expression

$$N_e = \frac{Q_m\, e^{1+\Gamma} (H_m/H)^{1+1/\Gamma} \exp[-(1+\Gamma)(H_m/H)^{1/\Gamma} \sec\chi]}{k_5\, n_0(N_2)(H_0/H)^{1+7/4\Gamma} + k_7\, n_0(O_2)(H_0/H)^{1+2/\Gamma}}. \tag{23.13}$$

In this expression the exponential function in the numerator on the right-hand side monotonously increases and tends to 1 for $H \rightarrow \infty$ and the order of magnitude of electron density is given by

$$N_e = O(H^{3/4\Gamma}). \tag{23.14}$$

[8] See the contribution of THOMAS to this volume, his Sect. 12, p. 44

Therefore electron density increases unboundedly with increasing height ($H \to \infty$, cf. Eq. (23.1)) and never reaches a maximum; this means that an ionized layer cannot appear in the steady state and it becomes impossible to interpret the stratification of electrons and ions in the F2-region.

δ) However, such an increase in electron density with height is not observed but electrons and ions are redistributed by ambipolar diffusion and finally reach a stratified distribution with a maximum at a certain height. As ambipolar diffusion is influenced by the Earth's magnetic field, the electron density distribution is also subject to its effect and this explains the geomagnetic control of the F2-layer. In the next two sections we shall obtain a more realistic shape for the electron density distribution, taking into account the ambipolar diffusion of electrons and ions.

24. Electron density distribution in the daytime. Taking account of *vertical transport by diffusion* the equation of continuity of electrons in the F2-region is given by Eq. (9.29a) or by

$$\frac{\partial N_e}{\partial t} = Q - L + \frac{\partial}{\partial z}\left\{\sin^2 I \cdot D_a \left[\frac{\partial N_e}{\partial z} + (1+\alpha_{aT})\frac{N_e}{T_p}\frac{\partial}{\partial z}T_p + \frac{N_e}{H_p}\right]\right\}. \tag{24.1}$$

α) *Simplifications.* If the same assumptions as in the previous section are adopted, Q is given by Eq. (23.4) and L by Eq. (23.10). For simplicity, however, we assume further that L is approximately proportional to the density of O_2 or N_2 and then from Eqs. (23.11) and (23.12) we get

$$L = B N_e, \tag{24.2}$$

$$B = B_0 (H_0/H)^{1+\delta/\Gamma}, \tag{24.3}$$

where δ is taken as 2 or 7/4 according to whether B is proportional to $n(O_2)$ or to $n(N_2)$. As regards the temperature dependence of D_a, Banks and Holzer[1] showed that when O^+ ions diffuse through an atomic oxygen gas, charge exchange collisions between O^+ ions and O atoms rather than their elastic collisions play an important role and the coefficient of ambipolar diffusion, D_a, depends on temperature in the following way:

$$D_a \propto T_i/(T_i + T_n)^{\frac{1}{2}}, \tag{24.4}$$

where T_n is the temperature of the atomic oxygen gas. In normal conditions T_n and T_i are not very different. If these are not distinguished, D_a is approximately proportional to $H^{\frac{1}{2}}/[O]$ and we get

$$D_a = D_{a0} (H/H_0)^{3/2 + 1/\Gamma}. \tag{24.5}$$

[1] Banks, P.M., Holzer, T.E. (1968): Planet. Space Sci. *16*, 1019

Lastly the ratio of plasma scale height to the neutral scale height for atomic oxygen is denoted by κ and this is assumed to take a constant value independent of height:

$$\kappa = H_p/H. \tag{24.6}$$

We shall also disregard the variation of gravity with height.

β) *New variables.* On these assumptions the differential equation (24.1) for a steady state ($\partial N/\partial t = 0$) can be transformed by changing the independent variable from z to H according to Eq. (23.1) and using Eqs. (23.4), (24.3, 5):

$$\begin{aligned} -Q_m e^{1+\Gamma}(H_m/H)^{1+1/\Gamma} \exp[-(1+\Gamma)(H_m/H)^{1/\Gamma} \sec\chi] \\ = -B_0 (H_0/H)^{1+\delta/\Gamma} N_e + \Gamma \frac{d}{dH}\left\{\sin^2 I\, D_{a0}(H/H_0)^{3/2+1/\Gamma}\right. \\ \left.\cdot\left[\Gamma \frac{dN_e}{dH} + \frac{1}{\kappa H}(1+\kappa\Gamma(1+\alpha_{aT}))N_e\right]\right\}. \end{aligned} \tag{24.7}$$

In order to simplify the expression we set

$$S \equiv Q_m e^{1+\Gamma}(H_m/H_0)^{1+1/\Gamma}, \tag{24.8}$$

$$R \equiv (1+\Gamma)\sec\chi (H_m/H_0)^{1/\Gamma}, \tag{24.9}$$

and introduce new variables[0]

$$y = (H_0/H)^p, \tag{24.10}$$

$$p = 1 + \frac{2}{3}\alpha_{aT} + \frac{2}{3\Gamma}\left(1+\frac{1}{\kappa}\right). \tag{24.11}$$

Then Eq. (24.7) is reduced to a differential equation of the following form:

$$\frac{d^2 N_e}{dy^2} - \frac{1}{2y}\frac{dN_e}{dy} + \left(\frac{\mu^2}{y^2} - \frac{B_0}{a^2} y^\nu\right) N_e = -\frac{S}{a^2} y^{\nu+\lambda(1-\delta)} \exp(-R y^\lambda), \tag{24.12}$$

where

$$\lambda \equiv \frac{1}{\Gamma p} = \frac{1}{\left(1+\frac{2}{3}\alpha_{aT}\right)\Gamma + \frac{2}{3}\left(1+\frac{1}{\kappa}\right)}, \tag{24.13}$$

$$\mu^2 \equiv \lambda^2 \left(1 + \frac{1}{2}\Gamma\right)\left[\frac{1}{\kappa} + \Gamma(1+\alpha_{aT})\right], \tag{24.14}$$

$$\nu \equiv -2 + \frac{1}{2p} + \lambda(1+\delta), \tag{24.15}$$

$$a^2 \equiv \frac{D_{a0}\sin^2 I}{\lambda^2 H_0^2}. \tag{24.16}$$

---

[0] In Sects. 24 and 25 a, A, c, d, p, α, β, γ, λ, μ, ν are used independently from their significations in other sections

γ) In order to find the *general solution* of the differential Eq. (24.12) we state that the corresponding homogeneous differential equation has two independent solutions, namely

$$y^{\frac{3}{4}}\,\mathrm{I}_{-\alpha}(A\,y^{\beta}) \quad\text{and}\quad y^{\frac{3}{4}}\,K_{\alpha}(A\,y^{\beta}),$$

where $\mathrm{I}_{-\alpha}(x)$ and $\mathrm{K}_{\alpha}(x)$ are Bessel functions with imaginary arguments[2,3] and

$$\alpha \equiv \left[\left(\frac{3}{4}-\mu\right)\left(\frac{3}{4}+\mu\right)\right]^{\frac{1}{2}} \Big/ \left(1+\frac{\nu}{2}\right), \tag{24.17}$$

$$\beta \equiv 1+\frac{\nu}{2}, \tag{24.18}$$

and

$$A \equiv B_0^{\frac{1}{2}}/(a\,\beta) = \left(\frac{B_0 H_0^2}{D_{a0}\sin^2 I}\right)^{\frac{1}{2}} \frac{\lambda}{\beta}. \tag{24.19}$$

Using the method of variation of constants the general solution of Eq. (24.12) can be expressed in the following form[4]:

$$N_e = \frac{S}{\beta a^2}\, y^{\frac{3}{4}} \left[\mathrm{K}_{\alpha}(A\,y^{\beta}) \int_0^y \exp(-R\,y^{\lambda})\cdot y^{\gamma}\,\mathrm{I}_{-\alpha}(A\,y^{\beta})\,\mathrm{d}y + \mathrm{I}_{-\alpha}(A\,y^{\beta}) \int_y^{\infty} \exp(-R\,y^{\lambda})\cdot y^{\gamma}\,\mathrm{K}_{\alpha}(A\,y^{\beta})\,\mathrm{d}y\right] + y^{\frac{3}{4}}[c\,\mathrm{K}_{\alpha}(A\,y^{\beta}) + d\,\mathrm{I}_{-\alpha}(A\,y^{\beta})], \tag{24.20}$$

where c and d are integration constants and

$$\gamma = \tfrac{1}{4} + \nu + (1-\delta)\,\lambda. \tag{24.21}$$

In order to determine the integration constant d let us consider the behavior of the above solution at sufficiently low heights, i.e. for $y \to \infty$. It can be shown that the solution (24.20) excluding the last group of terms containing c and d tends asymptotically to Q/B, i.e. the chemical equilibrium distribution of electron density[4]. This is the condition which a correct electron density distribution must satisfy because the effect of diffusion is negligible at sufficiently low heights. Since both β and A are positive as will be seen from Eqs. (24.15, 18, 19), $y^{3/4}\,\mathrm{K}_{\alpha}(A\,y^{\beta}) \to 0$ and $y^{3/4}\,\mathrm{I}_{-\alpha}(A\,y^{\beta}) \to \infty$ for $y \to \infty$.

Hence, N_e given by Eq. (24.20) does not satisfy the above-mentioned condition unless d vanishes. In order to determine the integration constant c, let us evaluate the value of the vertical flux of electrons (and ions) at sufficiently high altitudes. Using Eqs. (9.31, 32), (23.2) and (24.6), and neglecting the height change of gravity g, it is seen from the form of Eq. (24.1) that this flux F can be expressed as

$$F = -\sin^2 I\cdot D_a \left\{\frac{\mathrm{d}N_e}{\mathrm{d}z} + \frac{N_e}{H}\left[\frac{1}{K} + (1+\alpha_{aT})\Gamma\right]\right\}, \tag{24.22}$$

or by Eqs. (23.2) and (24.5, 10)

$$F = -\frac{\sin^2 I\cdot D_{a0}}{H_0}\, y^{-\frac{1}{p}\left(\frac{1}{2}+\frac{1}{\Gamma}\right)} \left\{\left[\frac{1}{K} + (1+\alpha_{aT})\Gamma\right] N_e - p\,\Gamma\,y\,\frac{\mathrm{d}N_e}{\mathrm{d}y}\right\}. \tag{24.22a}$$

---

[2] We adopt G.N. Watson's definition of $\mathrm{K}_{\alpha}(x)$, i.e. $\mathrm{K}_{\alpha}(x) = [\pi/(2\sin\alpha\pi)]\,(\mathrm{I}_{-\alpha}(x) - \mathrm{I}_{\alpha}(x))$.

[3] Erdelyi, A., Magnus, W., Oberhettinger, F., Tricomi, F.G. (1953): Higher transcendental functions, vol. II, p. 13. New York: McGraw-Hill

[4] Yonezawa, T., Takahashi, H. (1960): J. Radio Res. Lab. (Jpn.) 7, 335

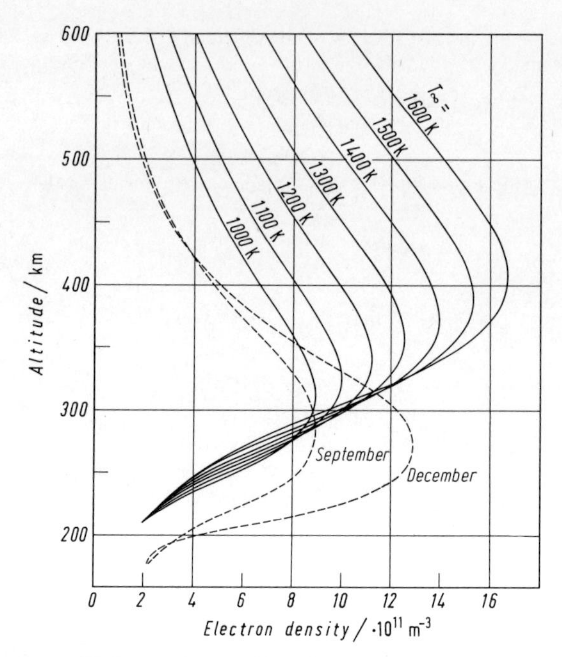

Fig. 12. Electron density profiles in the daytime $F2$-region calculated for JACCHIA's 1971 Model Atmospheres (CIRA 1972) with various exospheric temperatures. Standard noon electron density profiles for the sunspot number 100 at a 45° latitude location in September and December given by RAWER, RAMAKRISHNAN and BILITZA (1975) are also indicated by broken lines

Inserting Eq. (24.20) with $d=0$ into this equation and letting y tend to zero, the vertical flux of electrons at sufficiently high altitudes F_∞ can be seen to be given by[5]

$$F_\infty = -c\lambda\beta a^2 H_0 \left(\frac{A}{2}\right)^\alpha \Gamma(1-\alpha), \tag{24.22b}$$

where $\Gamma(x)$ is the gamma function.

Therefore the steady-state electron density distribution for a given vertical flux of electrons (and ions) at very great heights, F_∞, is represented by Eq. (24.20) with d equal to zero and c derived from Eq. (24.22b)

δ) *In the actual upper atmosphere*, the height gradient of the scale height for atomic oxygen Γ is not constant but varies with height.

In such a case we assume that the atmosphere is composed of a number of thin regions in each of which Γ may be regarded as constant; then Eq. (24.20) can be considered to represent the general solution in each region. If the integration constants c and d are so determined that N_e and dN_e/dz become continuous at the boundaries of neighboring regions, apart from the first (lowest) $d=0$ and the last (highest) c given by Eq. (24.22b), the electron density distribution over the whole atmosphere is obtained as the ensemble of the N_e's given by Eq. (24.20).

In this fashion we have calculated electron density profiles for some of the CIRA 1972[6] model atmospheres ($T_\infty = 1{,}000 \ldots 1{,}600$ K); the results are shown in Fig. 12.

---

[5] Yonezawa, T. (1972): J. Radio Res. Lab. (Jpn.) *19*, 109

[6] – (1972): COSPAR International Reference Atmosphere. Berlin: Akademie-Verlag

The intensity of ionizing radiation was assumed to be a linear function of T_∞ with a solar photon flux of $1.2 \cdot 10^{14}\,\mathrm{m}^{-2}\,\mathrm{s}^{-1}$ for $T_\infty = 1{,}000$ K and $2.4 \cdot 10^{14}\,\mathrm{m}^{-2}\,\mathrm{s}^{-1}$ for $T_\infty = 1{,}600$ K. As for other parameters we took $z_0 = 300$ km, $\sigma = 8 \cdot 10^{-22}\,\mathrm{m}^2$, and

$$B_0 = ([\mathrm{N}_2]_0 + 10 \cdot [\mathrm{O}_2]_0) \cdot 5 \cdot 10^{-13}\,\mathrm{s}^{-1},$$

$$D_{a0} = \frac{8 \cdot 10^{14}}{n_0} \left(\frac{T_{n0}}{1{,}000\,\mathrm{K}} \right)^{\frac{1}{2}} \mathrm{m}^2\,\mathrm{s}^{-1},$$

$$\chi = 30°, \quad I = 50°, \quad \kappa = 3, \quad \delta = 2, \quad F_\infty = 0.$$

n_0 is the total number density of the atmosphere at the height of 300 km and σ is the ionization cross section of an oxygen atom which is taken to be equal to the absorption cross section.

For comparison we have depicted in the same figure the mid-latitude standard electron density profiles at noon in September and December for a sunspot number of 100 according to RAWER, RAMAKRISHNAN and BILITZA[7]. The agreement between this and calculated profiles is not satisfactory; also the calculated layer peaks are situated at higher levels. In order to bring the calculated levels down to lower heights, we must take a smaller value of B_0 than given above or a larger value of D_{a0}. However, in view of the results of laboratory measurement of the rate coefficients for reactions (23.5) and (23.7)[8] and of the diffusion coefficient of O^+ in an atomic oxygen gas[9], it seems unreasonable to make such alterations to the values of the parameters. The true cause of the above discrepancy is probably transport due to neutral winds[10].

25. Electron density distribution in the nighttime. If there is no production of electrons and ions during the night, electrons and ions simply diminish in number density through the loss processes of Eqs. (23.5...8) and never attain a steady state, so that this phenomenon should be treated as being transient. However, during the night in winter at middle latitudes, electron density does not usually decrease with time after about midnight but is maintained at almost a constant value. For the explanation of this fact some process of replenishment of electrons and ions must be assumed to be going on.

α) *Precipitation of charged particles* from the protonosphere may conceivably be one of these processes[1,2]. If, by this mechanism, a downward flux of electrons and ions exists at sufficiently high altitudes, the electron density distribution in a steady state is given by Eq. (24.20) with S and d taken as equal to zero and c

---

[7] Rawer, K., Ramakrishnan, S., Bilitza, D. (1975): Preliminary reference profiles for electron and ion densities and temperatures proposed for the International Reference Ionosphere. Institut für physikalische Weltraumforschung, Freiburg

[8] Ferguson, E.E., Fehsenfeld, F.C., Goldan, P.D., Schmeltekopf, A.L. (1965): J. Geophys. Res. *70*, 4323

[9] Dalgarno, A. (1964): J. Atmos. Terr. Phys. *26*, 939

[10] See the contribution by STUBBE to this volume, p. 247

[1] Park, C.G. (1972): J. Geophys. Res. *75*, 4249

[2] Evans, J.V. (1972): J. Atmos. Terr. Phys. *34*, 175

derived from Eq. (24.22b), i.e.

$$N = \frac{|F_\infty|}{\lambda \beta a^2 H_0} \left(\frac{2}{A}\right)^\alpha \frac{1}{\Gamma(1-\alpha)} y^{3/4} \mathrm{K}_\alpha(A y^\beta). \tag{25.1}$$

In particular, if the atmosphere is isothermal and there is no difference among neutral, ion, and electron temperatures, we have $\Gamma \to 0$ and $\kappa = 2$; further, taking δ as 2, we find from Eqs. (24.13), (24.11) and (24.15), (24.18), (24.14), (24.17), (24.10), and (23.1) that the values of λ, ν, β, μ^2, and α are 1, 1, 3/2, 1/2, and 1/6, respectively and

$$y = \exp\left(-\frac{z - z_0}{H}\right). \tag{25.2}$$

Therefore Eq. (25.1) in this case becomes as follows if Eqs. (24.16) and (24.19) are noted:

$$N = \frac{2^{1/2}}{3^{1/3}\Gamma(\frac{5}{6})} \left(\frac{H}{D_{a0}^2 B_0 \sin^4 I}\right)^{1/3} |F_\infty| u^{1/2} \mathrm{K}_{1/6}(u), \tag{25.3}$$

where

$$u \equiv \frac{2}{3} \left(\frac{B_0 H^2}{D_{a0} \sin^2 I}\right)^{1/2} \exp\left[-\frac{3(z - z_0)}{2H}\right]; \tag{25.4}$$

here the subscript 0 to be attached to H has been omitted because H is now independent of height. Essentially the same expressions as Eq. (25.3) have been obtained by [3,4]. An example of the electron density profile given by Eq. (25.3) is shown in Fig. 13 together with the well-known Elias-Chapman distribution indicated by a broken line; this latter is given by

$$N_{ch} = N_m \exp\left\{\frac{1}{2}\left[1 - \frac{z - z_m}{H} - \exp\left(-\frac{z - z_m}{H}\right)\right]\right\}, \tag{25.5}$$

where N_m is the peak electron density and z_m is the height of the peak. It is seen that both curves are very similar in form. As is well known empirically, observed electron density profiles during the night can be closely represented by a Elias-Chapman distribution. This observational fact can naturally be accounted for if actual electron density distributions are to be expressed by Eq. (25.3).

β) Numerical calculations show that $u^{1/2} \mathrm{K}_{1/6}(u)$ takes a maximum value of 0.815 for u equal to 0.151. Therefore, the *peak electron density* N_m and the *height of the peak* z_m of the distribution of N given by Eq. (25.3) can be expressed in the following way:

$$N_m = 0.708 \left(\frac{H}{D_{a0}^2 B_0 \sin^4 I}\right)^{1/3} |F_\infty|, \tag{25.6}$$

[3] Geisler, J.E., Bowhill, S.A. (1965): University of Illinois Aeronomy Report No. 5. Aeronomy Laboratory, Department of Electrical Engineering. Urbana, Ill.: University of Illinois

[4] Yonezawa, T. (1965): J. Radio Res. Lab. (Jpn.) *12*, 65; Yonezawa, T. (1965): Space Res. 5, 49

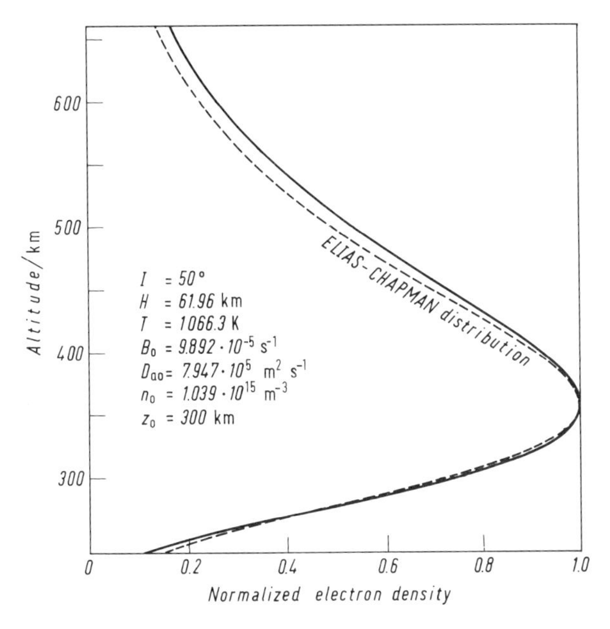

Fig. 13. A normalized electron density profile in the *F*2-region during the night when there is an influx of electrons and ions from the exosphere. The Elias-Chapman distribution of electron density with the same peak value and peak height is also indicated

$$\frac{z_m - z_0}{H} = \frac{1}{3} \ln \left(\frac{B_0 H^2}{D_{a0} \sin^2 I} \right) + 0.990. \tag{25.7}$$

γ) Using Eq. (25.6) it is possible to *estimate* the vertical *influx* of electrons (and ions) which is required for the maintenance of the F2-layer peak electron density. If JACCHIA's 1971 model atmosphere (CIRA 1972) for $T_\infty = 1{,}100$ K is assumed and z_0 is taken as 300 km, $H = 62$ km and from the statements in Subsect. 24δ $B_0 = 9.9 \cdot 10^{-5}\,\mathrm{s}^{-1}$ and $D_{a0} = 7.9 \cdot 10^5\,\mathrm{m}^2\,\mathrm{s}^{-1}$. Then, if I is taken as 50°, Eq. (25.6) indicates that a vertical electron influx of $1.0 \cdot 10^{12}\,\mathrm{m}^{-2}\,\mathrm{s}^{-1}$ is needed to maintain the peak electron density of $10^{11}\,\mathrm{m}^{-3}$. This is the influx of electrons (and O^+ ions) of an order of magnitude which can exist at very high altitudes, confirmed by observations of whistler atmospherics and by the incoherent scatter technique. It must, however, be noted that this value represents that of the vertical component of the flux so that the actual flux flowing along field lines must be larger than this value by a factor of $1/\sin I$ and that the original flux of protons in the magnetosphere which causes that of O^+ ions by charge exchange collisions must be considerably larger than the above value because the ion flow is subjected to considerable deceleration at the time of charge exchange collision when it changes from a proton to an O^+ ion flow. Incidentally from Eq. (25.7) the height of the peak is 360 km; this is reasonable for hours after midnight in winter at temperate latitudes.

δ) Some workers have dealt with the problem of theoretically deriving the *diurnal variation in electron density* including its transient change immediately after sunset by solving the continuity equation of electrons containing the diffusion term[5–8]. However, it seems difficult to express the solution analyti-

[5] Gliddon, J.E.C. (1959): Q. J. Mech. Appl. Math. *12*, 340, 347
[6] Ferraro, V.C.A. (1961): U.S. Air Force Technical Report, Contract No. AF 61 (052)-299
[7] Gliddon, J.E.C., Kendall, P.C. (1962): J. Atmos. Terr. Phys. *24*, 1073
[8] Gliddon, J.E.C., Kendall, P.C. (1964): J. Atmos. Terr. Phys. *26*, 721

cally if the electron loss rate is correctly assumed to be proportional to the densities of O_2 or N_2 as given by Eq. (24.3). Hence this rate has mostly been assumed to be either proportional to the density of atomic oxygen (the main constituent) or to be constant. Of course there is no such limitation when the differential equation is numerically integrated. This is the procedure now currently used[9].

ε) DUNCAN[10] considered the *layer shape*, however under much simplifying assumptions. He took the atmosphere as isothermal and assumed the loss coefficient of electrons, B, to be proportional to $n(\mathrm{O})$. Then δ in Eq. (24.3) is unity. If further thermal diffusion is neglected, it can readily be shown, using Eqs. (24.7) and (24.10...12) and the values of the parameters given by Eqs. (24.13...15), that the continuity equation for electrons (24.1) becomes as follows for the nighttime condition:

$$\frac{\partial N}{\partial t} = a^2 y \left[\frac{\partial^2 N}{\partial y^2} - \frac{1}{2y}\frac{\partial N}{\partial y} + \left(\frac{1}{2y^2} - \frac{B_0}{a^2}\right) N\right]. \tag{25.8}$$

As y is given by Eq. (25.2) in the present case, the Elias-Chapman distribution (25.5) is expressed as

$$N_{\mathrm{ch}} = N_{\mathrm{m}} \mathrm{e}^{1/2} (y/y_{\mathrm{m}})^{1/2} \exp(-y/2y_{\mathrm{m}}), \tag{25.9}$$

where y_{m} is the value of y corresponding to the height where $z = z_{\mathrm{m}}$. Now, suppose that such distribution of electron density exists at zero initial time. If N_{ch} is inserted into N in Eq. (25.8), its right-hand side, after some manipulation, becomes as follows:

$$2H y_{\mathrm{m}} \left(\frac{a^2}{4y_{\mathrm{m}}^2} - B_0\right) \frac{\partial N_{\mathrm{ch}}}{\partial z} - B_{\mathrm{m}} N_{\mathrm{ch}},$$

where B_{m} is the value of B at the height of z_{m} and is equal to $B_0 y_{\mathrm{m}}$. This expression represents the variation of electron density per unit time at a constant height just after the initial time. Its form suggests that the variation is composed of two parts, one a bodily movement of the distribution of N_{ch} with a velocity of $-2H y_{\mathrm{m}}(a^2/4y_{\mathrm{m}}^2 - B_0)$ which is independent of height and the other its exponential decay with a time constant of B_{m}^{-1} also independent of height. Thus, if a distribution of electron density is initially an Elias-Chapman distribution and undergoes a change through a process of diffusion and decay, it will still remain so at a moment later, and the same is true for each subsequent short interval. Therefore this kind of distribution will be maintained during the whole process of diffusion and decay which takes place according to Eq. (25.8).

Let U be the velocity of the bodily movement of the distribution mentioned above; this can be expressed as follows if Eq. (24.16) is taken into account:

$$U \equiv -\frac{D_{\mathrm{am}} \sin^2 I}{2H} + 2H B_{\mathrm{m}}. \tag{25.10}$$

---

[9] See the contributions by STUBBE in this volume and by RAWER and by SCHMIDTKE in vol. 49/7

[10] Duncan, R.A. (1956): Aust. J. Phys. 9, 436

This shows that if z_m is large, U is negative and if z_m is small, U is positive. Hence an Elias-Chapman layer situated at high levels will move down with the lapse of time while one at low levels will move up. The peaks of layers converge to the level where U is zero, i.e. where the following equation holds:

$$B = \frac{D_a \sin^2 I}{4H^2}. \tag{25.11}$$

ζ) Physically speaking, if a layer is situated at high levels, the effect of diffusion in the gravitational field predominates and it will fall; however, at lower and lower heights, the rate of electron disappearance rapidly increases and ultimately the decay of the lower edge of a layer becomes the most important process, making the *height of the peak* go up with the lapse of time at sufficiently low altitudes. These two opposing actions balance each other at the height determined by Eq. (25.11) where a layer is stabilized. It can readily be seen that this height is given by

$$z_{stab} = z_0 + \frac{H}{2} \ln[4B_0 H^2/(D_{a0} \sin^2 I)]. \tag{25.12}$$

As a special case of the solution for the problem of the diurnal variation in electron density, it can be proved under the same assumptions as above that, even if the initial distribution is not that of Eq. (25.9) it will tend to such distribution with the lapse of time. In view of this characteristic of the electron density distribution, we can well understand the empirical fact referred to earlier that the nocturnal electron density distribution in the F2-region can be represented by an Elias-Chapman distribution to a very good approximation if it is noted that the time constant is roughly given by $1/B$ and is of the order of two or three hours. It must, however, be noted that in reality δ is not equal to 1 as assumed here but is $7/4 \sim 2$ so that what has been stated hitherto is not strictly valid.

From the above considerations, the peak of the F2-layer is expected to be stabilized roughly at the height given by Eq. (25.12) several hours after sunset. On the other hand, the height of the peak of the layer produced by the inflow of charged particles from the exosphere is given by Eq. (25.7). When these two heights are compared, it is readily seen that the height given by Eq. (25.7) is higher than that given by Eq. (25.12) by $0.528\,H$, i.e. there is a difference of a few tens of kilometres. This may explain the observed fact that the height of the F2-layer in winter in middle latitudes is considerably higher near dawn than a few hours after sunset [4]. However, the situation is not so simple as above because winds and electric and magnetic fields in the upper atmosphere cause the F2-layer to undergo drift motion in the vertical direction and detailed studies of these effects are needed.

Here we only wish to add a simple estimation. If the velocity of the vertical drift is denoted by W, a steady-state layer will appear when this is balanced with the velocity given by Eq. (25.10). Then the increment in the layer height, Δz_{stab}, is found to be the following if it is noted that $(D_a)_{stab}$ and B_{stab} defined below satisfy the relation (25.11):

$$\Delta z_{stab} = H \operatorname{arcsinh}\left(\frac{HW}{(D_a)_{stab} \sin^2 I}\right) = H \operatorname{arcsinh}\left(\frac{W}{4HB_{stab}}\right), \tag{25.13}$$

where

$$(D_a)_{\text{stab}} = D_{a0} \exp\left(\frac{z_{\text{stab}} - z_0}{H}\right), \tag{25.14}$$

$$B_{\text{stab}} = B_0 \exp\left(-\frac{z_{\text{stab}} - z_0}{H}\right). \tag{25.15}$$

26. Diffusion of an ion cloud. The problem of diffusion of an ion cloud in relation to the reflection of radio waves from meteor trails has been the subject of study for decades. After a meteor has passed, electrons and ions of high density remain along the trail in a long needle-like form and they gradually diffuse outwards in a cylindrical form thereby enlarging their region of existence but diminishing in their number density. Thus it is expected that radio waves begin to be reflected by this cylindrical plasma cloud when its radius has grown sufficiently large but electron and ion densities are still not so low. This reflection will last for some time until plasma particle density becomes so low that reflection can no longer take place. An attempt was made to interpret the temporal variation in intensity of radio waves reflected from meteor trails, particularly the duration of their reflection and its frequency characteristics by formulating the above process mathematically[1]. However, such early-stage theories do not take into account the effects of the geomagnetic field and upper atmospheric electric fields and it was only in relatively recent years that genuine theoretical calculations were begun which took these effects into consideration. On the other hand, to measure upper atmospheric electric fields, experiments have been conducted to pursue the drift motion of visible ion clouds, such as of Ba^+ ions, which are ejected from rockets during twilight hours and the problem of plasma diffusion in the presence of electric and magnetic fields has been taken up from this point of view. It is, however, very difficult to deal mathematically with a truly realistic model and hence we shall confine our consideration mainly to the case in which there exist internal electric fields, due to charge separation, but not external electric fields. We shall also confine our attention to phenomena occurring near the 100 km level in the E-region.

α) In the E-region collisions between charged particles can be neglected compared with those between neutral and charged particles because the number density of neutral particles is sufficiently larger than that of charged particles. Then from Eqs. (4.2) and (4.5,6) the *equation of motion* of electrons or ions is seen to be the following if the inertia term is ignored (neutral particles are regarded to be at rest as a whole):

$$0 = \varepsilon_j q N(\boldsymbol{E} + \boldsymbol{v}_j \times \boldsymbol{B}) - \nabla p_j - N \frac{m_j m_n}{m_j + m_n} \nu_{jn} \boldsymbol{v}_j; \quad j = e \text{ or } i. \tag{26.1}$$

$\boldsymbol{B}$ is the geomagnetic field, $\boldsymbol{E}$ the internal space charge electric field in an ion cloud, N the electron or ion density assumed to be the same owing to the assumption of quasi-neutrality, m_j, $\boldsymbol{v}_j$, p_j, and ν_{jn} are the mass, velocity, partial pressure, and collision frequency with neutral particles of

[1] Yonezawa, T. (1946): Proc. Jpn. Acad. *22*, 213

electrons (j=e) or ions (j=i), respectively, m_n the mass of neutral particles, ε_j equal to -1 or 1 for j=e or i, respectively, and q is the absolute value of the charge on an electron. In the following we shall for simplicity[2] write $m_j \nu_j$ for $(m_j m_n/(m_j+m_n))\,\nu_{jn}$:

$$\nu_j = \frac{m_n}{m_j+m_n}\,\nu_{jn}. \tag{26.2}$$

It is readily seen that ν_e is almost equal to ν_{en}.

From the equation of continuity (4.1) we get also

$$\frac{\partial N}{\partial t} = -\nabla\cdot(N\,\boldsymbol{v}_j). \tag{26.3}$$

By taking the difference of Eq. (26.3) for j=e and i side by side, one gets:

$$\nabla\cdot N(v_i - v_e) = 0. \tag{26.4}$$

The space charge electric field $\boldsymbol{E}$ may be written in terms of a scalar potential function Ω, such that

$$\boldsymbol{E} = -\nabla\Omega. \tag{26.5}$$

If electron, ion, and neutral temperatures are assumed to be equal to one another, taking a temporally and spatially constant value T, we have from Eq. (5.9)

$$P_e = P_i = N\,k\,T. \tag{26.6}$$

β) Taking a coordinate system (x, y, z) whose z axis is parallel to the direction of the geomagnetic field, we obtain by *solving* Eq. (26.1) for $\boldsymbol{v}_j$ with the aid of Eqs. (26.2) and (26.5 ... 6)

$$N\boldsymbol{v}_j = -\tilde{D}_j(\nabla N + \varepsilon_j\, N\nabla\phi), \tag{26.7}$$

where

$$\tilde{D}_j = \begin{pmatrix} D_j^{\perp} & D_{H_j} & 0 \\ -D_{H_j} & D_j^{\perp} & 0 \\ 0 & 0 & D_j^{\parallel} \end{pmatrix}. \tag{26.8}$$

$$D_j^{\parallel} = \frac{kT}{m_j\,\nu_j}, \tag{26.9}$$

$$D_j^{\perp} = \frac{\nu_j^2}{\nu_j^2+\omega_j^2}\,D_j^{\parallel}, \tag{26.10}$$

$$D_{H_j} = \frac{\nu_j\,\omega_j}{\nu_j^2+\omega_j^2}\,D_j^{\parallel}, \tag{26.11}$$

---

[2] For a more detailed discussion see SUCHY's contribution in vol. 49/7, p. 1

$$\omega_{\mathrm{j}}=\frac{qB}{\varepsilon_{\mathrm{j}}m_{\mathrm{j}}}, \tag{26.12}$$

$$\phi=\frac{q\Omega}{kT}. \tag{26.13}$$

Substitution of $\boldsymbol{v}_{\mathrm{j}}$ from Eq. (26.7) into Eq. (26.3) gives

$$\begin{aligned}\frac{\partial N}{\partial t}&=D_{\mathrm{j}}^{\perp}\left(\frac{\partial^2 N}{\partial x^2}+\frac{\partial^2 N}{\partial y^2}\right)+D_{\mathrm{j}}^{\parallel}\frac{\partial^2 N}{\partial z^2}\\&+\varepsilon_{\mathrm{j}}\left\{D_{\mathrm{j}}^{\perp}\left[\frac{\partial}{\partial x}\left(N\frac{\partial\phi}{\partial x}\right)+\frac{\partial}{\partial y}\left(N\frac{\partial\phi}{\partial y}\right)\right]\right.\\&\left.+D_{\mathrm{j}}^{\parallel}\frac{\partial}{\partial z}\left(N\frac{\partial\phi}{\partial z}\right)+D_{H_{\mathrm{j}}}\left(\frac{\partial N}{\partial x}\frac{\partial\phi}{\partial y}-\frac{\partial N}{\partial y}\frac{\partial\phi}{\partial x}\right)\right\}.\end{aligned} \tag{26.14}$$

If we write this equation for $\mathrm{j}=\mathrm{i}$ and e and subtract one from the other we get:

$$\begin{aligned}&(D_{\mathrm{e}}^{\perp}-D_{\mathrm{i}}^{\perp})\left(\frac{\partial^2 N}{\partial x^2}+\frac{\partial^2 N}{\partial y^2}\right)+(D_{\mathrm{e}}^{\parallel}-D_{\mathrm{i}}^{\parallel})\frac{\partial^2 N}{\partial z^2}\\&=(D_{\mathrm{e}}^{\perp}+D_{\mathrm{i}}^{\perp})\left[\frac{\partial}{\partial x}\left(N\frac{\partial\phi}{\partial x}\right)+\frac{\partial}{\partial y}\left(N\frac{\partial\phi}{\partial y}\right)\right]\\&+(D_{\mathrm{e}}^{\parallel}+D_{\mathrm{i}}^{\parallel})\frac{\partial}{\partial z}\left(N\frac{\partial\phi}{\partial z}\right)+(D_{H_{\mathrm{e}}}+D_{H_{\mathrm{i}}})\left(\frac{\partial N}{\partial x}\frac{\partial\phi}{\partial y}-\frac{\partial N}{\partial y}\frac{\partial\phi}{\partial x}\right).\end{aligned} \tag{26.15}$$

If Eqs. (26.14, 15) for $\mathrm{j}=\mathrm{i}$ or e, regarded as simultaneous partial differential equations for N and ϕ, are solved under given initial and boundary conditions, the distributions of electron density and internal electric field potential are obtained as functions of time and position.

27. Diffusion of an initially spherical ion cloud[1]. It is convenient to use the cylindrical coordinate system (r,θ,z) with the z-axis in the direction of the geomagnetic field for treating the problem of the diffusion of an initially spherical ion cloud. It can be assumed that the situation is symmetrical around the z-axis so that all derivatives of any quantity with respect to θ vanish and Eqs. (26.14, 15) for $\mathrm{j}=\mathrm{i}$ become as follows:

$$\begin{aligned}\frac{\partial N}{\partial t}&=D_{\mathrm{i}}^{\perp}\left(\frac{\partial^2 N}{\partial r^2}+\frac{1}{r}\frac{\partial N}{\partial r}\right)+D_{\mathrm{i}}^{\parallel}\frac{\partial^2 N}{\partial z^2}\\&+D_{\mathrm{i}}^{\perp}\left[\frac{\partial}{\partial r}\left(N\frac{\partial\phi}{\partial r}\right)+\frac{N}{r}\frac{\partial\phi}{\partial r}\right]+D_{\mathrm{i}}^{\parallel}\frac{\partial}{\partial z}\left(N\frac{\partial\phi}{\partial z}\right),\end{aligned} \tag{27.1}$$

[1] Pickering, W.M. (1972): Planet. Space Sci. *20*, 149

$$(D_e^\perp - D_i^\perp)\left(\frac{\partial^2 N}{\partial r^2}+\frac{1}{r}\frac{\partial N}{\partial r}\right)+(D_e^\parallel - D_i^\parallel)\frac{\partial^2 N}{\partial z^2}$$
$$=(D_e^\perp + D_i^\perp)\left[\frac{\partial}{\partial r}\left(N\frac{\partial \phi}{\partial r}\right)+\frac{N}{r}\frac{\partial \phi}{\partial r}\right]+(D_e^\parallel + D_i^\parallel)\frac{\partial}{\partial z}\left(N\frac{\partial \phi}{\partial z}\right). \tag{27.2}$$

α) An exact solution of these equations can be obtained in the following way. Putting

$$N=d\ t^{-3/2}\exp[-t^{-1}(\lambda_1 z^2+\mu_1 r^2)] \tag{27.3}$$

and

$$\phi=a\ln\{b\,d\exp[-t^{-1}(\lambda_2 z^2+\mu_2 r^2)]\}, \tag{27.4}$$

where a, b, d, λ_1, μ_1, λ_2 and μ_2 are constants, and inserting these into Eqs. (27.1, 2), we find that, if

$$\lambda_1=\frac{1}{8}\left(\frac{1}{D_i^\parallel}+\frac{1}{D_e^\parallel}\right), \tag{27.5}$$

$$\mu_1=\frac{1}{8}\left(\frac{1}{D_i^\perp}+\frac{1}{D_e^\perp}\right), \tag{27.6}$$

$$\lambda_2=\frac{1}{8a}\left(\frac{1}{D_i^\parallel}-\frac{1}{D_e^\parallel}\right), \tag{27.7}$$

$$\mu_2=\frac{1}{8a}\left(\frac{1}{D_i^\perp}-\frac{1}{D_e^\perp}\right), \tag{27.8}$$

Eqs. (27.3, 4) are the exact solution of Eqs. (27.1, 2).

However, a solution of this form cannot be the one corresponding to the initial distribution of perfect sphere as seen from the reasoning below. It is evident from the functional form of Eqs. (27.3, 4) that in this solution iso-density surfaces of electrons and ions and equipotential surfaces of the space charge electric field, respectively, are always of similar geometric form. As may be seen by inserting Eqs. (26.9, 10) into Eqs. (27.5, 6)[2], μ_1 is greater than λ_1 so that the iso-electron- or iso-ion-density surface at a fixed time given by Eq. (27.3) is always an ellipsoid prolonged to the direction of the magnetic field and can never be a sphere. Thus a numerical method must be invoked to obtain solutions corresponding to an initial distribution of a form other than the one of the special ellipsoid mentioned above.

β) Pickering[1] solved Eqs. (27.1, 2) numerically for a spherical initial distribution and obtained the temporal variations of the distribution of N and ϕ. In the actual calculation, length and ion density are converted into non-dimensional quantities by dividing them by the typical dimension of an ion cloud R and by its typical initial volume ionization density n_0, respectively; time is also made non-dimensional by dividing by R^2/D_i.

In order to make the numerical calculation feasible it is necessary to limit the range of integration to a finite volume, and it is performed over the whole right cylinder with a non-dimensional height of $2L$ and a non-dimensional radius L. Let us denote a non-dimentionalized quantity by drawing a bar on it. The initial distribution was given by

$$\bar{N}_{\bar{t}=0}=\exp[-0.6(\bar{z}^2+\bar{r}^2)]+0.5, \tag{27.9}$$

---

[2] Cf. also the values of the parameters given in Eqs. (27.10) and (28.5) below

that is, the constant background ionization was taken as 1/3 of the initial electron density at the origin. The boundary conditions were taken to be zero value for the potential and zero normal derivative for electron density on $\bar{z}=\pm L$, $\bar{r}=L$. The values of the diffusion coefficient and other parameters are taken from those calculated by KAISER et al.[3] for a height of 97.5 km:

$$\begin{aligned} &\nu_{en}=10^5\,\mathrm{s}^{-1}, && T=203\,\mathrm{K}, && D_e^{\parallel}=2.62\cdot 10^4\,\mathrm{m}^2\,\mathrm{s}^{-1}, && D_i^{\parallel}=8\cdot 10\,\mathrm{m}^2\,\mathrm{s}^{-1},\\ &D_e^{\perp}=4.10\,\mathrm{m}^2\,\mathrm{s}^{-1}, && D_i^{\perp}=8.10\,\mathrm{m}^2\,\mathrm{s}^{-1}, && D_{H_e}=-328\,\mathrm{m}^2\,\mathrm{s}^{-1}, && D_{H_i}=0.201\,\mathrm{m}^2\,\mathrm{s}^{-1}. \end{aligned} \tag{27.10}$$

The results of numerical integration show that by diffusion the iso-electron-density surface comes to take an ellipsoid-like form prolonged in the direction of the magnetic field with the lapse of time particularly near the center of the region under consideration and the electric field is directed outward from the center of the ion cloud apart from what seems to be the effects of the boundaries of the finite region. The electron density at the center $\bar{N}_m(\bar{t})$ for $0\leqq\bar{t}\leqq 1$ is expressed very well by the following equation if the constant A_1 is appropriately chosen:

$$\bar{N}_m(\bar{t})=(A_1\bar{t}+1)^{-3/2}+0.5, \tag{27.11}$$

and the electron density in general $\bar{N}$ and the non-dimensionalized potential ϕ (cf. Eq. (26.13)) can be expressed in a similar form as Eqs. (27.3, 4):

$$\bar{N}=(\bar{N}_m(t)-0.5)\exp(-\bar{\lambda}_1\bar{z}^2-\bar{\mu}_1\bar{r}^2)+0.5, \tag{27.12}$$

$$\phi=a\ln b+a\ln[(\bar{N}_m(\bar{t})-0.5)\exp(-\bar{\lambda}_2\bar{z}^2-\bar{\mu}_2\bar{r}^2)+0.5]. \tag{27.13}$$

However, as $\bar{\lambda}_1$, $\bar{\mu}_1$, $\bar{\lambda}_2$ and $\bar{\mu}_2$ are now functions of time and a and b also change to some degree with time, the numerical solution just obtained probably cannot be said to exhibit the simple similarity behavior of the rigorous solution, Eqs. (27.3, 4). It is not entirely unreasonable to expect that if the volume of integration is taken wider and the background ionization lower, the numerical solution obtained above will ultimately approach the similarity solution for sufficiently large $\bar{t}$.

28. Diffusion of a cylindrical ion cloud[1]. In this Section we consider the diffusion of a cylindrical ion cloud with the axis making an angle Θ with the direction of the magnetic field (z-axis). Let the axis of the cylinder be contained in the zx-plane and all quantities have uniform values in the direction of the axis so that the problem can be treated as a two-dimensional one. If the axis of the cylinder is taken as the ζ-axis and the ξ-axis is taken perpendicular to it in the zx-plane, all derivatives with respect to ζ vanish. Thus Eqs. (26.14, 15) are transformed into

$$\begin{aligned} \frac{\partial N}{\partial t}=&D_j^{\perp}\frac{\partial^2 N}{\partial y^2}+(D_j^{\perp}\cos^2\Theta+D_j^{\parallel}\sin^2\Theta)\frac{\partial^2 N}{\partial\xi^2}\\ &+\varepsilon_j\left[D_j^{\perp}\frac{\partial}{\partial y}\left(N\frac{\partial\phi}{\partial y}\right)+(D_j^{\perp}\cos^2\Theta+D_j^{\parallel}\sin^2\Theta)\frac{\partial}{\partial\xi}\left(N\frac{\partial\phi}{\partial\xi}\right)\right.\\ &\left.+D_{Hj}\cos\Theta\left(\frac{\partial N}{\partial\xi}\frac{\partial\phi}{\partial y}-\frac{\partial N}{\partial y}\frac{\partial\phi}{\partial\xi}\right)\right], \end{aligned} \tag{28.1}$$

[3] Kaiser, T.R., Pickering, W.M., Watkins, C.D. (1969): Planet. Space Sci. *17*, 519

[1] Pickering, W.M., Windle, D.W. (1970): Planet. Space Sci. *18*, 1153

$$[(D_e^\perp - D_i^\perp)\cos^2\Theta + (D_e^\| - D_i^\|)\sin^2\Theta]\frac{\partial^2 N}{\partial\xi^2} + (D_e^\| - D_i^\|)\frac{\partial^2 N}{\partial y^2}$$
$$= [(D_e^\perp + D_i^\perp)\cos^2\Theta + (D_e^\| + D_i^\|)\sin^2\Theta]\frac{\partial}{\partial\xi}\left(N\frac{\partial\phi}{\partial\xi}\right)$$
$$+ (D_e^\perp + D_i^\perp)\frac{\partial}{\partial y}\left(N\frac{\partial\phi}{\partial y}\right) + (D_{He} + D_{Hi})\cos\Theta\left(\frac{\partial N}{\partial\xi}\frac{\partial\phi}{\partial y} - \frac{\partial N}{\partial y}\frac{\partial\phi}{\partial\xi}\right). \quad (28.2)$$

α) PICKERING and WINDLE[1], after having non-dimensionalized each quantity in a similar way as stated in the previous section, made a *numerical integration* over the finite region $-L \leqq \bar{\xi} \leqq L$, $-L \leqq \bar{y} \leqq L$ and obtained the solution of Eqs. (28.1, 2).

The boundary conditions are as above

$$\phi = 0, \quad \frac{\partial \bar{N}}{\partial \bar{\xi}} = 0 \quad \text{on } \bar{\xi} = \pm L$$

and (28.3)

$$\phi = 0, \quad \frac{\partial \bar{N}}{\partial \bar{y}} = 0 \quad \text{on } \bar{y} = \pm L,$$

and the initial condition is also as above

$$\bar{N}_{\bar{t}=0} = \exp[-0.6(\bar{\xi}^2 + \bar{y}^2)] + 0.5. \quad (28.4)$$

The values of the parameters such as the diffusion coefficient are taken from the results of calculation by KAISER et al.[2] for a height of 102 km:

$$\begin{array}{llll} \nu_{en} = 5\cdot 10^4\,\mathrm{s}^{-1}, & T = 222\,\mathrm{K}, & D_e^\| = 5.74\cdot 10^4\,\mathrm{m^2\,s^{-1}}, & D_i^\| = 17.8\,\mathrm{m^2\,s^{-1}}, \\ D_e^\perp = 2.24\,\mathrm{m^2\,s^{-1}}, & D_i^\perp = 17.8\,\mathrm{m^2\,s^{-1}}, & D_{H_e} = -358\,\mathrm{m^2\,s^{-1}}, & D_{H_i} = 0.880\,\mathrm{m^2\,s^{-1}}. \end{array} \quad (28.5)$$

β) Part of the *results* of calculation is shown in Figs. 14 to 16. Figure 14 gives the electron density on the axis of the ionization column, $\bar{N}_m$ (maximum electron density), at the altitude of 102 km as a function of $\bar{t}$; if Θ changes slightly from zero, the rate of decay of $\bar{N}_m$ by diffusion increases steeply, but after Θ has reached a value of about 5°, the rate of decay ceases to increase rapidly. This may qualitatively be understood if one considers that the plasma can move relatively freely in the direction parallel to the magnetic field and the velocity of this movement has a component in the radial direction which contributes largely to the diffusion of the plasma in this direction (perpendicular to the axis of the ionization column), and that this velocity component increases rapidly with increasing Θ. Figures 15 and 16 show isolines of electron density and equipotential contours at $\bar{t} = 1$ for $\Theta = 1°10'$ and $1°50'$ and altitude 102 km. In order to avoid the confusion with the values of electron density, the values of potential are increased by a factor of 10^4. The magnetic field is directed into the plane of the diagrams at angle Θ to the normal and is parallel to the plane determined by the $\bar{\xi}$-axis and the normal to the plane of the diagrams.

The important feature of Fig. 15 is that, apart from the regions which seem to be influenced by the boundaries, the space charge electric field is directed inward toward the axis of the ionization column. In this case the diffusive motion perpendicular to the column is impeded by the action of the space-charge electric field. On the other hand, in Fig. 16 showing the case where Θ is slightly augmented to $1°\,50'$ the space-charge electric field is directed mainly

[2] Kaiser, T.R., Pickering, W.M., Watkins, C.D. (1969): Planet. Space Sci. *17*, 519

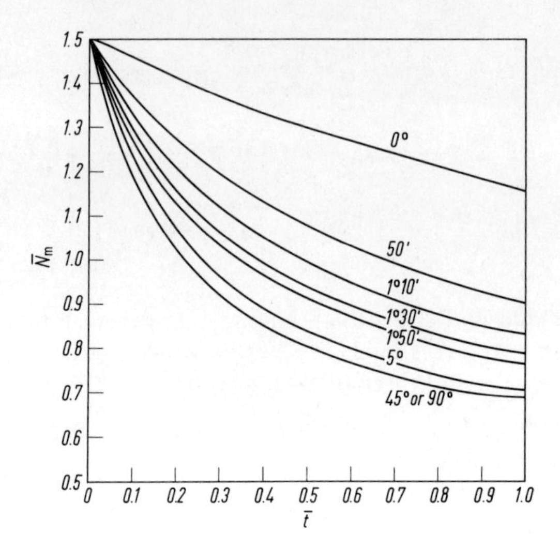

Fig. 14*. Normalized electron density on the axis of the ionization column $\bar{N}_m$ at an altitude of 102 km as a function of the normalized time $\bar{t}$ during the diffusion process of a cylindrical ion cloud. Numbers attached to the curves show the angle between the axis of the cloud and the direction of the magnetic field

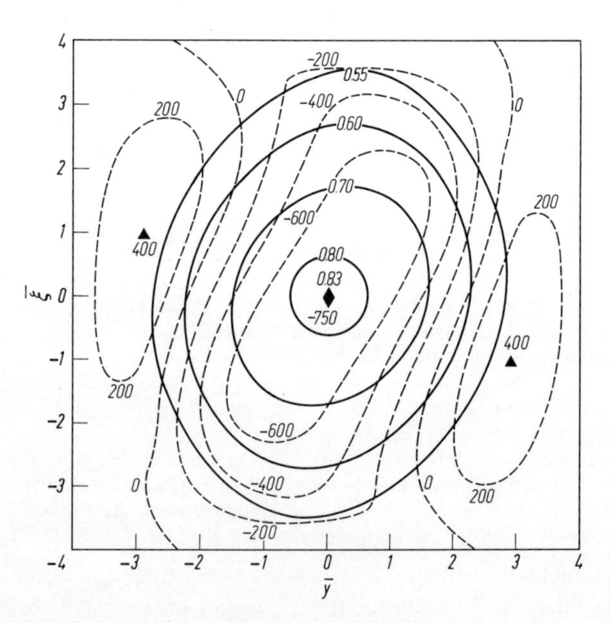

Fig. 15**. Isolines of electron density (solid curves) and equipotential contours (broken curves) for $\Theta = 1°\,10'$, $\bar{t} = 1$, and altitude 102 km during the diffusion process of a cylindrical ion cloud. See text for the direction of the magnetic field and coordinates. Potential unit is kT/q (cf. Eq. (26.13))

* From Pickering, Windle (1970): Planet. Space Sci. *18* (No. 8), 1158, Fig. 2

** From Pickering, Windle (1970): Planet. Space Sci. *18* (No. 8), 1159, Fig. 4

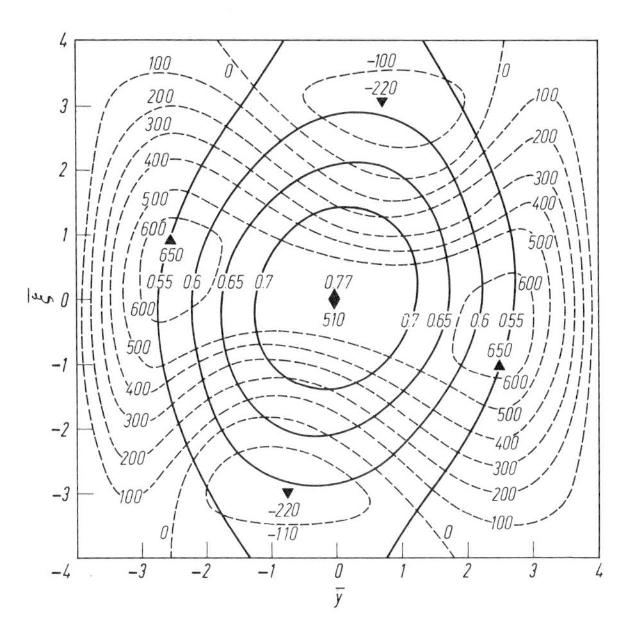

Fig. 16*. Isolines of electron density (solid curves) and equipotential contours (broken curves) for $\Theta = 1° 50'$, $\bar{t} = 1$, and altitude 102 km during the diffusion process of a cylindrical ion cloud. See text for the direction of the magnetic field and coordinates. Potential units is kT/q (cf. Eq. (26.13))

outward away from the column axis and thus enhances the rate of diffusion. If the value of Θ which separates the above two cases is denoted by Θ_c and called the "critical angle", Θ_c is essentially independent of $\bar{t}$ and estimated at $1° 24'$. This is the value of Θ which makes the left-hand side of Eq. (28.2) roughly equal to zero. As seen from Eq. (28.5), $D_i^{\perp} > D_e^{\perp}$ so that if Θ is less than Θ_c, the ions tend to escape from electrons, thus provoking a negative space charge on and near the column axis and hence an electric field directed inward. On the other hand, as is also seen from Eq. (28.5), $D_e^{\parallel}$ is overwhelmingly larger than other coefficients of that kind, so that when Θ becomes larger than Θ_c, the radial component of the diffusive motion of electrons parallel to the magnetic field has the greatest influence and electrons tend to escape from ions resulting in the accumulation of positive charge on and near the column axis and hence the space-charge electric field is directed away from the axis. These two situations may be referred to as "*electron-controlled*" *(slow) and* "*ion-controlled*" *(fast) diffusion.*

γ) PICKERING[3] applied the theoretical results obtained above to the study of the diffusion of *meteor trails* and compared them with the experimental work of WATKINS et al.[4] Although precise comparisons were not possible, it seemed reasonable to propose that the calculations of his study group support the suggestion of WATKINS et al. that field-aligned elongations of plasma may be produced by meteor trains.

* From Pickering, Windle (1970): Planet. Space Sci. *18* (No. 8), 1160, Fig. 6

[3] Pickering, W.M. (1973): Planet. Space Sci. *21*, 1671

[4] Watkins, C.D., Eames, R., Nicholson, T.F. (1971): J. Atmos. Terr. Phys. *33*, 1907

PICKERING and WINDLE[5] considered further the possible effects of electron-ion collisions on the diffusion of a cylindrical plasma irregularity oriented at right angles to the geomagnetic field lines, at altitude 300 km. The results obtained indicate that the diffusion may proceed at a rate rather greater than for a case for which these effects are assumed to be negligible. Furthermore, it was suggested that if the plasma density is great enough, the production of a "fin" of ionization along the magnetic field direction may be inhibited by the action of electron-ion collisions.

29. Diffusion of an ion cloud with an initial distribution of ellipsoidal form. PICKERING[1] made a numerical integration similar to that in the two foregoing sections for an initial ion-cloud distribution of the ellipsoidal form whose major axis is directed parallel to the magnetic field. Calculations were made for the values of the parameters corresponding to the conditions at the levels of 210 km and 310 km in the F-region as well as in the E-region. In the F-region collisions between electrons and ions cannot be ignored, and this effect was also taken into account. In the case of an ellipsoidal form of distribution its major to minor axis ratio R is a very important parameter. If it is smaller than a certain critical value R_c, the space charge electric field is directed outward and the fast diffusion of the ion-controlled type occurs; but if R is larger than R_c, the electric field is directed inward and the slow diffusion of the electron-controlled type occurs. The values of R_c have been estimated roughly at 70, 35, 30, 450, and 490 for the levels of 97.5, 102, 114, 210, and 300 km, respectively. These values are based on the results of numerical integration, but R_c may be calculated to a rough approximation by the following equation:

$$R_c^2 = (D_e^{\|} - D_i^{\|})/2(D_i^{\perp} - D_e^{\perp}). \tag{29.1}$$

The validity of this equation can be guessed from Eq. (27.2) if N and ϕ are assumed to have a functional form similar to those given by Eqs. (27.3, 4):

$$N \propto \exp[-(r^2/r_0^2) - (z^2/z_0^2)].$$

For varying $R(=z_0/r_0)$, the left-hand side of Eq. (27.2) changes sign at $R \simeq R_c$ so that the derivatives of ϕ on the right-hand side change sign also, or the direction of electric field is reversed at $R \simeq R_c$; it must however be noted that this is strictly correct only for small values of z/z_0 and r/r_0.

30. Effects of the electric field and wind. The conclusions reached in the foregoing sections are only valid when the effects of the electric field and wind in the ionosphere are disregarded. However, as stated at the beginning of Sect. 26, impetus to the study of the diffusion of ion clouds in the upper atmosphere has come not only from the problem of the meteor reflection of radio waves but also from the measurement of electric fields in the ionosphere; thus the above treatment of the diffusive motion of ion clouds ignoring

[5] Pickering, W.M., Windle, D.W. (1974): Planet. Space Sci. *22*, 833

[1] Pickering, W.M. (1973): Planet. Space Sci. *21*, 1073

external fields can by no means be satisfactory. Therefore we shall deal briefly with the effects of electric fields and winds[1].

α) Assuming that there are *weak inhomogeneities* in an infinitely wide magneto-active plasma, GUREVIČ and TSEDILINA[2] have made a theoretical study on how the inhomogeneities move and undergo spreading under the action of an external electric field and/or of the wind of neutral particles with a uniform speed. They have given the following expression for the plasma density at any instant of an irregularity which is of small dimensions at the initial instant[3]:

$$n(\boldsymbol{r},t)=\frac{n_0}{(2\pi)^3}\int \exp[i\boldsymbol{k}\cdot(\boldsymbol{r}-\boldsymbol{v}_a(\Theta)\,t)-D_a(\Theta)\,\boldsymbol{k}^2\,t]\,d\boldsymbol{k}, \tag{30.1}$$

where

$$D_a(\Theta)=\frac{2D_i(\Theta)\,D_e(\Theta)}{D_i(\Theta)+D_e(\Theta)}, \tag{30.2}$$

$$\boldsymbol{v}_a(\Theta)=\frac{D_i(\Theta)\,\boldsymbol{v}_e+D_e(\Theta)\,\boldsymbol{v}_i}{D_i(\Theta)+D_e(\Theta)}, \tag{30.3}$$

$$D_j(\Theta)=D_j^{\parallel}\cos^2\Theta+D_j^{\perp}\sin^2\Theta, \qquad (j=i,e). \tag{30.4}$$

Θ is the angle between vectors $\boldsymbol{k}$ and $\boldsymbol{B}$ while $D_j^{\parallel}$, $D_j^{\perp}$, and ω_j are given by Eqs. (26.9, 10), and (26.12); n_0 is the total number of ions in the irregularity and $\boldsymbol{v}_i$ and $\boldsymbol{v}_e$ are the drift velocities of ions and electrons, respectively. If the irregularity has initially an arbitrary distribution of plasma density $n(\boldsymbol{r},0)$, the integrand in Eq. (30.1) should be multiplied by the Fourier transform of $n(\boldsymbol{r},0)/n_0$ as an additional factor.

Here we shall be concerned with the drift motion of the irregularity, $\boldsymbol{v}$, due to the neutral wind $\boldsymbol{u}$ and the electric field $\boldsymbol{E}$. Then we have

$$\begin{aligned}\boldsymbol{v}_j=&\frac{1}{1+\gamma_j^2}[\boldsymbol{u}+\varepsilon_j\gamma_j\boldsymbol{u}\times\boldsymbol{e}+\gamma_j^2(\boldsymbol{u}\,\boldsymbol{e})\,\boldsymbol{e}]\\&+\frac{\varepsilon_j\gamma_j}{1+\gamma_j^2}[\boldsymbol{W}+\varepsilon_j\gamma_j\boldsymbol{W}\times\boldsymbol{e}+\gamma_j^2(\boldsymbol{W}\cdot\boldsymbol{e})\,\boldsymbol{e}], \qquad (j=i,e)\end{aligned} \tag{30.5}$$

where

$$\boldsymbol{e}=\boldsymbol{B}/B, \qquad \boldsymbol{W}=\boldsymbol{E}/B, \tag{30.6}$$

$$\gamma_j=|\omega_j|/\nu_j \qquad (j=i,e). \tag{30.7}$$

β) It can be shown that the *asymptotic expression* of Eq. (30.1) for sufficiently large t has two maxima and these make a drift motion with velocities approximately equal to $\boldsymbol{v}_a(\Theta=90°)$ and $\boldsymbol{v}_a(\Theta=0°)$. If these velocities are de-

---

[1] For experimental details see for instance Föppl, H., Haerendel, G., Haser, L., Loidl, J., Lütjens, P., Lüst, R., Melzner, F., Meyer, B., Neuss, H., Rieger, E. (1967): Planet. Space Sci. *15*, 357

[2] Gurevič, A.V., Tsedilina, E.E. (1967): Space Sci. Rev. 7, 407

[3] Giles, M., Martelli, G. (1971): Planet. Space Sci. *19*, 1

noted by v_1 and v_2, we get from Eqs. (30.3...7), and (26.9, 10):

$$\boldsymbol{v}_1=\frac{1}{1+\gamma_i\gamma_e}[\boldsymbol{u}+\gamma_i\gamma_e(\boldsymbol{u}\cdot\boldsymbol{e})\,\boldsymbol{e}+\gamma_i\gamma_e\,\boldsymbol{W}\times\boldsymbol{e}-\gamma_i\gamma_e(\gamma_e-\gamma_i)(\boldsymbol{W}\cdot\boldsymbol{e})\,\boldsymbol{e}] \tag{30.8}$$

$$\begin{aligned}\boldsymbol{v}_2=&\frac{1}{(1+\gamma_i^2)(1+\gamma_e^2)}[(1-\gamma_i\gamma_e+\gamma_i^2+\gamma_e^2)\,\boldsymbol{u}\\&+\gamma_i\gamma_e(\gamma_e-\gamma_i)\,\boldsymbol{u}\times\boldsymbol{e}+\gamma_i\gamma_e(1+\gamma_i\gamma_e)(\boldsymbol{u}\cdot\boldsymbol{e})\,\boldsymbol{e}+\gamma_i\gamma_e(\gamma_e-\gamma_i)\,\boldsymbol{W}\\&+\gamma_i\gamma_e(1+\gamma_i\gamma_e)\,\boldsymbol{W}\times\boldsymbol{e}-\gamma_i\gamma_e(\gamma_e-\gamma_i)(\boldsymbol{W}\cdot\boldsymbol{e})\,\boldsymbol{e}].\end{aligned} \tag{30.9}$$

From these results it is inferred that the irregularity will be split into two parts after a sufficient time has elapsed.

Regarding the time necessary for reaching such a condition, GUREVIČ and TSEDILINA[2] have made a calculation for the case in which a neutral wind of velocity u_m is blowing in the direction perpendicular to the magnetic field. The results show that, after the lapse of time, t, of $2D_i^{\parallel}/u_m^2$, the effect of drift on the shape of the inhomogeneity is already very considerable, but no sign of its splitting into two parts is yet seen. When t becomes as long as $20D_i^{\parallel}/u_m^2$, it has already been split clearly and the first maximum is displaced in the direction of u_m with a velocity much lower than u_m, while the second maximum is displaced in a different direction from that of the first one with a velocity more than 20 times higher than that of the first one, that is, with a velocity of the same order of magnitude as u_m. Therefore, it seems that it is not until the lapse of time t becomes much larger than $D_i^{\parallel}/u_m^2$ that the phenomenon of splitting is noticed. As γ_i for barium ions is of the order of 10 at a level of 150 km[4], their $D_i^{\parallel}(=kT/m_i\nu_i)$ is about $1\cdot10^4\,\mathrm{m^2\,s^{-1}}$ there. Hence, if a wind of $100\,\mathrm{m\,s^{-1}}$ is assumed to be blowing, it takes a few tens of seconds for an ion cloud to be split. In reality, electric fields exist in the upper atmosphere which in some cases will act to strengthen the action of the wind but in others to hinder it, making the time of onset of ion-cloud splitting earlier or later as the case may be. As seen from Eqs. (30.5, 6), the electric field $\boldsymbol{E}$ exerts the same effect on the drift velocity of ions as does the wind of velocity $\gamma_i E/B$; using this fact it is possible to estimate the time necessary for the splitting phenomenon to appear by the action of the electric field.

GILES and MARTELLI[3] reported that they observed the ionized component of a barium cloud to separate into two structures, one taking rapidly a field-aligned form and the other expanding in the direction approximately perpendicular to the magnetic field. They developed a method to evaluate the electric field in the ionosphere utilizing this phenomenon.

γ) *At greater heights*[4], above about 200 km, $\nu_j \ll |\omega_j|$ and γ_j given by Eq. (30.7) becomes so large that the Hall drift velocity due to the electric field (the second term in the second pair of brackets on the right-hand side of Eq. (30.5)) comes to be equal for both ions and electrons. v_1 and v_2 given by Eqs. (30.8, 9) becomes approximately equal to each other also if, as is plausible, electric fields in the upper atmosphere do not have a component parallel to the geomagnetic field. Therefore the splitting phenomenon of an ion cloud is not necessarily expected to occur at these heights but the component of the ionospheric electric field perpendicular to the magnetic field can be estimated by Bv, where v is the magnitude of the velocity component of the ion cloud perpendicular to both $\boldsymbol{E}$ and $\boldsymbol{B}$. For example, FÖPPL et al.[5] succeeded in

[4] Haerendel, G., Lüst, R., Rieger, E. (1967): Planet. Space Sci. *15*, 1

[5] Föppl, H., Haerendel, G., Haser, L., Lüst, R., Melzner, F., Meyer, B., Neuss, H., Rabben, H.-H., Rieger, E., Stöcker, J., Stoffregen, W. (1968): J. Geophys. Res. *73*, 21

measuring electric fields at 210 to 240 km heights by following barium ion clouds ejected from rockets for periods from a few tens of minutes to as long as two hours.

GILES and MARTELLI[6] made a theoretical calculation on the behavior of an infinitely long filament ion cloud in the presence of winds and electric fields. Although the analysis is rather complicated, the conclusions reached are similar to those stated above and we do not wish to enter into the details of this study here.

V. Vertical distributions of minor constituents in the high atmosphere

31. Escape of hydrogen and helium from the high atmosphere. STONEY pointed out as early as 1898 and 1900 that the atmospheric lighter atoms tend to escape from the upper atmosphere of the Earth and other inner planets. But, as is well known, it was not until the appearance of JEANS' "Dynamical Theory of Gases" that the quantitative expression for the escape became widely known. Jeans' formula can be written in the following form originally used by NICOLET[1]:

$$F_c = n_c (g_c/2\pi)^{1/2} r_c H_c^{-1/2} \left(1 + \frac{H_c}{r_c}\right) e^{-r_c/H_c}, \tag{31.1}$$

where F_c is the flux of atoms released outside the atmosphere per unit time through the unit area at the base of the exosphere or at the critical level, r the distance from the center of the Earth, n and H the number density and scale height of the atoms concerned, g the acceleration due to gravity and the subscript c designates the value of a quantity at the critical level. Eq. (31.1) can readily be derived by integrating at the critical level the number of those atoms which have an upward velocity exceeding the escape velocity, assuming that the velocities of the atoms are distributed according to the Maxwellian law; however, since the escape itself must make the distribution function deviate from the Maxwellian, Jeans' formula cannot be strictly valid.

Using Monte Carlo analysis, some workers have investigated to what extent Eq. (31.1) can be regarded as true. Although their results are not consistent with one another, it seems reasonable to assume the actual escape rate for hydrogen atoms to be about 70% of the rate given by Eq. (31.1) and that for helium atoms to be only a few percent below the rate given by Eq. (31.1)[2-4].

---

[6] Giles, M., Martelli, G. (1969): Planet. Space Sci. *17*, 1693
[1] Kockarts, G., Nicolet, M. (1962): Ann. Geophys. *18*, 269
[2] Brinkmann, R.T. (1970): Planet. Space Sci. *18*, 449
[3] Venkateswaran, S.V. (1971): Planet. Space Sci. *19* 275
[4] Chamberlain, J.W., Smith, G.R. (1971): Planet. Space Sci. *19*, 675

If the exospheric temperature T_∞ changes, the height of the critical level changes also and this results in changes of various quantities on the right-hand side of Eq. (31.1). Hence this expression is not convenient for calculating fluxes for various values of the exospheric temperature. KOCKARTS and NICOLET[1], assuming that the constituent under consideration is in *isothermal* diffusive equilibrium below the critical level, expressed F_c in terms of the values of various quantities at a certain level. Let us denote by a the distance of this level from the Earth's center and by z_c the difference in height between the critical level and this level; further let the subscript a refer to the values of the parameters at this level. Then, since

$$n_c = n_a \exp\left(-\frac{a}{a+z_c}\,\frac{z_c}{H_a}\right), \tag{31.2}$$

Eq. (31.1) can be expressed in the following way:

$$F_c = n_a (g_a/2\pi)^{1/2} \frac{a}{1+\frac{z_c}{a}} \, \frac{1 + H_a\left(1+\frac{z_c}{a}\right)\Big/ a}{H_a^{1/2}\, e^{a/H_a}}. \tag{31.3}$$

If the above-mentioned level is taken at the height of 500 km and the relation $z_c/a \ll 1$ is taken into account, by inserting numerical values into the parameters, Eq. (31.3) is brought into the following form:

$$F_c = 7.96 \cdot 10^6\, \mathrm{m}^{-2}\,\mathrm{s}^{-1}\, n_{500} \frac{1 + H_{500}/6.87\cdot 10^6}{H_{500}^{1/2}\, e^{6.87\cdot 10^6/H_{500}}}, \tag{31.4}$$

where the subscript 500 refers to the values of the parameters at the 500 km level. F_c divided by n_{500} gives approximately the effusion velocity v_E at the 500 km level:

$$v_E = F_c/n_{500}. \tag{31.5}$$

It should be noted that this depends on H_{500} only. The mean lifetime of an atom against effusion, τ_E, is defined as the time necessary for its concentration at 500 km to be reduced to 50%. Considering that the total number of atoms contained in a vertical column of unit cross section above 500 km is approximately given by $n_{500} H_{500}$, n_{500} is seen to decrease with time in proportion to $\exp(-v_E t/H_{500})$ if there is no replenishment of atoms; τ_E is given by

$$\tau_E = \frac{\ln 2 \cdot H_{500}}{v_E} = \frac{0.693\, H_{500}}{v_E}. \tag{31.6}$$

In Table 3 are shown some escape parameters for hydrogen and helium at 500 km calculated by Eqs. (31.4), (31.5), and (31.6) for several temperatures. In the case of hydrogen, however, the values of F_c have been adopted which are

30% less than those given by Eq. (31.4) for the reason stated above. From the table it is seen that a time shorter than one day is sufficient to reduce the concentration of atomic hydrogen significantly at a temperature not less than 1,000 K, if there is no replenishment of atoms. Therefore, a steady state cannot be maintained unless diffusion transport of hydrogen from lower levels constantly compensates its loss due to escape. The photodissociation of water vapor taking place above the mesopause is considered as the main source of hydrogen replenishment in this case[5]. For helium the effusion velocity is much slower; however, for temperatures not less than 1,000 K it will be lost from the atmosphere in a time much less than the age of the earth if there is no replenishment of atoms. Its existence in the present atmosphere is accounted for by its replenishment with that produced in the interior of the Earth by radioactive decay.

Table 3. Escape parameters for hydrogen and helium at 500 km

Temperature	Scale height/km		Effusion velocity/m s^{-1}		Loss-	
T/K	H	He	H	He	time/h H	time/a He
750	734	185	$6.19\cdot10^{-1}$	$1.36\cdot10^{-12}$	$2.28\cdot10^{2}$	$2.99\cdot10^{9}$
1,000	978	246	$5.75\cdot10^{0}$	$1.29\cdot10^{-8}$	$3.28\cdot10^{1}$	$4.20\cdot10^{5}$
1,250	1,223	308	$2.16\cdot10^{1}$	$3.07\cdot10^{-6}$	$1.09\cdot10^{1}$	$2.20\cdot10^{3}$
1,500	1,468	370	$5.18\cdot10^{1}$	$1.17\cdot10^{-4}$	$5.46\cdot10^{0}$	$6.96\cdot10^{1}$
1,750	1,712	431	$9.63\cdot10^{1}$	$1.55\cdot10^{-3}$	$3.42\cdot10^{0}$	$6.11\cdot10^{0}$
2,000	1,957	493	$1.53\cdot10^{2}$	$1.07\cdot10^{-2}$	$2.46\cdot10^{0}$	$1.01\cdot10^{0}$

32. Distributions of hydrogen and helium. We can use the same method explained in Sect. 19 to derive the expression for the vertical flux of hydrogen or helium atoms. In the case of these atoms, however, we cannot ignore the effect of thermal diffusion so that we must add a thermal diffusion term in the expression corresponding to Eq. (19.1). We assume here that hydrogen and helium atoms exist in the thermosphere in a sufficiently small proportion and their interaction with each other can be neglected. Then each species can be dealt with separately as one component in a binary mixture with the main constituent of the atmosphere. In this case thermal diffusion is expressed by the last term on the right-hand side of Eq. (3.21), which can be written as $-D_{12}\alpha_T\nabla\ln T$ if Eq. (5.6) is taken into account. Now let us take the subscript 1 in Eq. (19.9) as referring to hydrogen or helium; neglecting the terms containing the small quantity n_1/n in this equation and adding the above-mentioned thermal diffusion term, we obtain for the vertical velocity w_1 $(=\phi_1/n_1)$ of hydrogen or helium atoms the following expression:

$$w_1=-D_1\left[\frac{1}{n_1}\frac{dn_1}{dz}+\frac{1}{H_1}+(1+\alpha_T)\frac{1}{T}\frac{dT}{dz}\right]-K\left(\frac{1}{n_1}\frac{dn_1}{dz}+\frac{1}{H}+\frac{1}{T}\frac{dT}{dz}\right). \quad (32.1)$$

[5] Nicolet, M. (1970): Ann. Geophys. *26*, 531

Here D_1 is originally given by Eq. (19.8) but, on the assumption made above, it can be replaced by the diffusion coefficient for the binary mixture composed of hydrogen or helium and the atmospheric main constituent; this will be denoted by D_1 hereafter; H in Eq. (32.1) is scale height for the main atmosphere. As regards the thermal diffusion factor α_T, -0.27 is its appropriate value for helium as explained in Sect. 7, but in the case of hydrogen it changes with temperature, as shown in Table 2 in Sect. 7. Suitable values should be selected for respective cases in harmony with the temperatures under consideration. However, as its values are generally smaller for hydrogen than for helium, neglect of thermal diffusion in the case of hydrogen will not bring about serious error unless the temperature is very high.

Introducing the ratio Λ defined by

$$\Lambda = \frac{K}{D_1}, \tag{32.2}$$

we can write Eq. (32.1) in the following form:

$$\frac{\mathrm{d}n_1}{\mathrm{d}z} + \left[\frac{1}{1+\Lambda}\left(\frac{1}{H_1} + \frac{\Lambda}{H}\right) + \left(1 + \frac{\alpha_T}{1+\Lambda}\right)\frac{1}{T}\frac{\mathrm{d}T}{\mathrm{d}z}\right] n_1 = -\frac{\phi_1}{D_1(1+\Lambda)}. \tag{32.3}$$

This is the differential equation for the concentration n_1 in a steady state, ϕ_1 being the flux of atoms $(=n_1 w_1)$. When $\phi_1 = 0$, the corresponding distribution $n_{1\mathrm{eq}}(z)$ is given by

$$n_{1\mathrm{eq}}(z) = n_1(z_0)\frac{T_0}{T}\exp\left\{-\left[\int_{z_0}^{z}\left(\frac{1}{H_1} + \frac{\Lambda}{H}\right)\frac{\mathrm{d}z}{1+\Lambda} + \int_{z_0}^{z}\alpha_T(1+\Lambda)^{-1}\frac{\mathrm{d}T}{T}\right]\right\}. \tag{32.4}$$

When $\Lambda = 0$ there is no turbulence and the expression is reduced to that for the ordinary diffusive equilibrium distribution. When $\Lambda = \infty$ molecular diffusion is negligible and Eq. (32.4) leads to the mixing distribution characterized by the scale height H.

When $\phi_1 \neq 0$, we get by integrating Eq. (32.3)

$$n_1(z) = n_{1\mathrm{eq}}(z)\left[1 - \int_{z_0}^{z}\frac{\phi_1}{D_1(1+\Lambda)\,n_{1\mathrm{eq}}(z)}\,\mathrm{d}z\right]. \tag{32.5}$$

It will be seen from Eqs. (32.4) and (32.5) that any departure of the hydrogen or helium distribution from that given by ordinary diffusive equilibrium is due to eddy diffusion and/or a transport flow.

The vertical transport velocity given by Eq. (32.1) vanishes when the distribution of atoms is given by $n_{1\mathrm{eq}}$ of (32.4), and takes the greatest value in the upward direction when atoms are in a condition of complete mixing. In the latter case we have

$$\frac{1}{n_1}\frac{\mathrm{d}n_1}{\mathrm{d}z} = \frac{1}{n}\frac{\mathrm{d}n}{\mathrm{d}z} = -\frac{1}{H} - \frac{1}{T}\frac{\mathrm{d}T}{\mathrm{d}z} \tag{32.6}$$

so that w_1 is given by

$$w_{1\mathrm{mix}} = \frac{D_1}{H}\left(1 - \frac{H}{H_1} - \frac{\alpha_T H}{T}\frac{\mathrm{d}T}{\mathrm{d}z}\right). \tag{32.7}$$

It should be noted that this velocity is a physical parameter of the atmosphere independent of the abundance of the minor constituent under consideration. If the vertical transport velocity of a minor constituent is close to that given by Eq. (32.7) it is nearly in a condition of complete mixing; and if its vertical velocity is close to zero, it is nearly in the equilibrium condition given by Eq. (32.4).

33. The distribution of hydrogen atoms

α) BOWMAN et al.[1] and THOMAS and BOWMAN[2] have made a theoretical estimation of the concentrations of atmospheric minor constituents in the mesosphere and lower thermosphere (cf. Sect. 37). A diagram in the second paper gives $7.4 \cdot 10^{13}\,\mathrm{m}^{-3}$ for the *concentration* of hydrogen atoms *near 100 km* at noon. On the other hand, the same number density inferred from measurements of the Lyman-α line intensity is[3] about 1 to $3 \cdot 10^{13}\,\mathrm{m}^{-3}$, which makes us believe that the above-mentioned values of THOMAS and BOWMAN are too large. From the same follows a scale height of 7.5 km at 100 km for atomic hydrogen while the (empirical) CIRA 1972 model gives 6.2 km. Thus, the distribution of atomic hydrogen is seen to be fairly close to the condition of *complete mixing* since the scale height for hydrogen atoms in diffusive equilibrium, with the temperature at 100 km, would be about 180 km. Hence the vertical flux should be fairly near to that given by Eq. (32.7). According to KOCKARTS[3] the velocity $w_{1\,\mathrm{mix}}$ at 100 km is about $0.038\,\mathrm{m\,s}^{-1}$ so that if the hydrogen-atom concentration is taken as 1 to $3 \cdot 10^{13}\,\mathrm{m}^{-3}$, its maximum flux comes to be about 4 to $11 \cdot 10^{11}\,\mathrm{m}^{-2}\,\mathrm{s}^{-1}$; the actual *vertical flux* of H atoms will be somewhat smaller than this figure.

When the vertical flux takes such a large value as the above, the value of K or Λ has no great influence on the calculated distribution of concentration. This is because the right-hand side of Eq. (32.4) decreases while the expression in the brackets in Eq. (32.5) increases with increasing Λ, with the result that the product of these two factors, n_1, does not change to any great extent owing to their counteraction on each other. Thus, turbulence does not exert any great effect on the distribution of H atoms. As Λ decreases rapidly with increasing height, turbulence has no influence on the distribution of atoms at sufficiently high altitudes in any case.

β) It is also seen from Eq. (32.5) that the influence of the vertical flux on the distribution of atoms becomes small *at greater heights* because the decrease in $n_{1\mathrm{eq}}$ is not as rapid as the increase in D_1. Indeed, above about 500 km the number density profile of hydrogen atoms comes near to that of *diffusive equilibrium.* In Fig. 17 are shown the hydrogen-atom number density profiles given by KOCKARTS and NICOLET[4] in the height range 500 to 2,000 km for several exospheric temperatures.

The calculations are made on the assumption that the number density of H atoms at 100 km is $1 \cdot 10^{13}\,\mathrm{m}^{-3}$ and the turbopause is situated at the level of 105 km. If the values of effusion velocity at 500 km given by KOCKARTS[3] (without the correction of 30% stated in Sect. 31) are employed, the values of the vertical flux at 500 km are calculated to be about 2.0 to $3.5 \cdot 10^{11}\,\mathrm{m}^{-2}\,\mathrm{s}^{-1}$ and these correspond to flux values of about 2.2 to $4.0 \cdot 10^{11}\,\mathrm{m}^{-2}\,\mathrm{s}^{-1}$ at 100 km if

---

[1] Bowman, M.R., Thomas, L., Geisler, J.E. (1970): J. Atmos. Terr. Phys. *32*, 1661
[2] Thomas, L., Bowman, M.R. (1972): J. Atmos. Terr. Phys. *34*, 1843
[3] Kockarts, G. (1972): J. Atmos. Terr. Phys. *34*, 1729
[4] Kockarts, G., Nicolet, M. (1963): Ann. Geophys. *19*, 370

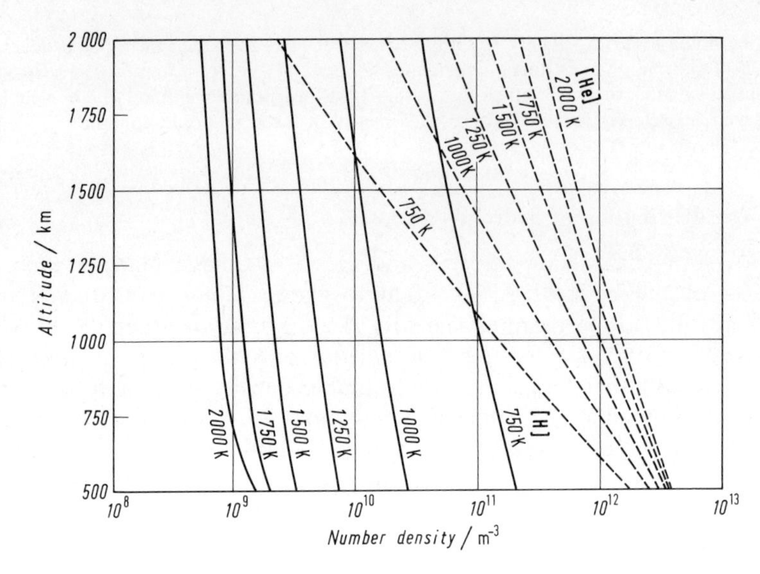

Fig. 17. Vertical distributions of hydrogen and helium atoms between 500 and 2,000 km for some exospheric temperatures

the geometrical fanning-out effect of the vertical flow is taken into account. The H-atom concentrations given in JACCHIA's 1971 model atmosphere[5] are almost the same as those shown in Fig. 17 for exospheric temperatures of not more than 1,300 K; but for higher temperatures they become somewhat smaller, and nearer to those given in the first paper of KOCKARTS and NICOLET[6]. At 1,900 K, JACCHIA's values are about 75% of those in the figure.

γ) The same Fig. 17 shows that the hydrogen-atom concentration increases rapidly with decreasing *exospheric temperature.* For an increase in temperature from 750 K to 2,000 K there is a reduction in concentration by a factor of about 1/130 at 500 km and even at 2,000 km still of nearly 1/70, the rate of reduction being especially marked at low temperatures. This is because the number of atoms that escape outside the atmosphere increases greatly with a rise in temperature. As a result of this the shape of the concentration profile deviates from that of diffusive equilibrium even at levels higher than 500 km when temperature is high; indeed we can see in Fig. 17 that this deviation appears in the height range 500–750 km if exospheric temperature considerably exceeds 1,500 K.

In order to see the situation at relatively low levels we have shown in Fig. 18 the profile in the height range 100 to 200 km given by KOCKARTS and NICOLET[4]. For exospheric temperatures of not less than 1,250 K it does not change to any great extent; the H-atom concentration at 200 km changes by a factor of only 2.5 for a temperature increase from 1,000 K to 2,000 K, but it increases by a factor of 10 for a temperature decrease from 1,000 K to 700 K.

In regard to the difference between day and night, PATTERSON[7] made a calculation taking into account the effect of a lateral flow around the Earth due to difference in day and night tempera-

[5] Jacchia, L.G. (1971): Smithson. Astrophys. Observat. Spec. Rep. *332*. The results were taken over into CIRA 1972

[6] Kockarts, G., Nicolet, M. (1962): Ann. Geophys. *18*, 269

[7] Patterson, T.N.L. (1966): Planet. Space Sci. *14*, 425

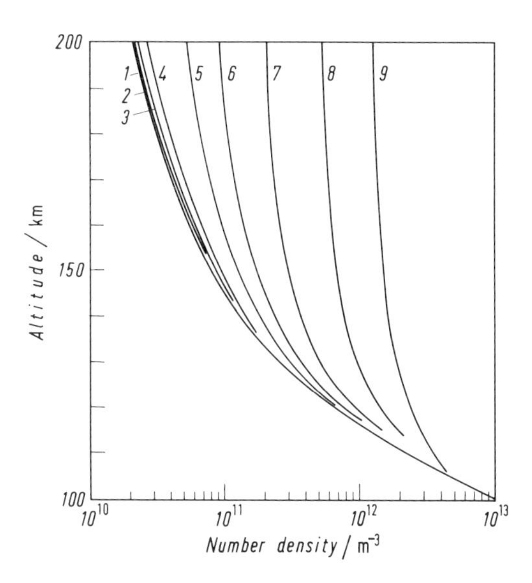

Fig. 18*. Vertical distributions of hydrogen atoms between 100 and 200 km for a number of exospheric temperatures from 600 K to 2,000 K calculated on the assumption that the turbopause is situated at a level of 105 km and that the number density at 100 km is $10^{13}\,m^{-3}$. Designation (1)...(9) of the different curves corresponds to the following exospheric temperatures: 2,000, 1,750, 1,500, 1,250, 1,000, 900, 800, 700, 600 K

tures. If exospheric temperature undergoes a diurnal variation changing sinusoidally between 700 K and 1,000 K or between 1,000 K and 1,500 K, the maximum-to-minimum ratio of the H-atom concentration at the base of the exosphere is obtained as 1.8 or 2.6, respectively.

34. The distribution of helium atoms

Helium is *produced* in the crust and mantle of the Earth *by radiative decay* of uranium and thorium. Of the atoms produced, the flux of those released into the atmosphere was estimated to be $7.7 \cdot 10^{9}\,m^{-2}\,s^{-1}$ by CRAIG and CLARKE[1] on the basis of their deep-sea measurement. This is considerably smaller than the corresponding figure for hydrogen atoms. According to KOCKARTS[2] $w_{1\,mix}$ given by Eq. (32.7) is about $1.8 \cdot 10^{-2}\,ms^{-1}$ at 100 km for helium; supposing that the He-atom concentration there is about $6 \cdot 10^{13}\,m^{-3}$ (estimated on the basis of the same mixing ratio as near the ground, $5.24 \cdot 10^{-6}$), we see that the above value of the helium flux is much lower than that corresponding to the maximum molecular diffusion transport velocity.

α) Thus, *near the 100 km level*, helium is in a state far from complete mixing, its distribution being given by Eq. (32.4). In other words, the vertical flux of helium atoms is so small that it has little influence on the distribution; this situation is entirely different from that of hydrogen atoms. On the other hand, as may be seen from Eq. (32.4), the effect of eddy diffusion could be very great, particularly when the exospheric temperature is low.

Figure 19 shows profiles of He-atom concentration calculated by KOCKARTS[2] for the values of the eddy-diffusion coefficient K of 0, 10, 10^2 and $10^3\,m^2\,s^{-1}$ and for a constant exospheric

* From Kockarts, Nicolet (1963): Ann. Geophys. *19* (No. 4), 373, Fig. 1

[1] Craig, H., Clarke, W.B. (1970): Earth Planet. Sci. Lett. 9, 45

[2] Kockarts, G. (1972): J. Atmos. Terr. Phys. *34*, 1729

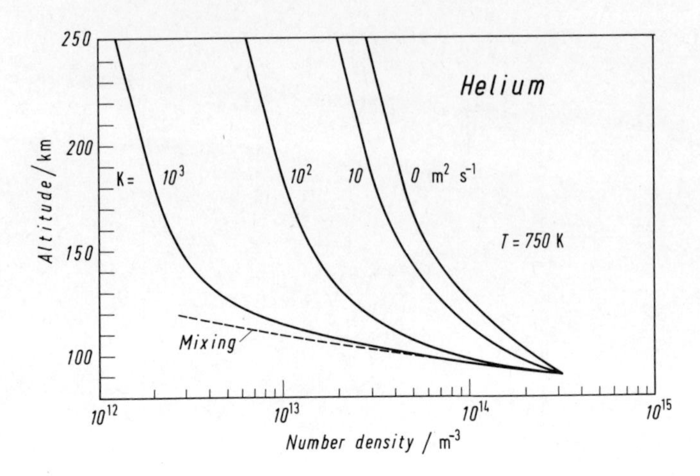

Fig. 19*. Vertical distributions of helium atoms between 90 and 250 km for different values of the eddy diffusion coefficient K in an atmospheric model with an exospheric temperature of 750 K

temperature of 750 K. It is seen that the profiles are very different for different K's; starting from the same concentration at 90 km, assumed values of K of 10 and $10^3\,m^2\,s^{-1}$ bring about a difference in the concentration at 250 km of not less than one order of magnitude. However, since the values of K are not quite certain, it is not advisable to determine the He-atom concentrations for increasing height from below upward by repeated use of Eq. (32.4).

β) As stated in the previous section, the *influence of eddy diffusion* becomes small at sufficiently *high altitudes* and the distribution tends to that given by diffusive equilibrium; hence the He-atom concentration profile can be determined if *one* value of He concentration at a sufficiently high level is obtained, except at low levels where diffusive equilibrium no longer prevails.

As will be seen from Fig. 21 below, He for the greater part of exospheric temperatures might be considered as the main constituent of the atmosphere at levels higher than about 1,000 km (and not too high). For example, according to JACCHIA and SLOWEY[3] helium occupies 80 and 67% (by mass) of the atmosphere at 700 km for temperatures of 700 K and 800 K, respectively, and the corresponding figures are 84 and 86% at 800 km. In such a case an approximate He concentration can be obtained by first deriving the atmospheric density from the atmospheric drag which an artificial satellite with a sufficiently high perigee suffers and then dividing this by the mass of a helium atom, adding a small correction appropriately, if necessary. JACCHIA, based on such values and neglecting eddy diffusion, has determined the distribution of helium on the assumption of diffusive equilibrium and has used the result in his model atmosphere. Some of the values of He-atom concentration taken from the model atmosphere are depicted in Fig. 17 for the height range 500 to 2,000 km, together with the H-atom concentrations referred to in the previous section. It should be noted that the neglect of eddy diffusion ($K=0$) may lead to an underestimation of He-atom concentration at low levels when the determination is started at a high level (cf. Fig. 19).

γ) *The vertical profile of helium* concentration in the lower thermosphere has rather often been measured by rocket techniques using mass spectrometers. Figure 20 shows the results of observations made in middle latitudes (White

* From Kockarts (1972): J. Atmos. Terr. Phys. *34* (No. 10), 1734, Fig. 1

[3] Jacchia, L.G., Slowey, J.W. (1968): Planet. Space Sci. *16*, 509

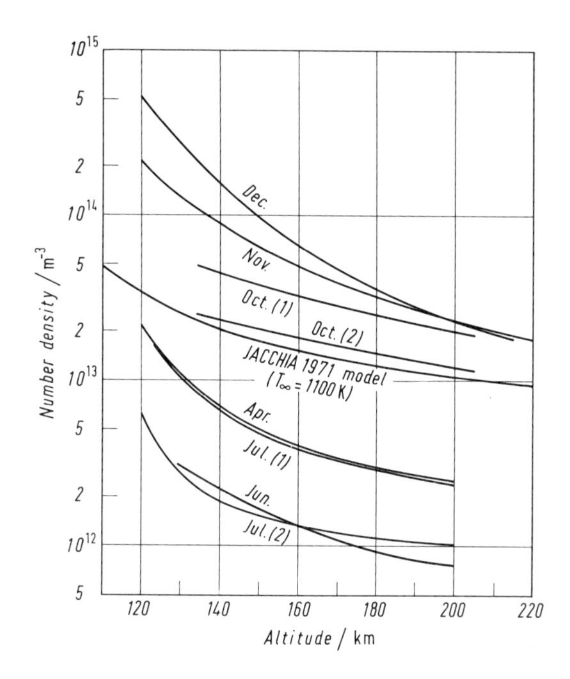

Fig. 20. Vertical distributions of helium atoms in the lower thermosphere measured by rocket technique together with that given by JACCHIA's 1971 Model Atmosphere ($T_{\infty}=1{,}100$ K). Times, places, and performers of the rocket measurements are as follows:

APR: 0345 MST, 15 Apr. 1965, White Sands, HEDIN and NIER

NOV: 0445 MST, 30 Nov. 1966; DEC: 1409 MST, 2 Dec. 1966 — White Sands, KASPRZAK et al.

JUN: 1249 MST, 21 June 1967; JUL (1): 1224 MST, 20 July 1967; JUL (2): 0200 MST, 20 July 1967 — White Sands, KRANKOWSKY et al.

OCT (1): 1533 CET, 10 Oct. 1967; OCT (2): 1423 CET, 4 Oct. 1967 — Salto di Quirra, Sardinia, BITTERBERG et al.

Sands (32.4 °N) or Sardinia (40 °N))[4-7]. Although there are great differences among measured values, it is seen that small values occur in summer, larger ones in winter, and intermediate ones in spring and autumn, such that there is a systematic seasonal variation. The maximum-to-minimum ratio amounts to about 30 to 80, though these figures themselves may not be very reliable.

For comparison we may look at the He-atom distribution given in JACCHIA's 1971 model atmosphere ($T_{\infty}=1{,}100$ K)[8]; it agrees fairly well with one of the October curves and so may correspond to a yearly average. In applying it, it must, however, be taken into account that the seasonal variation can be considerable. As the situation should be similar at higher levels, the curves of Fig. 17 should also be regarded as representing average conditions only.

[4] Hedin, A.E., Nier, A.O. (1966): J. Geophys. Res. *71*, 4121
[5] Kasprzak, W.T., Krankowsky, D., Nier, A.O. (1968): J. Geophys. Res. *73*, 6765
[6] Krankowsky, D., Kasprzak, W.T., Nier, A.O. (1968): J. Geophys. Res. *73*, 7291
[7] Bitterberg, W., Brutchhausen, K., Offermann, D., von Zahn, U. (1970): J. Geophys. Res. *75*, 5528
[8] Jacchia, L.G. (1971): Smithson. Astrophys. Observat. Spec. Rep. *332*

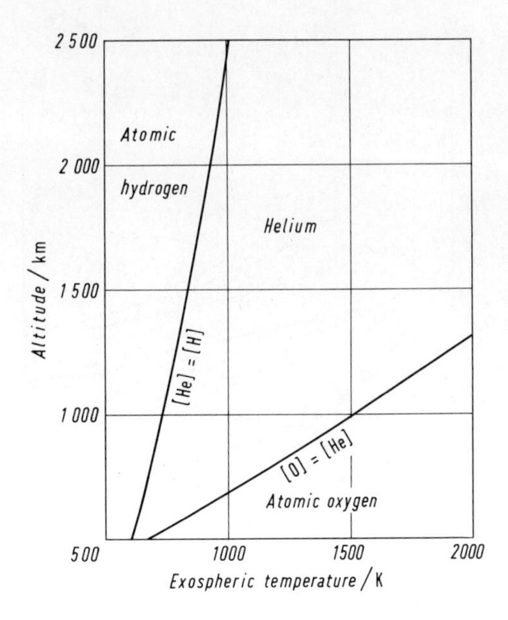

Fig. 21. Ranges of altitudes in which atomic hydrogen, helium, or atomic oxygen is the predominant constituent of the atmosphere between 500 and 2,500 km

Among the observed curves in Fig. 20 there are a few with a very steep slope at low heights but rapidly becoming gentle with increasing height, thus suggesting that eddy diffusion plays an important role (cf. Fig. 19). There are also others for which the slope varies with height in a way similar to JACCHIA's. Thus we cannot draw a definite conclusion from Fig. 20 as to whether eddy diffusion is important or not.

Let us for the moment consider the average conditions shown in Fig. 17. Using this figure and JACCHIA's 1971 model atmosphere[8] we have determined the so-called *transition heights*, namely the levels where the concentrations of hydrogen and helium atoms become equal, or where those of oxygen and helium atoms become equal. These are shown as a function of the exospheric temperature in Fig. 21. It is seen that, for a wide range of exospheric temperatures, helium becomes the main neutral constituent; in particular, when the temperature is low, this may happen even at a height as low as 500 km. Indeed, during the minimum of solar activity, helium may be observed at such low heights.

δ) The *seasonal variation* in helium concentration mentioned above is especially marked in high latitudes; this was demonstrated by drag data of artificial satellites such as Explorer 9, 14, 19, and, in particular, the balloon satellite Echo 2. The helium concentration over the winter pole is larger than that over the summer pole by a factor of approximately 3 to 4. This phenomenon is called the "*winter helium bulge*". According to observations by the artificial satellite OGO 6 in June 1969, the position of the maximum density depends somewhat on the longitude but is always found in the range between $-40°$ and $-70°$ geographic latitude, and in all cases fairly close to $-53°$

geomagnetic latitude[9]. The existence of the helium bulge has been independently inferred from observations of the intensity of the infrared 1,083 nm line of airglow which is emitted by helium atoms[10]. JOHNSON and GOTTLIEB[11] have attributed the helium bulge to a wind system blowing from the summer to the winter polar region; helium is transported on this wind system and concentrated in high latitudes on the winter hemisphere, resulting in the helium bulge there. Although this theory seems most promising, there are also other opinions that ascribe it to the latitudinal variation in the height of the helium turbopause[12,3] or to that in the coefficient of eddy diffusion[2].

ε) As stated above, the vertical flow of He atoms is estimated to be of the order of $10^{10}\,\mathrm{m}^{-2}\,\mathrm{s}^{-1}$. On the other hand, if their *rate of escape* is calculated using Fig. 17 and the effusion velocities listed in Table 3, the fluxes are $4.2\cdot 10^{10}$, $5.8\cdot 10^{9}$, $4.1\cdot 10^{8}$, $9.8\cdot 10^{6}$, $3.4\cdot 10^{4}$, and $2.5\,\mathrm{m}^{-2}\,\mathrm{s}^{-1}$ for exospheric temperatures of 2,000, 1,750, 1,500 1,250, 1,000 and 750 K, respectively. Therefore, only if the exospheric temperature were constantly around 2,000 K, the vertical flow of helium could just compensate the escape flow and a steady state should be maintained. In reality, however, during periods of minimum solar activity, the exospheric temperature becomes much lower than this value so that the escape flux is on the average much smaller than its vertical flux and helium should be accumulated in the atmosphere in a larger amount than actually observed. Thus, we must invoke some other mechanism which assists in releasing helium atoms outside the atmosphere in order to account for its amount presently found in the atmosphere.

One explanation starts with the plausible assumption that a part of the helium atoms are ionized by solar radiation and by collisions with energetic particles so that helium ions are produced which could be carried away by the "polar wind" (see Sect. 38)[13].

This mechanism is also conceivable in the case of hydrogen. In particular, when the temperature is low, this process should become relatively important. It may therefore be that the distribution of hydrogen atoms in the polar thermosphere is entirely altered during solar minimum.

35. Distributions of hydrogen, helium and other minor ions

α) When we consider the distribution of ions, we must take into account the *electromagnetic force* as well as gravity. However, as we are presently concerned with their vertical distribution in a quasi-static condition, only the vertical component of the electric field need be considered. This is here denoted by E. It is further assumed that other kinds of force can be derived

---

[9] Reber, C.A., Harpold, D.N., Horowitz, R., Hedin, A.E. (1971): J. Geophys. Res. *76*, 1845

[10] Tinsley, B.A. (1968): Planet. Space Sci. *16*, 91; Shefov, N.N. (1968): Planet. Space Sci. *16*, 1103; Christensen, A.B., Patterson, T.N.L., Tinsley, B.A. (1971): J. Geophys. Res. *76*, 1764

[11] Johnson, F.S., Gottlieb, B. (1970): Planet. Space Sci. *18*, 1707

[12] Cook, G.E. (1967): Planet. Space Sci. *15*, 627; Keating, G.M., Prior, E.J. (1968): Space Res. *8*, 982

[13] Patterson, T.N.L. (1968): Rev. Geophys. *6*, 553

from the potential ϕ. If gravity and the centrifugal force only are taken into account, we have

$$\phi = -g_0 \frac{r_0^2}{r} - \frac{1}{2} \omega^2 \sin^2\theta \cdot r^2, \tag{35.1}$$

where ω is the angular velocity of the Earth's rotation, θ colatitude, and the subscript 0 refers to the value of a variable at a datum level.

If the equation of motion (4.2) is applied to electrons and ions in a quasi-static condition, we get

$$\frac{1}{n_e} \frac{d}{dr}(n_e k T_e) = -m_e \frac{d\phi}{dr} - qE, \tag{35.2}$$

$$\frac{1}{n_i^+} \frac{d}{dr}(n_i^+ k T^+) = -m_i^+ \frac{d\phi}{dr} + qE, \tag{35.3}$$

where q is the unitary charge, the subscript e and the superscript + refer to electrons and positive ions, respectively, and the subscript i distinguishes different kinds of ions; we have further assumed a common temperature, denoted by T^+, for all kinds of ions; it is very often different from the temperature T_e of the electrons. The condition of neutrality of total charge is

$$n_e = \sum_i n_i^+. \tag{35.4}$$

From Eqs. (35.2)–(35.4) the vertical distributions of electrons and ions are found to be:

$$n_e = n_{e0} \frac{T_{p0}}{T_p} \exp\left[-\int_{r_0}^{r} \frac{m_e + m^+}{2k T_p} \frac{d\phi}{dr} dr\right] \tag{35.5}$$

and

$$n_i^+ = n_{i0}^+ \frac{T_{p0}}{T_p} \exp\left[-\int_{r_0}^{r} \frac{m_i^+}{kT^+} \frac{d\phi}{dr} dr + \int_{r_0}^{r} \frac{1}{kT^+} \frac{T_e m^+ - T^+ m_e}{2T_p} \frac{d\phi}{dr} dr\right], \tag{35.6}$$

where T_p is the plasma temperature defined by Eq. (9.31) and m^+ is the mean mass of ions given by

$$m^+ = \frac{\sum_i n_i^+ m_i^+}{\sum_i n_i^+}. \tag{35.7}$$

Eq. (35.6) shows that the concentration ratio of any two kinds of ions depends on T^+ only and not on T_e:

$$\frac{n_j^+}{n_k^+} \propto \exp\left[-\int_{r_0}^{r} \frac{(m_j^+ - m_k^+)}{kT^+} \frac{d\phi}{dr} dr\right]. \tag{35.8}$$

The electric field is obtained from Eqs. (35.2)...(35.4) by eliminating n_e and n_i^+:

$$qE = \frac{m^+ T_e - m_e T^+}{2T_p} \frac{d\phi}{dr} + k T_p \frac{d}{dr} \frac{T^+}{T_p}. \tag{35.9}$$

β) If we now assume that the *charged temperature ratio*, T_e/T^+, takes a constant value *independent of height*, we get from Eqs. (35.5) and (35.6) the following expressions:

$$\left(\frac{n_e T_p}{n_{e0} T_{p0}}\right)^{\frac{T_e+T^+}{T^+}} = \frac{\sum_i n_{i0}^+ \exp\left(-\int_{r_0}^{r} \frac{m_e+m_i^+}{kT^+} \frac{d\phi}{dr} dr\right)}{\sum_i n_{i0}^+} \tag{35.10}$$

and

$$n_i^+ = n_{i0}^+ \left(\frac{T_{p0}}{T_p}\right)^{\frac{T_e+T^+}{T^+}} \exp\left(-\int_{r_0}^{r} \frac{m_e+m_i^+}{kT^+} \frac{d\phi}{dr} dr\right) \cdot \left(\frac{n_{e0}}{n_e}\right)^{T_e/T^+}. \tag{35.11}$$

In the above equations (35.5)–(35.11), we can of course neglect m_e against m_i or m^+, but we have preserved it in order to show the symmetry or antisymmetry of the expressions with respect to electrons and ions.

γ) Below we shall further make the special assumption that the *ion temperatures* are *equal to* the *neutral temperature* denoted by T. We use suffix i also to identify the neutral species corresponding to the ion so denoted. Then, we have $m_i = m_e + m_i^+$ and

$$n_i = n_{i0} \frac{T_0}{T} \exp\left(-\int_{r_0}^{r} \frac{m_i}{kT} \frac{d\phi}{dr} dr\right), \tag{35.12}$$

and Eq. (35.10) can be written as follows (since now $T_e/T_{e0} = T^+/T_0^+ = T/T_0$):

$$\left(\frac{n_e}{n_{e0}}\right)^{\frac{T_e+T^+}{T^+}} = \left(\frac{T_0^+}{T^+}\right)^{\frac{T_e}{T^+}} \frac{\sum_i n_{i0}^+ \cdot (n_i/n_{i0})}{\sum_i n_{i0}^+}. \tag{35.13}$$

We additionally have the following expression from Eqs. (35.11) and (35.12):

$$\frac{n_i^+}{n_{i0}^+} = \frac{n_i}{n_{i0}} \left(\frac{n_{e0} T_0^+}{n_e T^+}\right)^{\frac{T_e}{T^+}}. \tag{35.14}$$

By differentiating Eq. (35.6) with respect to r, it is seen that, if the temperature gradient can be neglected, n_i^+ reaches a maximum at the level where

$$m^+ = \left(1 + \frac{T^+}{T_e}\right) m_i^+ \tag{35.15}$$

is satisfied. Since m^+ diminishes with increasing height, the peak in the vertical distribution of an ion should be situated at higher levels if the ion is lighter.

As the most important kinds of ions in the high atmosphere are O^+, He^+, and H^+, let us now consider these kinds of ions only. Applying Eq. (35.14) to these ions, eliminating the second factor, and designating the densities of charged and neutral species by N and n, respectively, we get:

$$\frac{N(\mathrm{H}^+)}{N(\mathrm{O}^+)}=\frac{n(\mathrm{H})\cdot n_0(\mathrm{O})}{n(\mathrm{O})\cdot n_0(\mathrm{H})}\cdot\frac{N_0(\mathrm{H}^+)}{N_0(\mathrm{O}^+)} \tag{35.16}$$

$$\frac{N(\mathrm{He}^+)}{N(\mathrm{O}^+)}=\frac{n(\mathrm{He})\cdot n_0(\mathrm{O})}{n(\mathrm{O})\cdot n_0(\mathrm{He})}\cdot\frac{N_0(\mathrm{He}^+)}{N_0(\mathrm{O}^+)}. \tag{35.17}$$

Thus, if the ratio of the concentrations of two kinds of ions at any datum level is given, its values above are determined by the distributions of their parent neutral atoms.

δ) As the ionization potentials of hydrogen and oxygen atoms are nearly equal, the *reacting system* of the charge exchange reaction

$$\mathrm{H}+\mathrm{O}^+ \rightleftarrows \mathrm{H}^+ + \mathrm{O} \tag{35.18}$$

are nearly in the state of resonance before and after the reaction so that this reaction proceeds in both directions very rapidly, promptly reaching an equilibrium condition. In this condition, if the statistical weights of the ground states of the atoms and ions are taken into consideration, the following relation holds:

$$\frac{N(\mathrm{H}^+)}{N(\mathrm{O}^+)}=\frac{9\cdot n(\mathrm{H})}{8\cdot n(\mathrm{O})}, \tag{35.19}$$

showing that the ratio of ion concentrations $n(\mathrm{X}^+)$ are completely determined by that of neutral concentrations for hydrogen and oxygen atoms.

In the case of He^+ and O^+, it is not a simple matter to obtain the ratio of their concentrations at a datum level because the mechanism of their production and loss as well as their diffusion must be taken into account (as for O^+ see Chapt. IV). Therefore, KOCKARTS and NICOLET[1] have assumed the ratio of the degrees of ionization of He and O at the datum level to be Y and made further calculations:

$$Y=\frac{N_0(\mathrm{He}^+)}{n_0(\mathrm{He})}\bigg/\frac{N_0(\mathrm{O}^+)}{n_0(\mathrm{O})}. \tag{35.20}$$

From Eqs. (35.17), (35.19), and (35.20) we have

$$N(\mathrm{H}^+)=\frac{N_e}{1+\frac{8}{9}(Yn(\mathrm{He})+n(\mathrm{O}))/n(\mathrm{H})}, \tag{35.21}$$

$$N(\mathrm{He}^+)=\frac{N_e}{1+\frac{9}{8}(n(\mathrm{H})+\frac{8}{9}n(\mathrm{O}))/(Yn(\mathrm{He}))}, \tag{35.22}$$

$$N(\mathrm{O}^+)=\frac{N_e}{1+\frac{9}{8}(n(\mathrm{H})+\frac{8}{9}Yn(\mathrm{He}))/n(\mathrm{O})}. \tag{35.23}$$

[1] Kockarts, G., Nicolet, M. (1963): Ann. Geophys. *19*, 370

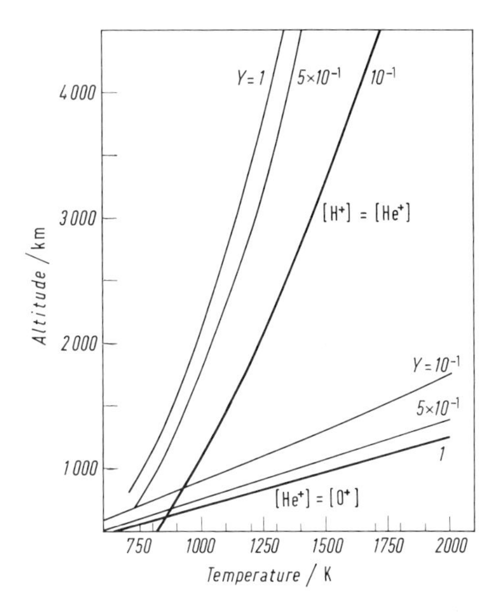

Fig. 22*. "Transition heights", i.e., levels where after the model computation the number density of H^+ ions becomes equal to that of He^+ ions or where the number density of He^+ ions becomes equal to that of O^+ ions as a function of the exospheric temperature for a few values of Y, the assumed ratio of the degree of ionization of He and O at 500 km

ε) An electron-density distribution is given by Eq. (35.5) and those of the neutrals can be regarded as known from the account given hitherto, so that *ion-density height distributions* can be obtained using Eqs. (35.21)...(35.23), provided the value of Y can be appropriately determined.

Kockarts and Nicolet[1] have assumed three values (1, 0.5, and 0.1) for Y and calculated ion density distributions in the height range of 500 to 3,000 km for several exospheric temperatures. They assume electron and ion temperatures to be equal to the exospheric temperature. Part of their results of calculation is shown in Fig. 22; the altitudes of transition levels for H^+ to He^+ or for He^+ to O^+ are expressed as a function of the exospheric temperature. The results are somewhat different for the three values of Y. Considering that Y is probably closer to 1 rather than to 0.1, at least in normal conditions[1], one can roughly see from the figure what kind of ion should be predominant for a given temperature and altitude.

When the temperature becomes lower, the area in which H^+ is the predominant ion extends to lower heights while, at higher temperature the areas in which O^+ and He^+ are predominant, extend to higher altitudes, but the lower boundary of the area of He^+ predominance also shifts upward.

The calculations also show that, if T_e and T^+ are not less than about 1,000 K, the ion population at 500 km can be regarded as being almost composed of O^+ only; if temperatures become lower, the upper limit of the range where O^+ is largely prevailing ($N(O^+) \simeq N_e$) comes down, e.g. to 350 km for T^+ and T_e equal to 700 K. It is remarkable that, for temperatures below 1,000 K, the ratio $N(O^+)/N(H^+)$ decreases greatly with increasing height. For $T^+ = T_e = 700$ K this ratio falls off from 1 to about 10^{-4} when the altitude increases from 500 km to 1,000 km.

---

* From Kockarts, Nicolet (1963): Ann. Geophys. *19* (No. 4), 384, Fig. 16

As seen from Eqs. (35.21)...(35.23), or Eq. (35.8), the *relative* concentrations of different kinds of ions do not depend on a difference in electron and ion temperatures. However, the electron concentration itself depends on it, as appears from Eq. (35.5), and hence the same is true for the absolute values of ion concentrations. For constant T^+ the decrease in electron concentration with height becomes more gradual as the ratio T_e/T^+ increases.

A summary of recent experimental evidence on ion composition is given in a short paper by K. RAWER in this volume.

36. The effect of the flow of main ions. Hitherto we considered the distribution of minor ions under quasi-static conditions. If, however, there is a vertical flow of the main constituent ions, other species of ions suffer from a drag resulting from collisions with the main ions, due to Coulomb interaction. Finally, all ions move together and their distributions will be influenced. This fact was first discussed quantitatively by SCHUNK and WALKER [5][1]. We shall now briefly outline the method of their study and their results. The effect becomes important mainly in the upper ionosphere.

α) According to BANKS[2] the following relationship holds among the *collision frequency* of a minor ion with neutral particles[3], ν_{in}, that with the major ions, O^+, $\nu_{ii'}$, the number density of neutral particles, n, that of atomic-oxygen ions, $N(O^+)$, and the ion temperature T_i (assumed to be common for the minor ion and the major ion O^+). The author finds:

$$\frac{\nu_{in}}{\nu_{ii'}}=2.3\cdot 10^{-9}\left(\frac{T_i}{K}\right)^{3/2}\frac{n}{N(O^+)}. \tag{36.1}$$

From this equation it can be seen that, if T_i is of the order of 1,000 K and $N(O^+)/n>7\cdot 10^{-4}$, the ratio becomes less than 0.1. These conditions may be regarded as roughly satisfied at heights above the F2-layer peak. Then, if we restrict our considerations to this region only, we may be able to ignore collisions between minor ions and neutral particles and to deal with the ternary system composed of electrons, O^+-ions, and the minor ions[0]. In the following, electrons, O^+-ions, and the minor ions will be indicated by the subscripts 1, 2, and 4, respectively.

β) In the following, we establish the *fundamental equations* finally determining the different density profiles. From Eqs. (3.1) and (3.2) the following three equations are obtained:

$$\frac{n^2}{\rho}(m_2 D_{12}\boldsymbol{d}_2+m_4 D_{14}\boldsymbol{d}_4)-\frac{D_1^T}{m_1}\frac{\nabla T_e}{T_p}=n_1\bar{\boldsymbol{V}}_1, \tag{36.2}$$

$$\frac{n^2}{\rho}(m_2 D_{42}\boldsymbol{d}_2+m_1 D_{41}\boldsymbol{d}_1)-\frac{D_4^T}{m_4}\frac{\nabla T_i}{T_p}=n_4\bar{\boldsymbol{V}}_4, \tag{36.3}$$

$$\boldsymbol{d}_1+\boldsymbol{d}_2+\boldsymbol{d}_4=0, \tag{36.4}$$

[1] Schunk, R.W., Walker, J.C.G. (1970): Planet. Space Sci. *18*, 1319

[2] Banks, P. (1966): Planet. Space Sci. *14*, 1105

[3] See SUCHY's contribution in vol. 49/7

[0] It is possible to discuss in a similar way the four-component system that includes neutral atomic oxygen as well, but the expressions derived will become very complicated [5]

where $\boldsymbol{d}_i(i=1, 2 \text{ or } 4)$ is given by Eq. (3.2); assuming that electron and ion temperatures are different, we have replaced the original temperature T by plasma temperature T_p and the original common temperature gradient ∇T by electron temperature gradient ∇T_e in Eq. (36.2) and by ion temperature gradient ∇T_i in Eq. (36.3), though the validity of this procedure is not warranted. When gravity and electric force are introduced in the force terms in Eq. (3.2) and Eq. (9.24) is taken into account, the following equations are derived:

$$\boldsymbol{d}_i = \frac{1}{p}(\nabla(n_i k T_p) - n_i m_i \boldsymbol{g} - n_i q_i \boldsymbol{E}, \quad i=1,2,4. \tag{36.5}$$

Multiplying both sides of Eq. (36.2) by $(m_1 D_{41} - m_2 D_{42})/(m_2 D_{12})$, adding the resulting equation to Eq. (36.3) side by side, and eliminating $\boldsymbol{d}_1$ and $\boldsymbol{d}_2$ using Eq. (36.4), we get

$$\begin{aligned} n_4 \bar{V}_4 + r n_1 \bar{V}_1 = &-\frac{n^2}{\rho} \boldsymbol{d}_4 (m_1 D_{41} - r m_4 D_{14}) \\ &- \frac{D_4^T T_i}{m_4 T_p} \nabla \ln T_i - r \frac{D_1^T T_e}{m_1 T_p} \nabla \ln T_e, \end{aligned} \tag{36.6}$$

where

$$r \equiv \frac{m_1 D_{41} - m_2 D_{42}}{m_2 D_{12}}. \tag{36.7}$$

Using the equations of charge neutrality and no electric current flow, the mass-velocity of this ternary system given by Eqs. (2.3) and (2.4) can be expressed approximately as

$$\boldsymbol{v}_0 \simeq \bar{\boldsymbol{v}}_1 + \frac{n_4(m_4 - m_2)}{n_1(m_1 + m_2)} (\bar{\boldsymbol{v}}_4 - \bar{\boldsymbol{v}}_1) \tag{36.8}$$

if $(n_4/n_1)^2$ is neglected compared with unity. Using this equation and Eq. (2.5) and considering that $m_1 D_{41}/m_2 D_{12}$ and $m_2 D_{42}/m_2 D_{12}$ are small quantities of the order of $(m_1/m_2)^{1/2}$ or less[0], Eq. (36.6) can be brought into the following form:

$$\begin{aligned} n_4 \bar{\boldsymbol{v}}_4 = n_4 \bar{\boldsymbol{v}}_1 - \frac{n n_4 m_1}{\rho} D_{41} \Big[&\nabla \ln n_4 - \frac{m_4}{k T_p} \boldsymbol{g} + \frac{T_e}{T_p} \nabla \ln n_1 \\ &+ \nabla \ln T_p - \frac{\beta}{T_p} \nabla T_i + (1 + \alpha - \gamma) \frac{1}{T_p} \nabla T_e \Big]; \end{aligned} \tag{36.9}$$

in deriving this equation Eqs. (9.33a) (with $T = T_e$) and (36.5) have also been used; α is given by Eq. (9.34) or (9.34a) and β and γ by the following

[0] This can be seen from Inequality (9.8) or can be demonstrated by a reasoning similar to that given in Sect. 9 with the aid of Eqs. (3.20) and (5.11) (cf. Eq. (9.13))

expressions:

$$\beta = -\frac{\rho}{n n_4 m_1 m_4} \frac{D_4^T}{D_{41}}, \tag{36.10}$$

$$\gamma = \frac{m_2 D_{42} - m_1 D_{41}}{m_2 D_{12}} \frac{\rho}{n n_4 m_1^2} \frac{D_1^T}{D_{41}}. \tag{36.11}$$

Eq. (36.9) can be applied separately to respective minor ions. γ takes the values of 0.764 for all singly charged minor ions [5]. β takes different values from one ion to another; for H^+, He^+, and N^+ values of -1.05, -0.98, and -0.15, respectively, were obtained by the method described by HIRSCHFELDER et al. [5].

γ) Assuming, for the sake of simplicity, *uniform temperature*, i.e. that neutral, electron, and ion temperatures are equal to one another and independent of height, let us consider the vertical distribution of minor ions. Under these particular assumptions we get from Eq. (36.9)

$$\bar{v}_4 = \bar{v}_1 - D\left(\frac{1}{n_4}\frac{dn_4}{dz} + \frac{m_4 g}{kT} + \frac{1}{n_1}\frac{dn_1}{dz}\right), \tag{36.12}$$

where $\bar{v}_i$ is the average speed of particles of species i and

$$D = \frac{n m_1}{\rho} D_{41}. \tag{36.13}$$

In Eq. (36.12) the first term on the right-hand side is representative of the drag effect due to the main ions, and the last term in the parentheses expresses the influence of the electric field (cf. Eq. (9.33a)). Since the electron or main ion-density distribution can be obtained as a function of the altitude by the method described in the previous chapter and their drift velocity $v_1(=F_1/n_1)$ can be calculated by Eq. (24.22) or (24.22a), Eq. (36.12) can be regarded as an equation for the unknown functions n_4 and v_4. If this equation is solved for n_4 simultaneously with the equation of continuity (4.1) and other equations of reaction expressing the production and loss of the minor ions, their number-density profile can be obtained.

δ) Just as an example we report on the work of SHUNK and WALKER [5][1].

These authors, taking into consideration the following reactions of production and loss of H^+, He^+, and N^+, have calculated the *number density profiles* of these minor ions. The values of the rate coefficients adopted are indicated in the parentheses on the right side of the reaction equations:

$$N_2 + h\nu \rightarrow N + N^+ + e^-, \quad (4.36\cdot 10^{-8}\,s^{-1}), \tag{36.14}$$

$$He^+ + N_2 \rightarrow He + N + N^+, \quad (1.5\cdot 10^{-15}\,m^3 s^{-1}), \tag{36.15}$$

$$N^+ + O_2 \rightarrow NO^+ + O, \quad (3.5\cdot 10^{-16}\,m^3 s^{-1}), \tag{36.16}$$

$$N^+ + O_2 \rightarrow O_2^+ + N, \quad (4.5\cdot 10^{-16}\,m^3 s^{-1}), \tag{36.17}$$

$$He + h\nu \rightarrow He^+ + e^-, \quad (8.16\cdot 10^{-8}\,s^{-1}), \tag{36.18}$$

$$H + O^+ \rightleftarrows H^+ + O, \quad (4\cdot 10^{-16}\,m^3 s^{-1}). \tag{36.19}$$

The results of this calculation are illustrated in Fig. 23 together with the profile of O^+-ion number density assumed for its establishment. For simplification it has been assumed that the atmospheric

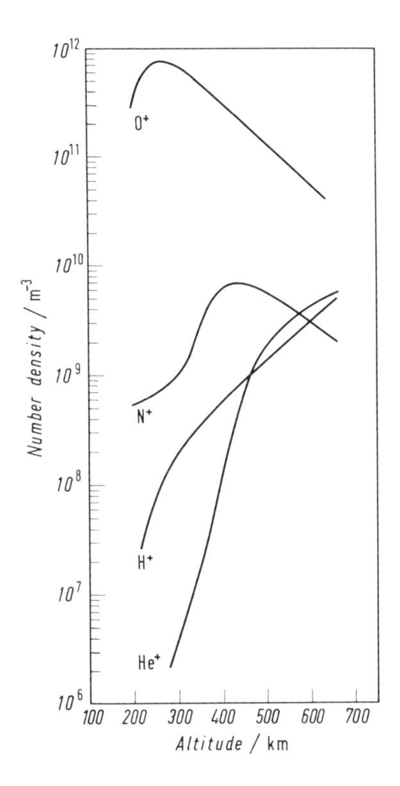

Fig. 23*. Number density profiles of H^+, He^+, and N^+ ions calculated on the basis of the assumed O^+ distribution given on top and the reactions as shown in the text. The atmospheric temperature is assumed to be 1,000 K and the flux of O^+ ions as well as that of minor ions are assumed to be zero at the top of the ionosphere

temperature is uniform and equal to 1,000 K and that the flux of O^+ ions as well as that of minor ions is zero at the top of the ionosphere.

ε) SCHUNK and WALKER have then given consideration to the effects exerted on the distribution of minor ions by the *flux* of the major or minor ions *at the top of the ionosphere*. In this case there is a restriction on the values of the upward flux of minor ions in order to avoid that their number density, at any height, may become negative. The maximum value is called the critical flux. Assuming the ambipolar flux of the main ions at the top of the ionosphere to be zero, the critical flux was found to be 25, 4.3 and $0.95 \cdot 10^{10}\,m^{-2}\,s^{-1}$ for H^+, He^+, and N^+, respectively. We show in Fig. 24 one such result valid for H^+. When the flux of minor ions at the top of the ionosphere comes near to the critical one, the profile of their number density abruptly undergoes a marked change and deviates widely from the original profile obtained for zero flux; the corresponding scale height comes then near to that of the main ion. The behavior is similar for He^+ and N^+, but in the latter case the distortion of the

* From Schunk, Walker (1970): Planet. Space Sci. *18* (No. 9), 1325, Fig. 2

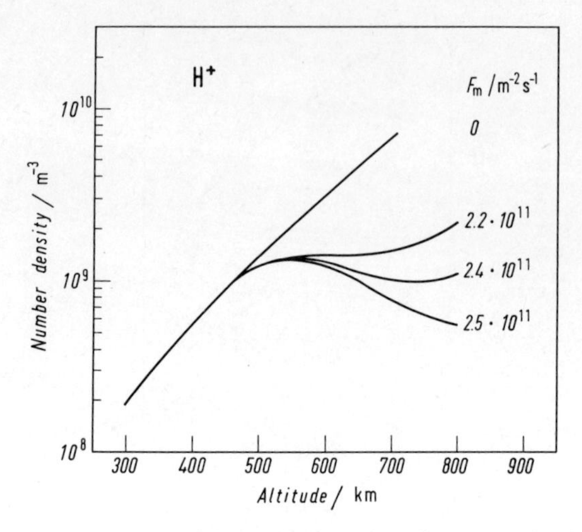

Fig. 24*. Number density profiles of H^+ ions calculated for a few assumed values of its flux at the top of the ionosphere, F_m. The ambipolar flux of the main ions is assumed to be zero at the top

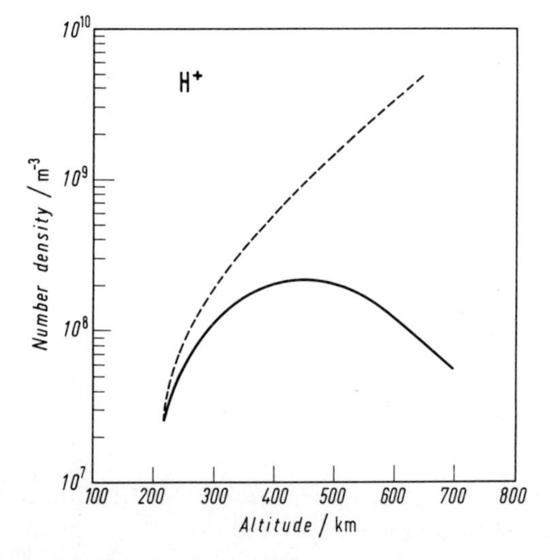

Fig. 25.** Comparison of the number density profiles of H^+ ions calculated for the zero ambipolar flux of the main ions at the top of the ionosphere (dashed line) and for that of $10^{13}\,m^{-2}\,s^{-1}$ (solid line)

profile is not very remarkable, owing to the relatively small difference in mass of O^+ and N^+ ions.

Another calculation clarified the effect of a large topside ambipolar flux of main ions on the distribution of minor ions. It was assumed that, at high altitudes, the first term on the right-hand side of Eq. (36.12) completely dominates other terms; $\bar{v}_4$ is set equal to $\bar{v}_1$ as the upper boundary condition. Fig. 25 compares the result with that when there is no ambipolar flux in the case of H^+

* From Schunk, Walker (1970): Planet. Space Sci. *18* (No. 9), 1329, Fig. 5
** From Schunk, Walker (1970): Planet. Space Sci. *18* (No. 9), 1330, Fig. 8

ions. The ambipolar flux is taken as $1 \cdot 10^{13}\,m^{-2}\,s^{-1}$ and the minor ion flux as $1.6 \cdot 10^{11}\,m^{-2}\,s^{-1}$. As $\bar{v}_4 = \bar{v}_1$ at high altitudes, $n_4(=F_4/v_4)$ keeps a constant ratio to $n_1(=F_1/v_1)$ there so that the minor ion scale height tends to that of the main ions; again, the minor ion-density profile deviates markedly from the original one for zero ambipolar flux, as is seen in Fig. 25.

37. Vertical distributions of other minor constituents. The method explained in Chapter III can be applied in principle to the discussion of the vertical profiles of minor constituents in the high atmosphere such as O_3, OH, H, H_2, H_2O, HO_2, H_2O_2, CH_4, N, NO, NO_2, N_2O, Ar, but also oxygen atoms and molecules in an excited state, $O(^1D_2)$ and $O_2(a^1\Delta_g)$, metallic atoms etc. This sort of studies has been made extensively in recent years. However, in almost all of these cases, complicated chemical reactions play a predominant role in determining the distribution of the constituent rather than the process of diffusion so that its discussion may lead to that of chemical reactions themselves. This is not the purpose of the present article[1]. For details, we refer the reader to the review paper by STROBEL[2]; only a cursory glance at the problem is given here.

α) In recent years studies on different *influences governing the occurrence* were made by several authors[3-7]. Consideration was given to a number of photochemical and chemical reactions, molecular and eddy diffusion and the distributions of minor constituents. With the exeption of [6] the eddy-diffusion coefficient, K, has always been assumed to be height-dependent; but a comparison of height-dependent and -independent cases was made in [3].

At 100 km, values of K between 200 and $450\,m^2\,s^{-1}$ were generally assumed (this is also the maximum value when K is height-dependent). However, the case where K takes a constant value of $1{,}000\,m^2\,s^{-1}$ was also considered [3], and [5] assumed three cases where K takes a maximum value of about $1{,}000\,m^2\,s^{-1}$ in the height range 80 to 100 km. As minor constituents oxygen, hydrogen and their compounds have most often been considered, but compounds of oxygen and nitrogen also were discussed[3]. As for time-dependence SHIMAZAKI and LAIRD[3] were first to compute time-dependent distributions of minor constituents[8]. In [6] only the case of photochemical equilibrium was considered but [4] as well as [7] deals with time-dependent cases. In particular, HUNT[4] has calculated the composition for photochemical equilibrium, diffusive-photochemical equilibrium and for diurnally varying diffusive-photochemical conditions. He examined in detail in what proportions the photochemical production of constituents, their loss through chemical reactions, and the divergence and convergence of their vertical flux contribute to the rate of change of their number density at various levels. We have already referred to the relevant work of THOMAS and BOWMAN[7] in Sect. 33.

[1] See THOMAS' contribution in this volume, p. 7

[2] Strobel, D.F. (1972): Radio Sci. *7*, 1

[3] Shimazaki, T., Laird, A.R. (1970): J. Geophys. Res. *75*, 3221; Shimazaki, T., Laird, A.R. (1972): Radio Sci. *7*, 23

[4] Hunt, B.G. (1971): J. Atmos. Terr. Phys. *33*, 1869

[5] George, J.D., Zimmerman, S.P., Keneshea, T.J. (1972): Space Res. *12/1*, 695

[6] Bowman, M.R., Thomas, L., Geisler, J.E. (1970): J. Atmos. Terr. Phys. *32*, 1661

[7] Thomas, L., Bowman, M.R. (1972): J. Atmos. Terr. Phys. *34*, 1843

[8] There is one inadequate point in their selection of the lower boundary condition, as explained in Sect. 22, and we do not know to what extent their results reproduce the real conditions of the high atmosphere. For some constituents, however, such as O_3, $O_2(a^1\Delta_g)$, OH, H, and $O(^3P)$, photochemical equilibrium was assumed at the lower boundary of 40 km and the corresponding distributions are free from the above-mentioned weakness. The same defect is, unfortunately, found in [5]

β) HUNT[4] assumed different *conditions at* the upper *boundary* (at 160 km), namely photochemical equilibrium for chemically fast-reacting constituents, but diffusive equilibrium for slowly reacting ones. At the lower boundary (at 60 km), photochemical equilibrium values was adopted or constant mixing ratios were assumed for respective constituents, with the exception of molecular hydrogen for which a downward flux was admitted. When a diurnal variation was admitted, the same boundary conditions as above were imposed, except that the photochemically active species were permitted to vary diurnally in sympathy with their diffusive counterparts inside the height range under consideration. BOWMAN et al.[6] have taken the downward vertical flux of O_2 at the upper boundary (120 km) to be equal to JHn, where J is the rate of photodissociation per molecule and n and H are the number density and scale height of molecular oxygen at the upper boundary.

For other constituents the number density at the same boundary was inferred on the basis of observations (e.g. $10^{12}\,\mathrm{m}^{-3}$ for atomic hydrogen), or photochemical equilibrium values were used, or diffusive equilibrium, i.e. zero vertical flux was assumed. At the lower boundary (at 60 km) appropriate values were adopted for constituents H_2, O_2 and H_2O, mostly $6\cdot 10^{16}\,\mathrm{m}^{-3}$ for the latter, corresponding to a mixing ratio of about 10^{-5}. Photochemical equilibrium values were taken for other constituents. THOMAS and BOWMAN[7] have assumed zero vertical flux at the upper boundary (140 km) except for $O(^3P)$, O_2 and H for which an upward vertical flux JHn, a downward one $-2JHn$, and an upward flux of $2\cdot 10^{11}\,\mathrm{m}^{-2}\,\mathrm{s}^{-1}$, respectively, were assumed. At the lower boundary (at 60 km) suitable density values were adopted for H_2, O_2, and H_2O but zero vertical flux was assumed for other constituents. Although these boundary conditions do not always seem adequate, we do not intend to discuss this point further.

γ) An interesting test can be obtained from the *distribution* of *atomic hydrogen* near the 100 km level[4,6,7]. From the number densities of H-atoms given by BOWMAN et al.[6] a scale height of about 5 km is found, thus smaller than the mean scale height of the atmosphere there which is about 6.2 km (see Sect. 33). Since the removal of H-atoms by chemical reactions does not occur rapidly different scale heights are not admissible[9]. In fact, THOMAS and BOWMAN[7] find a scale height of 7.5 km (see Sect. 33) and this is larger than the mean atmospheric scale height, thus admissible. While HUNT's[4] density value of $2.6\cdot 10^{13}$ at 100 km is in good agreement with observed values, the decrease with height is too slow, since it corresponds to an exospheric temperature of about 600 K only (cf. Fig. 18).

38. The effect of the polar wind on the ion-density distribution. The phenomenon that, in the polar region, thermal ions escape along geomagnetic field lines into the geomagnetic tail is called the polar wind. Concerning this phenomenon, we refer to the review paper due to LEMAIRE and SCHERER[1], and discuss here only the question of how the polar wind could influence the distribution of ionized species.

α) The *action of an electric field* is described by Eq. (35.9). In the case we consider here, there exists an upward field in the upper atmosphere; light positive ions are accelerated upward and tend to be set in motion outward along geomagnetic field lines. Since, in the polar region, such ions, after having eventually acquired sufficient velocity to overcome the gravitational force, will

---

[9] Probably the density of $10^{12}\,\mathrm{m}^{-3}$ adopted at 120 km[6] is too small

[1] Lemaire, J., Scherer, M. (1973): Rev. Geophys. Space Phys. *11*, 427

escape into the geomagnetic tail, they induce a decrease in ionospheric ion density[2].

The polar wind was first discussed by BANKS and HOLZER[3] using hydrodynamic equations such as those derived in Sect. 4. These equations are useful in a relatively low region where the collision frequencies between ions and other particles are high, but they are not applicable at sufficiently high altitudes where collisions practically do not occur and the gas-kinetic method is best employed. This method uses the Boltzmann equation for low-density plasmas (with the two-body collision term neglected), or the Vlasov equation[1]. Another method of calculation was first indicated by JEANS when estimating the rate of escape of neutral particles outside the atmosphere; this way has been taken by some authors[4,5].

β) The boundary between the two regions where the two methods mentioned above can be used respectively, i.e. the level separating the *collision-dominated and collisionless regions*, is called the *exobase* or *baropause*. Its height is determined by the condition

$$l_i = H_e, \tag{38.1}$$

where $H_e = -(\mathrm{d}\ln n_e/\mathrm{d}r)^{-1}$ is the electron-density scale height[6]; this latter is approximately given by

$$H_e \simeq \frac{2kT_p}{m_i g} \tag{38.2}$$

if Eq. (35.5) is noted; l_i is the Coulomb mean free path (against deflection) for ions of species i now under consideration and is given by

$$l_i = \left(w_i^2 + \frac{3kT_i}{m_i}\right)^{1/2} \Big/ \nu_i, \tag{38.3}$$

where w_i is the bulk velocity of the ions of species i and ν_i the relevant collision frequency.

γ) *Below the baropause* the hydrodynamic equations can be used. Let us, for the sake of simplicity, consider only three kinds of charged particles, namely electrons and two ionic species, H^+ and O^+. Denoting electrons by the subscript e, H^+ ions by 1, and O^+ ions by 2, we may regard H^+ ions as minor ions and O^+ ions as major ones. This is the situation in the main part of

---

[2] Besides this mechanism, it is also conceivable that ions get an outward velocity component by the drag effect of electrons ejected by photoionization during the day

[3] Banks, P.M., Holzer, T.E. (1968): J. Geophys. Res. *73*, 6846

[4] Dessler, A.J., Cloutier, P.A. (1969): J. Geophys. Res. *74*, 3730

[5] Marubashi, K. (1970): Rep. Ionos. Space Res. Jpn. *24*, 322

[6] When the whole atmosphere above a given level were compressed to a layer of uniform density equal to that at the level, then the thickness of that hypothetic layer is the scale height H at that level. When the mean free path, l, is greater than H, molecules moving vertically upward will normally suffer no collision and can leave the atmosphere

the ionosphere upto at least 800 km. We get from Eqs. (4.2), (4.5), and (5.9) the following equations[5]

$$m_1 N_1 w_1 \frac{\partial w_1}{\partial s} + \frac{\partial}{\partial s}(N_1 k T_1) + m_1 N_1 g_s = N_1 P_1' + N_1 q E_s, \tag{38.4}$$

$$\frac{\partial}{\partial s}(N_2 k T_2) + m_2 N_2 g_s = N_2 P_2' + N_2 q E_s, \tag{38.5}$$

$$\frac{\partial}{\partial s}(N_e k T_e) = N_e P_e' - N_e q E_s, \tag{38.6}$$

where s is a coordinate along the local geomagnetic field line, w_i the bulk velocity of ions of species i and P_i' is the net momentum gain which an ion of species i acquires per unit time, i.e. P_i' is equal to P_i given by Eq. (4.6) divided by n_i. Thus we have

$$P_1' = -\frac{m_1 m_2}{m_1 + m_2}(\nu_{12} + \nu_{1n}) w_1 - m_e \nu_{1e}(w_1 - w_e), \tag{38.7}$$

$$P_2' = \frac{m_1 m_2}{m_1 + m_2} \nu_{21} w_1 + m_e \nu_{2e} w_e, \tag{38.8}$$

$$P_e' = -m_e(\nu_{e1} + \nu_{e2}) w_e. \tag{38.9}$$

In deriving Eqs. (38.4) to (38.9) it has been assumed that the neutral particles are at rest and that the velocity of O^+ ions is so slow that it can be neglected compared with that of hydrogen ions, v_1, so that the inertia term due to the motion of O^+ ions can be dropped. In Eq. (38.6) the inertia term and the gravitational term are ignored owing to the light mass of electrons. As ν_{ij} is the gas-kinetic collision frequency, i.e. the average number of collisions one i-particle makes with any j-particle per unit time, it is proportional to N_j, and so we have the following relation:

$$N_i \nu_{ij} = N_j \nu_{ji}. \tag{38.10}$$

(For neutral particles i or j is replaced by n.)

δ) If the rates of production and loss of H^+ ions are denoted by Q_1 and L_1, the *equation of continuity* is

$$\frac{1}{A}\frac{\partial}{\partial s}(N_1 w_1 A) = Q_1 - L_1, \tag{38.11}$$

where A is the cross sectional area of a geomagnetic tube of force. We have from the quasi-neutrality of charge

$$N_1 + N_2 = N_e, \tag{38.12}$$

and from the condition of no accumulation of charge (no electric current)

$$N_1 w_1 - N_e w_e = 0. \tag{38.13}$$

Six unknown variables N_1, N_2, N_e, w_1, w_e, and E_s are determined in principle by the six equations (38.4...6) and (38.11...13).

ε) Let us first consider the *region below the baropause*. As will be made clear later (cf. Fig. 27, below), we have $N_1 \ll N_2 \simeq N_e$ below the baropause and hence $w_e \ll w_1$ from Eq. (38.13). Under these conditions the second terms on the right-hand side of Eqs. (38.7) and (38.8) and P_e' given by Eq. (38.9) are seen to be negligible if Eq. (38.10) is taken into account and we get from Eqs. (38.4...6), (38.11, 12)

$$I \equiv N_1 w_1 A = N_{10} w_{10} A_0 + \int_{s0}^{s} (Q_1 - L_1) A \, ds, \tag{38.14}$$

$$\begin{aligned} w_1 \left(1 - \frac{c_1^2}{w_1^2}\right) \frac{\partial w_1}{\partial s} = & \left(\frac{1}{A} \frac{\partial A}{\partial s} - \frac{1}{I} \frac{\partial I}{\partial s}\right) c_1^2 - \left(1 - \frac{m_2}{m_1} \frac{T_e}{T_2 + T_e}\right) g_s \\ & - \frac{m_2}{m_1 + m_2} (\nu_{12} + \nu_{1n}) w_1 + \frac{k}{m_1} \frac{\partial}{\partial s} (T_2 - T_1) \\ & - \frac{T_2}{T_1} \frac{c_1^2}{T_2 + T_e} \frac{\partial}{\partial s} (T_2 + T_e) \end{aligned} \tag{38.15}$$

$$N_2 \simeq N_{20} \frac{T_{20} + T_{e0}}{T_2 + T_e} \exp\left[- \int_{s0}^{s} \frac{m_2 g_s}{k(T_2 + T_e)} \left(1 - \frac{m_1}{m_1 + m_2} \frac{\nu_{21}}{g_s} w_1\right) ds \right], \tag{38.16}$$

$$c_1^2 = \frac{k T_1}{m_1} \tag{38.17}$$

(which may be called the sonic velocity of hydrogen ions). The subscript 0 means the value of a variable at a datum level and I is the flux of H^+ ions contained in a geomagnetic tube of force.

A, Q_1, L_1, T_1, T_2, T_e, ν_{1n}, and g_s can be regarded as given functions of height because their values can in principle be estimated on the basis of observations or some appropriate theories. Hence the same is true with I given by Eq. (38.14). $\nu_{21} w_1$ in Eq. (38.16) is proportional to the cross section for collisions between H^+ and O^+ ions, the mean relative velocity of these two kinds of ions and $N_1 w_1$. The collision cross section can in principle be estimated [5]; $N_1 w_1$ is given by Eq. (38.14) now that I can be regarded as known. The mean relative velocity can be calculated, if w_1 is not large as is usually the case below the baropause, by replacing it with the average of relative random velocities (if w_1 is not so small, some device such as the method of successive approximation should be used). In this way $\nu_{21} w_1$ can be taken as a known quantity and N_2 can be obtained using Eq. (38.16). Then ν_{12} which is the product of the collision cross section, relative velocity and N_2 can also be regarded as known. After all, variables on the right-hand side of Eq. (38.15) can be considered as known functions of height except w_1, and the bulk velocity of H^+ ions, w_1, can be found by solving this non-linear differential equation under appropriate boundary conditions. Once w_1 is known, N_1 can be calculated by Eq. (38.14); N_2 is already given by Eq. (38.16).

ζ) Next let us consider the *region above the baropause*. In this collisionless region the velocity distribution function of particles of species i, $f_i(\boldsymbol{v}_i, s)$, satisfies the Boltzmann equation (with vanishing collision term). Now this equation shows that the distribution function keeps the same value if its arguments ($\boldsymbol{v}_i$ and s) are changed along a trajectory of a particle in phase space so that

$$f_i(\boldsymbol{v}_i, s) = f_i(\boldsymbol{v}_{ib}, s_b), \tag{38.18}$$

where the subscript b identifies the value of a variable at the baropause and the quantities (variables) on both sides of this equation refer to a particle moving along the same trajectory. There are some relationships among these variables such as the law of conservation of energy and that of conservation of magnetic moment (the latter holds only when the magnetic field varies so slowly with the space coordinates and with time as to permit the guiding center approximation[7]); and hence, if the functional form of the distribution function is known at the baropause, it can be determined at any level above the baropause using Eq. (38.18) and the above-mentioned relationships.

As the baropause is the upper boundary of the collision-dominated region, the velocity distribution there can be regarded as approximately Maxwellian. LEMAIRE[8] has assumed the following simplified velocity-distribution function at this level:

$$f_i(\boldsymbol{v}_{ib}, s_b) = \begin{cases} 0 & \text{for } v_{ib\parallel} < 0, \\ N_i^* \left(\dfrac{m_i}{2\pi k T_i}\right)^{3/2} \cdot \exp\left(-\dfrac{m_i \boldsymbol{v}_{ib}^2}{2kT_i}\right), & \text{otherwise,} \end{cases} \tag{38.19}$$

where the subscript $\parallel$ denotes the components parallel to the magnetic field and N_i^* is a constant. Then the number density of the particle of species i just above the baropause, $N_i(s_{b+})$, is given by

$$N_i(s_{b+}) = \int_{v_{ib\parallel} > 0} f_i(\boldsymbol{v}_{ib}, s_b)\, \mathrm{d}\boldsymbol{v}_{ib} = \frac{N_i^*}{2} \tag{38.20}$$

and the upward flux of the particles along magnetic field lines is given by

$$F_i(s_{b+}) = \int_{v_{ib\parallel} > 0} v_{ib\parallel} f_i(\boldsymbol{v}_{ib}, s_b)\, \mathrm{d}\boldsymbol{v}_{ib} = \frac{N_i^*}{4} \sqrt{\frac{8kT_i}{\pi m_i}}. \tag{38.21}$$

From these two equations the velocity of effusion or bulk velocity is found to be

$$w_i(s_{b+}) = \sqrt{\frac{2kT_i}{\pi m_i}} \tag{38.22}$$

and for H^+ ions this is 0.80 times c_1 given by Eq. (38.17).

---

[7] Alfvén, H., Fälthammar, C.-G. (1963): Cosmical electrodynamics, § 2.3. Oxford: Clarendon
[8] Lemaire, J. (1972): J. Atmos. Terr. Phys. *34*, 1647

Considering that the particles have a bulk velocity, the form of the velocity distribution function assumed above seems too artificial and it may be more reasonable to express it as

$$f_i(\mathbf{v}_{ib}, s_b) = N_{ib}\left(\frac{m_i}{2\pi k T_i}\right)^{3/2} \exp\left\{-\frac{m_i}{2kT_i}\left[(v_{ib\parallel} - w_{ib})^2 + v_{ib\perp}^2\right]\right\}, \tag{38.23}$$

where the subscript $\perp$ denotes the component perpendicular to the magnetic field. However, LEMAIRE[8] has shown that adoption of Eq. (38.23) instead of the second equation (38.19) has only a small influence on the ion-density distribution. Further, the velocity distribution function (38.23) makes the mathematical analysis considerably more complicated and cannot make both the number density and the flux of particles continuous at the baropause[5]. Therefore we content ourselves here with adopting the simplest velocity distribution function given by Eq. (38.19).

LEMAIRE[8] has taken a datum level at the height of 950 km and assumed the number densities of H^+ and O^+ to be $N_1 = 3.2 \cdot 10^8\,\mathrm{m}^{-3}$, $N_2 = 7 \cdot 10^9\,\mathrm{m}^{-3}$; these are values measured at 85 °S dip latitude in the dusk region of the sunlit polar cap using a mass spectrometer on board the satellite OGO 2[9]. Electron, ion, and neutral temperatures have been assumed to take constant values 3,000 K, 3,000 K, and 1,000 K, respectively, and the NICOLET and KOCKARTS' (unpublished) model atmosphere has been used. A differential equation similar to Eq. (38.15) has been integrated upward to the baropause, taking various starting values of the H^+ ion flux at 950 km ranging between $6.4 \cdot 10^{10}$ and $3.2 \cdot 10^{11}\,\mathrm{m}^{-2}\,\mathrm{s}^{-1}$. The height of the baropause determined by Eqs. (38.1 ... 3) varies between about 1600 km and 1200 km corresponding to the above-mentioned values of the H^+ ion flux (cf. Fig. 26). Among a number of solutions for w_1 or the flux of H^+ ions obtained in this way, LEMAIRE has selected as the polar wind model the one which makes the H^+ ion flux value just below the baropause equal to that just above it given by Eq. (38.21); N_i^* in this equation ($i=1$) should be taken as twice the value of atomic hydrogen ion density just below the baropause calculated by Eq. (38.14) in order to ensure continuity at the baropause (cf. Eq. (38.20)). The solution makes both the number density and the flux of H^+ continuous at the baropause. It is found at a height of 1250 km and the electron and ion densities and their flux at this level are as follows:

$$N_e = 3.75 \cdot 10^9\,\mathrm{m}^{-3}, \quad N_1 = 1.4 \cdot 10^8\,\mathrm{m}^{-3}, \quad N_2 = 3.61 \cdot 10^9\,\mathrm{m}^{-3},$$

and the fluxes are

$$F_{H^+} = F_e = 5.44 \cdot 10^{11}\,\mathrm{m}^{-2}\,\mathrm{s}^{-1}.$$

Above the baropause, ion densities, effusion velocity, and so forth can be determined by utilizing the relation (38.18) and conservation laws.

η) *Results:* Fig. 26 shows the bulk velocity of H^+ as a function of height obtained with the above-mentioned model (dotted line). It is seen that the acceleration of ions becomes large just below the level of the baropause indicated by a vertical line; this is because the frictional force decreases rapidly when the velocity approaches sonic velocity. The solid line shows the case in which electron and ion temperatures are 2,500 K both, and the bulk velocity is seen to diminish for a lower temperature.

Figure 27 illustrates the number densities of H^+ and O^+ as functions of height obtained in the above-mentioned model and compares them with those of a static ionospheric model in diffusive equilibrium. Contrary to the case of diffusive equilibrium, the H^+ ion density decreases with increasing height. Therefore protons become a dominant constituent of the atmosphere only at heights not less than about 5,000 km. If they were in diffusive equilibrium this height should be approximately 1,700 km. As for O^+ ions, their rate of de-

[9] Taylor, Jr., H.A., Brinton, H.C., Pharo, III, M.W., Rahman, N.K. (1968): J. Geophys. Res. *73*, 5521

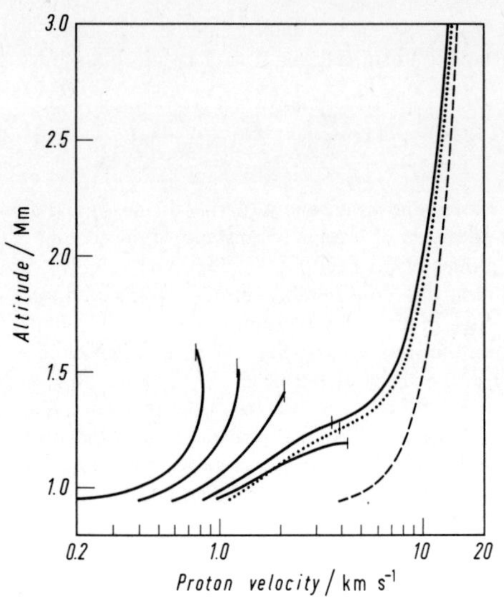

Fig. 26*. Proton bulk velocity vs. altitude. Dotted line corresponds to the model described in the text. Below the baropause altitudes shown by vertical lines the solid lines indicate a family of solutions of a hydrodynamical equation similar to Eq. (38.15) with different starting values of the bulk velocity at 950 km for T_e and T_i equal to 2,500 K and neutral temperature of 1,000 K. Only one of these solutions fits the kinetic solution above the baropause. The dashed line gives the bulk velocity in a kinetic model whose baropause would be at 950 km and $T_e = T_i = 3{,}000$ K

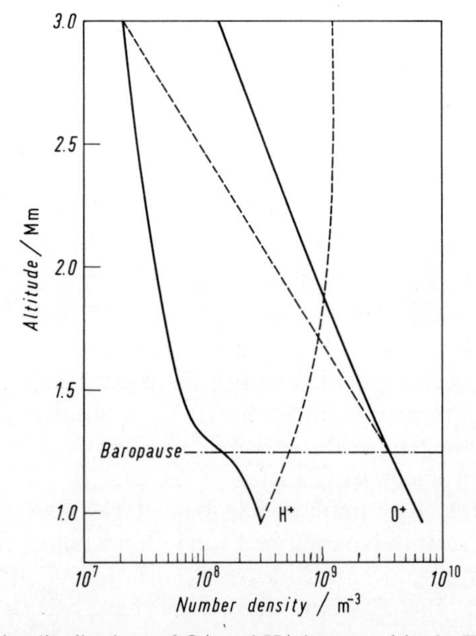

Fig. 27.** Number density distributions of O^+ and H^+ ions vs. altitude. The solid lines correspond to a polar wind model with electron and ion temperature equal to 3,000 K and neutral temperature to 1,000 K. The dashed lines show the distributions in the diffusive equilibrium condition

* From Lemaire (1972): J. Atmos. Terr. Phys. *34* (No. 10), 1650, Fig. 1

** From Lemaire (1972): J. Atmos. Terr. Phys. *34* (No. 10), 1652, Fig. 2

crease with height becomes less steep. This is the consequence of the intensification of the electric field accompanying an increase in the mean mass of ions as will be seen from Eq. (35.9).

In the model described above we have considered H^+ ions to be the minor constituent. It is also possible to view He^+ ions instead as the minor one and apply similar considerations.

One unsatisfactory point of the above treatment is the assumption of equal electron and ion temperatures which does not correspond to the actual conditions of the high atmosphere. Hence LEMAIRE[8] has extended the above calculations to the case where three kinds of ions O^+, H^+, and He^+ ions as well as electrons exist with constant ion temperatures of 1,500 K, 4,000 K, and 3,750 K, respectively and constant electron temperature of 4,500 K. The number density of He^+ at 950 km has been taken as $7 \cdot 10^6\,m^{-3}$. The results of this calculation at the level of 3000 km height are summarized in Table 4 together with those obtained above and those computed by Marubashi[5] under the assumption of $n_2 \equiv N(O^+) = 5 \cdot 10^{11}\,m^{-3}$ at 500 km and $T_e = T_i = 2{,}000$ K; results observed by HOFFMAN[1,10] are given for comparison. It will be seen from this table that the agreement between theory and observation is satisfactory.

Table 4. Polar wind characteristics at 3,000 km altitude

Model	LEMAIRE and SCHERER[1]	LEMAIRE[8]	MARUBASHI[5]	Observations[1,10]
N/m^{-3}				
O^+	$1.5 \cdot 10^8$	$1.3 \cdot 10^8$	$3 \cdot 10^8$	$(0.8 \ldots 2) \cdot 10^8$
H^+	$2.1 \cdot 10^7$	$3.6 \cdot 10^7$	$6 \cdot 10^7$	$(3 \ldots 5) \cdot 10^7$
He^+		$8.7 \cdot 10^5$		
$F/m^{-2}\,s^{-1}$				
O^+	$3.7 \cdot 10^2$	$3.3 \cdot 10^{-6}$		
H^+	$3.0 \cdot 10^{11}$	$6.7 \cdot 10^{11}$	$4 \cdot 10^{11}$	$5 \cdot 10^{11}$
He^+		$7.1 \cdot 10^9$		$5 \cdot 10^9$
$w/m\,s^{-1}$				
O^+	$2.5 \cdot 10^{-6}$	$2.6 \cdot 10^{-14}$		
H^+	$1.4 \cdot 10^4$	$1.8 \cdot 10^4$	$6 \cdot 10^3$	$(1 \ldots 1.5) \cdot 10^4$
He^+		$8.2 \cdot 10^3$		

39. Concluding remarks. From the above discussions it can be seen that the diffusion phenomena in the high atmosphere are in principle well understood and that no problem of fundamental importance seems to remain that must be solved before further progress in this field of atmospheric physics can be achieved. There remain, of course, a number of problems of less importance which must still be solved. In concluding this article we wish to mention some of them in the following.

As stated in Sect. 2, we have two methods of approach to the diffusion processes in the high atmosphere, namely the gas-kinetic and hydrodynamic

[10] Hoffman, J.H. (1971): Trans. Am. Geophys. Union 52, 301

one. The hydrodynamic method is not precise enough as to enable one to derive formulas concerning thermal diffusion, while the gas-kinetic method, in its usual form, can be applied only to the restricted case of a common temperature for all kinds of constituent gases. Some of our formulas were derived by making some hydrodynamic modifications in gas-kinetic ones (as in Sect. 36), or by adding a thermal diffusion term to hydrodynamic formulas (as in Sect. 9). Although this procedure may be approximately correct, we do not know at present to what degree of accuracy this is quantitatively allowed.

In the upper atmospheric problems atomic oxygen and atomic hydrogen play a very important part. However, as, in the laboratory, oxygen and hydrogen usually exist in molecular form, experiments are difficult to conduct for these atoms[1]. Reliable data are, however, needed in order to be able to make an adequate estimation of the transport properties of atomic oxygen and atomic hydrogen gases[2]. Some of these difficulties may be overcome by making detailed calculations on the intermolecular forces using quantum mechanical methods. Since atomic oxygen and hydrogen are quite liable to chemical reactions, this might bring about further complications to the problem.

Concerning the vertical distribution of oxygen atoms, it is seen in Sect. 21 that the one calculated does not agree with those observed. Apparently, the calculated effect of eddy diffusion makes the form of the distribution near the peak much broader than shown by the observations, though no such discrepancy is found with the distribution of molecular oxygen. It may be that the effect of eddy diffusion is not adequately taken into account in the current theory, or our values of the vertical eddy diffusion coefficient K were chosen too large. The latter may be particularly probable when one gives thought to the fact that (see Sect. 21) the value of K depends to a considerable extent on the assumed intensity of solar radiation, decreasing with a decrease in the latter; in fact, our solar radiation estimate used in that section may be seriously too large.

In Sect. 15 it has been pointed out that diffusive separation of atmospheric constituents sets in at a considerably lower level than expected from theory. This may be associated with existence of the turbopause, i.e. the abrupt termination of turbulence at about 110 km referred to in Sect. 13. The effects of atmospheric movements to cause atmospheric mixing appear to be much less effective above this level than a simple consideration suggests. At any rate the explanation for this apparent contradiction should be sought. It would also be advantageous if a more complete theory were developed to describe turbulent processes in the high atmosphere. At present too wide a variety of values ranging over two orders of magnitude have been obtained for ε, the rate of energy transfer to and among the eddies of different scales as described in Sect. 14.

In Sect. 24 it has been shown that the electron-density profiles theoretically expected in the daytime F2-region do not agree with those observed; rather often, the peak of the layer as obtained from this theory appears at too high an

[1] See THOMAS' contribution in this volume p. 7

[2] See SUCHY's contribution in vol. 49/7, p. 1

altitude. Here we arrive at the limit of applicability of pure diffusion theories. Effects of other agents such as winds and electric fields must be invoked even in an average steady state[3]. We should also bear in mind that the transport phenomena involving electrons and ions are more complicated than neutral ones and the first approximation may not be sufficient unlike the case of neutral gases[4]. Although some steps have been taken toward improving the degree of approximation [*4*, *5*], higher order approximation formulas may be needed in the case of plasma diffusion. Diffusion of an ion cloud in the presence of a wind and an external electric and magnetic field is a general problem which still deserves detailed study.

Lastly, there remains the problem of major constituent diffusion still to be solved as mentioned in Sect. 15.

General references

[*1*] Chapman, S., Cowling, T.G. (1952): The mathematical theory of non-uniform gases, 2nd edn. Cambridge: University Press (a) p. 252, (b) p. 248, (c) p. 249

[*2*] Spitzer, Jr., L. (1962): Physics of fully ionized gases, 2nd edn. New York, London: Interscience (a) p. 155 ff.

[*3*] Hirschfelder, J.O., Curtiss, C.F., Bird, R.B. (1954): Molecular theory of gases and liquids. New York: Wiley and Sons, London: Chapman and Hall (a) pp. 487, 516, (b) p. 517, (c) pp. 484, 486, (d) p. 480, (e) p. 477, (f) p. 488, (g) p. 716, (h) p. 518, (i) p. 579 Table 8.4–12, (j) p. 541, (k) § 8.4, (l) p. 30, (m) p. 946, (n) p. 1058

[*4*] Devoto, R.S. (1966): Transport properties of ionized monatomic gases. Phys. Fluids *9*, 1230

[*5*] Schunk, R.W., Walker, J.C.G. (1970): Thermal diffusion in the F2-region of the ionosphere. Planet. Space Sci. *18*, 535

[*6*] Ferziger, J.H., Kaper, H.G. (1972): Mathematical theory of transport processes in gases. Amsterdam: North-Holland (a) pp. 259 ff.

[*7*] Lettau, H. (1951) in: Compendium of meteorology. Malone, T.F. (ed.), p. 320. Boston, Mass.: American Meteorological Society

[*8*] Pasquill, F. (1962): Atmospheric diffusion. London: Van Nostrand (a) p. 70

---

[3] See STUBBE's contribution in this volume, p. 247

[4] See also the contribution of THOMAS in this volume, p. 7, and that of SUCHY in vol. 49/7, p. 1

Interaction of Neutral and Plasma Motions in the Ionosphere

By

P. STUBBE

With 25 Figures

A. Introduction

Up to the mid-1960's, winds were not considered a part of ionospheric theory. As a consequence, attempts to theoretically describe the ionospheric behavior generally gave unsatisfactory results. In particular the diurnal variation of the F-region electron density [*34*] and its maintenance at night could not be understood. HANSON and PATTERSON[1] suggested neutral winds to be one possible mechanism among others to account for the maintenance of the nighttime F-layer. Previously, DOUGHERTY[2] had discussed the effects of neutral winds, if present, on the F-region dynamics. A first estimate of the neutral wind velocities at F-region heights as well as a discussion of the forces by which they are determined was given by KING and KOHL [*25*]. This work turned out to be very fruitful in stimulating further investigations which soon led to an understanding of many ionospheric phenomena (e.g., [*35*], [*30*]).

Winds in the thermosphere affect the ionosphere in different ways. Since the neutral density in the thermosphere is much larger than the plasma density, winds carry the plasma along the magnetic field lines. Thereby, a vertical plasma velocity component is produced, provided the inclination angle is not just 0 or 90°. A wind blowing toward the pole causes a downward plasma motion, thereby lowering the layer height and reducing the electron-ion density since chemical losses increase with decreasing altitude. On the other hand, an equatorward wind raises the layer height and enhances the plasma density. This wind effect vanishes at the magnetic poles and the magnetic equator.

Another wind effect arises when the horizontal wind flow is not divergence free, as must generally be expected. In this case, a vertical motion is caused such as to reduce the total divergence. Since the resulting vertical velocities are different for the different neutral species, frictional forces are set up which

---

[1] Hanson, W.B., Patterson, T.N.L. (1964): Planet. Space Sci. *12*, 979

[2] Dougherty, J.P. (1961): J. Atmos. Terr. Phys. *20*, 167

enter into the equation of motion to give a third contribution besides the pressure gradient and gravity forces. They thus lead to a modification of the barometric law and an altered neutral composition, which in turn determines the ionospheric composition.

A third wind effect is also related to a vertical neutral motion; however, it does not require different velocities for the single neutral constituents. An upward motion corresponds to work done against gravity and thus cools the thermosphere. Conversely, a downward motion increases the thermospheric temperature. A temperature change leads to a changed thermospheric density and composition and thereby affects the ionosphere.

We shall confine our considerations in this article to thermospheric winds above about 150 km. Tidal winds in the dynamo region will not be discussed (see the contribution by [*32*] in this Encyclopedia), but we shall include electric fields resulting from tidal winds in our later discussion.

Thermospheric winds are driven by horizontal pressure gradients. These winds are subject to several other forces, e.g., inertial, Coriolis, frictional, and viscosity forces. The Coriolis force, although important, does not prevent the winds from blowing from high- to low-pressure regions. This distinguishes winds in the thermosphere from winds in the lower atmosphere. The wind velocity is strongly controlled by the ionospheric plasma density through friction. Thus the coupling between the ionosphere and thermospheric winds is mutual [*40*].

In this article we proceed in the following order. In Part B the fundamental equations for the wind velocity are established and their limits of applicability are discussed. The available methods of solving these equations as well as the characteristics of the calculated winds are reviewed in Part C. In Part D the experimental methods to determine thermospheric wind velocities are briefly outlined, and some experimental data are presented. In Part E we describe in some detail the effects of neutral winds on the F-region and in Part F the effects of plasma motions on the neutral wind. A short summary is given in Part G.

It should be noted that the term "wind" is used only for *horizontal neutral gas motions.*

B. Fundamental equations

1. General remarks. To obtain the wind velocity, the equation of motion for the neutral gas as a whole has to be solved. In order to perform this task, the force of friction with the plasma and the pressure gradient force have to be known. The first requires knowledge of the plasma density; therefore the continuity equations together with the equations of motion for the individual plasma constituents must be solved. For the second, the pressure gradient force, the horizontal temperature and density structures have to be known, necessitating solutions of the continuity equations for the individual neutral constituents and the energy equation for the neutral gas as a whole. We see, therefore, that a self-consistent description of the neutral wind requires the solution of the full set of transport equations for the zeroth moment (continuity equation), first moment (momentum equation or equation of motion), and second moment (energy equation or heat conduction equation) for all relevant neutral and charged constituents. A full solution of this set of equations is beyond our present possibilities. Solutions obtained so far have been based on different approximations with different degrees of realism. These solutions as well as the methods involved will be discussed in Sects. 14–18.

I. Equation of motion for the neutral gas

2. Navier-Stokes equation. The macroscopic velocity $\boldsymbol{v}_n$ of the neutral gas is described by the Navier-Stokes equation which reads

$$\rho_n \left[\frac{\partial \boldsymbol{v}_n}{\partial t} + (\boldsymbol{v}_n \operatorname{grad})\, \boldsymbol{v}_n\right] = -\operatorname{grad} p_n + \eta\, \Delta \boldsymbol{v}_n + \tfrac{1}{3}\eta \operatorname{grad} \operatorname{div} \boldsymbol{v}_n + \boldsymbol{F}, \tag{2.1}$$

where $p_n = n_n k T_n$ = hydrostatic pressure,
n_n = number density,
T_n = kinetic temperature,
k = Boltzmann's constant,
η = coefficient of viscosity,
F = external force per unit volume,
$\rho_n = m_n n_n$ = mass density,
m_n = average particle mass.

The subscript n refers to the neutral gas as a whole. It is assumed in Eq. (2.1) that η, the coefficient of viscosity, is constant in space.

The external force is essentially composed of a gravity, Coriolis, and frictional term:

$$\boldsymbol{F} = \rho_n \boldsymbol{g} + 2\rho_n (\boldsymbol{v}_n \times \boldsymbol{\Omega}) + \boldsymbol{P}, \tag{2.2}$$

where $\boldsymbol{g}$ is the acceleration due to gravity, $\boldsymbol{\Omega}$ is the Earth's angular velocity, and $\boldsymbol{P}$ is the frictional force per unit volume.

The centrifugal force may be neglected in the Earth's thermosphere. The gravity force, of course, vanishes in the equations of motion for the horizontal velocity components, but is one of the dominant forces in the equation for the vertical velocity.

3. The frictional force. Since at F-region heights motions of the ionospheric plasma are confined to the magnetic lines of force, the plasma cannot share the neutral gas motion. Therefore a frictional force is set up which is proportional to the plasma density and thus shows a strong height and time dependence.

α) In deriving an *expression for the frictional force*, the velocity distribution functions of the interacting gases have to be known. Assuming Maxwellian distributions displaced by the respective macroscopic velocity, the force acting on neutral species l due to friction with ion species j is given by [1]

$$\boldsymbol{P}_{lj} = -n_j \nu_{jl} \mu_{jl} (\boldsymbol{v}_{\rm n} - \boldsymbol{v}_j), \tag{3.1}$$

where $\mu_{jl} = m_j m_l/(m_j + m_l)$ = reduced mass,
ν_{jl} = averaged momentum transfer collision frequency of one ion of species j with neutral particles of species l. This definition takes account of Maxwellian velocity distributions in both species [2].

The collision frequency can be written as

$$\nu_{jl} = a_{jl} n_l, \tag{3.2}$$

where the factor a_{jl} is given only by the properties of the interacting particles, so that $a_{jl} = a_{lj}$. Therefore

$$\nu_{lj} = a_{jl} n_j = \frac{n_j}{n_l} \nu_{jl}, \tag{3.3}$$

with ν_{lj} the momentum transfer collision frequency [3] of one neutral particle of species l with ions of species j.

The total frictional force acting on the neutral gas is obtained from Eq. (3.1) by summation over j and l as

$$\boldsymbol{P} = -\sum_j n_j (\boldsymbol{v}_n - \boldsymbol{v}_j) \sum_l \nu_{jl} \mu_{jl}. \tag{3.4}$$

β) In using Eq. (3.1) or (3.4) together with the *collision frequencies* published in literature, care has to be applied. This is because two different definitions for the collision frequency are employed in the literature [2]. The collision frequencies given by STUBBE [1] and BANKS [4], [5] are properly defined to be used in conjunction with Eq. (3.1) or (3.4). On the other hand, the collision frequencies given by DALGARNO [*13*] are defined in such a way that they must be multiplied by m_l/μ_{jl} before inserting them in Eqs. (3.1) or (3.4).

γ) The *coefficients* a_{jl} for the ion and neutral species dominant at F-region altitudes are given in Table 1.

The ionospheric plasma determines the frictional force through the ion density and ion velocity. The contribution from the electron gas is negligible owing to the smallness of the electron mass.

δ) The *ion velocity* is caused by diffusion, electric fields, and winds. A common, but not correct, simplification is to neglect the diffusion and electric field contributions to obtain

$$\boldsymbol{v}_{\rm i} = (\boldsymbol{v}_{\rm n} \boldsymbol{B}) \boldsymbol{B}/B^2, \tag{3.5}$$

---

[1] Stubbe, P. (1968): J. Atmos. Terr. Phys. *30*, 1965
[2] See contribution by K. Suchy in vol. 49/7, p. 1
[3] Suchy, K., Rawer, K. (1971): J. Atmos. Terr. Phys. *33*, 1853
[4] Banks, P.M. (1966): Planet. Space. Sci. *14*, 1105

Table 1. $10^{15} a_{jl}/\mathrm{m}^3\,\mathrm{s}^{-1}$ for the ions O^+, NO^+, O_2^+ interacting with the neutral constituents O, N_2, O_2 after STUBBE[1]. T_{jl} is defined as $\mu_{jl}[(T_j/m_j)+(T_l/m_l)]$. For all practical purposes, T_j and T_l may be identified as the temperatures of the whole ion and neutral gases, respectively

j \ l	O	N_2	O_2
O^+	$1.86\left(\frac{T_{jl}}{1000\,\mathrm{K}}\right)^{0.37}$	1.08	1.00
NO^+	0.76	0.90	0.83
O_2^+	0.75	0.89	$1.17\left(\frac{T_{jl}}{1000\,\mathrm{K}}\right)^{0.28}$

where $\boldsymbol{B}$ is the induction of the geomagnetic field and the subscript i stands for the ion gas as a whole. The ion velocity given by Eq. (3.5) is directed parallel to the magnetic field lines. In a Cartesian coordinate system with the x-axis directed towards geographic south, the y-axis towards geographic east, and the z-axis vertically upwards, $\boldsymbol{B}$ has the components $(-B\cos I\cos D,\ B\cos I\sin D,\ -B\sin I)$, I being the inclination (positive in the northern hemisphere) and D the declination (positive towards the east). Hence it follows from Eq. (3.5) that the x- and y-components of $\boldsymbol{P}$ are given by

$$P_x = -\sum_j n_j \sum_l \nu_{jl}\mu_{jl}[v_{nx}(1-\cos^2 I\cos^2 D)+v_{ny}\cos^2 I\sin D\cos D], \tag{3.6}$$

$$P_y = -\sum_j n_j \sum_l \nu_{jl}\mu_{jl}[v_{ny}(1-\cos^2 I\sin^2 D)+v_{nx}\cos^2 I\sin D\cos D], \tag{3.7}$$

provided v_{nz} is small compared with v_{nx} and v_{ny}. Section 34 will discuss the extent to which Eqs. (3.6) and (3.7) are justified.

4. Thermospheric models and the pressure gradient force. It was outlined in Sect. 1 that in a self-consistent description of the thermospheric wind system the continuity and energy equations of the neutral gas have to be solved simultaneously with the momentum equation. This treatment automatically takes care of the pressure gradient force. Most authors, however, did not follow this way but instead externally specified the pressure gradient force by means of an empirical thermospheric model.

α) The official international model [11], due mainly to JACCHIA, is based on density measurements obtained from the orbital changes of satellites (see also RAWER's contribution in volume 49/7). In this model it is assumed that the neutral particle number densities follow from the barometric law

$$n_l(z) = n_l(z_0)\,\frac{T_n(z_0)}{T_n(z)}\exp\left(-\int_{z_0}^{z}\frac{dz}{H_l}\right), \tag{4.1}$$

where $H_l = kT_n/m_l g$ is the *scale height* of species l. The neutral temperature is analytically expressed as a function of altitude, subject to a set of parameters that in conjunction with the particle densities $n_l(z_0)$ are specified so that the total density $\rho_n = \sum_l m_l n_l$ closely agrees with the experimental density data obtained from satellite drag measurements.

In view of this procedure, the quantities $T_n(z)$ and $n_l(z_0)$ given by the JACCHIA model cannot be taken as the real temperature and particle densities, but rather as model parameters properly chosen to reproduce the total density. Therefore the pressure calculated from the JACCHIA model is far less accurate than the density. An even larger inaccuracy is introduced by the differentiation process required to obtain grad p_n from p_n. Even slight errors in the latitude and longitude dependence of p_n may thus result in appreciable errors in grad p_n. For these reasons the pressure gradient force is the least accurate among the forces determining the wind velocity.

β) Meanwhile, a few other thermospheric models have become available which are based on larger sets of data obtained with different techniques [1,2] [*6, 20, 21, 23, 24*]. These yield a better description of the particle densities n_1 and the neutral temperature T_n, but they underlie the same differentiation problem and thus are probably not much better with respect to the pressure gradient force.

For more details see RAWER's contribution in volume 49/7 and also [3,4] and [*1*]. Some of the new models depict the dependencies on hour [*18, 33*], geographic location [6], disturbance indices [5], describe composition variations [8-10], and high-latitude phenomena [7]. Heat sources and sinks play an important role. The most common source is heating by extreme ultraviolet (EUV) solar radiation [*12, 18, 33, 48*], but also corpuscular heating in the auroral oval is taken into account [*42*] (see also SCHMIDTKE's contribution in volume 49/7).

γ) An analytic representation for the pressure gradient can be obtained if an older version of the JACCHIA [11] model is used. In this model the particle densities at $z_0 = 120$ km are constant, and the temperature profile is described by

$$T_n(z) = T_\infty - (T_\infty - T_{120}) \exp[-s(z - 120\,\mathrm{km})] \tag{4.2}$$

with T_∞ the exospheric temperature, T_{120} the temperature at 120 km, and s a parameter which is related to T_∞ by

---

[1] Zahn, U. v., Köhnlein, W., Fricke, K.H., Laux, U., Trinks, H., Volland, H. (1977): Geophys. Res. Lett. *4*, 33

[2] Köhnlein, W., Krankowsky, D., Volland, H.: Kleinheubacher Ber. *19*, 539 (1976); Lämmerzahl, P.: J. Geomag. - Geoelectr. *31 S*, 85 (1979)

[3] Jacchia, L.G. (1979): Space Res. *XIX*, 179

[4] Zahn, U. v., Fricke, K.H. (1978): Rev. Geophys. and Space Phys. *16*, 169

[5] Jacchia, L.G., Slowey, J.W., Zahn, U. v. (1977): J. Geophys. Res. *82*, 684

[6] Hedin, A.E., Reber, C.A., Spencer, N.W., Brinton, H.C., Kayser, D.C. (1979): J. Geophys. Res. *84*, 1

[7] Roble, R.G., Dickinson, R.E. (1974): Planet. Space Sci. *22*, 623

[8] Forbes, J.M. (1978): J. Geophys. Res. *83*, 3691

[9] Mayr, H.G., Volland, H. (1976): J. Geophys. Res. *81*, 671

[10] Marubashi, K., Reber, C.A., Taylor Jr., H.A. (1976): Planet. Space Sci. *24*, 1031

[11] Jacchia, L.G. (1965): Smithson. Contr. Astrophys. *8*, 215

$$s = 0.0291\,\text{km}^{-1}\exp(-0.5q^2) \quad \text{and}$$
$$q = \frac{T_\infty/\text{K} - 800}{750 + 1.722 \cdot 10^{-4}(T_\infty/\text{K} - 800)^2}. \tag{4.3}$$

The temperature profile Eq. (4.2) permits the integration in Eq. (4.1) to be performed analytically. According to DICKINSON and GEISLER [*14*][12], the horizontal components of the pressure gradient are given by

$$(\text{grad}\, p_\text{n})_{x,y} = \rho_\text{n}\, g\, \frac{T_\text{n}}{T_\infty}\left[A\,\frac{\partial}{\partial x, y}\log_\text{e} T_\infty + B\,\frac{\partial}{\partial x, y}\log_\text{e} s\right], \tag{4.4}$$

where

$$A = (z/\text{km} - 120) - \frac{1}{s}\left[\frac{T_\infty}{T_\text{n}}\,\frac{T_\text{n} - T_{120}}{T_\infty - T_{120}} - \log_\text{e}\left(\frac{T_\text{n}}{T_{120}}\right)\right]$$

$$B = -(z/\text{km} - 120)\left[\frac{T_\infty}{T_\text{n}} - 1\right] + \frac{1}{s}\log_\text{e}\left(\frac{T_n}{T_{120}}\right).$$

5. Comparison of forces. The qualitative effects of the forces appearing in the Navier–Stokes equation may be described as follows. Normally, winds are driven by pressure gradients. In certain cases, e.g., in the presence of strong electric fields, the frictional force may also act as a driving force but usually is the main retarding force. The acceleration (or inertia) term tends to prevent temporal velocity changes, i.e., acts as a retarding force when the air is accelerated and as a driving forve when the air is decelerated. The viscosity term tends to prevent wind shears, i.e., it smoothes out the velocity-height profile by dimishing second derivatives with respect to height. Finally, the Coriolis term works to change the wind direction. In the northern hemisphere the sense of rotation is clockwise and opposite in the southern hemisphere.

In the following, the relative importance of the individual forces will be evaluated for altitudes around 300 km. This comparison is more easily visualized in terms of accelerations in units $10\,\text{m}\,\text{s}^{-2}$, i.e., compared with the gravitational acceleration. For the sake of simplicity, only the x-component will be discussed.

α) The first part of the *acceleration term,* $\partial v_{\text{n}x}/\partial t$, may be evaluated by expressing $v_{\text{n}x}$ in terms of harmonics. Thus

$$\frac{\partial v_{\text{n}x}}{\partial t} \approx n'\,\Omega\, v_{\text{n}x}, \tag{5.1}$$

where Ω = Earth's angular velocity $= 7.27 \cdot 10^{-5}\,\text{s}^{-1}$, n' = order of harmonic.

Of the second part of the acceleration term, $((\boldsymbol{v}_\text{n}\,\text{grad})\,\boldsymbol{v}_\text{n})_x$, we only consider the contribution $v_{\text{n}y}(\partial v_{\text{n}x}/\partial y)$ which is most easily tractable. We exchange

[12] Note that in [14] the expression for B contains a misprint

geographic longitude and local time and thereby obtain

$$v_{ny}\frac{\partial v_{nx}}{\partial y}=\frac{v_{ny}}{\Omega R_0\cos\varphi}\,\frac{\partial v_{nx}}{\partial t}\approx\frac{n'\,v_{nx}v_{ny}}{R_0\cos\varphi}, \tag{5.2}$$

where R_0 is the Earth's radius $=6370$ km, and φ is the geographic latitude.

Numerical values are given in Table 2 for both parts of the acceleration term up to the sixth harmonic.

Table 2. Numerical values for both parts of the acceleration term according to Eqs. (5.1) and (5.2) in units of $10\,\mathrm{m\,s^{-2}}$ (i.e., in g) for $v_{nx}=v_{ny}=100\,\mathrm{m\,s^{-1}}$ and $\varphi=45°$. First six harmonics

n'	1	2	3	4	5	6
First part	$7.3\cdot10^{-4}$	$1.5\cdot10^{-3}$	$2.2\cdot10^{-3}$	$2.9\cdot10^{-3}$	$3.6\cdot10^{-3}$	$4.4\cdot10^{-3}$
Second part	$2.2\cdot10^{-4}$	$4.4\cdot10^{-4}$	$6.7\cdot10^{-4}$	$8.9\cdot10^{-4}$	$1.1\cdot10^{-3}$	$1.3\cdot10^{-3}$

The diurnal variation of v_{nx} is best described by n′ between 1 and 2, but within ± 2 h around sunrise, 5 or 6 is more appropriate.

The x-component of the Coriolis acceleration is given by

$$2(\boldsymbol{v}_n\times\boldsymbol{\Omega})_x=2v_{ny}\Omega\sin\varphi. \tag{5.3}$$

For $v_{ny}=100$ m and $\varphi=45°$ the numerical value $1.0\cdot10^{-3}$ [$10\,\mathrm{m\,s^{-2}}$] is obtained.

β) For an estimate of the *frictional acceleration*, it is reasonable to consider O as the only neutral species and O^+ as the only ionic species at 300 km and to put the O^+ number density equal to $NmF2$, the electron number density at the F2 peak. We thus obtain from Eq. (3.6), with $D=0$ and $T_{jl}=1000$ K,

$$\frac{P_x}{\rho_n}\approx 9.3\cdot10^{-16}\,\mathrm{m^3\,s^{-1}}\sin^2 I\;NmF2\,v_{nx}. \tag{5.4}$$

Numerical values for P_x/ρ_n as a function of $NmF2$ are presented in Table 3.

Table 3. Numerical values of the acceleration due to friction for $v_{nx}=100\,\mathrm{m\,s^{-1}}$ and $I=45°$. The four $NmF2$ values correspond, from left to right, to high-solar-activity day, low-solar-activity day, high-solar-activity night, and low-solar-activity night

$NmF2/\mathrm{m^{-3}}$	$2\cdot10^{12}$	$6\cdot10^{11}$	$3\cdot10^{11}$	$8\cdot10^{10}$
P_x/ρ_n [$10\,\mathrm{m\,s^{-2}}$]	$9.3\cdot10^{-3}$	$2.8\cdot10^{-3}$	$1.4\cdot10^{-3}$	$3.7\cdot10^{-4}$

γ) The acceleration due to the *pressure gradient force*, $(1/\rho_n)(\mathrm{grad}\,p_n)_x$, can be evaluated by means of Eq. (4.4). Numerical values are given in Table 4.

Table 4. Numerical values for the acceleration due to the pressure gradient force for three assumed temperature gradients and $T_\infty = 1000$ K

$\frac{\partial T_\infty}{\partial x} \bigg/ \left(\frac{\text{K}}{10^4\,\text{km}}\right)$	100	200	300
$\frac{\text{grad}\, p_\text{n}}{\rho_\text{n}} \bigg/ (10\,\text{m}\,\text{s}^{-2})$	$1.5 \cdot 10^{-3}$	$3.0 \cdot 10^{-3}$	$4.5 \cdot 10^{-3}$

δ) The most difficult term to evaluate is the *viscosity term* because it involves a second derivative. Since the horizontal scale length is much larger than the vertical, we may write for the acceleration due to viscosity

$$\frac{1}{\rho_\text{n}} \eta (\Delta \boldsymbol{v}_\text{n} + \tfrac{1}{3} \,\text{grad div}\, \boldsymbol{v}_\text{n})_x \approx \frac{\eta}{\rho_\text{n}} \frac{\partial^2 v_{\text{n}x}}{\partial z^2}. \tag{5.5}$$

The coefficient of viscosity for atomic oxygen, the dominant neutral constituent at F-region heights, is [1]

$$\eta = 4.5 \cdot 10^{-5} \left(\frac{T_\text{n}}{1000\,\text{K}}\right)^{0.71} \text{kg}\,\text{m}^{-1}\,\text{s}^{-1}.$$

From the foregoing it is apparent that for noon conditions the frictional and pressure gradient forces are predominate. We therefore can approximately derive a $v_{\text{n}x}(z)$-profile by equating these two forces. For the ion density $n(\text{O}^+)$ a parabolic profile with a maximum at 300 km is assumed

$$n(\text{O}^+) \approx NmF2 \left[1 - \left(\frac{z - 300\,\text{km}}{Ym}\right)^2\right]$$

where Ym is the half-layer thickness. Around 300 km the pressure gradient force according to Eq. (4.4) can be further simplified with reasonable accuracy to give

$$(\text{grad}\, p_\text{n})_x \approx \rho_\text{n}\, g (z/\text{km} - 120) \frac{\partial}{\partial x} \log_\text{e} T_\infty.$$

With these assumptions we obtain as a rough estimate

$$\frac{\partial^2 v_{\text{n}x}}{\partial z} \approx \frac{2 v_{\text{n}x}}{Ym^2}$$

and

$$\frac{1}{\rho_\text{n}} \eta (\Delta \boldsymbol{v}_\text{n} + \tfrac{1}{3} \,\text{grad div}\, \boldsymbol{v}_\text{n})_x \approx \frac{\eta}{\rho_\text{n}} \frac{2 v_{\text{n}x}}{Ym^2}. \tag{5.6}$$

[1] Dalgarno, A., Smith, F.S. (1962): Planet. Space. Sci. 9, 1

For $T_\infty = 1000$ K, $v_{nx} = 100\,\mathrm{m\,s^{-1}}$, and the typical value $\rho_n = 2.5 \cdot 10^{-11}\,\mathrm{kg\,m^{-3}}$ [*11*] and $Ym = 80$ km (BECKER [2]), the numerical value $3 \cdot 10^{-3}/10\,\mathrm{m\,s^{-2}}$ is obtained for the acceleration due to viscosity.

In conclusion it may be said that among the forces discussed only the nonlinear part of the acceleration term may be neglected at F-region altitudes, except for a short period in the early morning hours.

ε) Something remains to be said about the *height and local time dependence* of the single terms in the Navier–Stokes equation. Two terms exhibit a strong height dependence, i.e., the friction and viscosity terms. The first depends on altitude mainly via the ion density. Consequently, friction is negligible in the E- and lower F-region as well as in the upper topside ionosphere. The acceleration due to viscosity depends on altitude via the neutral density ρ_n. This term is solely predominant above about 500 km but is negligible below about 200 km. A moderate height dependence is shown by the pressure gradient term which increases almost linearly with altitude. The Coriolis term depends weakly on height, being proportional to the wind velocity. The only term without any a priori height dependence is the linear part of the acceleration term.

The relative importance of the individual forces strongly depends on local time. Friction is important at day but is small and sometimes even negligible at night. As a consequence, the nighttime wind velocities are considerably larger than the daytime velocities, causing the Coriolis force to contribute significantly at night. The viscosity term is also greater at night than at day because of the diurnal variation of the total density which has a minimum at night. The acceleration term is maximum around midnight and sunrise, with the nonlinear part being noticeably only around sunrise [3].

6. The boundary conditions. If in Eq. (2.1) only the spatial derivatives with respect to z are taken into account, as is usually done, a differential equation of second order in z is obtained. Therefore two boundary conditions are required. These have been thoroughly discussed by KOHL [*29*] and BLUM and HARRIS [*9*].

Typical lower boundary altitudes used in wind calculations are 120 or 125 km in correspondence to the lower boundary altitudes of the thermospheric model [1] [*11*] employed. It is known that the wind structure is highly irregular up to 130...150 km [7]. Therefore, any lower boundary condition must necessarily be inadequate and one may as well use the simplest,

$$\boldsymbol{v}_n = 0. \tag{6.1}$$

The question arises: up to which altitudes are the solutions influenced by the obviously wrong boundary condition Eq. (6.1). The answer is given in Fig. 1 which shows the average zonal velocity v_{ny} as a function of altitude for two

[2] Becker, W. (1967): J. Geophys. Res. *72*, 2001
[3] Rüster, R., Dudeney, J.R. (1972): J. Atmos. Terr. Phys. *34*, 1075
[1] Jacchia, L.G. (1965): Smithson. Contr. Astrophys. *8*, 215

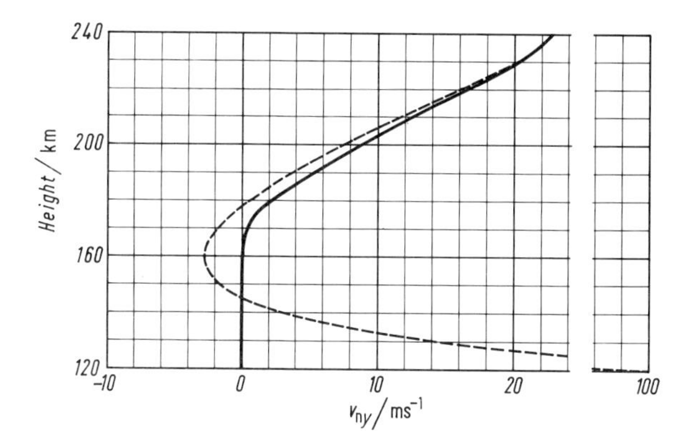

Fig. 1. Average zonal wind velocity v_{ny} as a function of altitude for latitude $\varphi=30°$, solar activity (Covington-) index $F_{10.7}=200$ and equinox (after BLUM and HARRIS [9]). ——— for $v_{ny}(120)=0$, ---- for $v_{ny}(120)=100\,m/s$

lower boundary values, viz., $v_{ny}(120)=0$ and $v_{ny}(120)=100\,\mathrm{m\,s^{-1}}$. We notice that above 180 km the two wind profiles are nearly identical, which is due to the fact that the locally acting forces are well in excess of the viscosity force, which is the only force effectively establishing a coupling between different altitude regions. We may conclude, therefore, that our lower boundary condition Eq. (6.1), although highly unrealistic, does not at all affect the wind velocity at F-region altitudes.

Owing to the exponential increases of the kinematic viscosity η/ρ_n with increasing altitude, the Navier-Stokes equation reduces to

$$\Delta \boldsymbol{v}_n + \tfrac{1}{3}\,\mathrm{grad}\,\mathrm{div}\,\boldsymbol{v}_n = 0 \tag{6.2}$$

for the transition $z\to\infty$. Equation (6.2) forms the basis for deriving an upper boundary condition. If we again neglect horizontal derivatives $\left(\frac{\partial}{\partial x, y} \ll \frac{\partial}{\partial z}\right)$, we have

$$\frac{\partial^2 v_{nx,y}}{\partial z^2} = 0, \tag{6.3}$$

which is solved by

$$\frac{\partial v_{nx,y}}{\partial z} = \mathrm{const.} \tag{6.4}$$

In order to prevent $v_{nx,y}\to\infty$ for $z\to\infty$, we have to put the constant equal to zero,

$$\frac{\partial v_{nx,y}}{\partial z} = 0. \tag{6.5}$$

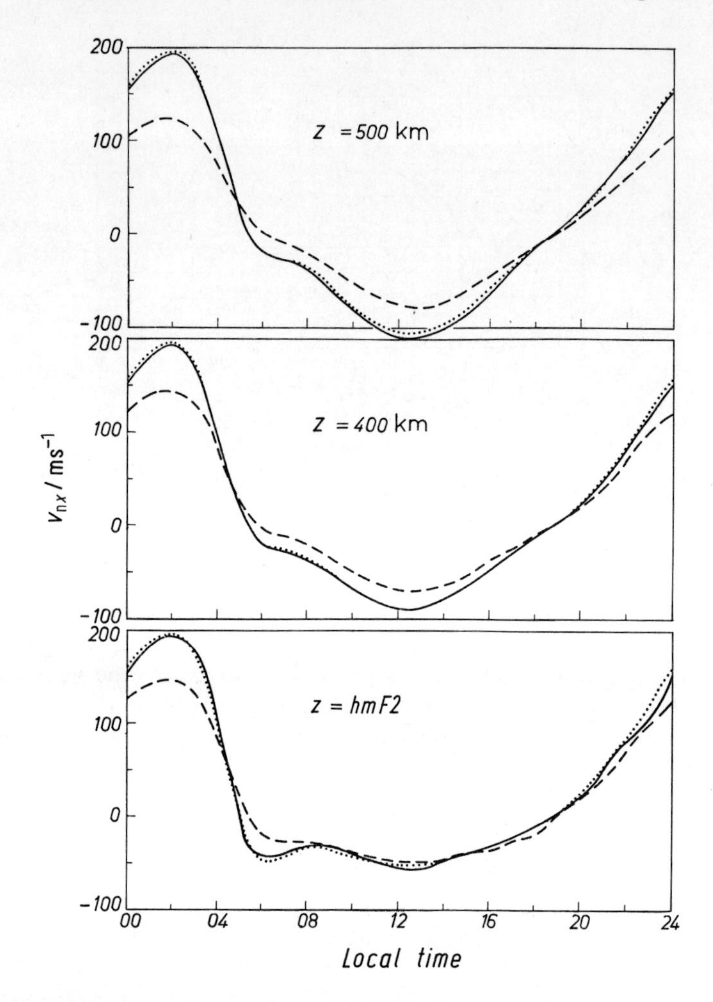

Fig. 2. North-south wind velocity v_{nx} as a function of local time for altitudes 500 km, 400 km, and $hmF2$ (height of F2-layer maximum) for latitude $\varphi=40°$ N, solar activity index $F_{10.7}=165$ and equinox (after KOHL [29]). ——— for $\frac{\partial v_{nx}}{\partial z}=0$ at 800 km, $\cdots\cdots$ for $\frac{\partial v_{nx}}{\partial z}=0$ at 600 km, - - - - for $v_{nx}=0$ at 1100 km

This is the common upper boundary condition, usually applied to boundary heights between 500 and 1000 km. The condition Eq. (6.5) has two deficiencies. The first is that this equation, although derived for $z\to\infty$, is applied to finite altitudes. Fortunately however, this does not markedly affect the results as can be seen from Fig. 2 where two solutions are shown with Eq. (6.5) applied to 600 and 800 km, respectively. The second deficiency of condition Eq. (6.5) is that it leads to solutions that above about 700 km no longer satisfy $\frac{\partial}{\partial x, y} \ll \frac{\partial}{\partial z}$, a relation needed to obtain Eq. (6.5). In order to see how strongly the solutions depend on the upper boundary condition, Fig. 2 also shows the results based on the specification $\boldsymbol{v}_n=0$ at 1100 km. We notice some deviation from the results obtained with Eq. (6.5), which however is small considering the extreme nature of the upper boundary condition $v_n = 0$. We may therefore expect to get realistic wind velocity profiles with the upper boundary condition Eq. (6.5) despite its inherent weaknesses.

II. Auxiliary transport equations for wind calculations

7. Energy equation for the neutral gas. Calculation of the wind velocity requires knowledge of the horizontal and vertical temperature structure. In turn, the temperature is affected by neutral gas motions. The neutral temperature is described by the energy equation (or heat-conduction equation)

$$\frac{\partial(C_v T_n)}{\partial t} = -v_n \operatorname{grad}(C_v T_n) + \frac{L}{n_n}\{Q + \operatorname{div}(\kappa \operatorname{grad} T_n) - p_n \operatorname{div} v_n + \eta[\tfrac{4}{3}(\operatorname{div} v_n)^2 - (\operatorname{rot} v_n)^2]\}, \tag{7.1}$$

where C_v is the specific heat per mole at constant volume, L is the Loschmidt's number, Q is the heat production or loss per unit volume and unit time, and κ is the thermal conductivity.

The right-hand terms correspond, from left to right, to convection, heat production or loss (due to UV absorption, IR irradiation, and heat exchange with the plasma), heat conduction, conversion of work into heat or vice versa, and viscosity heating.

Viscosity heating is generally small compared with the other terms in Eq. (7.1). The coupling between the neutral temperature and the wind velocity is manifested through the terms $v_n \operatorname{grad}(C_v T_n)$ and $(L/n_n) p_n \operatorname{div} v_n$, which in the lower thermosphere are of the same order of magnitude as the UV heat deposition.

The problem of specifying boundary conditions for Eq. (7.1) is similar to the corresponding problem for the Navier-Stokes equation. Typically, a temperature of the order 300 to 500 K is assumed at the lower boundary (120 or 125 km), while at the upper boundary $\partial T_n/\partial z = 0$.

8. Continuity equation for the neutral constituents. The particle densities of the neutral constituents enter into the Navier-Stokes equation mainly through the pressure-gradient term. The particle densities, on the other hand, are strongly affected by neutral gas motion, in particular by vertical motions which may be caused by a convergent or divergent horizontal flow. Therefore the continuity equations of the neutral constituents form an integral part of the complete set of transport equations by which the wind field is described.

The continuity equation for the lth neutral constituent reads

$$\frac{\partial n_l}{\partial t} = P_l - \operatorname{div}(n_l v_l), \tag{8.1}$$

where P_l is the particle production or loss per unit volume and unit time due to photochemical reactions. The div-term is favorably split into a contribution from the horizontal flow, $\mathrm{div}(n_l \boldsymbol{v}_l^{(H)})$, and a contribution from the vertical flow, $\partial/\partial z\,(n_l v_{lz})$. The latter can be analytically expressed in the following way.

In the equation of motion for the vertical velocity component v_{lz}, the inertial, viscosity, and Coriolis terms may be omitted because they are small compared with g, the acceleration due to gravity. The frictional force with the plasma is also negligible; however, the frictional force with the other neutral constituents, denoted by subscript m, has to be taken into account. We thus obtain from Eqs. (2.1), (2.2), and (3.1)

$$0=-\frac{1}{\rho_l}\frac{\partial p_l}{\partial z}-g-\frac{1}{m_l}\sum_m \nu_{lm}\,\mu_{lm}(v_{lz}-v_{mz}). \tag{8.2}$$

The velocity v_{lz} described by Eq. (8.2) has the physical meaning of a molecular diffusion velocity and will, for clarity, be denoted by $v_{lz}^{(M)}$. Employing the ideal gas law $p_l = k\,n_l\,T_n$, we obtain from Eq. (8.2)

$$v_{lz}^{(M)}=V_{lz}^{(M)}-D_l^{(M)}\left\{\frac{1}{n_l}\frac{\partial n_l}{\partial z}+\frac{1}{I_n}\frac{\partial T_n}{\partial z}+\frac{1}{H_l}\right\}, \tag{8.3}$$

where $D_l^{(M)}=k\,T_n/\sum_{m\neq l}\nu_{lm}\,\mu_{lm}$ = molecular diffusion coefficient,

$H_l = k\,T_n/m_l\,g$ = individual scale height of species l,

$V_{lz}^{(M)}=\sum_{m\neq l}\nu_{lm}\,\mu_{lm}\,v_{mz}^{(M)}/\sum_{m\neq l}\nu_{lm}\,\mu_{lm}$.

Below about 100 km, the vertical velocity is not governed by molecular diffusion but by turbulence. The average turbulent velocity is, according to LETTAU [*31*], given by

$$v_{lz}^{(T)}=-D^{(T)}\left\{\frac{1}{n_l}\frac{\partial n_l}{\partial z}+\frac{1}{T_n}\frac{\partial T_n}{\partial z}+\frac{1}{H_n}\right\}, \tag{8.4}$$

where $D^{(T)}$ = turbulent (or eddy) diffusion coefficient, and

$H_n = k\,T_n/m_n\,g$ = scale height of the neutral gas as a whole.

The eddy diffusion coefficient is an empirical quantity which can only be derived from observations in situ. Approximate expressions for $D^{(T)}$ are given by [1], based on [2].

The effect of molecular diffusion is to establish a density height distribution of species l according to the barometric law with the individual scale height H_l, whereas eddy diffusion tends to establish complete mixing, represented by a common scale height H_n for all neutral constituents.

---

[1] Shimazaki, T. (1971): J. Atmos. Terr. Phys. *33*, 1383

[2] Keneshea, T.J., Zimmerman, S.P. (1970): J. Atmos. Sci. *27*, 831

The total vertical velocity is the sum of the molecular and turbulent diffusion velocities[3],

$$v_{lz} = v_{lz}^{(\mathrm{M})} + v_{lz}^{(\mathrm{T})}. \tag{8.5}$$

Thus

$$v_{lz} = V_{lz}^{(\mathrm{M})} - (D_l^{(\mathrm{M})} + D^{(\mathrm{T})}) \left\{ \frac{1}{n_l} \frac{\partial n_l}{\partial z} + \frac{1}{T_\mathrm{n}} \frac{\partial T_\mathrm{n}}{\partial z} \right\} - \frac{D_l^{(\mathrm{M})}}{H_l} - \frac{D^{(\mathrm{T})}}{H_\mathrm{n}}. \tag{8.6}$$

Substitution of Eq. (8.6) in the continuity Eq. (8.1) yields a differential equation of second order in z. At a sufficiently high upper boundary, 300 km or more, the boundary condition can be obtained by assuming molecular diffusion equilibrium and neglecting $\partial T_\mathrm{n}/\partial z$, i.e.,

$$\frac{\partial n_l}{\partial z} = -\frac{n_l}{H_l}. \tag{8.7}$$

Proper lower boundary conditions can be specified in the mesosphere, below about 80 km. For the chemically inert constituents one may assume the composition at the lower boundary to be the same as at the Earth's surface, so that the total density determines all these particle densities. For the chemically active constituents it will, in general, be valid to assume chemical equilibrium at an appropriate height.

9. Continuity and momentum equations for the plasma constituents. The plasma density and velocity affect the neutral wind through friction.

α) The *continuity equation* for ion species j reads, if we neglect horizontal derivatives in the div-term and use N_j for the number densities of plasma components,

$$\frac{\partial N_j}{\partial t} = P_j - \frac{\partial}{\partial z}(N_j v_{jz}), \tag{9.1}$$

where P_j again is the particle effective production (or loss) per unit volume and unit time due to photochemical reactions. The continuity equation for the electron gas does not have to be considered separately because of quasi-neutrality,

$$N_\mathrm{e} = \sum_j N_j, \tag{9.2}$$

provided no negative ions are present. The neglect of horizontal derivatives in Eq. (9.1) is justified except for the equatorial anomaly region.

β) In the *equation of motion* for ion species j, we have as an additional external force the Lorentz force, but we may neglect the inertial, viscosity, and Coriolis terms which are small compared with the pressure gradient, gravity,

[3] See T. Yonezawa's contribution in this volume, p. 129

frictional, and Lorentz terms. The equation of motion thus reads

$$0=-\frac{1}{\rho_j}\operatorname{grad} p_j+\boldsymbol{g}-\frac{1}{m_j}\sum_l \nu_{jl}\,\mu_{jl}(\boldsymbol{v}_j-\boldsymbol{v}_\mathrm{n})+\frac{q}{m_j}(\boldsymbol{E}+\boldsymbol{v}_j\times\boldsymbol{B}) \tag{9.3}$$

where $\boldsymbol{E}$ is the electric field strength, q the unitary charge, and l denotes the individual neutral species. Restricting our considerations to the case that the gyrofrequency $\omega_j=qB/m_j$ is much larger than the collision frequencies ν_{jl}, which is satisfied above about 200 km, we obtain for v_{jz} [38]

$$\begin{aligned} v_{jz}=&\frac{E_x}{B}\cos I\sin D+\frac{E_y}{B}\cos I\cos D+v_{\mathrm{n}x}\sin I\cos I\cos D\\ &-v_{\mathrm{n}y}\sin I\cos I\sin D+v_{\mathrm{n}z}\sin^2 I-\frac{F_j}{\sum_l \nu_{jl}\,\mu_{jl}}\sin^2 I, \end{aligned} \tag{9.4}$$

where $F_j=k\left[\frac{1}{N_j}\frac{\partial}{\partial z}(N_j T_\mathrm{i})+\frac{1}{N_\mathrm{e}}\frac{\partial}{\partial z}(N_\mathrm{e} T_\mathrm{e})\right]+m_j g,$

T_i = ion temperature, T_e = electron temperature,

and N_e = electron (number) density.

γ) We notice that the diffusion velocity, represented by the last term in Eq. (9.4), depends on the electron density. This is because the ion and electron motions are coupled by means of an *electrostatic polarization field* which prevents charge separation. The condition $\omega_j \gg \nu_{jl}$ has the consequence that diffusion and wind-induced motions are possible only along the magnetic lines of force. In deriving Eq. (9.4) it was tacitly assumed that the frictional force between the different ionic species is small compared to the frictional force between the ions and neutral particles. This is no longer true in the topside ionosphere at altitudes where H^+ becomes a major ionic constituent.

The knowledge of v_{jz}, E_x, and E_y is sufficient to express all the other velocity components of interest [45]:

$$v_{jx}=v_{jz}\cot I\cos D-\frac{E_y}{B}\operatorname{cosec} I, \tag{9.5}$$

$$v_{jy}=-v_{jz}\cot I\sin D+\frac{E_x}{B}\operatorname{cosec} I, \tag{9.6}$$

$$v_{j\|}=v_{jz}\operatorname{cosec} I-\frac{E_x}{B}\cot I\sin D-\frac{E_y}{B}\cot I\cos D, \tag{9.7}$$

$$v_{j\xi}=v_{jz}\cot I-\frac{E_x}{B}\operatorname{cosec} I\sin D-\frac{E_y}{B}\operatorname{cosec} I\cos D, \tag{9.8}$$

$$v_{j\eta}=\frac{E_x}{B}\operatorname{cosec} I\cos D-\frac{E_y}{B}\operatorname{cosec} I\sin D. \tag{9.9}$$

$v_{j\|}$, $v_{j\xi}$, and $v_{j\eta}$ are the velocities upwards along the magnetic field lines, towards magnetic south and magnetic east, respectively.

Inserting Eq. (9.4) into Eq. (9.1) again yields a differential equation of second order in z. Boundary conditions typically applied are chemical equilibrium in the E- or lower F-region and a zero vertical flux ($N_j v_{jz}=0$) at the upper boundary in the diffusion controlled region (above about 600 km).

Equations (9.5) and (9.6) have to be used in place of Eq. (3.5) for an exact calculation of the x- and y-components of the frictional force between the plasma and the neutral gas.

III. Applicability of the transport equations in the upper atmosphere

10. General limits of applicability. A general limit of applicability of the transport equations is set by the natural requirement that the macroscopic quantities n, $\boldsymbol{v}$, and T be well defined. These quantities are well-defined when the following conditions are met:

(a) Volume elements in configuration space must exist large enough to contain a sufficiently large number of particles, but small enough that the change of the macroscopic quantities from one volume element to the next remains sufficiently small.

(b) Time intervals must exist, long compared with the mean time that a particle takes to cross the volume element, but short enough that the change of the macroscopic quantities between time intervals remains sufficiently small.

In order to have a statistical variability of less than 10^{-4}, the total number of molecules has to be greater than 10^8. Assuming an extremely low particle number density of 1 cm^{-3} ($10^6\ \mathrm{m}^{-3}$), our volume element has to have a minimum size of $10^2\ \mathrm{m}^3$, corresponding to a side length of 5 m. A typical macroscopic scale length is well in excess of 10^3 m, so that condition (a) is easily satisfied even for particle densities less than $10^6\ \mathrm{m}^{-3}$.

For a low particle velocity of 100 $\mathrm{m\,s}^{-1}$, the time to traverse the volume element is $5\cdot 10^{-2}$ s. A characteristic time for the macroscopic quantities to change is of the order 1 h for regular diurnal variations and 1 min for short scale processes.

We may conclude, therefore, that the conditions (a) and (b) do not constitute a practical limit of applicability of the transport equations in the upper atmosphere.

11. Rigorous limits of applicability. The statement that the conditions (a) and (b) of Sect. 10 are satisfied is analogous to saying that the Maxwellian transport equations are applicable. These are equations for the moments of the distribution function, derived from and physically equivalent to the Boltzmann equation. The Maxwellian transport equations are not directly usable for practical problems since in the equation for the nth moment, knowledge is required of the $(n+1)$th moment. The transport equations presented in Sects. 2, 7, 8, and 9 correspond to the first-order approximation of the Enskog–Chapman theory. They are valid only if the mean free path λ is small compared with a macroscopic scale length H, which may be interpreted as the scale height kT/mg, and if the mean collision time τ is small compared with a characteristic time t_0 for temporal variations of the macroscopic quantities to take place:

$$\lambda \ll H, \tag{11.1}$$

$$\tau \ll t_0. \tag{11.2}$$

If the conditions Eqs. (11.1) and (11.2) are relaxed to $\lambda \approx H$ and $\tau \approx t_0$, the upper limit of applicability of the transport equations lies at about 500 km. Strict application of Eqs. (11.1) and (11.2) yields an upper limit of 400 km at most.

Approximations of higher than first order, if manageable, lead to a more refined description of the physical system under consideration. They do not, however, extend the range of applicability and are useful only when the applicability of the first-order approximation is secured by fulfilment of the above conditions. Therefore a discussion of the limits of applicability of the transport equations in terms of a comparison of the Navier–Stokes with higher-order approximations, as given by [1] and referred to by [2], seems to miss the point.

We have thus far answered the question of up to which altitude the transport equations are valid. However, the relevant question is up to which altitude are the results derived from the transport equations correct, no matter whether or not the transport equations are valid in a strict physical sense. Considering that due to the upper boundary conditions imposed, $\boldsymbol{v}_\mathrm{n}$ and T_n are approximately constant with altitude above about 400 km while n_j and n_k approximately follow the barometric law there, it is easy to see that these questions are not necessarily equivalent. This aspect will be discussed in the following section.

12. Collisionless orbit considerations. An answer to the question raised at the end of Sect. 11 has to be found by solution of the Boltzmann equation from the collision-dominated thermosphere through a transitional zone characterized by $\lambda \approx H$ up to the collision-less exosphere. Even with numerical methods this task is beyond our present possibilities.

α) A widely used and well accepted way out of this difficulty is to introduce a *sharp boundary* at $\lambda = H$, termed the thermopause or exobase and to assume a discontinuous transition, i.e., that in the lower region collisions establish a Maxwell–Boltzmann distribution while in the upper region collisions are negligible. Correspondingly, the collisionless Boltzmann equation has to be solved above the thermopause, and from the multiplicity of solutions the one which becomes identical with the Maxwell distribution at the thermopause height z_0 has to be selected. This concept has been critically discussed in [1] and proved to be a valid approach except for hydrogen at temperatures above about 1500 K.

---

[1] Izakov, M.N. (1967): Space Sci. Rev. 7, 579

[2] Blum, P., Harris, I., Priester, W. (1972): COSPAR International Reference Atmosphere 1972, p. 399. Berlin: Akademie-Verlag

[1] Chamberlain, J.W. (1963): Planet. Space Sci. *11*, 901

In [1], the most simple case is assumed, i.e., constant temperature and particle density and zero horizontal velocity along the thermopause surface. The time-independent collisionless Boltzmann equation then reads

$$\dot{r}\frac{\partial f}{\partial r}+\dot{p}_r\frac{\partial f}{\partial p_r}=0,$$

where f is the distribution function in phase space, r the radial distance and p_r the radial momentum. This equation is solved by

$$f(r, p_r, p_\chi)=e^{\gamma-\gamma_0} f_M(r_0, p_r, p_\chi; T_0, n_0), \tag{12.1}$$

where $\gamma=\dfrac{GMm}{kT_0 r}$,

G the gravitational constant, M the Earth's mass, m the particle mass, p_χ the angular momentum, and f_M the Maxwell distribution belonging to the parameters T_0 and n_0. The subscript 0 indicates that the respective quantity refers to the thermopause at altitude z_0 or radial distance r_0 from the Earth's center. For the particle density n as a function of r we obtain from Eq. (12.1) by integration over the momentum space

$$n\equiv\int f\,d^3\boldsymbol{p}=n_0\,e^{\gamma-\gamma_0}. \tag{12.2}$$

This is the well-known *barometric law* in its general form. It thus appears that both the distribution function in phase space and the $n(r)$ distribution are the same in the collisionfree region $r>r_0$ and the collision-controlled region $r\leqq r_0$.

β) This, however, is not true since we have to rule out for $r>r_0$ those pairs p_r, p_χ which belong to incoming orbits with a velocity exceeding the escape velocity and to outgoing *escape orbits* not intersecting the thermopause surface. Thus for $r>r_0$ the particle density is necessarily smaller than that given by Eq. (12.2). Introducing a partition function

$$\zeta(\gamma)\equiv\int_a f\,d^3\boldsymbol{p}\Big/\int_t f\,d^3\boldsymbol{p} \tag{12.3}$$

where a and t indicate integration over the allowed and total momentum space, respectively, we have

$$n=\zeta(\gamma)\,n_0\,e^{\gamma-\gamma_0}. \tag{12.4}$$

A numerical evaluation shows that for $T_0=1500$ K the partition function drops to 0.9 at about $z=7500$ km for H, $z=40{,}000$ km for He and $z=200{,}000$ km for O and at even higher altitudes for $T_0<1500$ K. This means that the barometric law is correct to within 10% up to these altitudes, provided, of course, that the interplanetary space is populated only by particles originating from the Earth's atmosphere and that particle loss due to photoionization is negligible. This is a striking result considering that the macroscopic equations from which the barometric law follows are valid only up to 500 km at most.

With increasing height, the kinetic temperature,

$$T_{\rm kin} \equiv \frac{2T_0}{3U^2}(\langle v^2\rangle - \langle v_r\rangle^2), \tag{12.5}$$

$$U \equiv (2kT_0/m)^{1/2},$$

decreases since the distribution function becomes increasingly truncated and since the escape orbits gain increasing importance relative to the ballistic and satellite orbits. In the limit $\gamma \to 0$, when solely escaping particles give a contribution, the kinetic temperature approaches

$$T_{\rm kin} \approx 0.12\, T_0. \tag{12.6}$$

γ) CHAMBERLAIN[1] discusses only this asymptotic case but not the *height dependence* of $T_{\rm kin}$ in the vicinity of the thermopause. We can, however, readily evaluate $T_{\rm kin}$ as given by Eq. (12.5) if γ is sufficiently large for escape orbits to be disregarded so that only ballistic and satellite orbits need to be taken into account. We then obtain

$$T_{\rm kin} = \frac{T_0}{\zeta(\gamma)}\left[\operatorname{erf}(\sqrt{\gamma}) - \frac{2}{3\sqrt{\pi}}(3\gamma^{1/2} + 2\gamma^{3/2})\, e^{-\gamma}\right]. \tag{12.7}$$

Numerical results are shown in Fig. 3. We notice that for hydrogen a discontinuity appears at the thermopause for thermosphere temperatures around and above 1000 K. This is partly due to a general deficiency of the theoretical concept, and partly to the neglect of escape orbits in Eq. (12.7). Also shown in Fig. 3 is the height above which $T_{\rm kin}$ drops below $0.9\, T_0$ as a function of T_0. For hydrogen this height ranges from 17,000 km at $T_0 = 500$ K to 1200 km at $T_0 = 1500$ K and is by orders of magnitude greater for the other constituents. We may conclude, therefore, that the temperature results obtained from the energy Eq. (7.1) are valid far into the exosphere; although from a strict physical viewpoint the energy equation is valid up to about 400 to 500 km only. Care has to be applied for hydrogen at high temperatures.

The macroscopic horizontal velocity was not discussed in[1]. Since the effect of a high-energy truncation of the distribution function increases with increasing order of the moment of the distribution function, it is plausible to conclude that the macroscopic velocity is more affected by the truncation than the particle density, but less than the kinetic temperature. Thus the temperature limit appears as a lower, the particle density limit as an upper, limit to the validity of the velocity results obtained from the equation of motion.

δ) A more complex situation arises when n, $\boldsymbol{v}$, and T possess *gradients* in the thermopause surface. For this general case a solution has been given by HARTLE [*19*] but analysed only with respect to the particle density-height distribution. If we consider a point in the exosphere having the coordinates (r, φ, λ), we expect n, $\boldsymbol{v}$, and T at this point to be determined by the respective values at (r_0, φ, λ) if $(r - r_0)$ is small compared with r_0. With increasing height, however, orbits originating from more remote regions on the thermopause surface gain increasing importance on determining the properties at (r, φ, λ). Finally, if $(r - r_0)$ is large compared with r_0, the properties at (r, φ, λ) should be determined by the average values of n, $\boldsymbol{v}$, and T over the thermopause surface

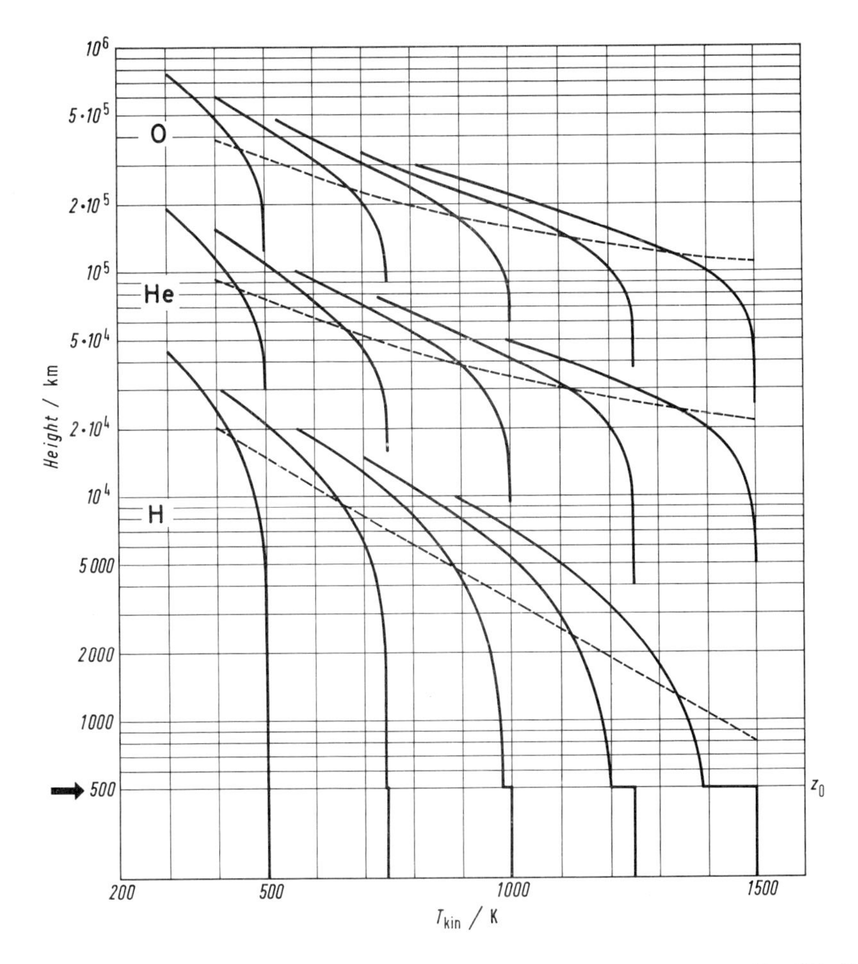

Fig. 3. Solid curves: T_{kin} as a function of altitude for thermopause temperatures $T_0=500$, 750, 1000, 1250, 1500 K and neutral constituents H, He, and O after Eq. (12.7). Dashed curves: Height at which $T_{\mathrm{kin}}=0.9\,T_0$ as a function of T_0

rather than by the particular values at (r_0, φ, λ). Consequently, $\boldsymbol{v}$ and T must vary with altitude and n must deviate from the barometric law in order for the transition from local to average behavior to be achieved. The height at which the transition occurs must depend on the horizontal scale length and should roughly be of the same order of magnitude. Disregarding local inhomogeneities, a typical horizontal scale length is of the order 10,000 km, corresponding to the distance pole-equator. We should expect, therefore, that up to altitudes of some thousand km above the thermopause the results at (r, φ, λ) should, for the case discussed, be the same as if n, $\boldsymbol{v}$, and T at (r_0, φ, λ) were representative for the whole thermopause. This qualitative conclusion is confirmed in [*19*].

ε) *Summarizing* we may state that in the high atmosphere the results obtained from the transport equations are valid up to much higher altitudes than the transport equations themselves, the validity range being larger for lower temperatures and higher particle masses.

C. Calculation of wind velocities

13. General remarks. According to our previous discussion, the equation of motion for the neutral gas is only a part within the system of coupled transport equations. Only the early papers on neutral winds dealt exclusively with the equation of motion, disregarding the coupling mechanisms (GEISLER [*17*], KOHL and KING [*30*], CHALLINOR [1]). In later works the mutual coupling between the wind velocity and the electron density (STUBBE [2], KOHL et al. [3], BAILEY et al. [*4*], RÜSTER [4], TORR and TORR [*47*], STROBEL and MCELROY [5]) and other ionospheric quantities (STUBBE [*43*], TANAKA and HIRAO [6]) was taken into account. All these attempts corresponded to one-dimensional solutions with respect to height or quasi-two-dimensional solutions by interchanging geographic longitude and local time.

Parallel to improving the one-dimensional solutions, work on two- or quasi-three-dimensional solutions was done (VOLLAND [7], VOLLAND and MAYR [*48*], BLUM and HARRIS [*9*], VEST [8]). True three-dimensional models, in which proper account is given of the magnetic control of the thermosphere, are not yet available, although some of the quasi-three-dimensional models are termed three dimensional by the respective authors.

A short description of the different methods used to solve the wind equation and the related transport equations and of the results thereby obtained will be given in the following sections.

I. Methods of calculating wind velocities

14. One- and quasi-two-dimensional solutions

α) In a *one-dimensional treatment*, the only spatial derivative is the altitude z. All derivatives with respect to x and y are neglected. However, derivatives with respect to y, the W-E coordinate, may approximately be taken into account by relating them to the corresponding derivatives with respect to time t by means of

$$\frac{\partial}{\partial y}=\frac{1}{\Omega R_0 \cos\varphi}\frac{\partial}{\partial t}=f(\varphi)\frac{\partial}{\partial t} \tag{14.1}$$

[1] Challinor, R.A. (1968): Planet. Space Sci. *16*, 557

[2] Stubbe, P. (1968): J. Atmos. Terr. Phys. *30*, 243

[3] Kohl, H., King, J.W., Eccles, D. (1968): J. Atmos. Terr. Phys. *30*, 1733

[4] Rüster, R. (1969): J. Atmos. Terr. Phys. *31*, 765

[5] Strobel, D.F., McElroy, M.B. (1970): Planet. Space Sci. *18*, 1181

[6] Tanaka, T., Hirao, K. (1973): J. Atmos. Terr. Phys. *35*, 1443

[7] Volland, H. (1967): Space Res. VII, p. 1193. Amsterdam: North-Holland Publishing Company

[8] Vest, R. (1973): A three-dimensional model of the thermosphere with auroral heating. Pennsylvania State University Report PSU-IRL-SCI-412

with R_0 the Earth's radius, Ω the Earth's angular velocity and φ the geographic latitude. Equation (14.1) is based on the idealization that the thermospheric properties are functions of latitude and local time only, thereby disregarding magnetic declination, magnetic inclination, and Universal Time effects.

By means of Eq. (14.1) we obtain from Eqs. (2.1), (2.2), and (3.4) the following differential equations for v_{nx} and v_{ny}:

$$\frac{\partial v_{nx}}{\partial t}=\frac{1}{1+v_{ny}f(\varphi)}\left\{\frac{\eta}{\rho_n}\frac{\partial^2 v_{nx}}{\partial z^2}-\frac{1}{\rho_n}\sum_j N_j(v_{nx}-v_{jx})\sum_l \nu_{jl}\mu_{jl}\right.$$
$$\left.+2v_{ny}\Omega\sin\varphi-\frac{1}{\rho_n}\frac{\partial p_n}{\partial x}\right\}, \tag{14.2}$$

$$\frac{\partial v_{ny}}{\partial t}=\frac{1}{1+v_{ny}f(\varphi)}\left\{\frac{\eta}{\rho_n}\frac{\partial^2 v_{ny}}{\partial z^2}-\frac{1}{\rho_n}\sum_j N_j(v_{ny}-v_{jy})\sum_l \nu_{jl}\mu_{jl}\right.$$
$$\left.-2v_{nx}\Omega\sin\varphi-\frac{1}{\rho_n}\frac{\partial p_n}{\partial y}\right\}, \tag{14.3}$$

where the subscripts j and l denote the ionic and neutral constituents, respectively.

These equations can be solved if the ion densities, the ion velocities, the neutral density, and the horizontal components of the neutral pressure gradient are known. The ion continuity Eqs. (9.1) can also be solved one or quasi-two dimensionally, except for the equatorial anomaly region where the ion velocity as given by Eqs. (9.4) to (9.9) is preferentially horizontal. The ionospheric plasma affects the wind velocity by giving rise to a frictional force. On the other hand, the neutral wind affects the plasma by setting up plasma motions parallel to the magnetic field lines, thereby leading to a redistribution of the plasma density. Thus the coupling is mutual, and it is therefore not sufficient to externally specify the ion density and insert it in Eqs. (14.2) and (14.3), as was done in the early papers[1] [*17*, *30*] and still is done in quasi three-dimensional modeling [*9*, *48*].

β) A simultaneous solution of the wind Eqs. (14.2), (14.3) and the continuity Eq. (9.1) is quite straightforward since the coupling, although important, is weak enough for allowing the equations to be mathematically decoupled by solving them independently and taking the other unknowns from the previous time step[2]. A time step width of the order 10 min is sufficiently small for this procedure. No prescription is required for the time dependence of the solutions. In particular it is not necessary for the solutions to be periodic. Within a transition time of the order 12 h the solutions become independent of the necessarily arbitrary initial profiles, and from there on are able to reproduce any kind of time dependence, subject to the given physical conditions. This is not the case for the methods used in quasi-three-dimensional modeling.

[1] Challinor, R.A. (1968): Planet. Space Sci. *16*, 557
[2] Stubbe, P. (1968): J. Atmos. Terr. Phys. *30*, 243

γ) The inherent weakness of the one-dimensional method is that the thermospheric quantities, in particular grad p_n, have to be externally fed in by using thermospheric models. This is because in a one-dimensional treatment of the energy and continuity equations of the neutral gas, the terms involving the divergence of the horizontal flux cannot fully be taken into consideration. Consequently the influence of the neutral wind on the neutral composition and temperature structure and thus on the pressure gradient cannot be described by the one- and quasi-two-dimensional method. This, however, must not necessarily mean a major restriction. If the thermospheric model used affords a realistic description of the observed thermospheric parameters, then the wind effects on these parameters are automatically included.

15. Quasi-three-dimensional solution with external thermospheric model. BLUM and HARRIS [*9*] extended the quasi-two-dimensional to a quasi-three-dimensional method, thereby taking into account velocity derivatives with respect to x, the N-S coordinate. They expanded the velocity components in Fourier series but neglected all terms of higher than second order.

This is a severe restriction for at least two reasons. First, the Fourier expansion forces the solution to be periodic, thereby not allowing study of other than regular diurnal variations. Second, it is known from quasi-two-dimensional solutions that harmonics of third and higher order are by no means negligible. This can be seen from Fig. 4 which shows, as a randomly chosen example, a Fourier analysis of the diurnal v_{nx} and v_{ny} variations for Adak, June and December 1959, based on a quasi-two-dimensional solution in [*45*], using the CIRA-72 thermospheric model [*11*]. In particular the v_{ny} results for June 1959 indicate that higher harmonics yield a large contribution, the amplitudes for the third to sixth harmonic amounting to 52, 32, 18, and 11% of the first harmonic, respectively.

The gain in accuracy by fully taking into account the nonlinear term is, according to BLUM and HARRIS [*9*], about 10% at middle and high latitudes, but much more than this in a zone around the equator. It is apparent, therefore, that the gain in accuracy by including the nonlinear term is greater than the loss in accuracy by truncating the Fourier series in the vicinity of the equator, but smaller at all other latitudes.

The method of [*9*] involves some more approximations which are not well justified. The ion density is taken from a semiempirical model[1], but is then also expanded in a Fourier series up to second order and averaged over longitude. The ion velocity is taken from Eq. (3.5) instead of Eqs. (9.4) to (9.9), and the declination angle is assumed to be zero. All these measures correspond to a smoothing process which leads to a background ionosphere far from reality. It appears, therefore, that the advantage of including the nonlinear term is overcompensated by the disadvantage of having to introduce all the mentioned approximations.

16. Quasi-three-dimensional solutions with self-consistent thermospheric model. VOLLAND and MAYR[1] [*48*] developed a self-consistent model by simultaneously solving the continuity, momentum, and energy equations for the neutral gas as a whole. This concept comes very close to the real physical

[1] Nisbet, J.S. (1970): On the construction and use of the Penn State MKI ionospheric model. Pennsylvania State University Report PSU-IRL-SCI-355

[1] Volland, H., Mayr, H.G. (1973): Ann. Geophys. *29*, 61

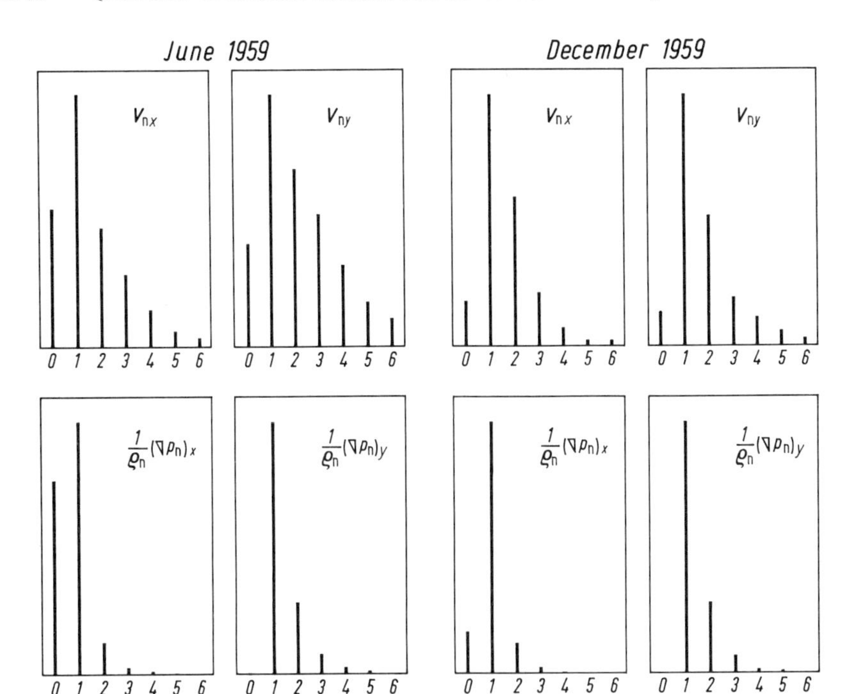

Fig. 4. Fourier analysis of the wind velocity and the pressure gradient at 300 km altitude for Adak ($\varphi = 52°$ N), June and December 1959. The graphs show the amplitudes of the zeroth to sixth harmonics in arbitrary units

situation since all variations of the thermospheric parameters are expressed in terms of the energy input only. In particular, it is not necessary in this treatment to introduce an external driving force for the wind system.

However, the simplifications needed to make this concept practically manageable are of an extreme nature. The most important *assumptions* are:

(1) Perturbation theory is applicable. This requires knowledge of the global average values T_0, ρ_0, p_0 which have to be taken from an empirical model. It furthermore requires that the temperature, density, and pressure perturbations T, ρ, p be small compared with T_0, ρ_0, p_0.

If this condition is satisfied, the total values (average + perturbation) may be replaced by the average values, except in the derivatives with respect to the independent variables where only the perturbations appear. As an example, the acceleration due to the pressure gradient reads

$$\frac{1}{\rho_0}\operatorname{grad} p \quad \text{instead of} \quad \frac{1}{\rho_0+\rho}\operatorname{grad} p.$$

In order to estimate how well this approximation is justified, we consult the CIRA-72 model [*11*] and find that for medium solar activity, ρ may amount to 40% of ρ_0 at 300 km and 60% at 400 km. This leads to an error in the equation of motion which cannot be tolerated if a claim is made for quantitative significance.

(2) The system thermosphere-ionosphere is linear. This condition requires that an energy source with frequency $\omega/2\pi$ leads to a response having the same frequency without any other harmonic. In[1], this assumption is slightly

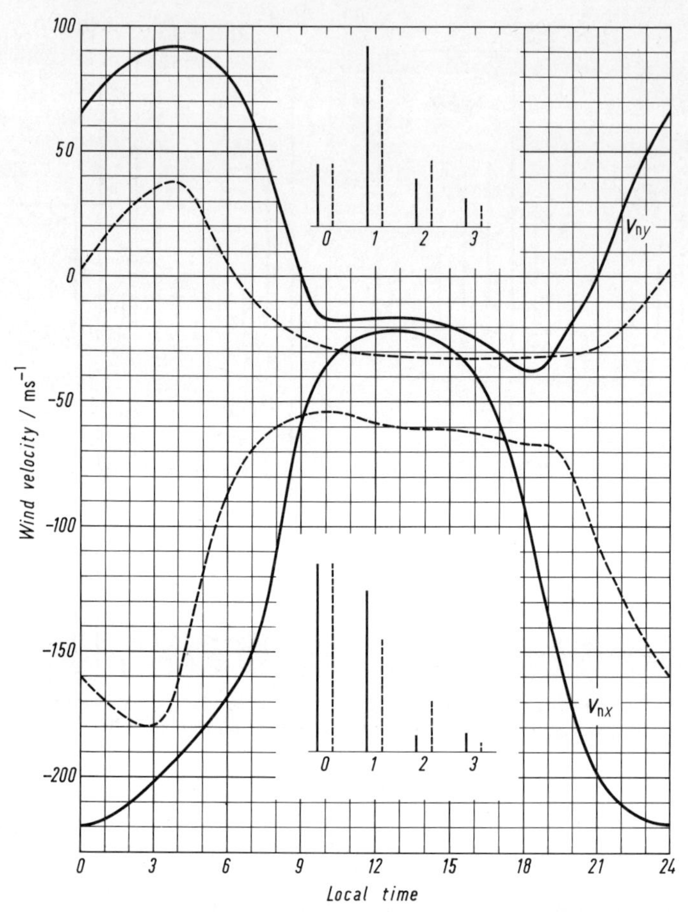

Fig. 5. Wind velocity components at 300 km altitude for Adak ($\varphi = 52°$ N) June and December 1959, based on the assumption that $\frac{1}{\rho}\frac{\partial p}{\partial x, y}$ is independent of local time and depends linearly on height, being 0 at 125 km and 2 cm s^{-2} at 300 km. Also shown is a Fourier analysis of the diurnal variation of the W-E velocity v_{ny} (upper) and N-S velocity v_{nx} (lower). Solid curves: December 1959. Dashed curves: June 1959

relaxed by allowing a coupling from the (1, 1) into the (2, 0) and (2, 2) modes but still neglecting other types of mode coupling. In order to see how linear, or rather nonlinear, the thermosphere-ionosphere system is, a one-dimensional calculation was performed assuming a time-independent pressure gradient and taking observed values for the ion density as a function of height and time of day. The results are shown in Fig. 5 and indicate an extremely high degree of nonlinearity. A linear system should react with a constant wind velocity. Actually, however, the first harmonic is by a factor of 3 higher than the zeroth harmonic for the W-E velocity and is about 75% of the zeroth harmonic for the N-S velocity. Another indication of the high nonlinearity is given by the results of Fig. 4 which show that a spectral decomposition of the velocity components and the corresponding driving force components yields largely different characteristics, although they should be identical for a linear system. Thus assumption (2) is probably even less justified than (1).

(3) The viscosity term is proportional to the velocity. This assumption is without any justification. It has been clearly demonstrated by several authors (e.g., RISHBETH [*36*], Figs. 7 and 8) how decisively the velocity profile depends on a proper treatment of the viscosity term. Assumption (3) is not even correct with respect to sign, so that it may be worse than neglecting viscosity. Physically, there is no need to introduce any restrictive assumption for the viscosity term since it belongs to the well-known terms in thermospheric physics.

The assumptions (1) to (3) as well as other assumptions not discussed here appear so drastic that the method probably fails to produce quantitatively correct results, at least with respect to the neutral wind. The method may be useful, however, to give semiquantitative support to certain qualitative concepts since it widely takes into account the coupling between the dynamic and thermodynamic aspects of the thermospheric behavior.

II. Characteristics of calculated wind velocities

17. Variation with local time. Figure 6 shows the wind field at 300 km for equinox conditions in a polar diagram, looking at the North pole. We notice

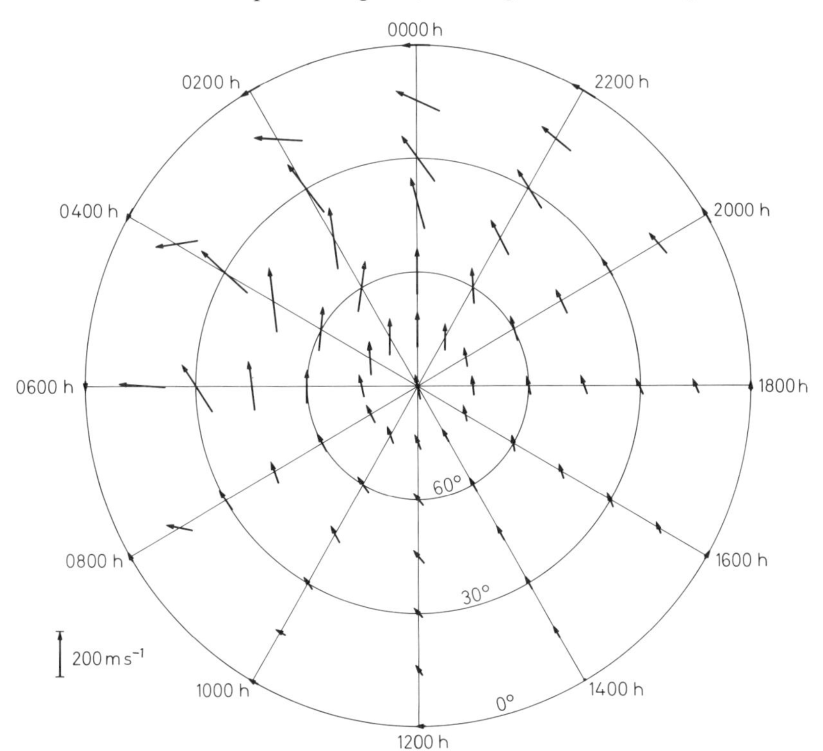

Fig. 6. Wind velocity at 300 km altitude as a function of latitude and local time for medium solar activity and equinox (after KOHL [*29*])

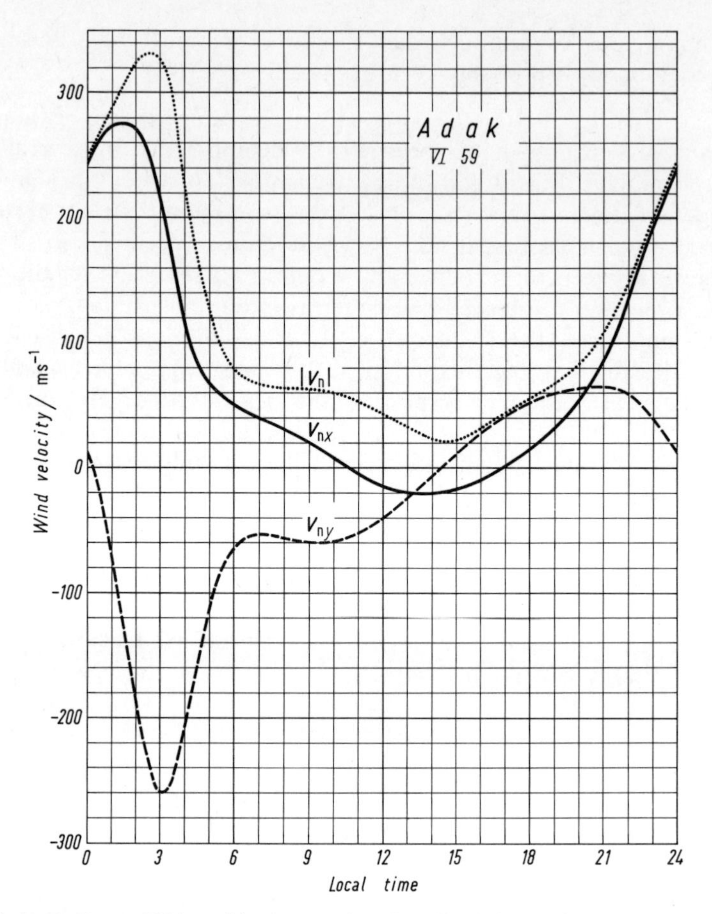

Fig. 7. Wind velocity at 300 km altitude as a function of local time for Adak ($\varphi = 52°$ N), June 1959, based on a modified JACCHIA model (after STUBBE [45]). Dotted curve: Total velocity $|\boldsymbol{v}_n|$. Solid curve: N-S component v_{nx}, positive towards south. Dashed curve: W-E component v_{ny}, positive towards east

that the wind velocity is higher on the night side than on the day side, which is mainly due to the difference in the ion drag. The direction is roughly from high to low pressure, but visibly modified by the Coriolis force as can be seen from the eastward rotation of the wind vector when progressing from lower to higher latitudes. The meridional velocity component is directed towards the pole at day and towards the equator at night, while the zonal component is directed such that the wind is blowing away from the 14 h meridian.

A more detailed picture for one particular station, Adak, is given in Figs. 7 and 8 for June and December 1959. These figures reveal one very artificial characteristic of the wind field, namely, the W-E component v_{ny} has a westward maximum close to 3 h and a zero close to 14 h local time independent of season. This is because the thermospheric properties according to the CIRA-72 model [11] do not depend on the day length, although actually they should.

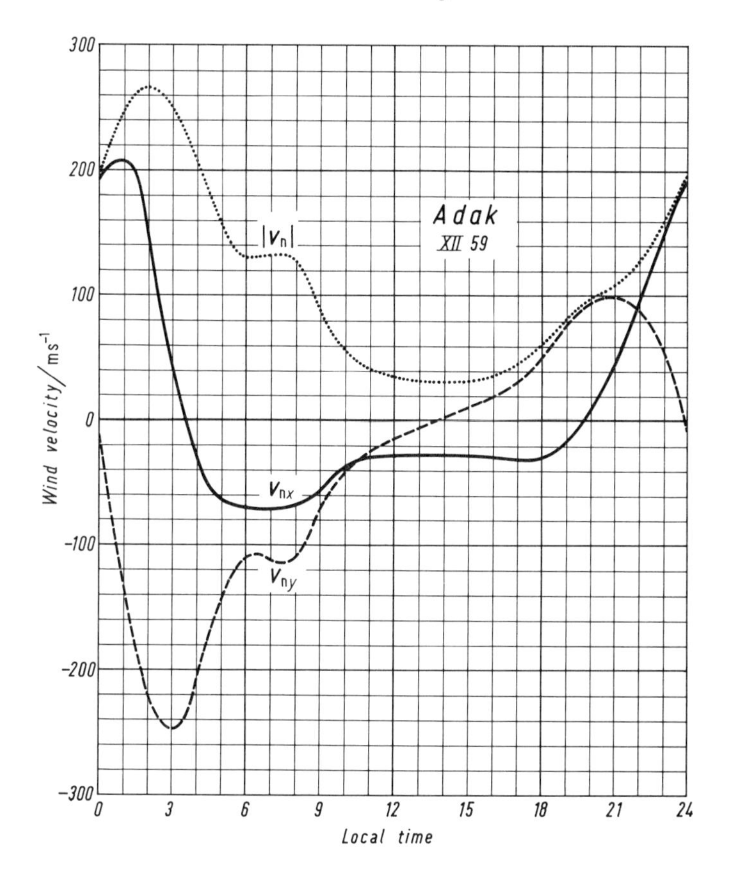

Fig. 8. Same as Fig. 7, but for December 1959

The prevailing wind is seen to be equatorward in the meridional and westward in zonal direction, while observations (KING-HELE [*26*]) indicate a prevailing eastward wind, corresponding to a thermospheric superrotation (for more information on superrotation see RISHBETH [*37*]).

The wind direction for the cases shown in Figs. 7 and 8 is plotted in Fig. 9. It can be seen that the wind vector rotates clockwise with a nonuniform rate. Around sunrise the wind direction is nearly constant for a period of about four hours.

18. Variation with height. The shape of the velocity-height profile is mainly controlled by three forces, i.e., pressure gradient, friction, and viscosity. The pressure-gradient force increases nearly linearly with height and thus tends to establish a wind velocity which also monotonously increases with height. Friction leads to an indentation of the velocity profile near the F2-layer maximum, while viscosity tends to minimize second derivatives with respect to height and thus smoothes out the profile and reduces the velocity above the

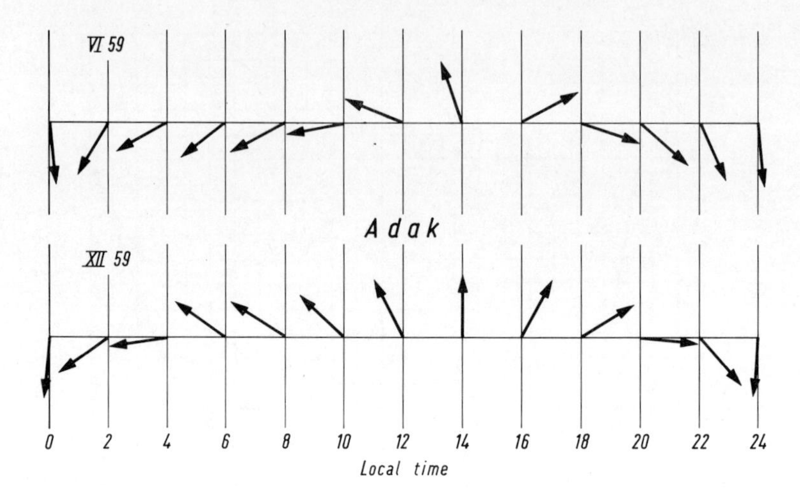

Fig. 9. Wind direction at 300 km altitude as a function of local time for Adak ($\varphi = 52°$ N), June 1959 (upper part) and December 1959 (lower part), based on a modified JACCHIA model (after STUBBE [45]). Note that the length of the arrow is not a measure of the wind velocity. Arrow directed upwards corresponds to poleward wind

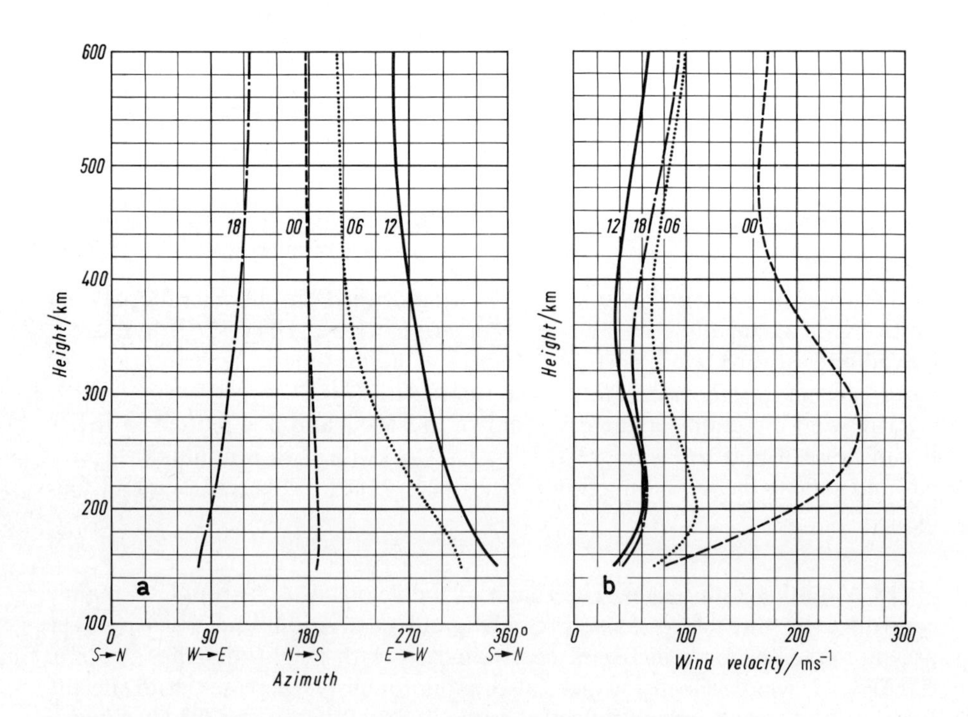

Fig. 10a, b. Total wind velocity and wind direction as a function of altitude for Adak ($\varphi = 52°$ N), June 1959, four local times, based on a modified JACCHIA model (after STUBBE [45]). **(a)** Azimuth, measured from N to E. **(b)** Total wind velocity $|\boldsymbol{v}_n|$

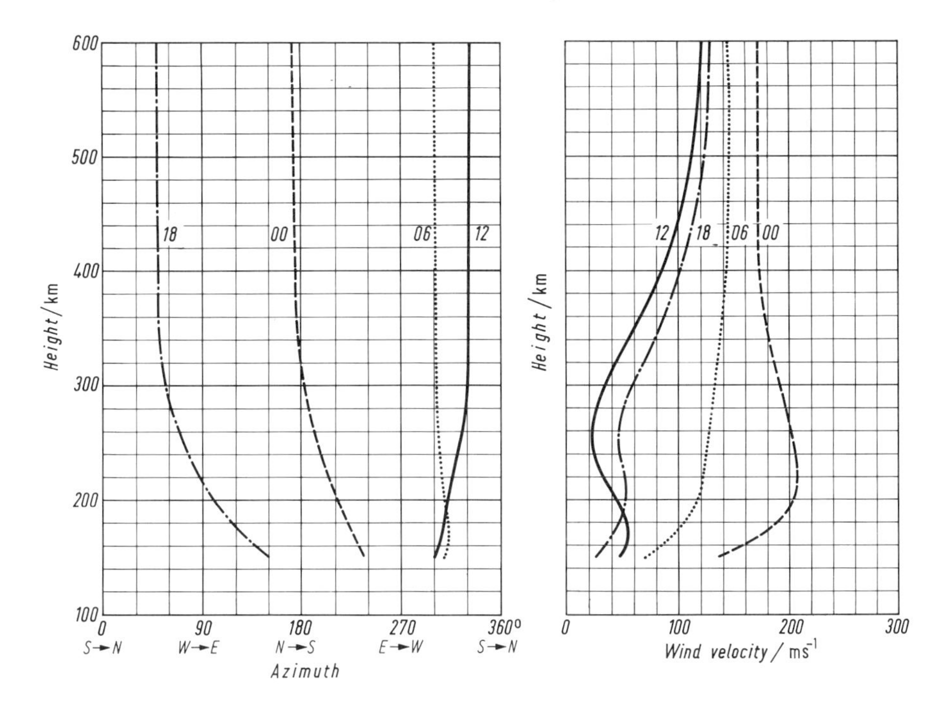

Fig. 11. Same as Fig. 10, but for December 1959

F2-layer peak. The depth of the valley in the velocity profile increases with increasing $NmF2$ (electron density at the F2-peak) and decreases with increasing $hmF2$ (height of F2-peak) since the effect of viscosity is larger at higher altitudes due to the exponential increases of the kinematic viscosity η/ρ_n.

In determining the wind direction as a function of height, the Coriolis force is important in addition to the pressure gradient, friction, and viscosity forces. In the lower thermosphere, where the pressure gradient and Coriolis force may be dominating (geostrophic case), the wind direction is perpendicular to the pressure gradient. With increasing altitude, friction, and/or viscosity become the direction-determining forces. In a first approximation, friction and viscosity lead to a wind direction coinciding with the direction of the pressure gradient. A closer look, however, shows that this is not quite true. Friction is stronger for the zonal component v_{ny} than for the meridional component v_{nx} since the F-region ions are immobile perpendicular to the magnetic field. Thus friction causes a twist towards the meridian. A stronger friction causes larger second derivatives in the velocity profile. Depending on their sign, viscosity therefore should either increase or decrease the amount of the meridional twist.

Quantitative results for both the wind velocity and direction, again for Adak, are shown in Figs. 10 and 11 for summer and winter and high solar activity. At low solar activity, when the frictional force is considerably smaller, the velocity profiles are much smoother and bear some resemblance with the neutral temperature profiles.

D. Observed wind velocities

19. Experimental methods

α) One method to directly measure wind velocities is to *release* a vapor cloud. By tracing the neutral and charged parts of the cloud, the wind velocity and electric field strength can be independently obtained. A description of this method is found in[1]. Its disadvantage is that it can be used during twilight only.

β) Another direct method is to determine the wind velocity from the Doppler shift of *airglow* lines[2]. In particular, the oxygen 6300 Å airglow has been used[3]. The results reported show that the meridional component is equatorward at night and that the latitudinal component is eastward after sunset and westward before sunrise, which is in agreement with the theoretical predictions (see Figs. 7 and 8). The Doppler-shift method is applicable at night-time only.

γ) Indirectly, wind velocities in the magnetic N-S direction can be derived from *incoherent scatter* (i.sc.) measurements of the ion velocity[4]. With the St. Santin–Nancay device, the ion velocity component parallel to the magnetic field, $v_i\|$, can be measured[5]. Knowing $v_i\|$, the wind velocity component in the magnetic N-S direction, $v_{n\xi}$, is, according to Eqs. (9.4) and (9.7), given by

$$v_{n\xi} = v_i\| \sec I + \frac{F_i \operatorname{tg} I}{\sum_l \nu_{il} \mu_{il}}, \tag{19.1}$$

where it is assumed that only one ionic species needs to be taken into account. At the F2-peak F_i is simply given by

$$F_i = k \frac{\partial}{\partial z}(T_i + T_e) + m_i g, \tag{19.2}$$

while at all other altitudes F_i involves height derivatives of the electron density. In most cases the derivative $\partial(T_i + T_e)/\partial z$ may also be omitted at the F2-peak. In this case it is sufficient to know the F2-peak height $hmF2$ as well as the neutral particle densities n_l at that height in order to convert $v_i\|$ into $v_{n\xi}$.

The Millstone facility measures the ion velocity $0.96 v_i\| + 0.28 v_i\perp$ [*15*] which cannot unambignously be used to determine $v_{n\xi}$. Usually, however, $0.28 v_i\perp \ll 0.96 v_i\|$ may be a reasonable assumption, so that Eq. (19.1) can approximately be applied to the Millstone results, too.

---

[1] Föppl, H., Haerendel, G., Haser, L., Loidl, J., Lütjens, P., Lüst, R., Melzner, F., Meyer, B., Neuss, H., Rieger, E. (1967): Planet. Space Sci. *15*, 357

[2] Vassy, A.T. and E.: This Encyclopedia, vol. 49/5, p. 5

[3] Armstrong, E.B. (1969): Planet. Space. Sci. *17*, 957

[4] Rawer, K., Suchy, K.: This Encyclopedia, vol. 49/2, Sect. 42; Rawer, K.: vol. 49/7, his Sect. 5

[5] Vasseur, G. (1969): J. Atmos. Terr. Phys. *31*, 397

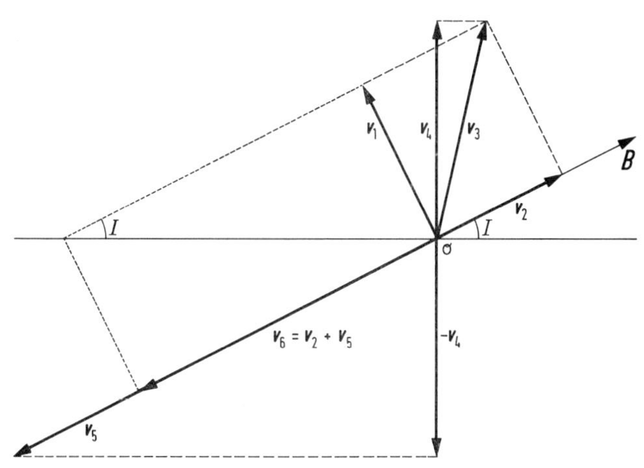

Fig. 12. Sketch to illustrate Eq. (19.3): *1* The electromagnetic drift velocity $\boldsymbol{v}_1$ and the wind induced velocity $\boldsymbol{v}_2$ add up to the resulting velocity $\boldsymbol{v}_3$ which has the z-component $\boldsymbol{v}_4$. *2* In order to compensate $\boldsymbol{v}_4$, a diffusion velocity $\boldsymbol{v}_5$ with z-component $-\boldsymbol{v}_4$ is built up. The resulting velocity parallel to the magnetic field line is $\boldsymbol{v}_6=\boldsymbol{v}_2+\boldsymbol{v}_5$. *3* It can be seen from the sketch that $v_6=-v_1\cot I$, which corresponds to Eq. (19.3)

It has been shown [*46*] that under nighttime conditions the vertical ion velocity component, v_{iz}, may vanish above about 300 km. With $v_{iz}=0$ we obtain from Eq. (9.7)

$$v_{i\|}=-\frac{\cot I}{B}(E_x\sin D+E_y\cos D)=\frac{E_\eta}{B}\cot I, \tag{19.3}$$

where E_η is the electric field strength component in the magnetic W-E direction. This is a rather surprising result since it means that the ion velocity parallel to the magnetic field lines may be a measure of the electric field strength, although the velocity $\boldsymbol{E}\times\boldsymbol{B}/B^2$ does not have a component along the magnetic field. A geometric illustration of the processes leading to Eq. (19.3) is given in Fig. 12. Applying Eq. (19.3) to the results of VASSEUR [3], E_η is found to be of the order ± 1 to 2 mV/m.

δ) Another indirect method is to *iteratively specify* the wind velocity component $v_{n\xi}$ at the F2-peak such that *hmF2* obtained from a solution of the ion continuity equations agrees with measured *hmF2* values [*44, 45*]. In applying this method it has to be made sure that not only the theoretical and experimental *hmF2* values, but also the *NmF2* values are in agreement since *hmF2* is not only determined by the wind velocity but also by photochemistry.

20. Experimental data. In this section some experimental wind data obtained with the methods outlined above will be presented.

α) Tables 5 and 6 give a summary of the wind measurements obtained by means of the *vapor cloud technique* [1] up to 1973.

[1] Rieger, E. (1973): Private communication

Table 5. Wind velocities, measured by vapor cloud technique for morning twilight. Bracketed values for azimuth correspond to transformation of southern hemisphere station into northern hemisphere (after RIEGER [1])

Season	Location	Geogr. lat.	Date	Local time	Height /km	Velocity /m s^{-1}	Azimuth (N→E)
Winter	Sahara	31°	15. 02. 64	05.57	142	59	175°
Equinox	Kiruna	68°	02. 04. 70	02.46	163	74	147°
					172	76	167°
					195	105	173°
			09. 04. 67	02.00	240	190	165°
	Sardinia	40°	30. 09. 65	05.20	236	92	233°
	Wallops	36°	25. 09. 66	04.51	253	45	166°
	Woomera	−31°	16. 10. 69	04.46	295	128	281° (259°)
Summer	Churchill	59°	05. 08. 67	01.36	268	147	198°
	Sardinia	40°	18. 06. 66	03.42	163	200	228°
					198	218	212°
					208	275	215°
					221	220	204°
			28. 06. 66	03.53	195	117	205°
					220	124	204°
	Chamical	−30°	11. 11. 72	04.26	177	46	314° (226°)
					225	78	314° (226°)
					248	78	4° (176°)

A comparison of some of these datum points with calculated winds [*28*] is shown in Fig. 13. Both the measured and calculated wind vectors rotate clockwise as local time progresses. There is some disagreement, however, with respect to magnitude and direction. Particularly in the second half of the night the theoretical winds are considerably larger in magnitude than the measured winds. Some of the disagreement may be due to the occurrence of electric fields at higher latitudes. It must also be kept in mind that the calculations describe an average behavior while the observations represent single values.

β) With the techniques explained in Subsect. 19γ and δ diurnal variations of the wind velocity in the magnetic N-S direction at altitude $hmF2$ were determined and are shown in Fig. 14 for St. Santin-Nancay (France)[2] and in Fig. 15 for Adak (Alaska) [*45*]. Figure 15 also presents a comparison with theoretical results. The agreement is relatively good with respect to phase but rather poor with respect to magnitude. There may be several reasons for the disagreement, but doubtlessly the main reason has to be sought in the thermospheric model from which the pressure gradient force is derived. Models based on satellite drag measurements[3] [*11*] as well as models based on neutral mass spectrometer measurements [*20*] involve a good deal of smoothing and averaging. It is inevitable that errors in the pressure are highly amplified when the pressure is differentiated in order to get the pressure gradient. Progress in the theoretical description of the thermospheric wind system appears possible only with an improved thermospheric model or a more realistic three-dimensional solution of the transport equations with a self-consistent thermospheric model.

[2] Ameyenc, P., Vasseur, G. (1971): Paper presented at the URSI/COSPAR Symposium on "Dynamics of the Thermosphere and Ionosphere above 120 km", held at Seattle

[3] Jacchia, L.G. (1965): Smithson. Contr. Astrophys. *8*, 215

Table 6. As Table 5 but for evening twilight

Season	Location	Geogr. lat.	Date	Local time	Height /km	Velocity /m s^{-1}	Azimuth (N→E)
Winter	Kiruna	68°	17. 01. 71	15.48	170	40	90°
					185	37	74°
			14. 02. 70	16.16	176	34	270°
					185	56	292°
					216	150	283°
			23. 02. 70	16.50	173	155	84°
	Sahara	31°	18. 11. 65	17.44	174	79	60°
			19. 11. 65	17.44	202	73	62°
			27. 11. 64	17.38	166	86	52°
			30. 11. 64	17.37	207	139	50°
Equinox	Andoya	71°	17. 04. 69	22.48	256	550	138°
	Kiruna	68°	15. 03. 69	18.55	140	408	270°
					196	348	282°
					234	293	289°
			15. 03. 71	18.55	191	140	260°
			17. 03. 69	19.10	146	440	270°
					192	350	260°
			20. 03. 68	19.23	172	60	255°
					211	26	194°
			23. 03. 68	19.35	167	60	126°
					219	70	118°
			24. 03. 71	19.32	185	280	260°
			04. 04. 70	20.12	209	12	22°
			07. 04. 67	20.57	235	160	173°
			08. 04. 67	21.03	237	120	185°
			10. 04. 67	21.14	230	120	280°
			11. 04. 67	21.21	236	100	180°
			23. 10. 67	17.05	149	70	77°
					206	67	90°
	India	8°	28. 03. 68	18.55	134	42	339°
					151	51	280°
					176	62	248°
					209	34	168°
			30. 03. 68	18.55	150	67	276°
					176	28	297°
					207	21	125°
	Woomera	−31°	02. 03. 71	19.38	205	32	69° (111°)
					218	38	91° (89°)
			17. 10. 69	19.15	250	122	95° (85°)
Summer	Churchill	59°	05. 08. 67	22.04	258	160	260°
			16. 08. 70	21.56	229	242	185°
					234	105	161°
	Sardinia	40°	16. 06. 66	20.29	154	29	161°
					174	82	199°
					190	113	176°
					230	96	172°
			06. 07. 64	20.23	150	75	180°
	Chamical	−30°	02. 11. 72	19.28	140	50	338° (202°)
			04. 11. 72	19.27	158	70	345° (195°)
					183	54	34° (146°)

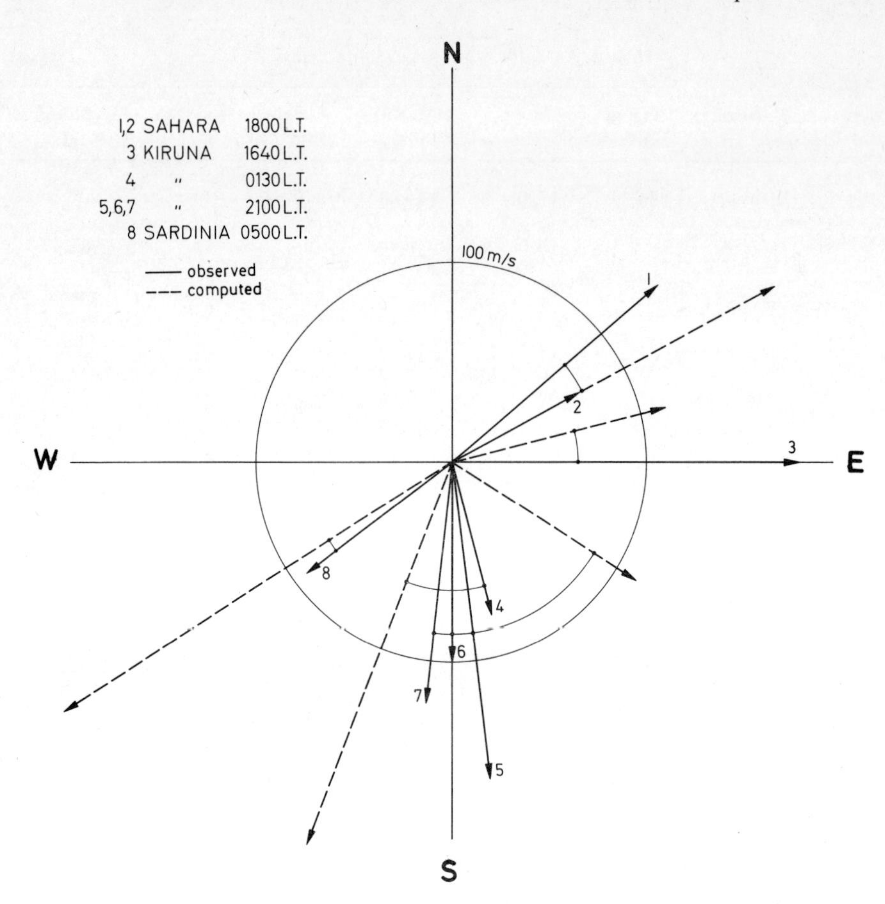

Fig. 13. Comparison of calculated and measured wind velocities for different locations and local times at heights between 200 and 240 km (after Kohl [*28*]). *1, 2* Sahara 18.00 LT. *3* Kiruna 16.40 LT. *4* Kiruna 01.30 LT. *5, 6, 7* Kiruna 21.00 LT. *8* Sahara 05.00 LT

The noon wind velocity in the magnetic N-S direction as a function of season and latitude is shown in Fig. 16 [*45*]. It can be seen that at low latitudes the wind velocity is always close to zero, being slightly positive, i.e., towards south, in summer and negative in winter. At higher latitudes the wind velocity is always negative, being larger in summer than in winter. There is a pronounced increase of the northward velocity from lower to higher latitudes.

γ) More recent results obtained with the incoherent scatter technique (see Sect. 19) refer to the North American station Millstone Hill, Mass., specifying diurnal and seasonal variations[4,5]. The semidiurnal tide was particularly studied at the low-latitude station Arecibo[6]. Nighttime measurements of winds and temperatures were described by Hernandez and Roble [*22*].

[4] Roble, R.G., Salah, J.E., Emery, B.A. (1977): J. Atmos. Terr. Phys. *39*, 503

[5] Emery, B.A. (1978): J. Geophys. Res. *83*, 5691; *83*, 5704

[6] Harper, R.M. (1979): J. Geophys. Res. *84*, 411

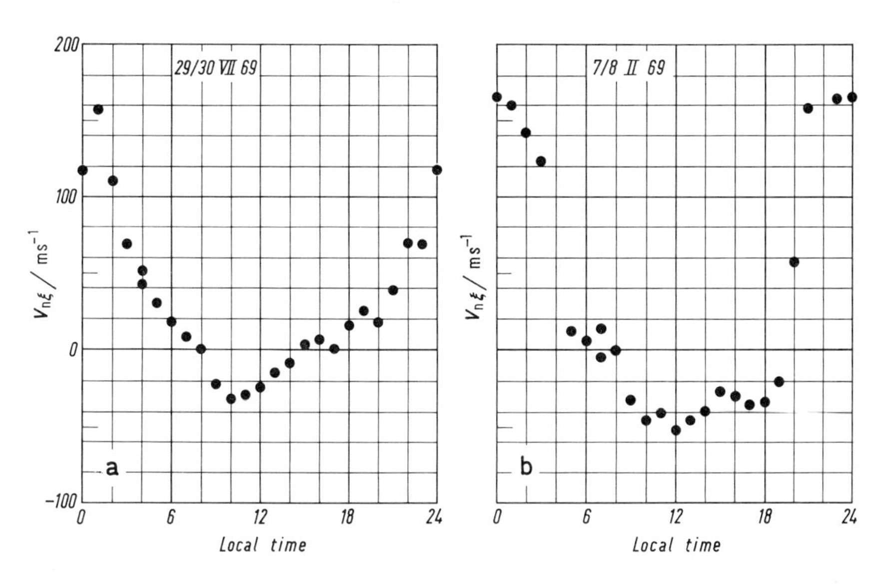

Fig. 14a, b. Wind velocity component in the magnetic N-S direction, positive towards south, at altitude *hmF2* for St. Santin-Nancy ($\varphi = 47°$ N) as a function of local time[2]

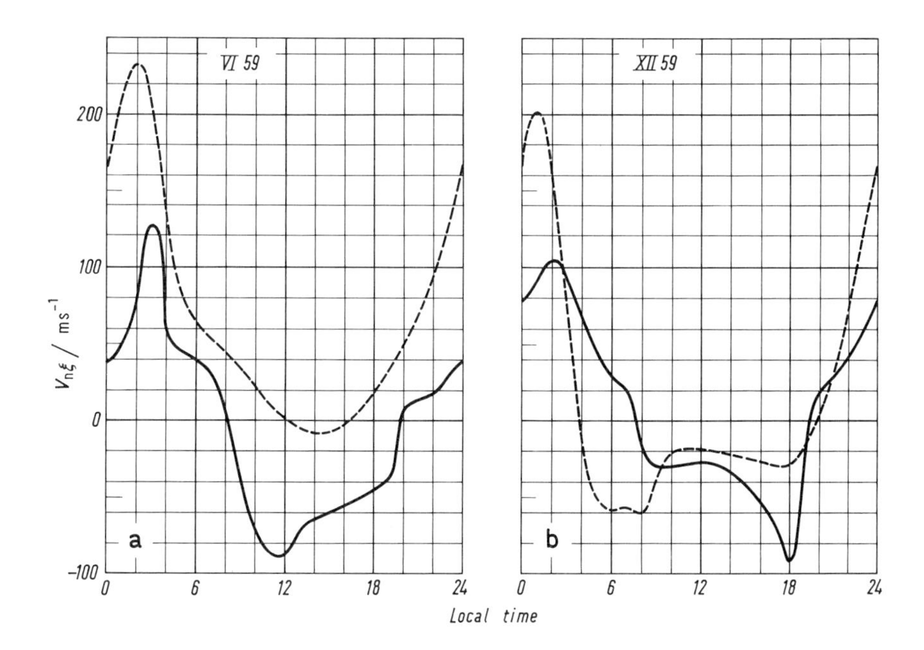

Fig. 15a, b. Wind velocity component in the magnetic N-S direction at altitude *hmF2* for Adak ($\varphi = 52°$ N) as a function of local time (after STUBBE [45]). Dashed curves: Calculated velocities. Solid curves: Velocities required to reproduce observed *hmF2* results[3]

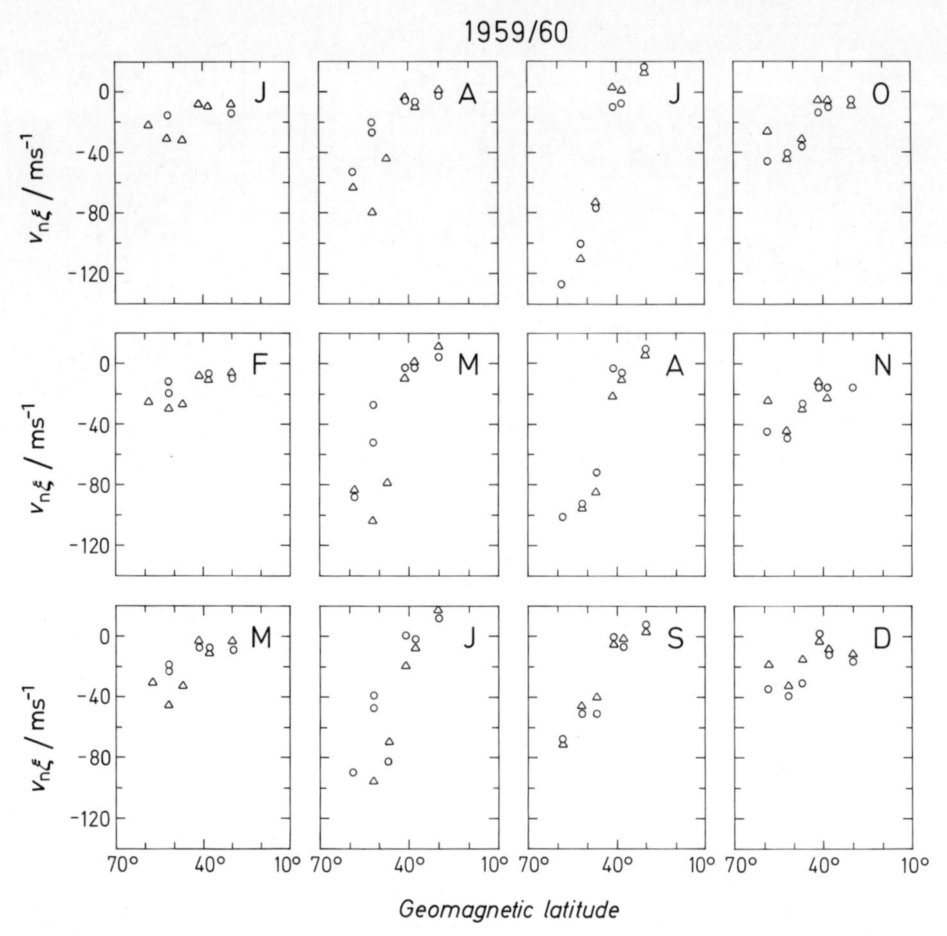

Fig. 16. Noon values of the wind velocity component in the magnetic N-S direction at altitude *hmF2* as a function of season and geomagnetic latitude for the years 1959/60 (○ 1959, △ 1960) (after STUBBE [*45*])

E. Effects of winds on the ionosphere

21. General remarks. Two different categories of wind effects on the ionosphere [*34*] can be distinguished. First, winds directly affect the ionosphere by causing plasma motions which are parallel to the magnetic field lines if the ion-neutral collision frequency is small in relation to the ion gyrofrequency. These plasma motions immediately influence the electron-ion distribution and thereby greatly control the regular ionospheric behavior. Second, winds affect the properties of the neutral atmosphere and indirectly also the ionospheric properties. In particular the neutral composition and temperature are strongly controlled by divergent or convergent horizontal wind flows.

I. Direct effects

22. Properties of the F 2-peak. In the absence of neutral winds, a "day equilibrium" or "night stationary layer" is formed when the apparent upward motion due to loss is equal to the real downward motion due to diffusion. RISHBETH[1] has shown that the F2-peak occurs at a height where

$$\frac{\beta H_{\mathrm{i}}^2}{D_{\mathrm{a}}} \approx C \approx \begin{cases} 0.60 \text{ at day} \\ 0.13 \text{ at night} \end{cases} \tag{22.1}$$

provided only the ionic constituent 0^+ has to be taken into account. Here,

$\beta = 0^+$ loss per unit volume and unit time divided by the 0^+-particle density,

H_i = scale height of ionizable neutral constituent, i.e., atomic oxygen,

$D_{\mathrm{a}} = \dfrac{k(T_{\mathrm{e}}+T_{\mathrm{i}})}{\nu(0^+,0)\,\mu(0^+,0)} \sin^2 I$ = ambipolar diffusion coefficient.

Now, if an additional vertical ion drift W is present, as caused by a neutral wind in an inclined magnetic field, the layer will either be raised or lowered depending on whether W is upward or downward. According to RISHBETH [*36*], the displacement of $hmF2$ is

$$\Delta hmF2 \approx 0.9\, W \frac{H_{\mathrm{i}}^2}{D_{\mathrm{a}}}, \tag{22.2}$$

where D_{a} and H_i are to be taken at the peak height given by Eq. (22.1), i.e., without vertical drift. A change of the peak height leads to a change of the daytime $NmF2$ according to

$$\Delta NmF2 \approx NmF2 \exp\{0.75\, \Delta hmF2/H_i\} \tag{22.3}$$

which is due to the fact that the loss coefficient β decreases with increasing altitude in proportion to the density of the molecular neutral constituents. For nighttime conditions, the decay of $NmF2$ may be roughly described in terms of an effective loss coefficient β_{eff} by

$$NmF2(t) \approx NmF2(t_0) \exp\{-\beta_{\mathrm{eff}}(t-t_0)\}. \tag{22.4}$$

Considering that the height dependence of β_{eff} may be approximately represented by

$$\beta_{\mathrm{eff}}(z) \approx \beta_{\mathrm{eff}}(z_0) \exp\{-(z-z_0)/H_l\}$$

where H_l is the scale height of the molecular constituents, a change in $hmF2$ leads to the following change in β_{eff}:

$$\Delta \beta_{\mathrm{eff}} \approx \beta_{\mathrm{eff}} \exp\{-\Delta hmF2/H_l\}. \tag{22.5}$$

[1] Rishbeth, H. (1966): J. Atmos. Terr. Phys. *28*, 911

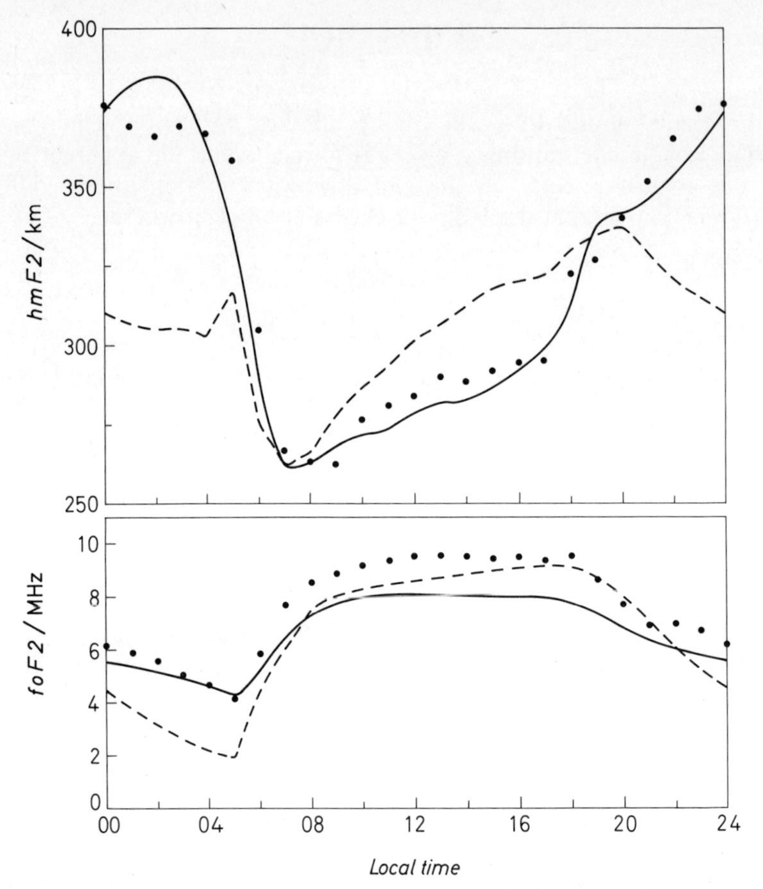

Fig. 17. Calculated and observed diurnal variations of $hmF2$ and $foF2$ for Ft. Monmouth ($\varphi = 40°$ N). September 1960 (after Kohl [29]). $foF2$, the critical frequency, is a measure of $NmF2/\text{m}^{-3} = 1.24 \cdot 10^{10}\,(foF2/\text{MHz})^2$. Dots: Measured value. Solid curves: Calculation including winds. Dashed curves: Calculation excluding winds

Equations (22.1) to (22.5) are only approximately valid, but they are very useful as rough guides to an understanding of the reaction of the F-layer to the action of neutral winds.

23. Diurnal variation of the F-layer. We have seen from Figs. 6–11 and 14–16 that the meridional wind component is poleward at day and equatorward at night. Therefore, the wind-induced vertical ion velocity is downward at day and upward at night [see Eq. (9.4)] so that according to Eqs. (22.2), (22.3) and (22.5), $hmF2$ and $NmF2$ are reduced at day and enhanced at night compared to the case without wind. Particularly the nighttime behavior of the F-layer could not be understood without winds.

A comparison of calculated diurnal variations of $NmF2$ and $hmF2$ with and without winds and the corresponding measured quantities is shown in Fig. 17. We notice that $hmF2$ without winds is much smaller at night than

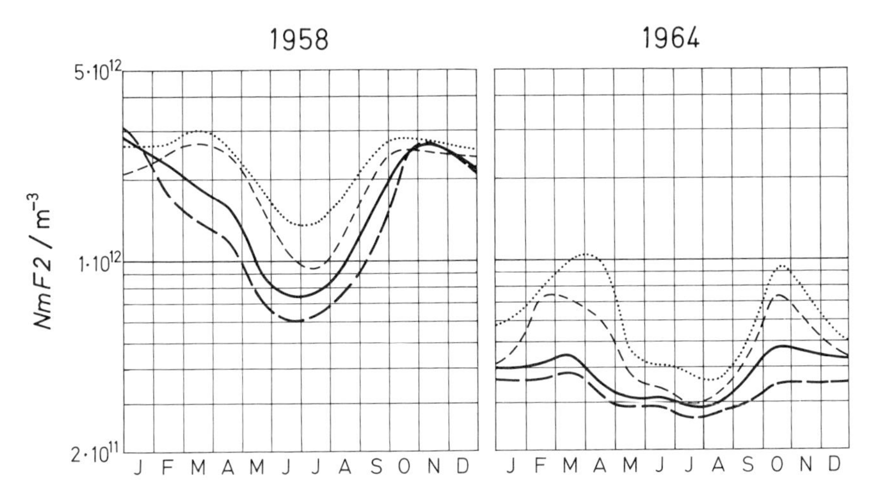

Fig. 18. Seasonal variation of $NmF2$ for four stations, · · · · · · Yamagawa ($\varphi=31°$ N), - - - - Akita ($\varphi=40°$ N), ——— Lindau ($\varphi=52°$ N), — — Uppsala ($\varphi=60°$ N), and the years 1958 (high solar activity) and 1964 (low solar activity)

observed. As a consequence, the decay of $NmF2$ after sunset is much too steep, while with inclusion of winds the agreement between the calculated and measured quantities is satisfactory. It was probably the greatest success of wind theory that it led to a resolution of the maintenance problem of the nighttime F-layer. Another local time effect, the evening enhancement of $NmF2$ at mid latitudes in summer and the occurrence of an appertaining secondary $NmF2$ minimum around noon, can also be understood only if winds are included in the calculations [1].

24. Seasonal variation of the F2-layer. From photochemical considerations one would expect that $NmF2$ is higher in summer than in winter. Just the opposite is the case. This so-called summer-winter anomaly of the F-region is most pronounced at high latitudes and high solar activity [*34*]. This can be seen from Fig. 18, which shows the seasonal variation of NmF2 for four latitudes and the years 1958 (high solar activity) and 1964 (low solar activity).

There is little doubt [1–4] [*39*, *44*] that a good portion of the summer-winter anomaly is to be explained in terms of a seasonal variation of the neutral composition with relatively more O in winter and O_2 in summer. Indeed, it has been observed by Hedin et al. [*20*] that the O/O_2 ratio increases from summer to winter. On the other hand, if we look at Fig. 16, we notice that the meridional wind velocity also exhibits a very pronounced seasonal variation, which is stronger for higher than for lower latitudes and thus resembles the

[1] Kohl, H., King, J.W., Eccles, D. (1968): J. Atmos. Terr. Phys. *30*, 1733

[1] Rishbeth, H., Setty, C.S.G.K. (1961): J. Atmos. Terr. Phys. *20*, 263

[2] Wright, J.W. (1963): J. Geophys. Res. *68*, 4379

[3] Duncan, R.A. (1969): J. Atmos. Terr. Phys. *31*, 59

[4] King, G.A.M. (1970): J. Atmos. Terr. Phys. *32*, 433

NmF2 behavior. Up to about 30° geomagnetic latitude, the wind tends to increase the summer ionization and decrease the winter ionization, thereby reducing or reverting the summer-winter anomaly. Above 30° geomagnetic latitude, however, the northward wind has a maximum in summer and consequently gives support to the summer-winter anomaly. It has also been shown [*44, 45*] that for low solar activity the maximum northward velocity occurs in winter, thereby reducing the summer-winter anomaly. This gives a natural explanation for the dependence on solar activity of the summer-winter anomaly.

It appears, therefore, that not only seasonal composition changes but also seasonal wind changes are responsible for the seasonal variation of the F-layer. Both effects are working together at higher latitudes and higher solar activity and are working against each other at lower latitudes and lower solar activity. Correspondingly, the seasonal F-layer variations are stronger in the first than in the second case.

25. Wind effects near the dip equator. The features of the equatorial F-region are mainly to be attributed to the action of electrical fields. By day the electric field strength possesses an eastward component which, according to Eq. (9.4), gives rise to an upward plasma motion. Thereby, the plasma density at high altitudes is enhanced over the value that would correspond to diffusive equilibrium along the magnetic field line. Consequently, plasma moves downward along the field line and leads to an increase of the plasma density at some distance from the dip equator with crests around 15° dip latitude. The total plasma velocity is a superposition of the field-aligned diffusion velocity and the perpendicular $\boldsymbol{E} \times \boldsymbol{B}$ drift velocity, exhibiting a "fountain-like" pattern.

The inverse effect of this plasma motion upon the neutral winds was studied by ANDERSON and ROBLE [*3*], and in [2].

Meridional winds are able to modify the shape of the latitudinal *NmF2* distribution. A qualitative illustration is shown in Fig. 19. A transequatorial wind, say southward, will raise the layer height in the northern and reduce it in the southern hemisphere, thereby increasing *NmF2* in the northern and decreasing it in the southern hemisphere. This effect is overcompensated by the plasma transport from the northern into the southern hemisphere which leads to an enhanced *NmF2* in the southern and reduced *NmF2* in the northern hemisphere [1]. The net effect, then, consists in an over-all decrease of *NmF2* in the equatorial zone which is greater in the northern than in the southern hemisphere.

A diverging wind, i.e., a poleward wind having a velocity increasing with latitude, will lead to a reduction of *hmF2* and *NmF2*, the amount of which increases with latitude [*41*]. A converging wind, on the other hand, will result in a corresponding enhancement of *hmF2* and *NmF2*. Diverging and converging winds are typical of the equinox, while transequatorial winds occur in summer and winter.

The combined effects of electric fields and winds were investigated at a low-latitude incoherent scatter station in Puerto Rico [*8*].

[1] Bramley, E.N., Young, M. (1968): J. Atmos. Terr. Phys. *30*, 99

[2] Ching, B.K., Straus, J.M. (1977): J. Atmos. Terr. Phys. *39*, 1389

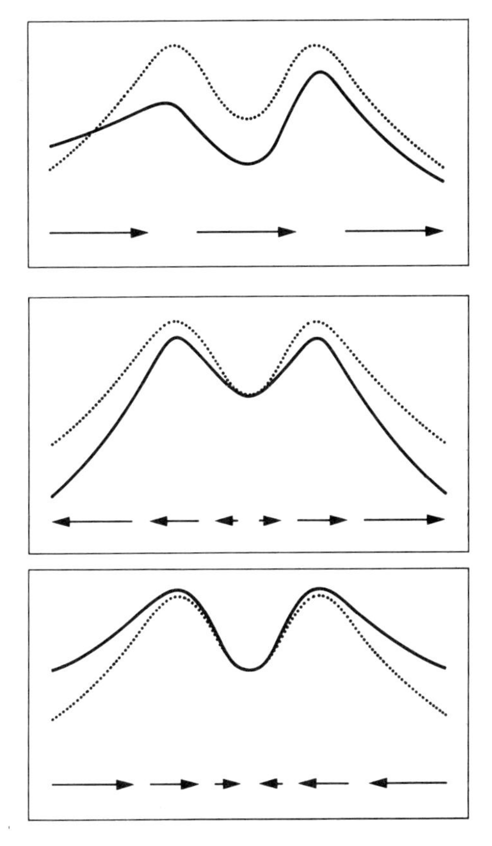

Fig. 19. Sketch to illustrate the effect of meridional winds on *NmF2* as a function of latitude in the equatorial zone. Dotted curve (identical in all three cases): Excluding winds. Solid curves: Including winds. Upper section: Transequatorial wind. Middle section: Diverging wind. Lower section: Converging wind

26. Wind effects depending on magnetic inclination. Equation (22.2) approximately describes the effect of a wind-induced vertical plasma drift on the height of the F-layer. The vertical velocity W appearing in Eq. (22.2) follows from Eq. (9.4):

$$W = \sin I \cos I (v_{nx} \cos D - v_{ny} \sin D) = v_{n\xi} \sin I \cos I. \tag{26.1}$$

W has a maximum at an inclination of $I = 45°$. It would appear, therefore, that the wind effect on the ionosphere is also maximum at $I = 45°$. However, inserting Eq. (26.1) in Eq. (22.2) gives

$$\Delta hmF2 \sim \frac{v_{n\xi}}{D_a} \sin I \cos I = \frac{v_{n\xi}}{D_{a0}} \cot I, \tag{26.2}$$

where D_{a0} is defined by $D_a = D_{a0} \sin^2 I$. Equation (26.2) can not, of course, be applied in the vicinity of the dip equator. Nonetheless, it seems to indicate that the effect of winds on the F-layer is more pronounced at lower than at higher latitudes. This, however, is not necessarily so since D_{a0} relates to an altitude z' which, by means of Eq. (22.1), is by itself a function of inclination I. For z' we obtain from Eq. (22.1)

$$z' = z_0 - \frac{H_i H_l}{H_i + H_l} \log_e \left[\sin^2 I \frac{C}{H_i^2} \frac{D_{a0}(z_0)}{\beta(z_0)} \right],$$

where z_0 is an arbitrary reference height. Since

$$D_{a0}(z') \approx D_{a0}(z_0) \exp\left\{ \frac{z' - z_0}{H_i} \right\},$$

we find

$$D_{a0}(z') \sim (\sin I)^{-\frac{2H_l}{H_i + H_l}}.$$

The exponent lies between -0.67 and -0.73, depending on whether loss by dissociative recombination[1,2] is controlled by O_2 or N_2. Thus we have approximately [34]

$$\Delta hmF2 \sim v_{n\xi} \sin^{0.7} I \cot I. \tag{26.3}$$

This function still increases with decreasing inclination if $v_{n\xi}$ is constant, which is a rather unexpected result considering that W has a pronounced maximum at $I = 45°$. It is, however, confirmed by more detailed calculations so that we may safely state that the effect of a given wind is higher for smaller inclinations.

27. Wind effects depending on magnetic declination. If we express the wind effect on the F-layer in a geographic rather than in a magnetic coordinate system, we obtain instead of Eq. (26.3)

$$\Delta hmF2 \sim \sin^{0.7} I \cot I (v_{nx} \cos D - v_{ny} \sin D), \tag{27.1}$$

where D is the magnetic declination, positive towards east. In the case $D = 0°$ or $D = 180°$ the ionosphere is influenced only by the meridional wind component, in the case $D = 90°$ or $D = 270°$ only by the zonal component.

We have seen in Sect. 17 that the wind vector rotates clockwise for an observer looking towards the pole. Let us, for the sake of simplicity, assume that the rate of rotation is constant and that the wind vector points towards the equator at midnight. Then we can easily demonstrate the effect of a varying declination angle on the ionosphere with the help of Fig. 20. The wind direction can be obtained from the graph by connecting the center of the circle with the respective local time.

First we consider D to be zero. In this case the wind is perpendicular to the magnetic meridian at 6 and 18 h, having a poleward component between 6 and 18 h and an equatorward component between 18 and 6 h. Thus the wind in-

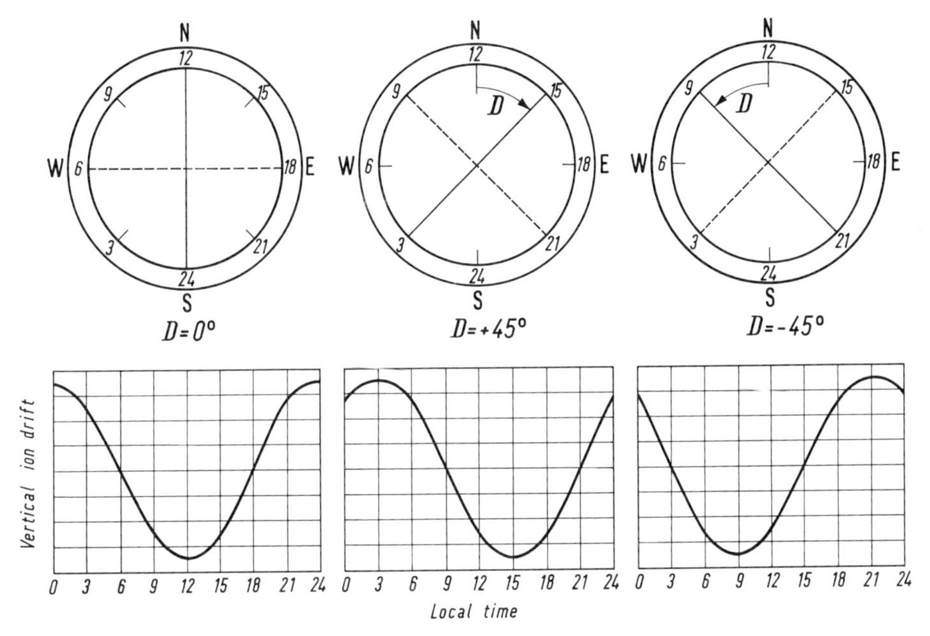

Fig. 20. Sketch to illustrate the effect of the magnetic declination on the wind induced vertical plasma drift. Upper row: Wind direction as a function of local time. The wind direction is obtained by connecting the center of the circle with the respective local time. The solid (dashed) lines indicate the local times at which the wind is parallel (perpendicular) to the magnetic meridian. Lower row: Wind induced vertical ion drift as a function of local time: Left: $D=0°$. Middle: $D=+45°$. Right: $D=-45°$

duced plasma drift W is upward between 18 and 6 h and downward between 6 and 18 h. Next we assume D to be positive, say 45°. For this case the wind is perpendicular to the magnetic meridian at 9 and 21 h, so that W is upward between 21 and 9 h and downward between 9 and 21 h. Thus the change from up to down occurs later in the morning and from down to up later in the evening than for $D=0$. As a consequence, the wind-induced reduction of $hmF2$ at day as well as its enhancement at night begins later, so that $NmF2$ is larger at day than for $D=0$ but exhibits a steeper decay in the afternoon and evening hours. The opposite is the case for negative D, i.e., $NmF2$ is smaller at day but larger at night than for $D=0$.

These merely qualitative conclusions are confirmed by observations[1] and theoretical calculations[2] as can be seen from Fig. 21, which shows the experimental and theoretical diurnal $foF2$ variations for two stations with nearly equal geographic latitudes and magnetic inclinations but different magnetic declinations.

The inclination and declination effects add up to a longitude effect which, however, is hard to predict on the basis of simple theoretical arguments since I

---

[1] Eyfrig, R.W. (1963): J. Geophys. Res. *68*, 2529

[2] Kohl, H., King, J.W., Eccles, D. (1969): J. Atmos. Terr. Phys. *31*, 1011

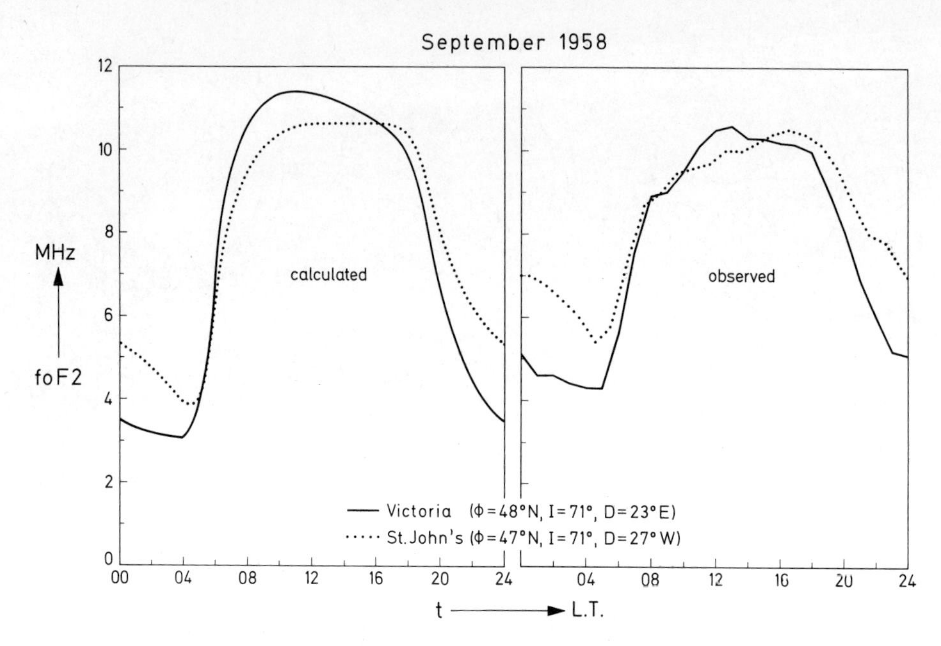

Fig. 21. Calculated (left) and observed (right) diurnal variations of $foF2$ for two stations, September 1958. Solid curves: Victoria ($\varphi=48°$ N, $I=71°$, $D=+23°$). Dashed curves: St. Johns ($\varphi=47°$ N, $I=71°$, $D=-27°$)[2]

and D are, for any given geographic latitude, complicated functions of geographic longitude.

28. Wind effects at high latitudes. At sufficiently high latitudes in winter the photoionization rate is zero throughout the day. Nonetheless, the ionosphere is maintained at rather high values for $NmF2$. In addition $NmF2$ exhibits a regular diurnal variation, although the sun never rises above the horizon. There is no doubt that the maintenance problem cannot be resolved in terms of thermospheric winds, but the modulation of the plasma density could be ascribed to the action of winds[1].

In summer the situation is just opposite in that the sun is always above the horizon, with a small diurnal variation of the solar zenith angle. The diurnal variation of $NmF2$ should be correspondingly small. The observations, however, reveal a rather pronounced diurnal $NmF2$-variation, having little relation to local time[2]. Again, an explanation can be given in terms of thermospheric winds[3].

Based on the foregoing sections, a qualitative prediction of the wind effect can be simply given for any station once its geometry is known.

---

[1] Eccles, D., King, J.W., Rüster, R., Slater, A. (1973): J. Atmos. Terr. Phys. *35*, 1285

[2] Dudeney, J.R. (1976): J. Atmos. Terr. Phys. *38*, 291

[3] King, J.W., Eccles, D., Kohl, H. (1971): J. Atmos. Terr. Phys. *33*, 1067

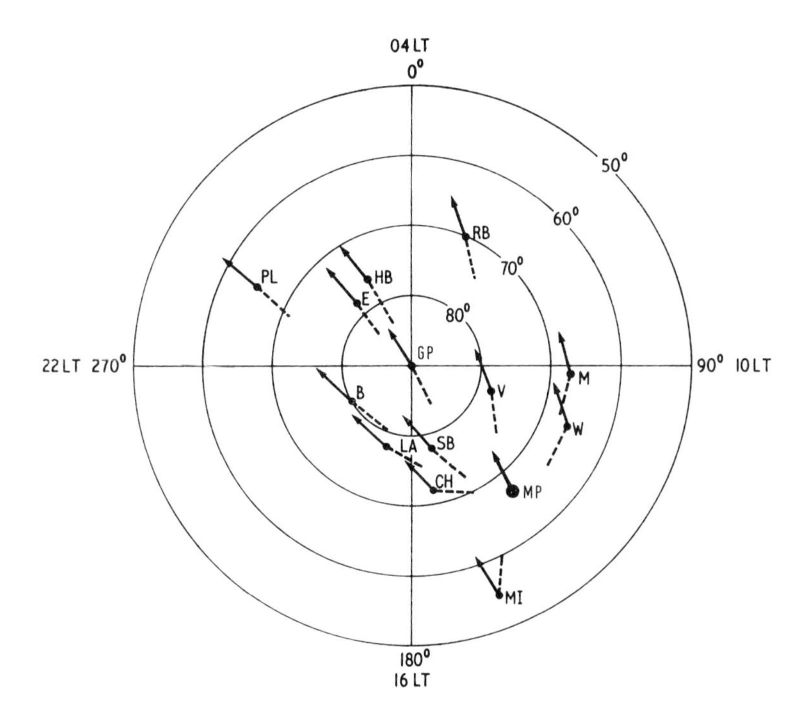

Fig. 22. Geographic position of Antarctic observatories together with direction of magnetic meridian (arrows) and wind direction at 400 km at 04 UT (dashed lines)[3]. Wind blows away from magnetic pole. GP = Geographic Pole, MP = Magnetic Pole, PL = Port Lockroy, E = Ellsworth, HB = Halley Bay, RB = Roi Baudouin, M = Mirny, V = Vostok, W = Wilkes, MI = Macquarie Island, SB = Scott Base, CH = Cape Hallett, LA = Little America, B = Byrd

Particular attention has been paid to the Antarctic stations since most of them exhibit a conspicuous Universal Time dependence. The reason can be seen from Fig. 22 which shows the geographic positions of the Antarctic stations together with the directions of the magnetic meridians and the wind directions at 04 UT. It is just by accident that most of the stations are located such that the wind direction coincides with the magnetic meridian at 04 UT. At this time the wind blows from the magnetic towards the geographic pole, thus leading to upward plasma motions and producing $NmF2$ maxima close to 04 UT.

Were the observatories in the Antarctic region centered around the magnetic pole rather than around the geographic pole, no UT. Effect would have been found. This situation is found in the Arctic region, so that no common UT effect can be seen there[4]. The effect of the displacement of geomagnetic and geographic poles was theoretically modelled in[5].

29. Wind effects during magnetic storms. It is well known that the thermospheric temperature is increased during magnetic storms, with preference to higher latitudes[1]. Consequently the equatorward wind at night will be increased and the poleward wind at day be decreased as a result of the ad-

---

[4] Challinor, R.A. (1970): J. Atmos. Terr. Phys. *32*, 1959

[5] Roble, R.G., Dickinson, R.E. (1974): Planet. Space Sci. *22*, 623

[1] Roemer, M. (1972): COSPAR International Reference Atmosphere, p. 341. Berlin: Akademie-Verlag

ditional pressure gradient. This implies that $NmF2$ is increased, giving rise to a "positive storm"[2].

α) Calculations of BURGE, ECCLES, KING and RÜSTER [*10*] have shown that the wind system is more strongly affected in winter than in summer, as can be seen from their Fig. 2, so that, except for the equinoxes, the additional storm-time wind is not symmetric about the equator. The results indicate that the additional storm-time wind can be represented by a converging plus a trans-equatorial component which is directed from the winter into the summer hemisphere. A detailed qualitative prediction as to the effect of such an additional wind is not easy to give, but at least one can say that Fig. 19 allows for the possibility of a negative storm in the equatorial region. It should be mentioned that winds are only partly responsible for the magnetic storm behavior of the ionosphere, at least as far as their direct effect, i.e., plasma transport effect, is concerned. Another contribution is expected to come from neutral composition changes, which may, however, also be related to storm-time winds. For this subject see Sect. 32.

β) Observational results obtained with different techniques were published[3-5]. More recently, attention is directed toward combining theoretical reasoning with observed facts so that basic assumptions of the theory are taken from measured data[6] [*2*] (see also RAWER's contribution in volume 49/7).

30. Wind effects on the ionospheric temperature. The direct wind effects described in Sect. 23 to 29 were related to momentum transfer. Another direct wind effect is connected with energy transfer and thus leads to a change of the ionospheric temperature. A net exchange of energy between two gases occurs when either the temperatures or the macroscopic velocities are different. The heat transferred from the neutral to the ion gas per unit volume and unit time is given by[1]

$$Q_T = 2N_\mathrm{i}\frac{\mu_\mathrm{in}^2}{m_\mathrm{i} m_\mathrm{n}}\left\{\frac{3}{2}k(T_\mathrm{n}-T_\mathrm{i})\,\nu_\mathrm{in}^{(\mathrm{E})} + \frac{1}{2}\frac{T_\mathrm{i}}{T_\mathrm{in}} m_\mathrm{n}(\boldsymbol{v}_\mathrm{i}-\boldsymbol{v}_\mathrm{n})^2\,\nu_\mathrm{in}^{(\mathrm{M})}\right\}, \tag{30.1}$$

where T_in, the reduced temperature, is defined in the legend to Table 1, and $\nu_\mathrm{in}^{(\mathrm{E})}$, $\nu_\mathrm{in}^{(\mathrm{M})}$ are two different transport collision frequencies[2] relating to energy and momentum transfer. For velocity differences of the order few hundred meters per second, $\nu_\mathrm{in}^{(\mathrm{E})}$ and $\nu_\mathrm{in}^{(\mathrm{M})}$ are nearly equal and may be replaced by the common collision frequency ν_in. In order to estimate to which extent the ion temperature T_i is increased by friction, we may approximately assume that the frictional heat gain is compensated by the first term in the brackets, i.e., $Q_T=0$.

---

[2] Jones, K.L., Rishbeth, H. (1971): J. Atmos. Terr. Phys. *33*, 391
[3] Rishbeth, H. (1975): J. Atmos. Terr. Phys. *37*, 1055
[4] Rüster, R., King, J.W. (1976): J. Atmos. Terr. Phys. *38*, 593
[5] Marubashi, K., Reber, C.A., Taylor Jr., H.A. (1976): Planet. Space Sci. *24*, 1031
[6] Richmond, A.D., Matsushita, S. (1975): J. Geophys. Res. *80*, 2839
[1] Stubbe, P. (1971): J. Sci. Ind. Res. *30*, 379
[2] See K. Suchy's contribution in vol. 49/7, p. 1

Values for $T_i - T_n$ as a function of $|\boldsymbol{v}_i - \boldsymbol{v}_n|$ are given in Table 7 when the neutral gas consists of 0 and $T_n = 1000$ K.

Table 7. $(T_i - T_n)$/K as a function of $|\boldsymbol{v}_i - \boldsymbol{v}_n|$ for $T_n = 1000$ K and neutral gas consisting of atomic oxygen

$\lvert\boldsymbol{v}_i - \boldsymbol{v}_n\rvert$/m s^{-1}	100	200	300	500	750	1000
$(T_i - T_n)$/K	6.5	26	60	174	426	833

We see that for the regular wind system the increase in T_i is rather modest. For the polar region, however, FEDDER and BANKS [*16*] have calculated ion temperature increases up to 500 K.

II. Indirect effects

31. Wind effects on the neutral temperature

α) Two terms in the energy Eq. (7.1) for the neutral gas contain the *neutral velocity*, namely

$$\boldsymbol{v}_n \operatorname{grad}(C_v T_n) \quad \text{and} \quad \frac{L}{n_n} p_n \operatorname{div} \boldsymbol{v}_n .$$

The main contributions to these terms stem form the vertical derivatives, so that the temporal temperature change due to the velocity dependent terms may be approximated by

$$\frac{\partial T_n}{\partial t} \approx -v_{nz} \frac{\partial T_n}{\partial z} - \frac{L}{C_v n_n} p_n \frac{\partial v_{nz}}{\partial z} . \tag{31.1}$$

Equation (31.1) does not explicitly involve the horizontal wind velocity. The vertical velocity, however, is strongly dependent on the thermospheric wind velocity as we shall see later.

The second term on the right-hand side of Eq. (31.1) can be put in a more convenient form. Using the continuity equation and the ideal gas law, we have

$$p_n \frac{\partial v_{nz}}{\partial z} = k n_n \left(\frac{\partial T_n}{\partial t} + v_{nz} \frac{\partial T_n}{\partial z} \right) - \frac{\mathrm{d} p_n}{\mathrm{d} t} . \tag{31.2}$$

β) At this point it is useful to *separate v_{nz} into two parts*[1], the first, $v_{nz}^{(B)}$, describing the "breathing" of the atmosphere due to thermal expansion or contraction, and the second, $v_{nz}^{(D)}$, representing the wind induced vertical velocity.

[1] Rishbeth, H., Moffet, R.J., Bailey, G.J. (1969): J. Atmos. Terr. Phys. *31*, 1035

By definition, $v_{nz}^{(B)}$ is the velocity of an isobar. Thus

$$\frac{\partial p_n}{\partial t} + v_{nz}^{(B)} \frac{\partial p_n}{\partial z} = 0 \tag{31.3}$$

and consequently $\frac{dp}{dt}$ is given by

$$\frac{dp_n}{dt} = v_{nz}^{(D)} \frac{\partial p_n}{\partial z}, \tag{31.4}$$

or upon employing the barometric law

$$\frac{dp_n}{dt} = -v_{nz}^{(D)} \frac{p_n}{H_n}. \tag{31.5}$$

Insertion of Eqs. (31.2) and (31.5) in Eq. (31.1) yields

$$\frac{\partial T_n}{\partial t} = -(v_{nz}^{(B)} + v_{nz}^{(D)}) \frac{\partial T_n}{\partial z} - \frac{L m_n g}{C_p} v_{nz}^{(D)}. \tag{31.6}$$

Some typical values for the temporal change of T_n caused by convection (first term) and adiabatic heating (second term) are given in Tables 8 and 9. Since $\partial T_n/\partial z$ is positive in the thermosphere, both terms are positive for negative, i.e., downward, and negative for positive, i.e., upward, motion.

Table 8. Numerical examples for convective heating. Temporal temperature change/K/h as a function of $v_{nz}^{(D)}$ and $\partial T_n/\partial z$. Bracketed values behind $\partial T_n/\partial z$ indicate height to which the respective value for $\partial T_n/\partial z$ typically applies. Positive value for $\partial T_n/\partial t$ belongs to negative value for $v_{nz}^{(D)}$ and vice versa

$v_{nz}^{(D)}/\mathrm{m\,s^{-1}}$ $\frac{\partial T_n}{\partial z} \Big/ \frac{\mathrm{K}}{\mathrm{km}}$	0.1	0.2	0.5	1	2	3
0.5 (290)	0.2	0.5	0.9	1.8	3.6	5.4
1 (260)	0.4	1.1	1.8	3.6	7.2	11
2 (220)	0.7	2.2	3.6	7.2	14	22
3 (200)	1.1	3.2	5.4	11	22	32
5 (170)	1.8	5.4	9.0	18	36	54
10 (120)	3.6	11	18	36	72	108

Table 9. Numerical examples for adiabatic heating. Temporal temperature change as a function of $v_{nz}^{(D)}$. Positive value for $\partial T_n/\partial t$ belongs to negative value of $v_{nz}^{(D)}$ and vice versa. N_2 is assumed to be the dominant neutral species

$v_{nz}^{(D)}/\mathrm{m\,s^{-1}}$	0.1	0.3	0.5	1	2	3
$\frac{\partial T_n}{\partial t} \Big/ \frac{\mathrm{K}}{\mathrm{h}}$	3	9	16	31	63	94

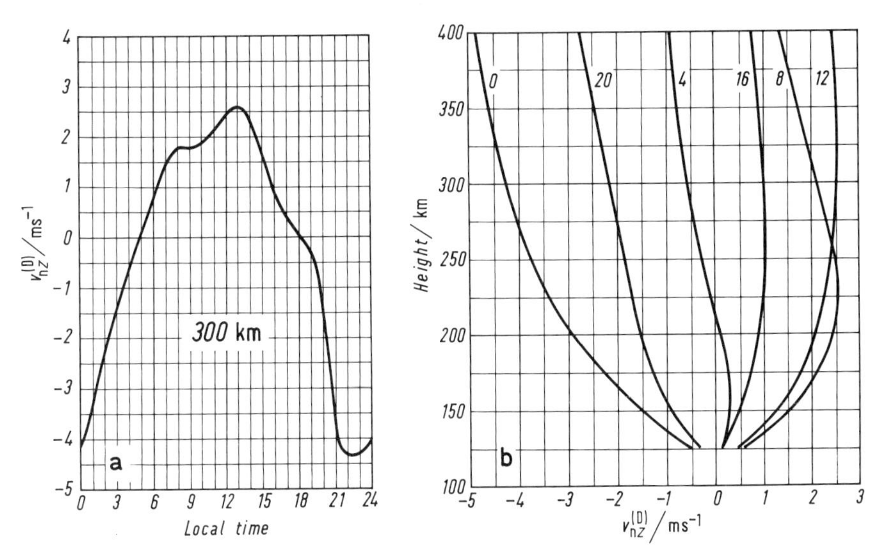

Fig. 23a, b. Vertical velocity $v_{nz}^{(D)}$ as a function of local time and height for low solar activity, equinox and latitude 45°. (**a**) $v_{nz}^{(D)}$ at 300 km as a function of local time. (**b**) $v_{nz}^{(D)}$ as a function of height for local times 0, 4, 8, 12, 16, 20 h[1]

In the isothermal region above about 300 km, a typical value for $\partial T_n/\partial t$ due to extreme ultraviolet (EUV) heat input is of the order 50 K/h[2]. We see that heating caused by vertical transport comes close to this value if $v_{nz}^{(D)}$ is of the order $1\ \mathrm{m\,s^{-1}}$.

γ) In order to evaluate $v_{nz}^{(D)}$, we assume[1] that the divergence of the horizontal neutral air flow is balanced by the divergence of a vertical flow so that *barometric equilibrium* is maintained. Assuming that the vertical flux is zero at infinity, we get

$$v_{nz}^{(D)}(z) = \frac{1}{n_n(z)} \int_z^\infty \left[\frac{\partial}{\partial x}(n_n v_{nx}) + \frac{\partial}{\partial y}(n_n v_{ny})\right] dz. \tag{31.7}$$

Numerical values for $v_{nz}^{(D)}$ as a function of local time and altitude have been calculated[1]. An example obtained for low solar activity is shown in Fig. 23.

δ) It is apparent from the results of Fig. 23 and Tables 8 and 9 that the thermospheric wind system has a paramount *influence on* the *thermospheric temperature*. We see from Fig. 23 that the wind-induced vertical velocity $v_{nz}^{(D)}$ is upward at day and downward at night. Thus the neutral temperature is decreased at day and increased at night, thereby reducing the amplitude of the diurnal temperature variation. The average global temperature is, of course, diminished by the wind system. This is because energy is required to maintain the wind field against friction.

It follows from Eq. (31.7) that a convergent wind flow (negative divergence) results in a downward motion and thus increases the temperature, while a div-

[2] Bailey, G.J., Moffett, R.J. (1972): Planet. Space Sci. *20*, 1085

ergent wind flow leads to a temperature decrease. This conclusion allows qualitative prediction of how the thermosphere reacts upon heating in a limited area, a situation which occurs during magnetic storms when an additional heat source is operating at high latitudes. At the location of maximum temperature increase the wind is necessarily divergent, so that the effect of the primary heat source will be reduced, due to adiabatic cooling. At some distance from the heat source, the wind will be retarded. Consequently the wind flow is convergent and thus gives rise to a temperature increase. In summary we may say, that the thermospheric wind field acts to smooth out the global temperature distribution. For the sake of completeness it should be added that thermospheric gravity waves [3] may have the same effect [27]. For more details on this subject see [2, 4, 5].

32. Wind effects on the neutral composition. In Sect. 31 it was assumed that the neutral constituents possess a common vertical velocity. For this case it has been shown [1] that vertical transport is not able to noticeably affect the particle density-height distribution and thus the neutral composition. If, on the other hand, each neutral constituent has its own vertical velocity, a frictional force is set up [see Eq. (8.2)] which in the lower thermosphere may be of the same order of magnitude as the pressure gradient and gravity force. As a result, the $n_l(z)$ profile will be markedly influenced, and in general, changes in the neutral composition will be caused.

A detailed knowledge of the individual vertical velocities is nearly impossible to obtain. What would be required for this purpose is a very realistic three-dimensional solution of the coupled transport equations from the upper thermosphere down to the mesosphere-thermosphere boundary, taking into account photochemistry and turbulent diffusion. Nobody has yet even tackled this task.

α) In order to demonstrate, in principle, the *effect of vertical transport on the neutral composition*, we may start from the following rather simple assumptions.

(1) T_n is constant with time and height.

(2) The neutral constituent under consideration, denoted by l, is a minor constituent in the sense that the major constituent, denoted by m, is not affected by friction with constituent l.

(3) The divergence of the horizontal wind flow is caused by a velocity divergence only.

(4) All constituents have the same horizontal velocity.

It follows from (2) that n_m obeys the barometric law, and from (1)+(2) that $\partial n_m/\partial t = 0$. Furthermore, it follows that in the absence of a divergent wind $v_{mz} = 0$. Thus, combining Eqs. (8.1) and (8.3), the continuity equation for con-

---

[3] Jones, W.L.: This Encyclopedia, vol. 49/5, p. 177

[4] Mayr, H.E., Volland, H. (1972): J. Geophys. Res. 77, 6774

[5] Mayr, H.E., Volland, H. (1973): J. Geophys. Res. 78, 2251

[1] Rishbeth, H., Moffett, R.J., Bailey, G.J. (1969): J. Atmos. Terr. Phys. *31*, 1035

stituent 1 in the absence of a divergent wind reads

$$\frac{\partial n_l}{\partial t} = P_l + \frac{\partial}{\partial z}\left\{D_l^{(M)}\left[\frac{\partial n_l}{\partial z} + \frac{1}{H_l}\right]\right\}. \tag{32.1}$$

P_l positive means production, negative means loss of species l per unit volume and time. In the presence of a divergent wind we have

$$\frac{\partial n_l}{\partial t} = P_l + \frac{\partial}{\partial z}\left\{D_l^{(M)}\left[\frac{\partial n_l}{\partial z} + \frac{1}{H_l}\right]\right\} - \frac{\partial}{\partial z}(n_l v_{mz}^{(D)}) - n_l\left(\frac{\partial v_{mx}}{\partial x} + \frac{\partial v_{my}}{\partial y}\right). \tag{32.2}$$

The continuity equation for the major constituent m yields

$$\frac{\partial}{\partial z}(n_m v_{mz}^{(D)}) = -n_m\left(\frac{\partial v_{mx}}{\partial x} + \frac{\partial v_{my}}{\partial y}\right). \tag{32.3}$$

Combining Eqs. (32.2) and (32.3) and using the barometric law for the major constituent, we obtain

$$\frac{\partial n_l}{\partial t} = P_l + \frac{\partial}{\partial z}\left\{D_l^{(M)}\left[\frac{\partial n_l}{\partial z} + \frac{1}{H_l}\right]\right\} - v_{mz}^{(D)}\left[\frac{\partial n_l}{\partial z} + \frac{n_l}{H_m}\right]. \tag{32.4}$$

Thus, if we assume that at a given time the minor constituent is in diffusive equilibrium[2], a comparison of Eqs. (32.4) and (32.1) shows that the effect of vertical transport may be expressed through

$$\frac{\partial n_l}{\partial t} = v_{mz}^{(D)} n_l\left[\frac{1}{H_l} - \frac{1}{H_m}\right] = v_{mz}^{(D)}\frac{n_l}{H_l}\left[1 - \frac{m_m}{m_l}\right]. \tag{32.5}$$

Equation (32.5) shows that an upward motion leads to an increase of n_l if $m_l > m_m$ and to a decrease if $m_l < m_m$. This is equivalent to saying that the light constituents are enhanced by a convergent wind and the heavy constituents by a divergent wind, where the terms "light" and "heavy" are used in relation to the mass of the major constituent or, more generally, to the average mass. It must be emphasized, however, that Eq. (32.5) is a rather crude approximation which can only be used for qualitative argumentation. Nonetheless, the conclusions drawn from it are fully confirmed by more elaborate investigations[3-5]. In particular it is confirmed that the effect of vertical transport is more efficient if the individual to average mass ratio is far from unity.

β) Although the derivation of Eq. (32.5) from the transport Eqs. (8.1) and (8.3) is quite elementary, it may still appear puzzling that a convergent wind, i.e., a wind leading to accumulation of air, can nonetheless cause a depletion of the heavy constituents. The reason is that the downward

[2] See T. Yonezawa's contribution in this volume, p. 129
[3] Mayr, H.E., Volland, H. (1972): J. Geophys. Res. 77, 6774
[4] Mayr, H.E., Volland, H. (1973): J. Geophys. Res. 78, 2251
[5] Reber, C.A., Hays, P.B. (1973): J. Geophys. Res. 78, 2977

flux, caused by the downward motion of the major constituent via friction, transports more particles into the mesospheric reservoir than are supplied by the convergent horizontal wind flux. On the other hand, if the wind is divergent, more heavy particles are extracted from the mesospheric reservoir than are carried away by the horizontal wind.

The *conclusions* that can be drawn from Eq. (32.5) in conjunction with Eqs. (31.6) and (31.7) are summarized in Table 10.

Table 10. Wind effects on neutral temperature and composition. The terms "heavy" and "light" are used in relation to the average mass. The terms "decreasing" and "increasing" relate to the case without wind. The total particle density variation is a superposition of column 3 with 4 or 5, respectively (e.g., for a converging wind the "heavy constituent" O_2 may decrease in the lower thermosphere, where the temperature effect is small, but may increase in the upper thermosphere where the temperature effect predominates)

Type of wind	Vertical velocity	Temperature	Heavy constituents	Light constituents
converging	downward	increasing	decreasing	increasing
diverging	upward	decreasing	increasing	decreasing

γ) Table 10 makes it easy to *predict* certain *thermospheric variations.*

(i) Seasonal variations [3, 5]

Type of wind: From summer to winter hemisphere. Thus, on the average, diverging in summer and converging in winter hemisphere.

Vertical velocity: Upward in summer and downward in winter hemisphere.

Neutral temperature: Decrease in summer and increase in winter hemisphere, thus smoothing out seasonal variation.

Neutral constituents: Modest increase of O and strong increase of He in winter and decrease in summer hemisphere. Modest decrease of O_2 and strong decrease of Ar in winter and increase in summer hemisphere. Little variation of N_2.

(ii) Magnetic storm variations [4]

Type of wind: Diverging at higher latitudes (due to additional heat source) and converging at lower latitudes.

Vertical velocity: Upward at higher latitudes and downward at lower latitudes.

Neutral temperature: Increase of temperature at higher latitudes due to the additional heat source, however, reduced by upward motion. Increase of temperature at lower latitudes due to downward motion.

Neutral constituents: Little change of N_2 in the lower thermosphere, but increase in the upper thermosphere due to temperature increase. Decrease of O at higher latitudes in lower, but slight increase in upper thermosphere. Increase of O at lower latitudes. Increase of O_2 and Ar at higher latitudes and decrease at lower latitudes. Decrease of He at higher latitudes and increase at lower latitudes.

Thus the wind-induced composition changes should give rise to a negative ionospheric storm at higher, and to a positive storm at lower, latitudes. Superimposed on this indirect wind effect is the direct wind effect dealt with in Sect. 29.

δ) A few words should be said about the *characteristic time* for wind-induced composition changes to take place. The real cause of the composition change is the vertical diffusion velocity set up by a converging or diverging wind. Thus it is the diffusion time constant

$$\tau_D = \frac{H_l^2}{D_l^{(M)}} \tag{32.6}$$

that determines the speed of the composition change. τ_D is of the order of a few days around 120 km and several minutes at F2-layer heights. The effect of the vertical velocity on the particle density is stronger at lower altitudes where the free vertical flow is more efficiently inhibited by friction. Owing to the exponential decrease of τ_D with altitude, any change in composition has sufficient time to propagate to higher altitudes while it is built up at lower altitudes. Thus one should expect that a diffusion time constant slightly above the turbopause is a characteristic measure for the whole thermosphere. This corresponds to a characteristic time of about 5 d.

F. Effects of plasma motions on the neutral wind

33. General remarks. The distinction between wind effects on the ionosphere and ionospheric effects on winds is not legitimate in a strict sense. Each of these effects is mutual. If a wind causes an ionospheric effect, either directly by momentum or energy transfer, or indirectly by changing the neutral temperature and composition, the ionosphere reacts with a change of the frictional force and thereby the wind velocity. On the other hand, if a plasma motion sets up a neutral motion, the latter reacts to the ionosphere by either of the means described in Sects. 23 to 32. A distinction, however, can be made with respect to the cause and effect sequence. In this sense we shall deal here with those motions which are first produced within the plasma, then transferred to the neutral gas, and subsequently act back on the ionosphere, thereby modifying the primary plasma motion.

34. Effect of plasma diffusion on the neutral wind. First, we recall some relationships that were derived in Sect. 9. The vertical diffusion velocity as a function of inclination is given by

$$v_{iz}^{(D)}(I) = v_{iz}^{(D)}(90°)\sin^2 I, \tag{34.1}$$

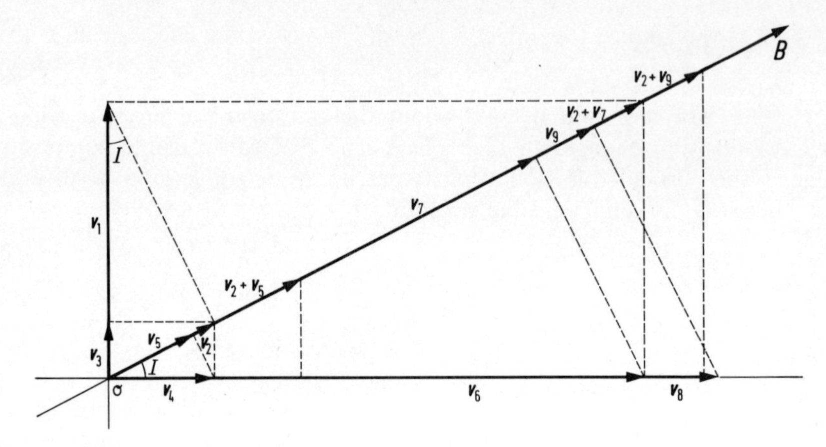

Fig. 24. Sketch to illustrate Eqs. (34.5) and (34.6). All vectors originate in O. *1* Let the ionospheric conditions be such that for $I=90°$ the plasma diffusion velocity $\boldsymbol{v}_1$ would be produced. $\boldsymbol{v}_1$ has the component $\boldsymbol{v}_2$ ($v_2=v_1 \sin I$) along the field line. $\boldsymbol{v}_2$ has the vertical component $\boldsymbol{v}_3$ ($v_3=v_2 \sin I =v_1 \sin^2 I$) and the horizontal component $\boldsymbol{v}_4$. *2* The plasma accelerates the neutral gas in the horizontal direction. When the neutral gas has attained the velocity $\boldsymbol{v}_4$, it gives rise to the additional plasma velocity $\boldsymbol{v}_5$ along $\boldsymbol{B}$. The total plasma velocity along $\boldsymbol{B}$ is $\boldsymbol{v}_2+\boldsymbol{v}_5$ which has a horizontal component in excess of $\boldsymbol{v}_4$. Thus the acceleration of the neutral gas is continued, coming to a close when the neutral velocity is $\boldsymbol{v}_6$. The reason for the acceleration to become zero for $v_{n\xi}=v_6$ is that $\boldsymbol{v}_6$ produces the field aligned plasma velocity $\boldsymbol{v}_7$ which, added to $\boldsymbol{v}_2$, has the horizontal component $\boldsymbol{v}_6$, so that the horizontal plasma velocity equals the horizontal neutral velocity. *3* $\boldsymbol{v}_2+\boldsymbol{v}_7$ has the vertical component $v_{iz}=v_1=v_{iz}^{(D)}(90°)$, and $v_{n\xi}=v_6=v_{iz}^{(D)}(90°)\cot I$. *4* In order to demonstrate that $\boldsymbol{v}_6$ is really the equilibrium velocity, we assume that $v_{n\xi}=v_8>v_6$. Then the total field aligned plasma velocity is $\boldsymbol{v}_2+\boldsymbol{v}_9$, which has a horizontal component smaller than v_8. Consequently the ion drag is negative, leading to a reduction of $v_{n\xi}$

so that the total vertical ion velocity, assuming no electric field, reads

$$v_{iz}=v_{n\xi} \sin I \cos I+v_{iz}^{(D)}(90°) \sin^2 I, \tag{34.2}$$

where $v_{n\xi}$ is the horizontal wind velocity towards magnetic south. For the corresponding ion velocity component we have

$$v_{i\xi}=v_{n\xi} \cos^2 I+v_{iz}^{(D)}(90°) \sin I \cos I. \tag{34.3}$$

Now we assume that at time t_0 the neutral air is at rest and that no horizontal pressure gradient force is present. At time t_0 the vertical ion velocity is

$$v_{iz}=v_{iz}^{(D)}(90°) \sin^2 I. \tag{34.4}$$

The meridional ion velocity at time t_0, viz., $v_{iz}^{(D)}(90°) \sin I \cos I$, exerts a drag on the neutral gas and thus causes a meridional neutral motion. The drag comes to an end when $v_{n\xi}=v_{i\xi}$, i.e., when $v_{n\xi}$ has reached the value

$$v_{n\xi}=v_{iz}^{(D)}(90°) \cot I. \tag{34.5}$$

The horizontal neutral motion modifies the initial plasma motion. Inserting Eq. (34.5) in Eq. (34.2) gives

$$v_{iz}=v_{iz}^{(D)}(90°). \tag{34.6}$$

Thus a diffusing plasma causes a meridional neutral gas motion which reacts on the plasma in such a way that the resulting vertical plasma motion corresponds to diffusion for $I=90°$ (or no magnetic field). The physical processes leading to the relations Eqs. (34.5) and (34.6) are illustrated in Fig. 24.

The time for Eqs. (34.5) and (34.6) to be reached is, according to Eq. (3.1), given by

$$\tau = \frac{m_n}{\mu_{in}} \frac{1}{a_{in} n_i}, \tag{34.7}$$

where a_{in} is the collision frequency coefficient defined by Eq. (3.2). τ amounts to 15 min for $n_i = 10^{12}\,\mathrm{m}^{-3}$ and 2.5 h for $n_i = 10^{11}\,\mathrm{m}^{-3}$. Therefore the processes described above should be quite effective in the F2-layer, at least for daytime conditions.

Equation (34.5) allows estimation of the error introduced by calculating the wind velocity with the ion velocity given by Eq. (3.5), i.e., without taking into account plasma diffusion (see Sect. 3). Typical values for $v_{iz}^{(D)}(90°)$ are listed in Table 11. These have to be multiplied by $\cot I$ in order to get the maximum error in the meridional wind velocity due to the neglect of plasma diffusion in the frictional force.

Table 11. $v_{iz}^{(D)}(90°)/\mathrm{m\,s^{-1}}$ as a function of height for three levels of solar activity

Height	Low solar act.	Medium solar act.	High solar act.
200 km	2.5	2	1.5
250 km	10	5	3.5
300 km	37	11	7
350 km	128	25	10
400 km	429	54	23
hmF2 (day)	≈10	≈10	≈10
hmF2 (night)	≈50	≈50	≈50

We see from Table 11 that at low latitudes the error in $v_{n\xi}$ introduced by neglecting the influence of diffusion may be quite noteworthy.

35. Effect of $E \times B$ plasma drift on the neutral wind. Next we assume that the plasma motion is caused by an electric field. In this case v_{iz} and $v_{i\xi}$ are given by

$$v_{iz} = \frac{E_\eta}{B} \cos I + v_{n\xi} \sin I \cos I \tag{35.1}$$

and

$$v_{i\xi} = -\frac{E_\eta}{B} \sin I + v_{n\xi} \cos^2 I. \tag{35.2}$$

Again we introduce a time t_0 such that at $t = t_0$ the neutral gas is at rest. Consequently,

$$v_{iz} = \frac{E_\eta}{B} \cos I. \tag{35.3}$$

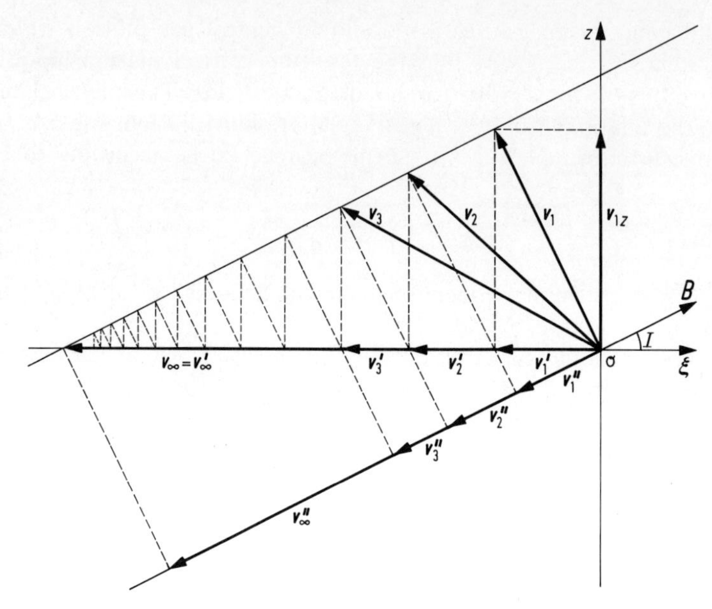

Fig. 25. Sketch to illustrate Eqs. (35.4) and (35.5). All vectors originate in O. *1* The electric field causes the velocity $\boldsymbol{v}_1$ ($v_1 = E_\eta/B$) perdendicular to $\boldsymbol{B}$. $\boldsymbol{v}_1$ has the vertical component $v_{1z} = v_1 \cos I$ and the horizontal component $v'_1 = -v_1 \sin I$. *2* The plasma accelerates the neutral gas in the horizontal direction. When the neutral gas has attained the velocity $\boldsymbol{v}'_1$, it gives rise to the additional plasma velocity $\boldsymbol{v}''_1$ along $\boldsymbol{B}$. The total plasma velocity then is $\boldsymbol{v}_2$. Its horizontal component $\boldsymbol{v}'_2$ is greater than $\boldsymbol{v}'_1$, so that the acceleration of the neutral gas continues. *3* The mutual acceleration of the neutral gas through the plasma and of the plasma through the neutral gas comes to an end when the total plasma velocity $\boldsymbol{v}_\infty$ is equal to the meridional neutral velocity $\boldsymbol{v}'_\infty$. Thus the equilibrium neutral velocity is $v_{n\xi} = v'_\infty = -v_1/\sin I$, while the vertical ion velocity is zero

As before, the neutral gas is set in motion horizontally until $v_{n\xi}$ and $v_{i\xi}$ are equal. Thus the meridional neutral velocity caused by the electromagnetic plasma drift is

$$v_{n\xi} = -\frac{E_\eta}{B} \frac{1}{\sin I}. \tag{35.4}$$

Inserting Eq. (35.4) in Eq. (35.1) yields the resulting vertical ion velocity

$$v_{iz} = 0. \tag{35.5}$$

Thus if an electric field is acting for a period comparable with or greater than τ [see Eq. (34.6)], a neutral velocity in the magnetic N–S direction as given by Eq. (35.4) is produced. The neutral motion reacts on the plasma in such a way that the initial vertical plasma drift given by Eq. (35.3) is reduced to zero. The effect of an electric field on the ionosphere and the thermosphere, therefore, strongly depends on its characteristic time compared with τ. In particular, a constant or slowly varying electric field does not have a marked effect on the vertical ion density distribution, except in the equatorial region where I is almost zero. A geometric illustration of the processes leading to Eqs. (35.4) and (35.5) is given in Fig. 25.

G. Conclusion

Neutral winds are caused by horizontal pressure gradients and modified by plasma motions or by the mere presence of an ionospheric plasma because of friction. At day the hottest region of the thermosphere is near the equator, at night near the poles. Consequently the meridional wind component is directed towards the equator at night and towards the poles at day.

Meridional winds strongly affect the ionospheric properties. In the F-region where the ion-neutral collision frequency is much smaller than the ion gyrofrequency, the wind-induced plasma motion is confined to the magnetic field lines. An equatorward wind thus causes an upward plasma motion which leads to an increase of the F-layer height, a reduction of chemical ion loss, and consequently to an increase of the F-layer plasma density. A poleward wind, on the other hand, gives rise to a reduction of the F-layer height and plasma density. This wind effect partly or fully explains a good deal of the observed F-region properties, e.g., the regular diurnal variation, the seasonal variation, the effect of magnetic declination and the Antarctic UT effect, to mention a few.

Thermospheric winds also affect the neutral temperature and neutral composition. Converging winds result in a temperature increase and a decrease of the particle density ratio of heavy to light neutral constituents. The helium (He) winter bulge and the increase of the O to O_2 number density ratio from summer to winter are two examples for the effects of winds on the neutral atmosphere.

In general one can say that the physical mechanisms controlling the wind field and the interaction of neutral and plasma motions are rather well understood. It would certainly come as a surprise if any basically new mechanism were to be found. Our knowledge is insufficient, however, with respect to the temporal and spatial structure of the thermosphere in general and of the thermospheric pressure in particular. Unfortunately, the key parameter, grad p_n, is a differential quantity. Its realistic description, therefore, requires extremely accurate experimental or theoretical methods. Still open questions are related with energy sources in the atmosphere, the thermospheric response to these sources, and magnetic control of the thermosphere.

List of main symbols

Symbol	Meaning	First appearance (Eq.)
	Coordinates	
x, y, z	Cartesian coordinates, pointing towards geographic south, east, and vertically upwards, respectively	(3.6)
ξ, η, z	Cartesian coordinates, pointing towards geomagnetic south, east, and vertically upwards, respectively	(9.8)
	Subscripts	
e	Electron gas	
i	Ion gas as a whole	

Symbol	Meaning	First appearance (Eq.)
j, k	Individual ion species	
l, m	Individual neutral species	
n	Neutral gas as a whole	
	Vectors	
$\mathbf{B}$	Geomagnetic induction	(3.5)
$\mathbf{E}$	Electric field strength	(9.3)
$\mathbf{F}$	External force per unit volume acting on a gas in the Earth's atmosphere	(2.1)
$\mathbf{g}$	Acceleration due to gravity	(2.2)
$\mathbf{P}$	Frictional force per unit volume	(2.2)
$\mathbf{v}$	Macroscopic velocity	(2.1)
$\mathbf{\Omega}$	Earth's angular velocity	(2.2)
	Scalars	
a	Collision frequency coefficient	(3.2)
C_p	Specific heat per mole at constant pressure	(31.6)
C_v	Specific heat per mole at constant volume	(7.1)
D	Magnetic declination, positive towards east	(3.6)
D_a	Ambipolar diffusion coefficient	(22.1)
$D^{(M)}$	Molecular diffusion coefficient	(8.3)
$D^{(T)}$	Turbulent (or eddy) diffusion coefficient	(8.4)
q	Proton charge	(9.3)
foF2	Critical frequency of F2-layer	Fig. 17
$hmF2$	Height of F2-peak	Fig. 2
H	Scale height	(4.1)
I	Magnetic inclination, positive in the Northern hemisphere	(3.6)
k	Boltzmann's constant	(2.1)
L	Loschmidt's number	(7.1)
m	Particle mass	(2.1)
n	Particle number density (N for charged particles)	(2.1)
$NmF2$	Electron number density at F2-peak	(5.4)
p	Pressure	(2.1)
P	Particle production or loss per unit volume and unit time due to photochemical reactions	(8.1)
Q	Heat production or loss per unit volume and unit time due to UV absorption, IR irradiation, and heat exchange	(7.1)
R_0	Earth's radius	(5.2)
T	Kinetic temperature	(2.1)
T_{jl}	Reduced temperature	Table 1
T_∞	Exospheric neutral temperature	(4.2)
$v_{nz}^{(B)}$	Vertical neutral velocity due to thermospheric "breathing"	(31.3)
$v_{nz}^{(D)}$	Vertical neutral velocity due to divergent or convergent wind flow	(31.4)
W	Wind-induced vertical ion velocity	(22.2)
Ym	Semi-thickness of parabolic F layer approximation	following Eq. (5.5)
β	Loss coefficient	(22.1)
η	Coefficient of viscosity	(2.1)
κ	Thermal conductivity	(7.1)
λ	Mean free path	(11.1)
μ	Reduced mass	(3.1)
ν	Collision frequency	(3.1)
φ	Geographic latitude	(5.2)
ρ	Mass density	(2.1)
τ	Characteristic time	(11.2)

General references

[1] Alcayde, A., Bauer, P., Hedin, A., Salah, J.E. (1978): Compatibility of seasonal variations in mid-latitude thermospheric models at solar maximum and low geomagnetic activity. J. Geophys. Res. *83*, 1141

[2] Anderson, D.N. (1976): Modeling the midlatitude F-region ionospheric storm using east-west drift and a meridional wind. Planet. Space. Sci. *24*, 69

[3] Anderson, D.N., Roble, R.G. (1974): The effect of vertical $E \times B$ ionospheric drifts on F region neutral winds in the low-latitude thermosphere. J. Geophys. Res. *79*, 5231

[4] Bailey, G.J., Moffett, R.J., Rishbeth, H. (1969): Solution of the coupled ion and neutral air equations of the mid-latitude ionospheric F2-layer. J. Atmos. Terr. Phys. *31*, 253

[5] Banks, P., Kockarts, G. (1973): Aeronomy. New York: Academic Press

[6] Barlier, F., Berger, C., Falin, J.L., Kockarts, G., Thuillier, G. (1978): A thermospheric model based on satellite drag data. Ann. Geophys. *34*, 9

[7] Bedinger, J.F. (1972): Thermospheric motions measured by chemical releases. Space Res. XII, p. 919. Berlin: Akademie-Verlag

[8] Behnke, R., Kohl, H. (1974): The effects of neutral winds and electric fields on the ionospheric F2-layer over Arecibo. J. Atmos. Terr. Phys. *36*, 325

[9] Blum, P.W., Harris, I. (1975): Full non-linear treatment of the global thermospheric wind system. Part 1: Mathematical method and analysis of forces. Part 2: Results and comparison with observations. J. Atmos. Terr. Phys. *37*, 193

[10] Burge, J.D., Eccles, D., King, J.W., Rüster, R. (1973): The effects of thermospheric winds on the ionosphere at low and middle latitudes during magnetic disturbances. J. Atmos. Terr. Phys. *35*, 617

[11] Jacchia, L.G. (1972): Atmospheric models in the region from 110 to 2000 km. COSPAR International Reference Atmosphere 1972, p. 227. Berlin: Akademie-Verlag

[12] Creekmore, S.P., Straus, J.M., Harris, R.M., Ching, B.K., Chiu, Y.T. (1975): A global model of thermospheric dynamics. I. Wind and density fields derived from a phenomenological temperature. J. Atmos. Terr. Phys. *37*, 491. II. Wind, density, and temperature fields generated by EUV heating. J. Atmos. Terr. Phys. *37*, 1245

[13] Dalgarno, A. (1964): Ambipolar diffusion in the F-region. J. Atmos. Terr. Phys. *26*, 939

[14] Dickinson, R.E., Geisler, J.E. (1968): Vertical motion field in the middle thermosphere from satellite drag densities. Mon. Weather Rev. *96*, 606

[15] Evans, J.E. (1972): Ionospheric movements measured by incoherent scatter: A review. J. Atmos. Terr. Phys. *34*, 175

[16] Fedder, J.A., Banks, P.M. (1972): Convection electric fields and polar thermospheric winds. J. Geophys. Res. *77*, 2328

[17] Geisler, J.E. (1966): Atmospheric winds in the middle latitude F-region. J. Atmos. Terr. Phys. *28*, 703

[18] Harris, I., Mayr, H.G. (1975, 1977): Diurnal variations in the thermosphere. 1. Theoretical formulation. J. Geophys. Res. *80*, 3925. 2. Temperature, composition, and winds. J. Geophys. Res. *82*, 2628

[19] Hartle, R.E. (1971): Model for rotating and non-uniform planetary exospheres. Phys. Fluids *14*, 2592

[20] Hedin, A.E., Mayr, H.G., Reber, C.A., Spencer, N.W., Carigan, G.R. (1974): Empirical model of global thermospheric temperature and composition based on data from the OGO-6 quadrupole mass spectrometer. J. Geophys. Res. *79*, 215

[21] Hedin, A.E., Reber, C.A., Newton, G.P., Spencer, N.W., Brinton, H.C., Mayr, H.G. (1977): A global thermospheric model based on mass spectrometer and incoherent scatter data. MSIS 2. Composition. J. Geophys. Res. *82*, 2148

[22] Hernandez, G., Roble, R.G.: Direct measurements of nighttime thermospheric winds and temperatures. 1. Seasonal variations during geomagnetic quiet periods. J. Geophys. Res. *81*, 2065 (1976). 2. Geomagnetic storms. J. Geophys. Res. *81*, 5173 (1976). 3. Monthly variations during solar minimum. J. Geophys. Res. *82*, 5505 (1977)

[23] Jacchia, L.G. (1974): Variations in thermospheric composition: A model based on mass spectrometer and satellite drag data. J. Geophys. Res. *79*, 1923

[24] Jacchia, L.G. (1977): Thermospheric temperature density and composition: New Models, Cambridge (Mass.). Smithson. Astrophys. Obs., Spec. Rept. 375
[25] King, J.W., Kohl, H. (1965): Upper atmospheric winds and ionospheric drifts caused by neutral air pressure gradients. Nature *206*, 699
[26] King-Hele, D.G. (1964): The rotational speed of the upper atmosphere, determined from changes in satellite orbits. Planet. Space. Sci. *12*, 835
[27] Klostermeyer, J. (1973): Thermospheric heating by atmospheric gravity waves. J. Atmos. Terr. Phys. *35*, 2267
[28] Kohl, H. (1970): Wind systems in the thermosphere. Space Res. X, p. 550. Amsterdam: North-Holland Publishing Company
[29] Kohl, H. (1972): Ein globales Windsystem in der Thermosphäre und sein Einfluß auf die F-Schicht der Ionosphäre. Habilitationsschrift, University of Göttingen
[30] Kohl, H., King, J.W. (1967): Atmospheric winds between 100 and 700 km and their effects on the ionosphere. J. Atmos. Terr. Phys. *29*, 1045
[31] Lettau, H. (1951): Diffusion in the upper atmosphere. Compendium of Meteorology, Amer. Meteor. Soc., Boston, p. 320
[32] Matsushita, S. (1967): Lunar tides in the ionosphere. This Encyclopedia, vol. 49/2, p. 547
[33] Mayr, H.G., Hedin, A.E., Reber, C.A., Carignan, G.R. (1974): Global characteristics in the diurnal variations of the thermospheric temperature and composition. J. Geophys. Res. *79*, 619
[34] Rawer, K., Suchy, K. (1967): This Encyclopedia, vol. 49/2, p. 1
[35] Rishbeth, H. (1967): The effect of winds on the ionospheric F2-peak. J. Atmos. Terr. Phys. *29*, 225
[36] Rishbeth, H. (1972): Thermospheric winds and the F-region: A review. J. Atmos. Terr. Phys. *34*, 1
[37] Rishbeth, H. (1972): Superrotation of the upper atmosphere. Rev. Geophys. and Space Phys. *10*, 799
[38] Rüster, R. (1971): Solution of the coupled ionospheric continuity equations and the equations of motion for the ions, electrons and neutral particles. J. Atmos. Terr. Phys. *33*, 137
[39] Rüster, R., King, J.W. (1973): Atmospheric composition changes and the F2-layer seasonal anomaly. J. Atmos. Terr. Phys. *35*, 1317
[40] Schunk, R.W. (1975): Transport equations for aeronomy. Planet. Space Sci. *23*, 437
[41] Sterling, D.L., Hanson, W.B., Moffett, R.J., Baxter, R.G. (1969): Influence of electromagnetic drifts and neutral air winds on some features of the F2 region. Radio Sci. *4*, 1005
[42] Straus, J.M., Creekmore, S.P., Harris, R.M., Ching, B.K. (1975): Effects of heating at high latitudes on global thermospheric dynamics. J. Atmos. Terr. Phys. *37*, 1545
[43] Stubbe, P. (1970): Simultaneous solution of ionospheric equations (short title). J. Atmos. Terr. Phys. *32*, 856
[44] Stubbe, P. (1973): Atmospheric parameters from ionospheric measurements. Z. Geophys. *39*, 1043
[45] Mitteilungen aus dem Max-Planck-Institut für Aeronomie, Nr. 52 (1974). Berlin, Heidelberg, New York: Springer. [English translation: Pennsylvania State University Report PSU-IRL-SCI-418 (1973)]
[46] Stubbe, P., Chandra, S. (1970): The effect of electric fields on the F-region behaviour as compared with neutral wind effects. J. Atmos. Terr. Phys. *32*, 1909
[47] Torr, M.R., Torr, D.G. (1969): A theoretical investigation of the ionosphere. CSIR Research Report 271, Pretoria, South Africa
[48] Volland, H., Mayr, H.G. (1972): A three-dimensional model of thermosphere dynamics – I. Heat input and eigenfunctions, II. Tidal waves, III. Planetary waves. J. Atmos. Terr. Phys. *34*, 1745

Extreme Ultraviolet Observational Data on the Solar Spectrum

By

G.M. NIKOL'SKIJ

With 22 Figures

I. Spectral observations

1. Early experiments. With the advent of rocket techniques, it became possible to extend records of the solar spectrum toward the wavelength region below 290 nm. As early as 1943 KIEPENHEUER and REGENER[1] prepared a spectrograph with optics of LiF and a guiding system for rocket experiments. Then in 1946 American investigators using a German V 2 rocket[2] recorded the solar spectrum down to the "ozone boundary" (290–210 nm), however, with rather low resolution, below 0.4 nm. In 1947 the UV (ultraviolet) spectrum was recorded to approximately the same wavelength by the staff of the Applied Physics Laboratory of John Hopkins University[3]. For the next three years the shortwave limit of the rocket spectra remained virtually unchanged (near 190 nm) while resolution improved.

The general findings of these experiments were 1) the intensity of the continuum decreases faster than had been expected; 2) color temperatures of UV emission were lower than in the visible: $T_c = 5000$ K at a wavelength of 300 nm and 4500 K at 200 nm. Along with spectrographic methods, various other methods were employed to record solar UV emission within wide spectral ranges. First, phosphors (generally $CaSO_4$:Mn) were used for this purpose. Combined with LiF and CaF_2 windows, these phosphors are sensitive in the wavelength ranges 100–134 and 123–134 nm, respectively. Such experiments indicated the essential difference between the energy distribution in the solar EUV (extreme ultraviolet) and Planck's law. It was found that the absorption in the terrestrial atmosphere of EUV radiation becomes negligible at heights

[1] Kiepenheuer, K.O. (1948): Solar-terrestrische Erscheinungen, Ten Bruggencate, P. (ed.): FIAT Review of German Science 1939–1946, Vol. I.10, pp. 228–284. Wiesbaden: Dieterich'sche Verb. Buchh.

[2] Baum, W.H., Johnson, F.S., Oberly, J.J., Rockwood, C.C., Strain, C.V., Tousey, R. (1946): Phys. Rev. **76**, 781

[3] Hopfield, J.J., Clearman jr., H.E. (1948): Phys. Rev. *73*, 877

above 20 km for wavelengths from 170 to 200 nm and above 100 km from 140 to 170 nm.

Solar astronomy made good progress by space techniques in 1952 when a spectrogram including the Lyman alpha line (121.57 nm), the brightest line in the whole solar spectrum below 279 nm, was obtained by the Colorado University group.

A great deal of analysis of the solar spectrum was then carried out within the wavelength ranges where terrestrial atmospheric absorption is important – from 290 nm down to 150–200 nm – and where the continuum intensity becomes negligible and the spectrum changes to a line spectrum. Along with numerous absorption lines, often blended, the brightest emission UV lines are situated within this spectral region, e.g., the resonance doublet of ionized Mg, 279.55 and 280.27 nm. These lines are similar to the well-known Ca II doublet – H and K – and also represent a reversal in the center of very broad absorption lines. The total radiation energy within the UV Mg I doublet was found to be[4] $18\ \mathrm{mW\,m^{-2}}$ ($1\ \mathrm{mW\,m^{-2}} = 1\ \mathrm{erg\,cm^{-2}\,s^{-1}}$) . Results concerning line identification in the spectral region in question are reported in[5, 6].

Both for geophysics and for astrophysics (though for different reasons) the most interesting subjects are the emission lines in the solar spectrum from a wavelength of 200 nm down to the x-ray range. Therefore we shall concentrate our attention on this spectral region, the extreme ultraviolet (EUV).

2. Instrumentation used

α) Generally *gratings* are used for recording the EUV spectrum because refraction optics is not feasable in the spectral region below 100 nm. Along with certain advantages, the gratings have some drawbacks: applied as a camera, they produce much scattered light which makes it difficult to obtain spectrograms of good quality. Two types of arrangement are used: a grazing-incidence system and a normal-incidence one. The latter type can be used successfully only for the spectral region below 30–50 nm, since it is difficult to get a shorter-wavelength spectrum without a background of scattered light of longer wavelengths. The scattered light can be reduced by one order of magnitude or more by the use of a predispersion system, which simultaneously plays the role of a collector. It produces the image of the solar disk on the entrance-slit jaw, which enables stigmatic spectra to be obtained. This type of equipment was used from 1955 on by the NRL group[1]. The "predispersion" grating was deformed in a special manner to compensate for an astigmatism of the principal system (Fig. 1).

For the purpose of reducing scattered light and enlarging the reflectivity in the EUV region, the grating surface is coated with one or sometimes several layers of e.g., quartz, Ge, Al_2O_3, or SiO. At a wavelength of 120 nm, for instance, good results are obtained with a grating coated with Au, Pt, or Ir.

[4] Ivanov-Holodnyj, G.S. (1958): Izv. Akad. Nauk SSSR, Ser. Geofiz., Nr. 9, 1104

[5] Malitson, H.H., Purcell, J.D., Tousey, R., Moore, C.F. (1960): Astrophys. J. *132*, 746

[6] Kačalov, V.P., Pavlenko, N.A., Jakovleva, A.B. (1958): Izv. Akad. Nauk SSSR, Ser. Geofiz., Nr. 9, 1099

[1] Purcell, J.D., Packer, D.M., Tousey, R. (1960): Space Res. *1*, 581

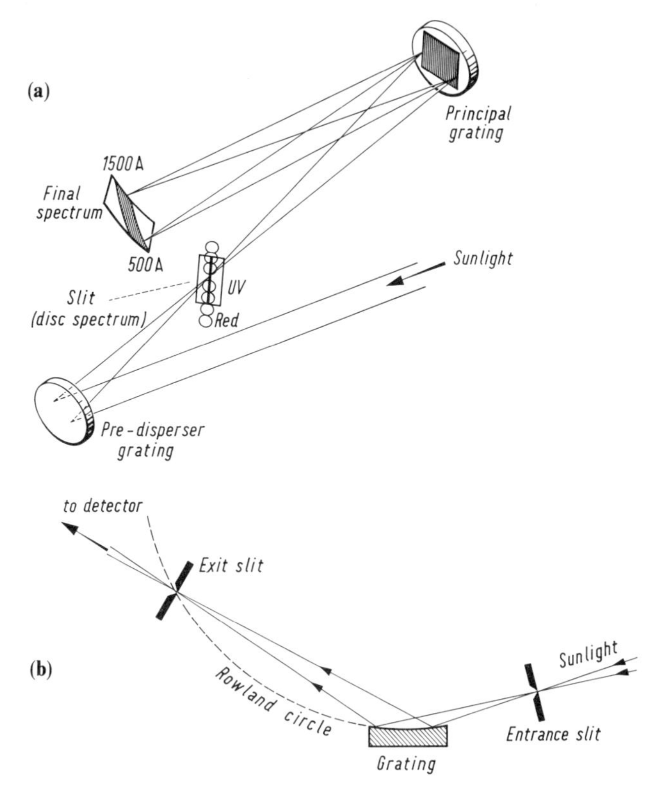

Fig. 1a, b. Two types of solar EUV rocket spectroscopes: (**a**) double-dispersion spectrograph (NRL, top) (**b**) grazing-incidence monochromator (AFCRL, bottom)

β) For wavelengths below 30–50 nm *grazing incidence* is generally applied to improve the reflectivity, most frequently the Rowland system. In this system the grating, slit, and spectrum are placed along a circle whose diameter is equal to the radius of the grating curvature (spectral lines are perpendicular to the circle plane).

In a grazing incidence system scattered light is reduced considerably along with increasing reflectivity. For example, according to EDLEN, the short-wave limit of a grating that extends down to 32 nm at a normal incidence is 5.3 nm at an incidence under 85.6°. The advantages of the grazing-incidence system are limited by considerable astigmatism and diminution of the aperture.

SCHUMANN emulsion is used for photographic recording of EUV spectra.

γ) The Air Force Cambridge Research Laboratory (AFCRL) group (H.E. Hinteregger) adopted *photoelectric detection* of the EUV spectrum. Their first spectrometer was of the Rowland type with grazing-incidence; wavelength scanning was obtained by moving the output slit along the Rowland circle, and a multiplier of open type combined with a telemetry system was used for detecting EUV spectral intensities. The first spectrometer covering the wavelength range 25–130 nm was flown on an Aerobee rocket on 12 March 1959.

δ) A good *pointing* and stabilization *system* is necessary for such space experiments. In early flights different pointing arrangements of large aperture were used: a diffuse reflector set in front of the entrance slit; a sphere of LiF combined with a diaphragm of smaller aperture instead of an entrance slit (aperture increased to 120°); a wedge-shaped slit with a wedge angle ≈10° pointed toward the Sun (the aperture in this case reaches 60°). The imperfection of such pointing systems is apparent.

The advent of the biaxial pointing control designed at Colorado University[2] and improved by other groups was the next advance in solar rocket astronomy. The guiding accuracy in the first experiments was one or two minutes of arc (′). There was no compensation for rotation around the Sun-rocket axis.

For many years now the guiding accuracy in solar rocket experiments has reached one second of arc (″).[3] The first three-axis stabilized rocket was designed by a British rocket group. To compensate for rotation about the Sun-rocket axis, a magnetograph was used as sensor. There are several types of three-axis pointing controls, two of them designed by NASA. A review on pointing control systems is included in [4].

ε) The procedure for absolute calibration in the EUV spectrum is very difficult, especially for wavelengths below 50 nm, as the reflectivity of the optical surfaces changes during flight. The problems of calibration and photocathode sensitivity variations are considered in a series of papers.[5]

The problem of photometric calibration of the EUV spectrum was considered at a special symposium. Data on some instrumentation used for recording solar EUV radiation is reported together with the observational results[6].

3. Photographic recording of the solar EUV spectrum

α) The solar EUV spectrum in the *neighborhood of the Lyman alpha* line has been frequently photographed after the first records obtained by different American authors[1–5].

The first flux estimates for 24 lines within the range 120.6–181.7 nm were reported in[5]; the spectrogram was taken at an altitude of 95–150 km on 6 August 1958.

β) The Colorado University group made a magnificent contribution to the first rocket solar experiments in recording, for the first time, the *Lyman α lines*

---

[2] Stacey, D.S., Stith, B.A., Nidey, R.A., Pietenpol, W.A. (1954): Electronics *27*, 149

[3] Burton, W.M., Ridgeley, A., Wilson, R. (1967): Mon. Not. R. Astron. Soc. *135*, 207

[4] Wilson, R. (1967): Paper at IAU Symp., Prague

[5] Reeves, E.M., Parkinson, W.H. (1970): Appl. Opt. *9*, 1201

[6] Reeves, E.M., Parkinson, W.H. (1969): Calibration Methods in the ultraviolet and X-ray regions of the spectrum (Munich Symposium, May 1968), Special publication SP-33. Paris: European Space Research Organisation

[1] Rense, W.A. (1953): Phys. Rev. *91*, 229

[2] Johnson, F.S., Purcell, J.D., Tousey, R. (1954): Phys. Rev. *95*, 621; Bull. Am. Phys. Soc. *29*, 33

[3] Johnson, F.S., Malitson, H.H., Purcell, J.D., Tousey, R. (1958): Astrophys. J. *127*, 80

[4] Jursa, A.S., Le Blanc, F.G., Tanaka, Y. (1955): J. Opt. Soc. Am. *45*, 1085

[5] Behring, W.E., Alister, H.Mc., Rense, W.A. Ap. J. *127*, 676; Aboud, A., Behring, W.E., Rense, W.A. (1959): Astrophys. J. *130*, 381

of helium He I and He II. At the launchings of 4 June 1958 and 30 March 1959, VIOLETT and RENSE, using a grazing-incidence spectrograph [6], recorded the EUV spectrum down to 8.4 nm, which was very good for that time.

The gratings used had areas of 1 cm × 2 cm and 1.8 cm × 3.6 cm ruled with 600 lines per mm and a 50-cm radius of curvature. The plate resolution was 1.26, 0.93, and 0.66 nm/mm at the wavelengths 120, 60, and 30 nm, respectively. The spectral resolution near the Lyman α line of hydrogen (121.6 nm) was 0.04 nm. The pointing control system was due to PIETENPOL et al. [7]

The wavelengths of 150 lines appearing on the spectrograms of both flights and rough visual estimates of line intensities were listed in a relative intensity scale taking 1000 for Ly α and 5 for the weakest of the lines. However, it was impossible to compare intensities but for adjacent lines because the variation with wavelength of the reflectivity of the optics, emulsion sensitivity, and absorption in the terrestrial atmosphere were not known. An attempt at an absolute calibration of the line intensities in the VIOLETT and RENSE list was made by IVANOV-HOLODNYJ and NIKOL'SKIJ. [8] The resonance line of ionized helium (30.38 nm) was discovered for the first time in these spectrograms.

Some of the identified lines were false ones, being either cracks in the photographic emulsion or lines of other orders of the grating. Three fourths of the 74 identified lines agree, however, with modern identifications.

γ) NRL performed a series of experiments to record the solar spectrum in a *wide spectral range.* Two normal-incidence spectrographs were arranged in the same container; one was designed for the spectral region 150–350 nm and the other for 50–250 nm. In 1954 a series of spectrograms were taken that overexposure for the Lyman α line of hydrogen. [9] On the 1955 spectrogram 40 lines appeared within the spectral range 121.5–154.2 nm, including the lines of lithium-like ions, C IV, N V, O VI, originating in the transition layer between the chromosphere and the corona. No absolute calibration was carried out. [10]

On a 30-s spectrogram obtained on 13 March 1959, approximately 100 lines, which included 8 lines of the Lyman series, and the continuum were discovered within the range 50–180 nm. A normal-incidence spectrograph with a dispersion of 4 nm/mm and an improved grating coated with Ge and Al_2O_3 was carried aboard an Aerobee-Hi rocket to an altitude of 198 km.

On 19 April 1960, the NRL group obtained a stigmatic spectrum that extended 49.9–155 nm and showed about 200 lines. [11] The improved apparatus of the previous experiments was used together with a predispersion grating. Both gratings were ruled with 600 lines per mm and had a radius of curvature of 400 mm. Since the directions of dispersion of these two gratings were perpendicular, the spectra were slightly tilted. The slit was about 0.1 nm wide. The spectrograms are shown in Fig. 2.

δ) The results obtained from a spectrogram photographed with 60-s exposure time at 157–206 km altitude are reported by [11]. In order to obtain absolute calibration the Lyman α *intensity* was also measured with an ion chamber aboard the same rocket. According to this experiment, the Lyman α

[6] Violett, T., Rense, W.A. (1959): Astrophys. J. *130*, 954
[7] Pietenpol, W.A. (1954): Electronics *27*, 149
[8] Ivanov-Holodnyj, G.S., Nikol'skij, G.M. (1961): Astron. Zh. *38*, 45
[9] Johnson, F.S., Pursell, J.D., Tousey, R. (1954): Phys. Rev. *95*, 621
[10] Johnson, F.S., Malitson, H.H., Purcell, J.D., Tousey, R. (1958): Astrophys. J. *127*, 80
[11] Detwiller, C.R., Garrett, D.L., Purcell, J.D., Tousey, R. (1961): Ann. Géophys. *17*, 263

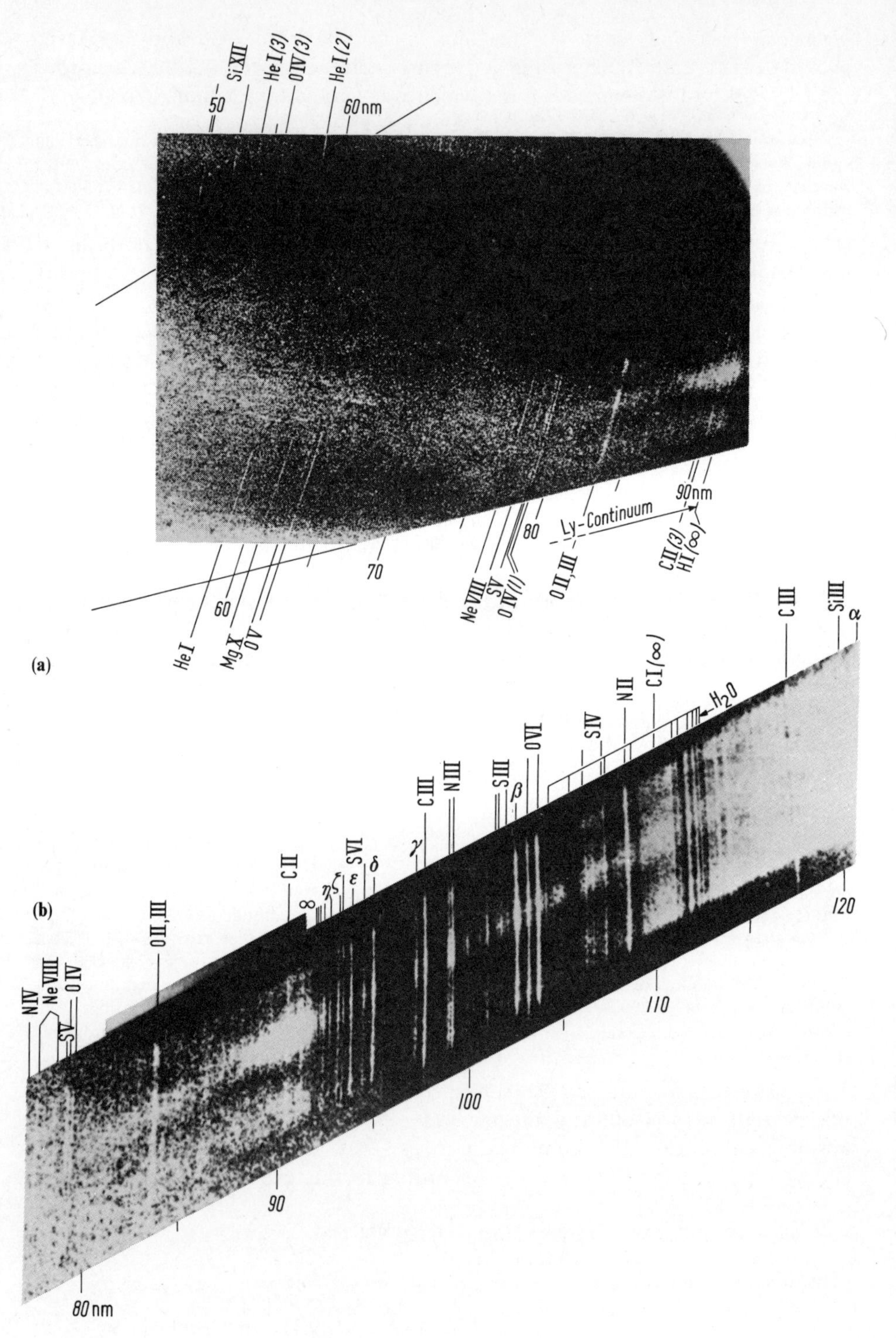

Fig. 2a, b (caption see opposite page)

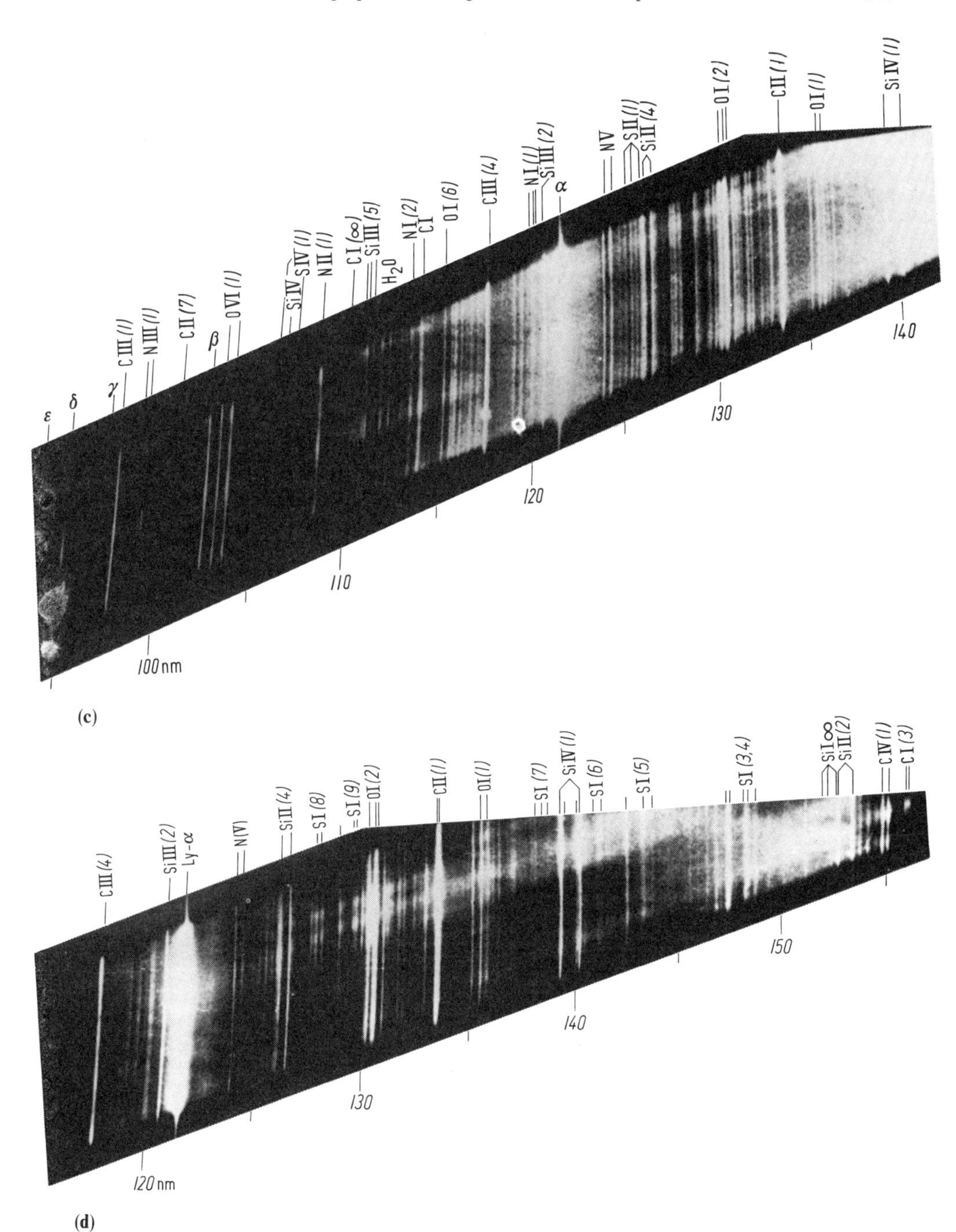

Fig. 2a–d. Double dispersion normal incidence spectrogram (four subranges) of the solar radiation in the range 50–150 nm (courtesy: GARRETT, PURCELL, TOUSEY, NRL) 19 April 1960

Table 1. Near Earth flux of bright solar EUV lines as obtained from photographic rocket observations [11]

[nm]	Ion	Flux [$mW m^{-2}$]	λ [nm]	Ion	Flux [$mW m^{-2}$]
189.209	Si III	0.10	126.504	Si II	0.020
181.742	Si II	0.45	126.066	Si II	0.010
180.801	Si II	0.15	124.278	N V	0.003
167.081	Al II	0.08	123,880	N V	0.004
163.700	C I	0.16	121.567	H I	5.1
164.047	He II	0.07	120.652	Si III	0.030
156.140	C I	0.09	117.570	C III	0.010
155.077	C IV	0.06	113.989	C I	0.003
154.819	C IV	0.11	108.570	N II	0.006
159.944	Si II	0.041	103.761	O IV	0.025
152.670	Si II	0.038	103.191	O VI	0.020
140.279	Si IV	0.013	102.572	H I	0.060
139.379	Si IV	0.030	99.158	N III	0.010
139.568	C II	0.050	98.979	N III	0.006
133.451	C II	0.050	97.703	C III	0.050
130.602	O I	0.025	94.974	H I	0.010
130.486	O I	0.020	93.780	H I	0.05
130.217	O I	0.013	83.5	{O II, O III	0.010

flux from the entire disk at a distance of 1 AU* is 5.1 $mW m^{-2}$. The absorption by molecular nitrogen (N_2) and oxygen (O_2) as well as by H_2O vapor carried with the rocket was discovered in the spectrum. The total column content of water vapor molecules within the light path was found to be $4.6 \cdot 10^{20} m^{-2}$. The total energy flux in the higher members of the Lyman series is about 0.12 $mW m^{-2}$. The intensity in the Lyman continuum below 91.2 nm can be described by the radiation of a black body at 6600 K thus slightly above the color temperature in the visible; total intensity is 0.24 $mW m^{-2}$. Line identifications and intensities given in [11] are reproduced in Table 1.

ε) Within the next several years the NRL group headed by TOUSEY considerably extended the UV spectra toward the *short-wave end*.

In the experiment of 21 June 1961 the short-wavelength limit approached 17 nm. The grating used was ruled with 600 lines per mm and had a 400-mm radius of curvature. In the experiments of 22 August 1962 and 10 May 1963, the short-wavelength limit was extended to 15 nm. On 20 September 1963 the spectrum within the range 3.3–18.8 nm was obtained. A grating ruled with 2400 lines per mm was used. The resolution on this spectrogram was 0.015 nm and the uncertainty of the wavelength determination was 0.002 to 0.01 nm. Estimated intensities of 30 lines within the range 3.3–7 nm were given. [12]

The NRL group recorded for the first time the bright C VI Lyman α line (3.37 nm). The soft x-ray spectrum has been recorded many times. [13–15]

* In the following we always indicate radiation flux values in free space taken at 1 AU, i.e., near the Earth, except where specially mentioned. [1 AU (astronomical unit) $= 1.5 \cdot 10^{11}$ m = 150 Gm = average distance between Earth and Sun.]

[12] Austin, W.E., Purcell, J.D., Tousey, R., Widing, K.G. (1966): Astrophys. J. *145*, 373

[13] Tousey, R. (1967): Astrophys. J. *149*, 239

[14] Austin, W.E., Purcell, J.D., Snider, C.B., Tousey, R., Widing, K.G. (1967): Space Res. 7, 1252

[15] Widing, K.G., Sandlin, G.D. (1968): Astrophys. J. *152*, 545

A survey covering the time up to 1971 was published by WALKER.[16] His Tables 1 and 2 cover most experiments made until this date.

ζ) A remarkable *improvement in sensitivity* was reached in 1971 by applying multigrid diffraction filters for preselection. Such filters[17], according to their dimensions, give strong preference to the chosen wavelength range so that radiation from longer wavelengths is very considerably reduced. This is an important advantage at the short-wave end of the spectrum where straylight from strong long-wave lines, in particular hydrogen Lyman α (121.6 nm), can be very disturbing. Photographic records obtained on board a rocket in the wavelength range 4–21 nm allowed the detection of some unknown lines and reached a resolution of 5 pm. For example, the splitting of the Fe α emission at 18.8 nm could be shown.[18] This result was later confirmed with a 3-m-radius spectrograph with a resolution of 4 pm.[19] After [5] the highest resolutions reached until 1976 were 10 pm absolute and about 3 pm relative value.

4. EUV-spectrum study by photoelectric techniques

α) Photoelectric detection of the EUV spectrum in *rocket experiments* has been used by HINTEREGGER's group at the Air Force Cambridge Laboratory (Bedford, Massachussets). The first record within the region 25–130 nm was obtained on 12 March 1959.[1] Four peaks were found, corresponding to the lines at 28.2, 33.5, 30.4, and 121.6 nm. The grating used had a 2-m radius of curvature and was ruled with 600 lines per mm. This experiment showed that radiation of wavelength below 30.4 nm is absorbed in the terrestrial atmosphere above the 200 km level. The use of a grating ruled with 1200 or 2400 lines per mm allowed the short-wave limit of the spectrum to be extended to 12.5 and 6 nm, respectively. Absolute calibration was done in the laboratory for wavelengths greater than 50 nm; below that limit an indirect calibration method had to be used, namely, by extrapolation because the laboratory spectrum was too faint.

At the beginning of 1960 the solar spectrum was recorded over the range 6–130 nm. Records obtained on 19 January (30–130 nm) and 29 January 1960 (6–30 nm) at 200 km altitude are reproduced in HINTEREGGER's paper[2], where he also calculated the resulting ionization of interplanetary gas. The resolution varied from 0.2 nm (near 6 nm) to 1.6 nm (near the H Lyman α 121.6 nm line). On the record one can see 60 peaks corresponding to the emission lines, including those due to other orders of the grating.

According to HINTEREGGER's data, apart from the very strong Lyman α line of hydrogen the total energy emitted within 6–130 nm is 10 mW m^{-2}. After correction for absorption in the terrestrial atmosphere, this value increases to 15 mW m^{-2}.

---

[16] Walker, A.B.C. (1972): Space Sci. Rev. *13*, 672
[17] Schmidtke, G. (1968): Optik *27*, 267; Appl. Optics (1970): *9*, No. 2
[18] Schmidtke, G., Schweizer, W. (1971): Astrophys. J. *169*, L27
[19] Behring, W.E. et al. (1972): Astrophys. J. *175*, 493
[1] Hinteregger, H.E., Damon, K.R., Heroux, L., Hall, L.A. (1960): Space Res. *1*, 615
[2] Hinteregger, H.E. (1960): Astrophys. J. *132*, 801

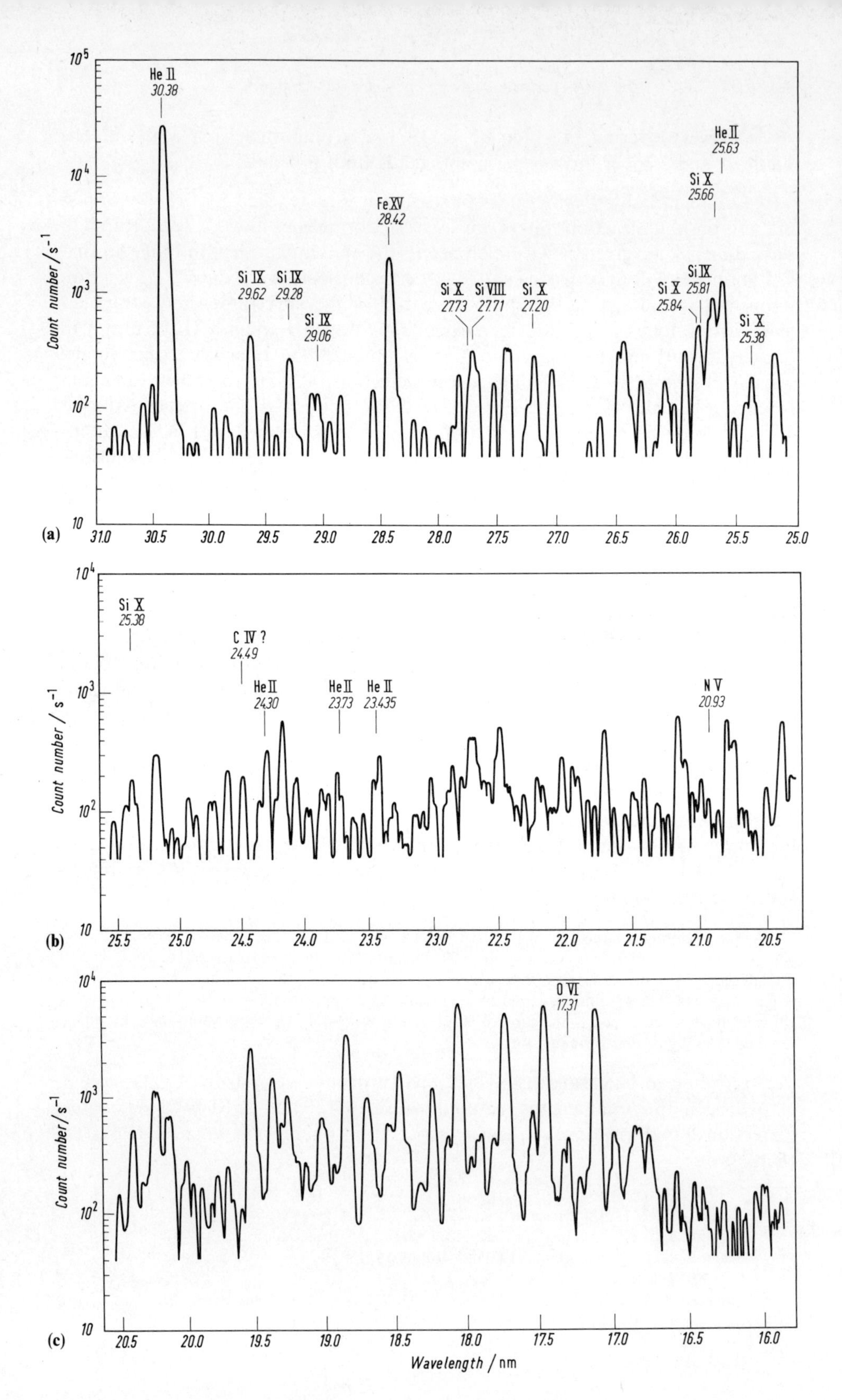

Fig. 3a–c (caption see opposite page)

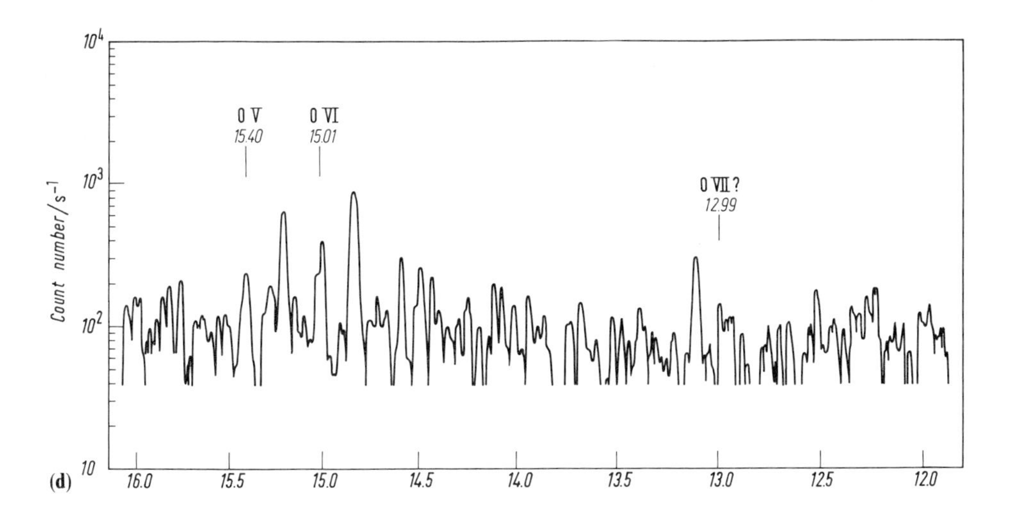

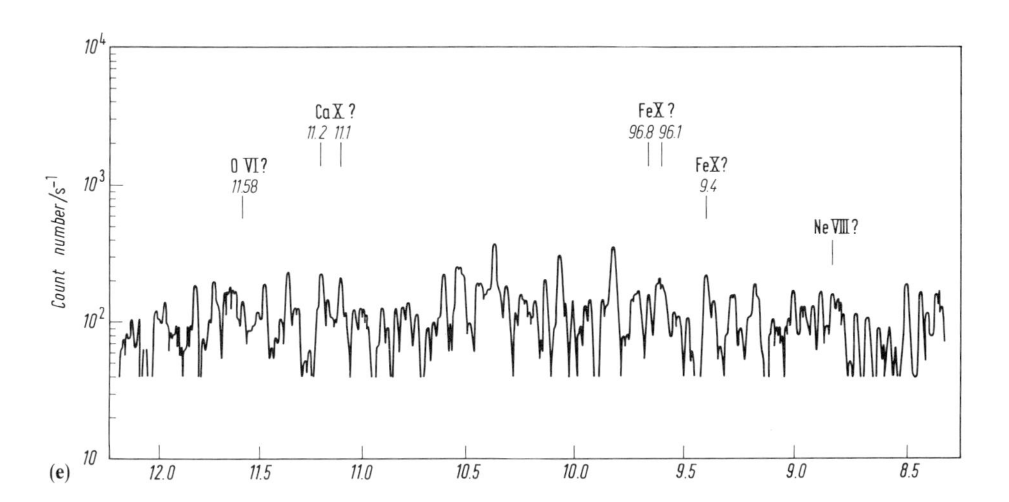

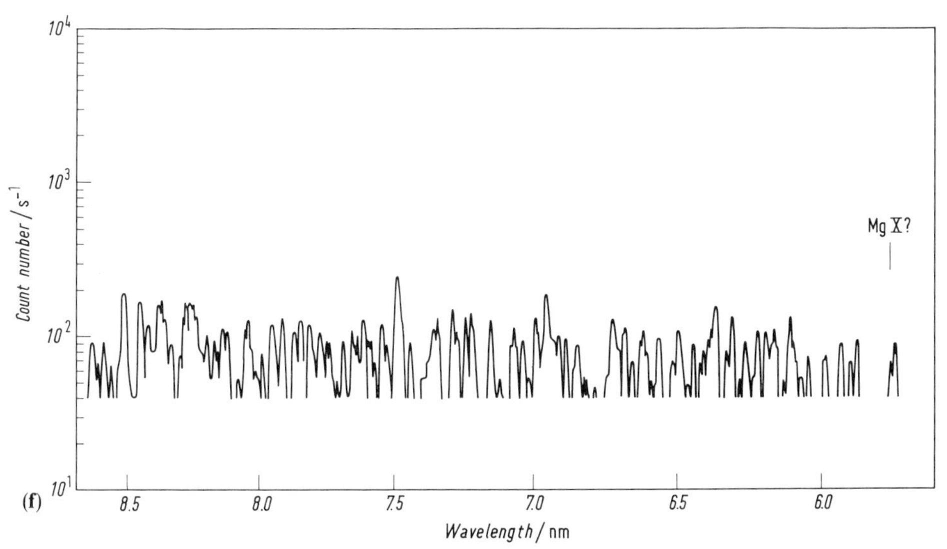

Fig. 3a–f. Solar spectrum (six subranges) in the range 5.5–31 nm from one 100-s scan of the AFCRL grazing-incidence monochromator aboard an Aerobee-Hi rocket[5]

In the summer of 1960, using the same method, HINTEREGGER[3] obtained a record with a better resolution of 0.3–8.5 nm within the region 25–130 nm. In 1961 the experiment was repeated[4] with even a better resolution of 0.15–0.3 nm. The total radiation flux appeared somewhat lower than might have been concluded from the observed decrease in solar activity (the solar radio wave flux at 10.7 cm, the 'COVINGTON index', was halved).

The monochromator used in the summer experiment had a grating with a 2-m radius of curvature, ruled with 300 lines per mm. The error of the wavelength determination was not greater than 0.2 nm. As estimated by[4], the absolute calibration uncertainties were $\pm 10\%$ near the H Lyman α line, and $\pm 25\%$ near 25 nm.

In 1963 the AFCRL obtained new records of the solar EUV spectrum upward from 5.5 nm. HINTEREGGER published a record of the range 5.5–31 nm obtained on 2 May 1963, the resolution of which reached 0.01 nm; a grating ruled with 1200 lines per mm and with a 2-m radius of curvature was used.

The records are shown[5] in Fig. 3a–f, where the ordinate is the logarithm of the counting rate. To obtain the photon flux density, the counting rate must be multiplied by the calibration factor $1.4 \cdot 10^9\ \mathrm{m}^{-2}$, the uncertainty being estimated to lie between a factor of 2 to 3.

In all the described experiments of AFCRL the monochromator was carried by an Aerobee rocket, thus the peak altitude could never be above 250 km.

β) An apparatus was first lifted to a *higher altitude* of 550 km, and thus for a longer time, aboard the satellite OSO-III in spring 1967.[6]

The concave grating, ruled with 600 lines per mm on a gold-coated surface and having a 2-m radius of curvature, was operated in grazing incidence at an angle of 86°. The entrance slit was 0.05 mm × 8 mm. Four output slits were moved along the Rowland circle.[7] Scanning was done in "steps", there were 2040 steps over the whole spectral region measured. The spectral resolution varied from 0.15 nm near the short-wave limit to 0.3 nm near the long-wave boundary of the EUV spectrum. Therefore each emission line was covered by an average of four steps. Ground commands could order the satellite to either perform scanning or fix the slit in any of the 2040 positions for a longer time.

Compared with previous experiments, scattered light was significantly reduced. Absolute calibration had been done in the laboratory for 10 wavelengths; the intensity was measured with a tungsten photocathode of an assumed efficiency of 1 electron per absorbed quantum. HALL and HINTEREGGER[8] presented a spectrum constructed by averaging 6 scans recorded on 11 March 1967, 0930–1002 h TU (Fig. 4). No single scan deviated from the average by more than 3%. The solar radio flux at 10.7 cm lead to a Covington activity index of 142 (ground observed), which is rather high. The OSO-III mission lasted about 6 months. Unfortunately, one week after the beginning of the experiment a steady decrease in the sensitivity of the apparatus began;

---

[3] Hinteregger, H.E. (1961): J. Géophys. Res. *66*, 2367

[4] Hall, L.A., Damon, K.R., Hinteregger, H.E. (1963): Space Res. *3*, 745

[5] Hall, L.A., Schweizer, W., Hinteregger, H.E. (1965): J. Geophys. Res. *70*, 2241

[6] Hall, L.A., Hinteregger, H.E. (1969): Solar Phys. *6*, 175

[7] The circle was in fact constructed as a beryllium-copper belt forming a continuous loop with four slits in it. Each scanning is followed by another one at the next slit

[8] Solar Radiation in the EUV and its Variation with solar Rotation, COSPAR, Leningrad 1970

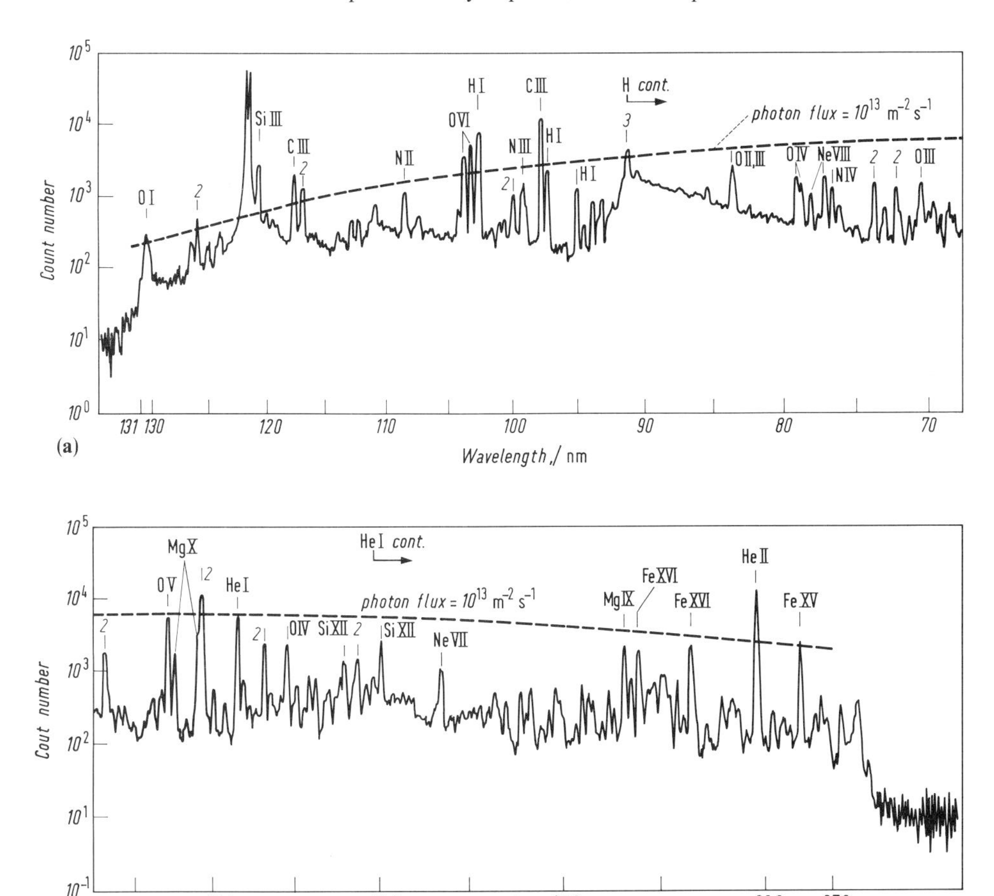

Fig. 4a, b. Solar spectrum in the range 27–131 nm observed aboard the satellite OSO-III (11 March 1967). Line widths are instrumental. HALL, HINTEREGGER [5]

after 2 months the sensitivity had decreased by a factor of 10. Fortunately, this did not interfere with a study of the 27-day variations of EUV fluxes on 7 lines. The absolute fluxes in individual lines and over wide spectral ranges (130–27 nm) are tabulated in the paper mentioned above. Table 2 contains data on absolute values of the flux in EUV lines emitted by ions of different stages of ionization 9 days after switching on of the experiment. During this time interval the solar radio emission flux at 10.7 cm had changed nearly by a factor of 2.

A special investigation of the spectrum of the quiet Sun was performed with the Harvard College Observatory spectrometer carried on OSO-IV. Results of this experiment are reported by [9].

[9] Dupree, A.K., Reeves, E.M. (1971): The EUV-spectrum of the quiet sun, I, 1971, TR-17; Astrophys. J.

Table 2. Solar flux in the 131–27 nm range.[a] OSO-III, 11 March 1967, 0945 UT (after HALL and HINTEREGGER [8])

Wave length λ [nm]	Identi-fication	Energy flux near Earth [$\mu W\,m^{-2}$]	Notes
1	2	3	4
130.60			
130.49, bl.	O I	15.0	
130.22	O I	8.2	
126.50	Si II	7.2	
126.07	Si II	3.6	
124.28	N V	4.6	
123.88	N V	6.9	
121.57	H I	5000.0	nonlinear range
120.65	Si III	59.0	
117.50	C III	36.0	
112.83	Si IV	4.7	Blended with C I
112.25	Si IV	3.8	
108.05	N II	10.8	
103.76	O VI	32.0	Blended with C II
103.19	O VI	45.0	
131.0–102.7	Un-resolved	67.0	
131.0–102.7	Integral	5303.8	
102.57	H I	67	
99.0	N III	12.1	
97.70	C III	90	
97.25	H I	16.4	
94.97	H I	7.1	
94.45	S VI	1.9	
93.78	H I	4.6	
93.34	S VI	2.8	
93.07	H I	2.8	
92.62	H I	2.8	
102.7–91.1	Un-resolved	25	
102.7–91.1	Integral	232.5	
91.1–89.0	H cont.	70	
90.4	C II	2.8	
89.0–86.0	H cont.	62	
86.0–83.0	H cont.	34	
83.5	O II, III	12.3	
83.0–80.0	H cont.	19	
91.1–80.0	Un-resolved	2.3	
91.1–80.0	Integral	202.4	
80.0–77.0	H cont.	9.8	
79.02			
79.01, bl.	O IV	6.5	
78.77	O IV	3.3	
78.65	S V	2.0	
78.03	Ne VIII	3.0	
77.04	Ne VIII	5.9	
77.0–74.0	N cont.	5.1	
76.51	H IV	4.7	
76.0	O V	1.9	
74.0–71.0	H cont.	2.4	
71.0–68.0	H cont.	1.2	
70.3	O III	6.5	
80.0–63.0	Un-resolved	11.2	
80.0–63.0	Integral	63.5	
62.97	O V	29.0	
62.53	Mg X	8.1	
60.98	Mg X	16.0	
59.96	O III	2.7	
58.43	He I	30.0	
554	O IV	11.0	
52.10	Si XII	7.4	
50.8	O III	3.0	
50.4	He I const.	20.0	b
49.93	Si XII	15.2	
46.52	Ne VII	6.7	
63.0–46.0	Un-resolved	16.7	
63.0–46.0	Integral	165.8	

[a] Single lines are listed to 0.01 nm (0.1 Å) and unresolved multiplets to the nearest Å (0.1 nm).
[b] Total photon flux $4.10\,m^{-2}s^{-1}$

Table 2. (continued)

Wave length λ [nm]	Identi-fication	Energy flux near Earth [μW m^{-2}]	Notes	Wave length λ [nm]	Identi-fication	Energy flux near Earth [μW m^{-2}]	Notes
1	2	3	4	1	2	3	4
41.7	S XIV (?)	4.8		37.0–27.0	Un-resolved	152.6	Many blended weak lines
46.0–37.0	Un-resolved	25.2		37.0–27.0	Whole range	684.7	
46.0–37.0	Integral	30.0		131.0–27.0	Whole range without Lyman-α	1680	
36.81	Mg IX	30.0					
36.5		9.1					
36.07	Fe XVI	20.0					
33.54	Fe XVI	43.0					
30.38	He II	353					
28.41	Fe XV	77.0					

A detailed description of the apparatus is given in [10]. The spectrometer used was of the normal-incidence type. The grating was ruled on a gold-coated surface and had 1800 lines per mm. The solar radiation was collected by a platinum-coated concave mirror having 0.5-m radius of curvature. The angular dimension of the entrance slit was $(1')^2$. The grating rotated step by step to provide scanning over the wavelength range 30–140 nm; each step was 0.01 nm lasting 0.08 s, total scanning time being 25 min. The output slit was 0.32 nm wide and resolution was somewhat less than 0.2 nm. A multiplier with a tungsten cathode was used as detector.

Absolute calibration uncertainty was apparently within a factor of 2. The uncertainty of the relative spectral sensitivity factor was given as about 20 % in the range 52–125 nm and 50 % beyong this region. Twenty-seven scans recorded on 26–27 October 1967 were selected for analysis. The spectrum presented in Fig. 5 was obtained by averaging all these scans. The list of 150 lines, with types of ions and transition for each line and line intensities (in mW m^{-2} sr^{-1}) and of photon flux from the solar surface (in m^{-2} s^{-1} sr^{-1}) is tabulated in the above-mentioned paper.[6] Table 3 reproduces the intensity distribution in the H continuum (near H Lyman α), and in the He I continuum. The detailed energy distribution in the He I recombination continuum is a subject of special interest.

Comparing their record of the solar spectrum with the spectrum of the entire solar disk, DUPREE and REEVES point out that the relative line intensities are different in both spectra. For example, the ratio of the lines H Lyman β to O VI is 2.5 in the center of the disk and 1.5 over the entire disk; this is due to different intensity distributions over the solar disk of both lines.

[10] Dower, R., Diamond, S., Hazen, N.: Harvard College Observatory OSO-61 Experiment Handbook, Tr-10, May, 1969

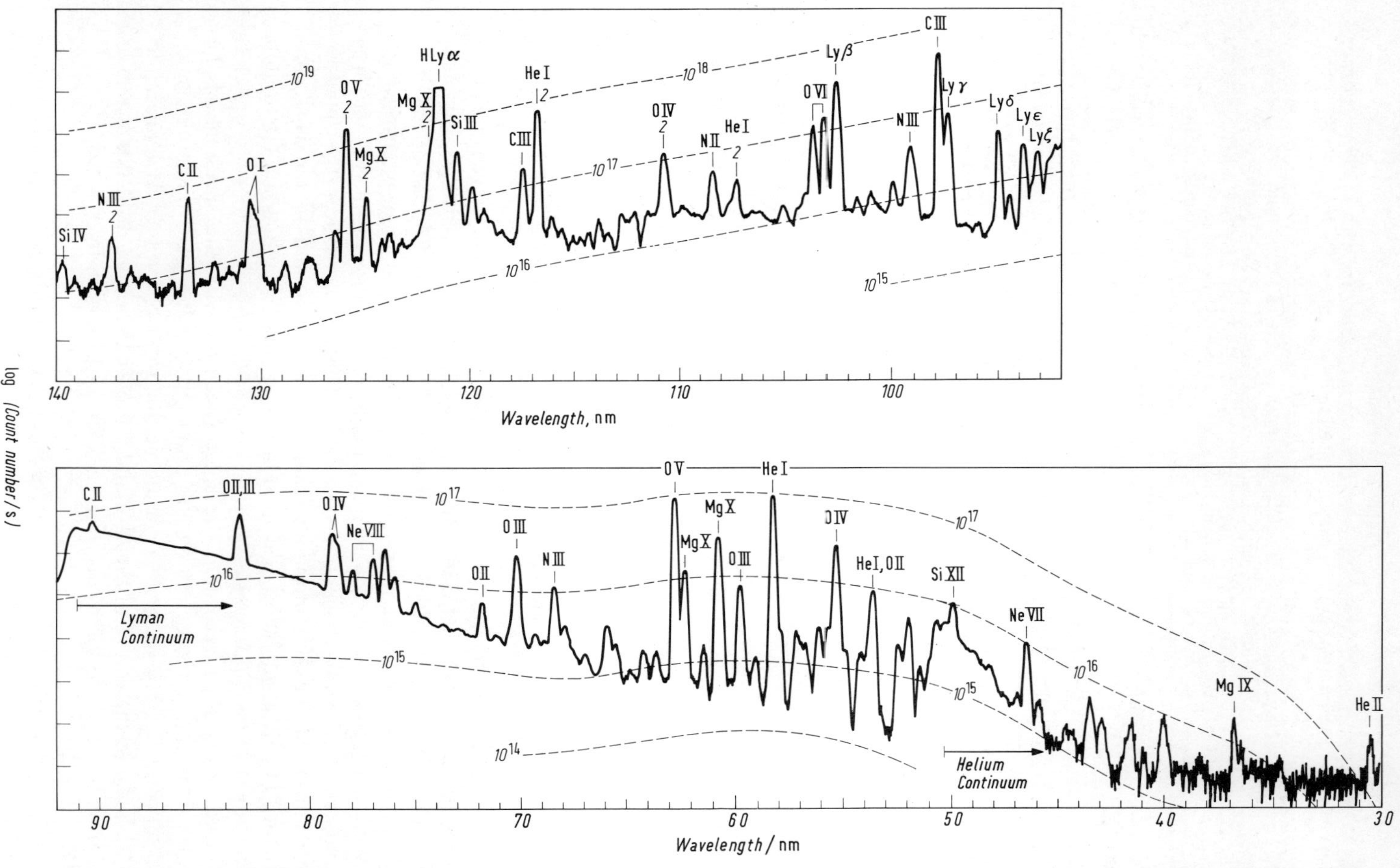

Fig. 5. Spectrum of the quiet Sun (near center of the disk) deduced from 27 spectral scans on 26–27 October 1967 by the Harvard spectrometer[6] aboard OSO-IV. Absolute calibration curves of specific intensity (photon flux in units $m^{-2}\,s^{-1}\,sr^{-1}\,nm^{-1}$) are shown as dotted lines

Table 3. OSO-IV quiet sun continuum intensities[9]

Wave-length λ [nm]	Flux density[a] photons [$m^{-2}s^{-1}nm^{-1}$]	energy [$mWm^{-2}nm^{-1}$]	Feature	Comments
125	6.47 E16	1.03 E2	L α wing	Red wing of L α;
124	8.47 E16	1.36 E2		? C continuum, recombination
123	1.11 E17	1.80 E2		to excited ^{1}D term (123.94 nm)
120.5	1.16 E17	2.65 E2		Blue wing of L α;
120	1.33 E17	2.19 E2		? S I continuum, recombination
118	6.44 E16	1.08 E2		to ^{3}P ground term (119.68 nm).
117	4.41 E16	7.47 E1		Recombination to ^{3}P ground term of C I (110.04 nm)
110	4.41 E16	1.03 E2	C I continuum	
108	5.68 E16	8.06 E1		
106	9.81 E16	5.81 E1		
104	3.09 E16	4.28 E1		
102	2.24 E16	3.13 E1		
100	1.61 E16	2.29 E1		
98	8.50 E15	1.72 E1		
96	6.31 E15	1.30 E1		
95.5	5.91 E15	1.23 E1		
91.2	2.13 E17	4.63 E2	Hydrogen Lyman continuum	Recombination to ground level of H I (91.17 nm)
90	1.65 E17	3.63 E2		
88	1.06 E17	2.39 E2		
86	7.00 E16	1.62 E2		
84	4.50 E16	1.07 E2		
82	3.02 F16	7.31 E1		
80	2.07 E16	5.16 E1		
78	1.43 E16	3.66 E1		
76	1.10 E16	2.86 E1		
74	8.38 E15	2.25 E1		
72	6.53 E15	1.80 E1		
70	4.91 E15	1.39 E1		
68	3.16 E15	9.92		
66	2.18 E15	6.56		
64	1.72 E15	5.34		
50.4	1.70 E16	6.72 E1	He I continuum	Recombination to ground level of He I (50.43 nm)
50	1.37 E16	5.44 E1		
49	8.75 E15	3.56 E1		
48	5.69 E15	2.35 E1		
47	3.63 E15	1.53 E1		
46.5	2.91 E15	1.24 E1		

[a] E a means factor of 10^a

5. The spectrum below 6 nm

α) In the short-wavelength spectrometer of FRIEDMAN's group (NRL) the grating was replaced by a potassium acid phthalate crystal, used as a Bragg

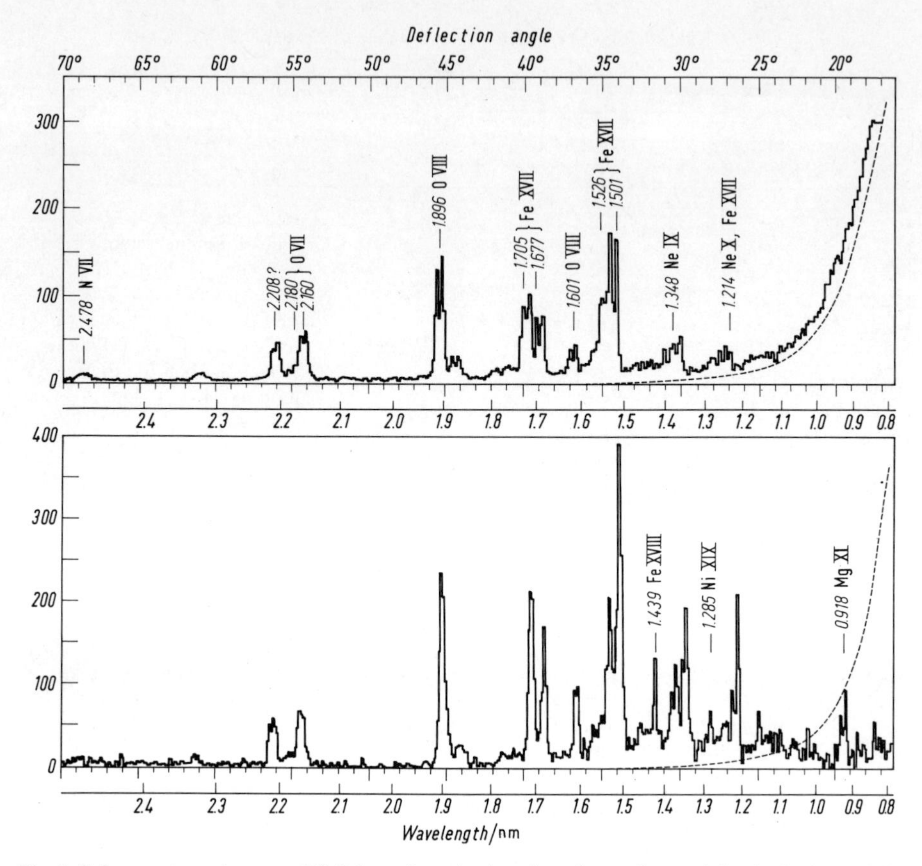

Fig. 6. Solar spectrum between 0.8–2.5 nm. Low (top) and moderate (bottom) level of solar activity. Observation with a crystal spectrometer aboard the OVI-IO satellite by RUGGE, WALKER (Aerospace Corp.)[7]

scattering crystal. The relation between scattering angle θ and wavelength λ is

$$n\lambda = 2d \sin\theta,$$

where $n = 1, 2, 3, \ldots$; $2d = 2.663$ nm. In solar EUV spectroscopy various crystals are used so that $2d$ can be chosen in the range between 0.4 and 6.3 nm.

The NRL spectrometer was carried by an Aerobee rocket on 25 July 1963. Fourteen lines were discovered within 1.37–2.48 nm. The uncertainty of wavelength measurement was 0.01 nm. The contribution of these 14 lines to the total radiation flux was estimated as being 80–90% of the total flux in that range. The remainder of the flux was caused by a continuum, which may, however, consist of weaker overlapping lines. A list of the recorded lines and line intensities is given in [1]. Some of the lines were identified as emissions of O VII, O VIII, Fe XVII, and N VII (by comparison with spectra obtained from a theta-pinch machine according to EDLEN and TYREN).

[1] Blake, R.L., Chubb, T.A., Friedman, H., Unzicker, A.E. (1965): Astrophys. J. *142*, 1

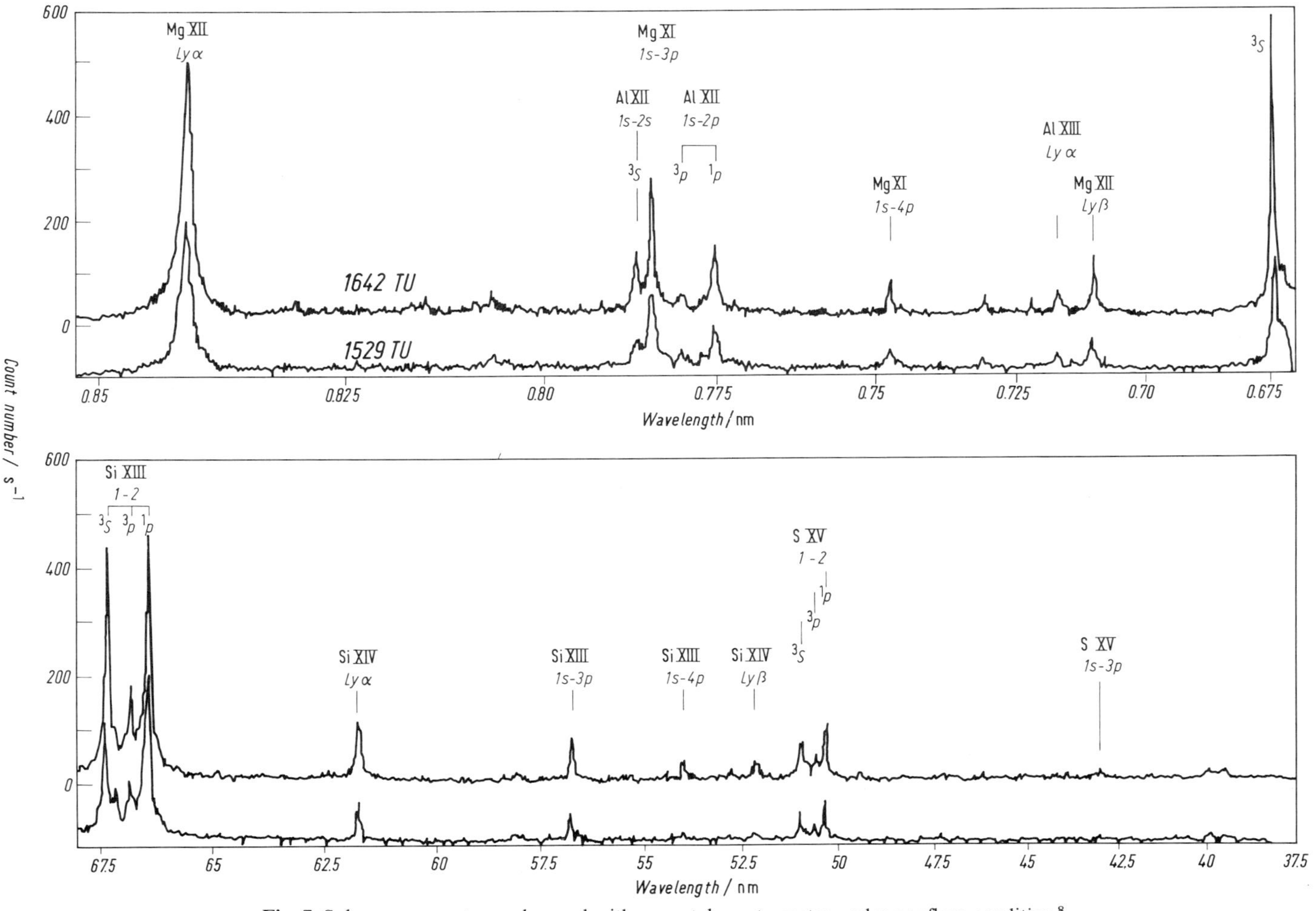

Fig. 7. Solar x-ray spectrum observed with a crystal spectrometer under nonflare condition[8]

β) Extension toward *shorter wavelengths*, however, was achieved with gratings. After TOUSEY's group had succeeded in recording the solar spectrum down to 3 nm, the next advance, down to 1 nm, was made by the groups directed by MANDEL'ŠTAM[2] and by WILSON[3,4]. In the autumn of 1965 MANDEL'ŠTAM and his colleagues with a spectrograph carried to 500 km obtained spectrograms extending down to a line at 0.95 nm, identified as Mg XI emission.

The grating used was ruled on a gold-coated surface with 600 lines per mm and 1-m radius of curvature; the grating had a blaze so that the light intensity was best near 6 nm. The slit was wedge shaped with maximal width 0.05 mm. Between 0.95 and 3.4 nm, 14 single and blended lines were recorded. Of these 14 lines, apart from the blend 1.09–1.19 nm (observed for the first time and identified with Ne IX, Na X, Ne X and the 2.85 nm line associated with C VI emission), 10 had been recorded earlier by FRIEDMAN's group.[5] The other 2 lines at 3.0 and 3.18 nm are perhaps lines of the second order, i.e., emissions at 1.49 and 1.6 nm.

γ) Many *photoelectric measurements* of the solar spectrum within 0.8–2.5 nm have been carried out. An NRL experiment aboard the OSO VI-10 satellite during 1966–1967 recorded more than 50 lines within this range.[6,7] Figure 6 presents the spectrum variation as a function of solar activity for nonflare conditions; in this latter case if there are sufficient active centers on the disk lines can be observed down to 0.43 nm. Figure 7 presents the spectrum obtained by British scientists.[8] Data on the EUV spectrum below 2.5 nm are summarized by POUNDS in a good review.[9]

6. EUV spectrum at the solar limb

α) Solar EUV spectra out of the solar disk were obtained for the first time by British investigators on 9 April 1965. BURTON et al.[1] using a Rowland-type spectrograph photographed the chromospheric spectrum in the range 97.7–280.3 nm.

The slit was directed tangentially to a solar altitude 7000 km above the limb. The uncertainty of the pointing was $\pm 2''$ corresponding to ± 1500 km. More than 90% of the 300 observed lines were identified. Unfortunately, no spectrum of the solar disk was simultaneously taken. Therefore the comparison of absolute line intensities at the limb was made using another limb spectrum obtained on 19 April 1960 by the NRL group.[2]

The limb-to-disk intensity ratio varies within a range of 0.01 to 0.1 and is enhanced according to the ionization potentials of the emitting ions. The line intensity distribution over the disk and at

---

[2] Žitnik, I.A., Krutov, V.V., Maljavkin, L.P., Mandel'štam, S.L., Čeremuhin, G.S. (1967): Kosm. Issled. 5, 276; Space Res. (1967): 7, 1263

[3] Burton, W.M., Wilson, R. (1965): Nature *207*, 61

[4] Jones, B.B., Freeman, F.F., Wilson, R. (1968): Nature *219*, 252

[5] Friedman, H. (1965): Astrophys. J. *142*, 1

[6] Fritz, G., Kreplin, R.W., Meerins, J.F., Unzicker, A.E., Friedman, H. (1967): Astrophys. J. *148*, 1133

[7] Rugge, H., Walker jr., A.B.C. (1968): Space Res. *8*, 439

[8] Walker, A.B.C., Rugge, H.R. (1969): Enhancement of the solar X-ray spectrum below 25 A during solar flares, Amsterdam: North-Holl. Co.

[9] Pounds, K.A. (1970): Ann. Géophys. *26*, 555

[1] Burton, W.M., Ridgeley, A., Wilson, R. (1967): Mon. Not. R. Astron. Soc. *135*, 207

[2] Detwiller, C.R., Garrett, D.L., Purcell, J.D., Tousey, R. (1961): Ann. Géophys. *17*, 263

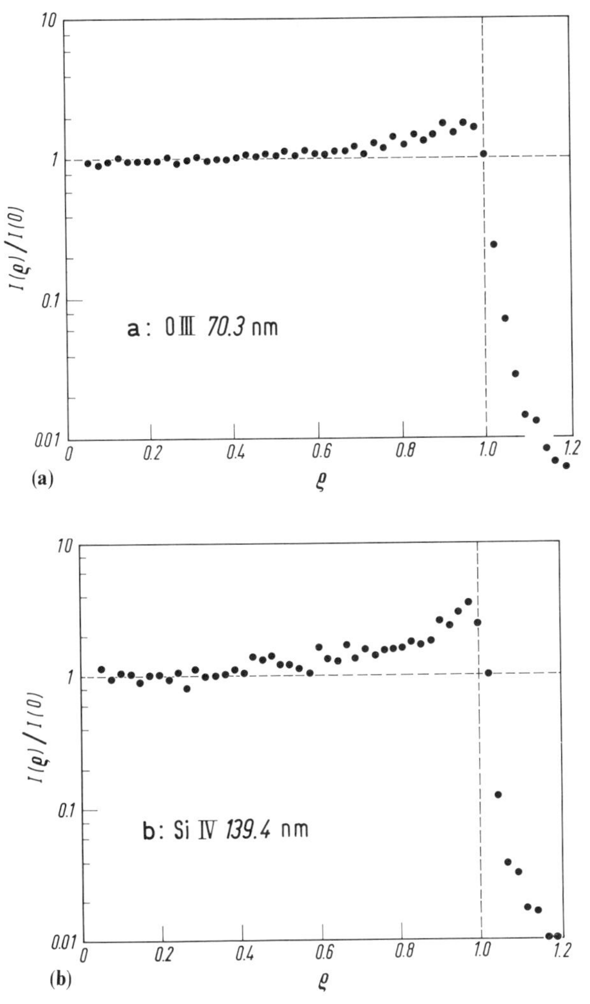

Fig. 8a–d. Center-to-limb intensity variation for the resonance lines of four different ions (OIII, SiIV, OVI, SiXII). Points represents data obtained by averaging over between 40 and 50 good quality spectroheliograms at solar latitudes between $+10°$ and $-10°$ only (WITHBROE[3]). The observations were obtained with the Harvard spectrometer-spectroheliometer aboard OSO-VI[4] (November 1967) (Fig. 8c, d see page 230)

the limb was studied by WITHBROE.[3] Twelve lines emitted by ions at various ionization stages were considered. The records obtained during the OSO-IV mission (Oct.–Nov. 1967) with the Harvard Observatory's spectrometer-spectroheliograph were used as input.[4] Spatial resolution was 1′. (For a more detailed description of these data, see the following section.)

Figure 8 shows the distribution of the relative line intensities over the solar disk. Each of the curves is the result of averaging several records. An analysis of these data is made below.

[3] Withbroe, G.L. (1970): Sol. Phys. *11*, 42, 208

[4] Goldberg, L. (1968): Science *162*, 95

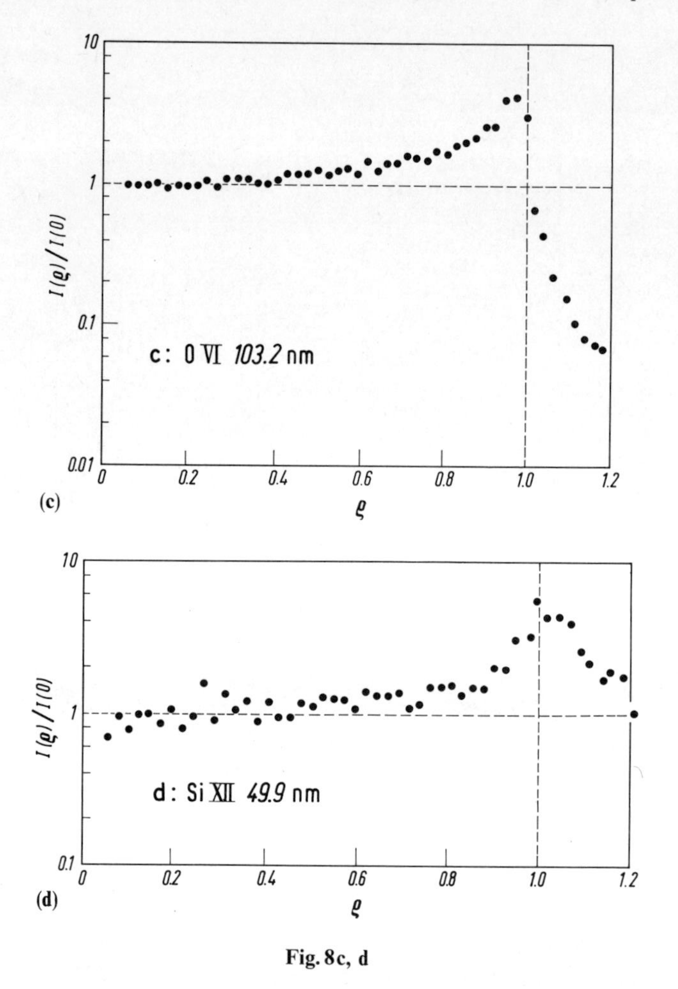

Fig. 8c, d

A series of solar limb spectra within 30–280 nm was obtained aboard British Skylark rockets in 1965–1969. The Sun's image was stabilized on a tangential slit with an accuracy of 5″.

The apparatus is described and the 1965 results reported in [5]. The limb spectrum differs from the disk spectrum in showing higher intensities of the intercombination lines. Neutral metal lines are absent. The lines of the limb spectrum are emitted in a highly ionized, hot plasma by ions with ionization potentials above *8* eV.

The authors list more than 600 lines, 507 of which have been identified (255 of these are emitted by Fe II ions). Wavelengths and estimated line intensities are also given. The uncertainty of the wavelength measurement is ±0.005 nm in the first order of the grating. The inaccuracy of the relative intensity determination is claimed to lie within a factor of 2 over the range of about 10 nm and a factor of 3 over the whole recorded spectrum of 75–250 nm. The absolute calibration was only approximately known.

---

[5] Burton, W.M., Ridgeley, A., Wilson, E. (1967): Mon. Not. R. Astron. Soc. *135*, 207

β) Tousey's group at NRL made observations of the *flash spectrum* during the eclipse of 7 March 1970, applying the well-known "slitless" method, with the image of the chromosphere used as a slit.

The focal length was 110 cm, ruling 1200 lines per mm. The guiding uncertainty was not greater than ±2.5″. Unfortunately, only one film covering 140–200 nm could be safely recorded.

The following results were obtained. Near a wavelength of 155.5±1 nm are many Si I lines originating in the lower chromosphere. Near 161±5 nm C I lines appear. The He II line at 164 nm shows intensity enhancement toward the limb, as does the Al II line at 167 nm. There are Si II lines seen above the 7″ layer in the chromosphere at 152.67, 153.34, 180.80, 181.69, and 181.74 nm; the brightest lines within this spectral region are 190.87 nm C III, 166.61 nm O III, 185.47 nm and 186.28 nm Al III, 189.20 nm Si III.

However, some weaker lines of He II extended higher up. The lines 151.62 and 155.08 nm of C IV and 140.11, 140.48, 139.38, and 140.28 nm of 0 IV reach 17″ altitude above the limb.

Three weak unidentified lines, 142.5±0.2, 144.8±0.2, and 146.7±1 nm, seen in the spectrum originate in the corona, since they cover about 180° of the position angle.

7. Linewidth in the EUV range

α) The procedure for EUV linewidth determination is rather complicated because the linewidths are generally small while the spectral resolution of the apparatus is low, usually not better than 0.01 nm. For instance, the expected width of the He II line 30.4 nm is approximately 0.004 nm and never more than 0.007 nm.[1] On the other hand the widths of the lines emitted by various ions would provide important information on the physical properties of the upper solar atmosphere responsible for the EUV radiation provided they can be measured. Hitherto, only the profiles of the *most prominent lines*, H Lyman α and H Lyman β, which are of the order of 0.1 nm and thus broad enough for measurement, have been studied. In one of the first photographs of H Lyman α taken by Rense and his colleagues[2], a very faint reversal has been discovered despite the low resolution in the core of the line profile.[3] The existence of a narrow reversal gap in the center of the H Lyman α line, caused by interplanetary hydrogen, was predicted theoretically in 1958.[4]

On 21 July 1959, using the NRL double-dispersion grating spectrograph operating in the 13th order, Purcell and Tousey obtained a spectrogram in the vicinity of H Lyman α with a resolution of about 0.003 nm; another series of similar spectra was taken on 19 April 1960.[5]

The microphotometer tracing of the line profile is shown in Fig. 9. The most prominent feature of this profile is the narrow absorption in the line core,

[1] Nikol'skij, G.M. (1962): Dokl. Akad. Nauk SSSR *147*, 809; Geomagn. Aeron. (1962): *2*, 1025
[2] Rense, W.A. (1953): Phys. Rev. *91*, 229
[3] Report at the IAU Assembly, Moskva 1958
[4] Nikol'skij, G.M. (1958): Astron. Zh. *35*, 657
[5] Purcell, J.D., Tousey, R. (1960): Space Res. *1*, 115; J. Geophys. Res. *65*, 370 (1960); Mém. Soc. R. Sci. Liège *4*, 283 (1961)

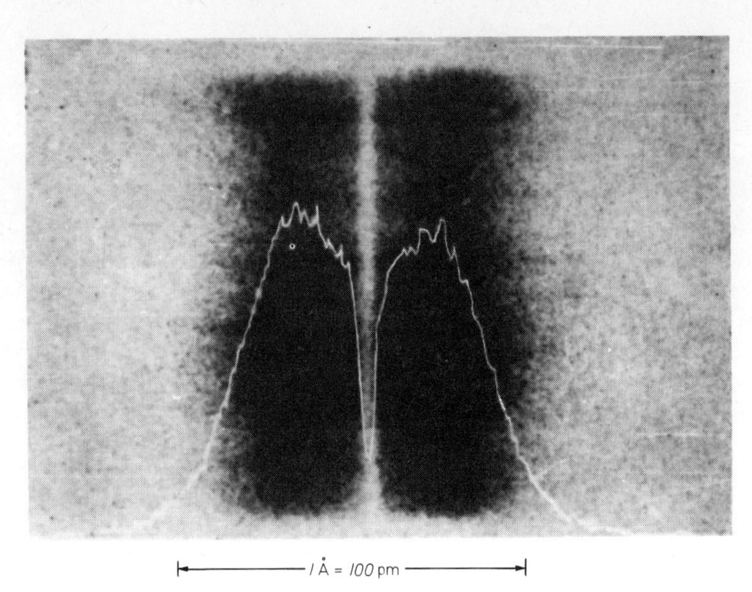

Fig. 9. Spectrum and profile of H Lyman α (21 July 1959). Spectral resolution about 3 pm (NRL, PURCELL, TOUSEY[5])

the width of which is between 0.0025 and 0.004 nm, according to PURCELL and TOUSEY. This value points to a kinetic temperature of about 800–2000 K for the absorption in question and takes place in the upper atmosphere that is called "the Earth's corona". The temperature of the interplanetary gas is not lower than 10,000 K. These theoretical calculations[4] were made on the basis of an overestimated value of the interplanetary hydrogen density and the "terrestrial corona" was left out of account.

For the H Lyman α line this narrow central absorption is superimposed on a shallow, broad minimum between two peaks. The separation of these peaks is about 0.04 nm. This wide, shallow depression is due to reversal in the solar atmosphere and is caused by the large optical depth of the chromosphere.

On 22 August 1962 the same group obtained solar spectrograms including H Lyman α and Lyman β lines with a resolution of only 0.007 nm so that the narrow terrestrial absorption was undetectable.[6] Figure 10 presents the H Lyman α profiles according to the 1959 and 1962 data for the active and quiet regions of the solar disk. The broad central reversal shows variations from point to point over the disk, as well as with time. The Lyman α profile is similar to that of Lyman β, but the central depression of the latter is somewhat shallower and the peak separation is less by a factor of 15, due to its 5-times-lower oscillator strength and shorter wavelength compared with Lyman α. The recorded lines and corresponding profiles are shown on Fig. 11.

---

[6] Tousey, R., Purcell, J.D., Austin, W.E., Garrett, D.L., Widing, K.G. (1964): Space Res. *4*, 703

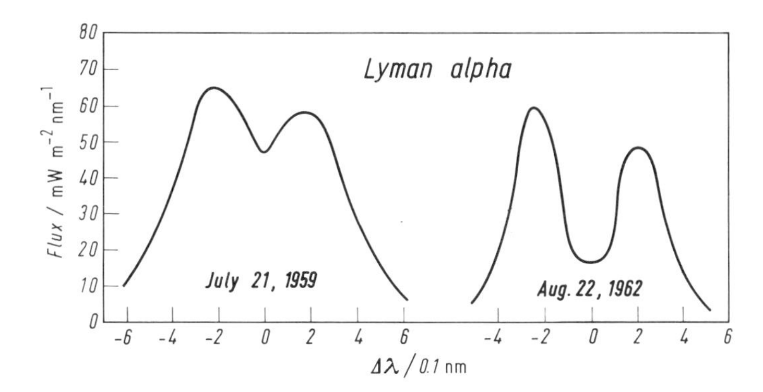

Fig. 10. Change in character of central verversal of Lyman α with solar activity. Active Sun – 1959, quiet Sun – 1962. Spectral resolution about 7 pm (NRL: TOUSEY et al.[6])

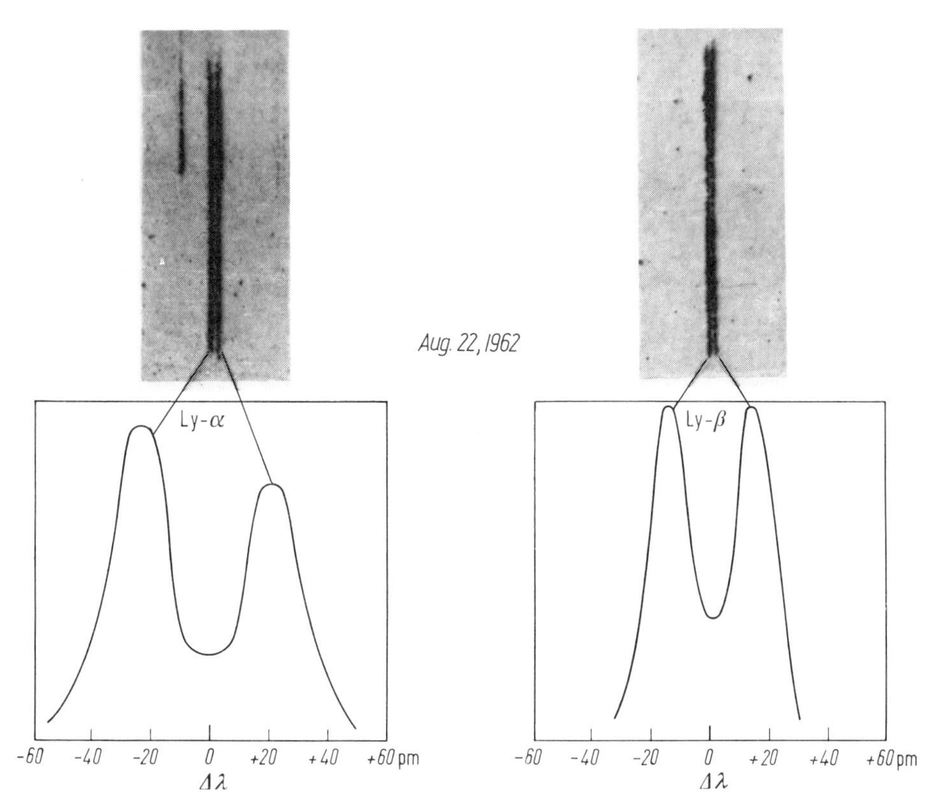

Fig. 11. Profiles of Lyman α and Lyman β (22 August 1962). Spectral resolution about 7 pm (NRL: TOUSEY et al.[6])

β) An *explanation* for the whole central reversal may be given qualitatively in terms of incoherent scattering in the optically dense medium[7-9]. After repeated absorption and reemission, a H Lyman α photon can move out of the emitting region or finally loose its energy by transformation into heat (true absorption). Each absorption may occur at a different frequency but close to the true resonance frequency. Such frequency deviations from the resonance result from a Doppler effect caused by the thermal velocities of atoms, since the absorption coefficient is greatest at the resonance frequency and decreases outward from the line core, the rate of photon escape is enhanced. Conversely, the rate of the photon emission falls further from the line center. It is the combination of these two processes that is responsible for the double-peak profile.

The observed H Lyman α profiles have been discussed by Morton and Widing[10] in terms of the Jefferies-Thomas theory of line formation by noncoherent scattering in the chromosphere with deviation from local thermodynamic equilibrium. Comparing the observed and calculated profiles, they derived a temperature between 5500 and 115,000 K. This value should be too high, keeping in mind that the greater part of the Lyman α flux arises within regions where the temperature is about 7000 K (black-body temperature). Some critical remarks to the derivation were made by Obridko[11], who estimated the upper temperature limit within the H Lyman α emission region to be between 10,000 and 20,000 K.

γ) Rense's group[12] obtained high-resolution spectrograms with the *oxygen* OI triplet lines in the range 130.2–130.6 nm. All these line profiles have a flat-top shape showing reversal.

An Aerobee rocket was used (launchings on 19 November 1967 and 21 November 1968). The instrumental broadening was 0.0035 and 0.002 nm in both flights in both cases the effects of film granulation were considerable. The OI profiles were constructed after averaging a few traces. The line profiles have broadened tops with an absorption core. From an analysis of the profiles the authors concluded that the absorption core, which is about 0.002 nm narrow and quite deep (almost 100%), was caused by telluric oxygen, while the broad shallow depression of about 10% was of solar origin. The half-width of the OI lines (after correcting for self-absorption) is about 0.002 nm. The latter value gives a thermal velocity of about 25 km/s at a temperature within the emitting region of about 10,000 K.

The high-resolution echellete-spectrograph flown on an Aerobee rocket[13] on 21 November 1968 recorded the resonance lines of OII at 133.452 and 133.569 nm (transition $2^2P^0-2^2D$). The brightness of these lines in active regions is 3 times higher than in undisturbed ones. The line profiles have a nearly Gaussian shape with half-widths of about 0.022 nm (corresponding to $6 \cdot 10^5$ K). Since the ionization temperature is about 20,000 K, the nonthermal velocity must be supersonic (30 km/s), apparently due to macroturbulent motions.

δ) Spectra with *very high resolution* were more recently obtained on board the manned vehicle Skylab with a spectrograph in the range 97–400 nm. The results of a profile study of the CIII resonance line (97.7 nm) were reported by[14].

---

[7] Ivanov, V.V. (1966): Theory of stellar spectra, Sci. 390

[8] Avrett, E.H., Hummer, D.G. (1965): Mon. Not. R. Astron. Soc. *130*, 295

[9] Jefferies, J.T., Thomas, R.N. (1959): Astrophys. J. *129*, 401; *131*, 695

[10] Morton, D.C., Widing, K.G. (1961): Astrophys. J. *133*, 596

[11] Obridko, V.N. (1963): Astron. Žh. *40*, 466

[12] Bruner jr., E.C., Jones, R.A., Rense, W.A., Thomas, G.F. (1970): Astrophys. J. *162*, 281

[13] Berger, R.A., Bruner jr., E.C., Stevens, R.J. (1970): Solar Phys. *12*, 370

[14] Moe, O.K., Nicolas, K.R., Bartoe, J.-D. (1976): Skylab/ATM preprint, June

The spectrograph slit determined a field of view on the Sun of 2″ by 60″. The exposure time was 44 min for a quiet Sun but only 10 min for active regions. After correction for the instrumental profile, the Doppler width ($\Delta\lambda_D$) was found at $\lambda = 97.7$ nm to be 0.007 nm in nondisturbed regions. In the line wings ($\Delta\lambda > 1.2\Delta\lambda_D$) the intensity is higher than in the Doppler profile.

In active regions the emission line is much wider, its profile deviating significantly from the regular Doppler shape. There is a central reversal of 10% depth, the long-wavelength peak being higher than the other one. In other cases no reversal is seen: the line top is flat with depression on the short wavelength side. The 50% width is about 0.03 nm.

The ionization temperature T_i of CIII is 50,000 K. The columnar density of C III ions in the solar atmosphere may be estimated from the radiation flux ($F \approx 0{,}05$ mW m^{-2}) or the emission measure of the regions with the above temperature (see Sect. 3). Our calculations give about 10^{17} m^{-2} as columnar density which gives an optical thickness (τ_c) of about 1. However, with pure scattering for a resonance line the self-absorption becomes significant only where $\tau \gtrsim 10^4$.

Thus, since thermal broadening at $5 \cdot 10^4$ K can account for only 20% of the observed line width of C III in quiet regions, this width should be caused by nonthermal or "turbulent", almost isotropic motion with velocities of about 20 km s^{-1}. The line deformation in the active regions may then be due to stronger, sometimes anisotropic motions.

II. XUV solar images

8. Summary of observations. Considerable information on EUV radiation properties can be obtained from an analysis of the distribution of the EUV line brightness over the solar disk. EUV solar images can be obtained by different methods: 1) scanning the spectrometer entrance slit over the disk, 2) using a pinhole camera, 3) with a grazing-incidence reflector, and 4) with a slitless grating spectrograph.

α) The first solar disk images in the soft x-ray region were obtained by the NRL group.[1] Images in some EUV emission lines such as He II 30.4 nm and He I 58.4 nm were later photographed by American, Russian, and British investigators.

As is well-known, x-ray experiments on wavelengths below 10 nm require special optics. The first useful x-ray solar disk images were obtained using a *pinhole camera.*

On April 1960, a battery of 8 pinhole cameras was flown on an Aerobee-Hi rocket. The pinholes were 0.13 mm in diameter and were placed 160 mm from the film, giving a 1.5-mm solar image at a resolution of about 3′. The cameras were kept pointed at the sun within 1′, but the pointing system did not compensate for the rotation of the cameras round the Sun-camera axis, which caused discrete features to be drawn into arcs in the picture extending to 160°. Nevertheless, after a "defolding" procedure BLACKE et al.[2] found a correlation between x-ray structure and H

[1] Chubb, T.A., Friedman, H., Kreplin, W.R., Black, R.L., Unzicker, A.E. (1961): Mém. Soc. R. Sci. Liège *4*, 235

[2] Blacke, R.L., Chubb, T.A., Friedman, H., Unzicker, A.E. (1963): Astrophys. J. *137*, 7

Table 4. XUV heliograms

No.	Day	Wavelength λ [nm]	Contrast	Spatial resolution	Methods and remarks	References
	1	2	3	4	5	6
1	8 v 56	L$\alpha \sim 121.6$	>1	2′	Slitless spectrograph with 15° prism. Qualitative identification of active regions with Ca^+ plages ($d_{\odot} = 2.8$ mm)	Mercure, R., Miller, Jr., S.C., Rense, W.A., Stuart, F. (1956): J. Geophys. Res. *61*, 573
2	13 iii 59	L$\alpha \sim 121.6$	≈ 5	0.5′	Two concave gratings. Coincidence with Ca^+ plage. Lα plages are greater. Prominences at the limb.	Purcell, J.D., Packer, D.M., Tousey, R. (1959): Nature *184*, 8
3	19 iv 60	2–6	≈ 55	3′	Pinhole camera ($d = 1.6$ mm). Images are smoothed by rotation over 160° angle	Chubb, T.A., Friedman, H., Kreplin, W.R., Blake, R.L., Unzicker, A.E. (1961): Mém. Soc. Roy. Sci. (Liège) *4*, 235
	21 vi 61	4–9	high	2′		
		2.3–4	higher	3′		
		1.2–2.5	more high	6′		
4	4 iv 63	0.8–15	(≈ 200) Contrast between bright and weak parts of active region	1.3′	Slit scanning across solar disk. Our estimation from reference data	Blake, R.L., Chubb, T.A., Friedman, H., Unzicker, A.E. (1965): Astrophys. J. *142*, 1
	25 vii 63	4.4–6	(35)	0.5′		
		0.8–2	(90)			
		4.4–6	(30)	0.5′		
5	10 v 63	Lα 121.6			Spectroheliogram of solar disk part of 7′. The limb brightening is greater in 550.3-nm coronal emission regions. Maximum contrast appears on O VI lines.	Tousey, R., Austion, W.E., Purcell, J.D., Widing, K.G. (1965): Ann. Astrophys. *28*, 755; Tousey, R. (1964): Quart. Bull. *5*, 123
		Lβ 102.5				
		O VI 103.2	low	$\approx 1'$		
		103.8				
		C III 97.7				
6	10 v 63	Fe XV 28.4	high		Slitless spectrograph with concave grating	Purcell, J.D., Garrett, D.L., Tousey, R. (1964): Astrophys. J. *69*, 147
		He II 30.4	middle			
	28 vi 63	Fe XVI 33.5 }	high	1′		
		36.1 }				
		Mg IX 36.8 }				
		Me I 58.4				
		O V 63				

7	6 vi 63	17–40 1–11 1– 9		high	3′	Pinhole camera, oscillation of image with 4 solar radii	Zitnik, I.A., Krutov, V.V., Maljavkin, L.P., Mandelstam, S.L. (1964): Kosmic issl. *2*, 920
8	25 vii 63	4.4	7	high	2′	Pinhole camera, used 19 iv 60 and 21 vi 61	Blake, R.L., Chubb, T.U., Friedman, H., Unzicker, A.E. (1965): J. Astrophys. *142*, 1
9	20 ix 63	Fe XV	28.4	high	1′	Slitless spectrograph with concave gravity. Limb brightening, especially within 550 nm coronal emission regions	Tousey, R., Austin, W.E., Purcell, J.D., Widing, K.G. (1965): Ann. Astrophys. *28*, 755; Purcell, J.D., Garrett, D.L., Tousey, R. (1964): J. Astrophys. *69*, 147
10	11 viii 64	2.5 nm 4.4–7		high		Pinhole camera. Limb brightening. Coincidence with Ca^+ plage	Russell, P.C. (1965): Nature *205*, 684
11	17 xii 64	6 nm		high		Pinhole camera. Limb brightening. Coincidence with Ca^+ plage	Black, W.S., Booker, D., Burton, W.M., Jones, B.B., Shenton, D.B., Wilson, R. (1965): Nature *206*, 654
12	16 ii 65 4 iii 65	L He II	121.6 30.4	2–5 2–3	1′ 1′	Spectroheliograph with telemetry at OSO-II. Quiet Sun	Tousey, R. (1967): Astrophys. J. *149*, 239
13	17 ii 65	0.8 1.2 4.4		very high	1′	Parabolic refractor with grazing incidence. Coronal emission is softer than that associated with plages on the solar disk	Giaconni, R., Reidy, W.P., Zehnpfennig, T., Lindsay, J.C., Muney, W.S. (1965): Astrophys. J *142*, 1274
14	9 iv 65	6–15 nm Fe IX He II	 17.1 30.4	very high (6–15 nm, Fe IX) low	3′ and 10′	Pinhole camera with flat grating within 6–15 nm great limb brightening occurs	Burton, W.W., Wilson, R. (1965): Nature *207*, 61
15	12 iv 65	Mg II	280.3 ±0.2	weak	1′	Cassegrain-Maksutov system with BF SOLZ filter. Slow contrast feature in absorption are visible	Fredga, K. (1966): Astrophys. J. *71*, 399
16	23 vi 65	0.8–2.4 2 0.8–1.8 1.2		very high	1.5′	Pinhole camera. Fluxes are measured.	Broadfoot, A.L. (1967): Astrophys. J. *149*, 675
17	20 ix 65	1 2 2.5 17–20		very high high 20	 1′ and 5′	Pinhole cameras. Oscillation 1.5 solar radius. Our contrast estimation from reference data.	Zitnik, I.A., Krutov, V.V., Maljavkin, L.P., Mandelstam, S.L., Ceremuhin, G.S. (1967): Kosmic. Iss. *5*, 276

Table 4. XUV heliograms (continued)

No.	Day	Wavelength λ [nm]	Contrast	Spatial resolution	Methods and remarks	References
	1	2	3	4	5	6
18	20 x 65	L 121.6	2	2″	Cassegrain telescope with ionization chamber. Resolution 2″	Sloan, W.A. (1966): Astrophys. J. *71*, 399
19	20 x 65	2.4 4 4.4–4.8 3.2	very high	2′ 4′	Pinhole cameras. Resolution 1′. Good quality images.	Russell, P.C., Pounds, K.A. (1966): Nature *209*, 490
20	9 iv 65 17 xii 64 20 x 65 2 ii 66	5 } 18 } He II 30.4	high moderate	2′	Pinhole camera with and without grating. Limb brightening. Darkening at poles. Coincidence with Ca^+ plages.	Burton, W.M. (1969): Sol. Phys. *8*, 53
21	1 ii 66	He II 30.4 Mg IX 36.8 } Fe XV 28.4 } Fe XVI 33.5 } 36.1 }	middle high		Slitless spectrograph. Quiet Sun. Mg IX limb brightening. Fe lines are underexposed: plages are visible only.	Purcell, J.D., Snider, R., Tousey, R. (1967): J. Astrophys. *149*, 239
22	6 iv 66	1.4 2.4 4.5	(30–50)	2′	Pinhole cameras. Our contrast estimate from reference data	Cuchois, Y., Senemand, G., Bonnelle, Ch., Montel, M., Senemand, Ch. (1966): C.R. Acad. Sci. (Paris) *263*, 1082
23	28 iv 66	105–135 (Lα)	≈0.1 dark features	2.5″	Cassegrain telescope. Scanning with gas chamber	Sloan, W.A. (1968): Sol. Phys. *5*, 329
24	28 iv 66	15–70 Mg IX 36.8 Ne VII 46.5 Si XII 49.9 52.2 O IV 55.5	high	10″	Slitless spectrograph. Much overlapping images of different brightness. Active Sun. All images, except He are rings	Purcell, J.D., Snider, R., Tousey, R. (1967): Astrophys. J. *149*, 239

		He I He II O V Fe IX -XVI	58.4 30.4 63 15- 36				
25	20 v 66	0.8 0.8–1.2 1.2–1.6 1.6–2		very high	20″	Two elements grazing incidence reflector. ($\lambda=5.8$ mm). Very good images. The emitted region extends beyond limb	Underwood, J.H., Muney, W.S. (1967): Sol. Phys. *1*, 129
26	4 x 66	5.1 3.4		very high	a few min. of arc.	Frensel zone plate. Limb brightening, darkening at the poles	Elwert, G. (1968): Symp. on Struc. and Develop. of Solar Act. Reg., p. 439
27	8 vi 68	0.35–1 0.35–1.4 4.4 –6		very high	2″	Slitless spectra. $F=132$ cm. Flare. The loops are visible. Dynamical features	Vaiana, G.S., Reidy, W.P., Zehnpfennig, T., Spreybroeck, K., van, Giacconi, R. (1968): Science *161*, 564
28	7 iii 70	17–50				Coronograph LYOT with occulting disk. The eclipse day. Emission extends to 3–9 solar radii	Koomen, M.J., Purcell, J.D., Tousey, R.: (1970): Nature *226*, 1135

and K Ca II plages and cm-radiowave emission. The contrast between the brightness of active and undisturbed regions lies between 37 and 108 with an average value of about 55, which agrees with the theoretical contrast of the order of 100 predicted earlier by IVANOV-HOLODNYJ and NIKOL'SKIJ[3]. The total solar flux within the wavelength range below 6 nm was 0.3 mW m^{-2} at the time of the experiment (19 April 1960).

Data on the distribution of EUV brightness over the solar disk from solar disk images obtained by some of the methods mentioned are listed in Table 4. Unfortunately, we have insufficient information on tne contrast between the active regions and the undisturbed background. The calibration procedure of such photographs is not complicated, and can be performed using combinations of different pinholes to form a series of small solar images. In any case, the contrast may be crudely estimated by indirect methods, utilizing, for instance, data from other experiments; in Table 4 the data for such cases are given in brackets.

β) The most advanced technique for observing XUV images seems to be the *slitless spectrograph* or the *reflector* operating in *grazing incidence*. Though a pinhole is the simplest equipment, its sensitivity is extremely low because of its small aperture. Grazing-incidence telescopes have a considerable advantage: their higher sensitivity enables solar images to be photographed at wavelengths near 1 nm with an exposure time of about 1 s. At a grazing incidence of 54′ and a wavelength of 0.5 nm, the use of stainless steel mirrors (with an effective aperture of 1.6 cm^2) enhanced the reflectivity[4] by up to 70%. A picture taken on wavelengths below 4.4 nm is shown on Fig. 12a. Several bright regions can be seen. For comparison Fig. 12b shows a heliogram on H Lyman alpha taken from a ground observatory at the same time. Interesting results were also obtained from corrected photographs of the solar disk image (Fig. 12a, b). Results derived from pictures taken with the reflector[5] on 13 March 1965 are shown in Table 4.

Isodense lines over the solar disk were constructed[6] for wavelengths of 0.8 and 4.4 nm. For the longer wavelength a detailed system of isodense lines could be established that showed the coronal x-ray emission extending up to $0.2 R_{\odot}$ above the solar limb ($R_{\odot}$ is the radius of the Sun). Using the calibration curve given for the shorter wavelength range at 0.8 nm, we found a contrast of about 30 between the brightest part of the active region, about 40″ in diameter, and the near background.

The same group obtained for the first time some x-ray photographs with a resolution as high as 1″. A flare that occurred near the solar limb on 8 June 1968 could be located with this accuracy.[7] In the wavelength range 0.35–1.4 nm the flare structure is especially well-seen and is similar to the H-alpha line structure. The most prominent x-ray feature of this flare seems to be

---

[3] Ivanov-Holodnyj, G.S., Nikol'skij, G.M. (1961): Astron. Zh. *38*, 45

[4] Underwood, J.H., Muney, W.S. (1967): Solar Phys. *1*, 129

[5] Giacconi, R., Reidly, W.P., Zehnpfennig, T., Lindsay, J.C., Muney, W.S. (1965): Astrophys. J. *142*, 1274

[6] Reidly, W.P., Vaiana, G.S., Zehnpfennig, T., Giacconi, R. (1968): Astrophys. J. *151*, 333

[7] Vaiana, G.S., Reidy, W.P., Zehnpfennig, T., van Speybroeck, L., Giacconi, R. (1968): Science *161*, 564

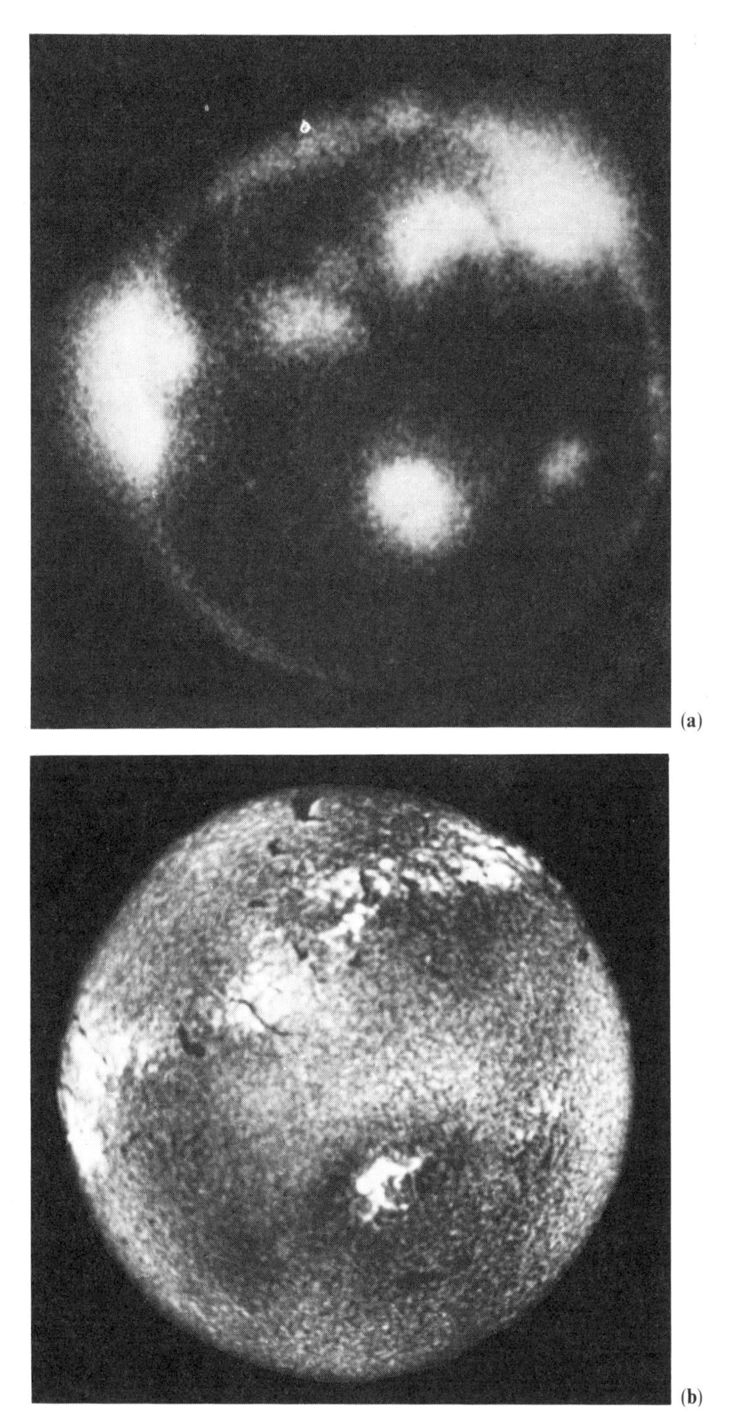

Fig. 12. (**a**) X-ray spectroheliogram ($\lambda \leqq 4.4$ nm) observed by UNDERWOOD and MUNEY[4] on 20 May 1966 (Solar diameter ≈ 5.8 mm, spectral resolution $\approx 20''$). (**b**) Corresponding Hα heliogram (for comparison)

associated with the active H-alpha filament that had disappeared 30 min before the flare event. Loop and arcs associated with condensations of the coronal material (white corona) can be seen.

YANG et al.[8] during the total solar eclipse of 7 March 1970 launched a slitless spectrograph on board an Aerobee-150 rocket. A wavelength range 90–220 nm was covered. The analysed emission lines of quiet prominences stemmed from ions with temperatures $3 \cdot 10^4$–$3 \cdot 10^5$ K. These lines are emitted in the very thin transition zone between the prominence and the corona. The density of the surrounding corona is significantly lower; this may be taken as a confirmation of the idea that prominences are condensations from the corona.

γ) Recently, many *heliograms* have been obtained from rockets and satellites. On 10 July 1972 aboard an Aerobee rocket the Sun was observed with a spatial resolution of 3″. The solar image diameter was 9.2 mm. The active regions, dark fibrils and supergranulae with sizes of about 40″ are well visible. The brightness of fibrils is only about 1/10 that of chromospheric features and 1/3 to 1/8 of the brightness of active regions[9]. A very large number of heliograms were obtained on board of the United States's manned orbiting USA station, Skylab.

Launch date was 14 May 1973, the orbit altitude was 435 km. The station was equipped with a set of instruments for solar observation known as the Apollo Telescope Mount (ATM). The orientation was controlled by three powerful gyroscopes and reached an accuracy of about 2.5″. Heliograms were obtained with the following three instruments:

1) Tousey's XUV spectroheliograph with focal length of 2 m, two gratings with 3600 mm^{-1} for the wavelength region 15–65 nm. The spatial resolution was 5″, the solar image diameter 18.5 mm, the spectral resolution 0.013 nm. About 700 photographic negatives were obtained (NRL).

2) Reeves's scanning UV spectroheliograph with an out-of-axis parabolic mirror mount producing a solar image on the slit which cuts out a field of 5″ by 5″. By the scanning motion a field of 5′ was covered. The spectral range was 30–135 nm. The data were transmitted by telemetry (Harvard Coll. Obs.).

3) Giacconi's x-ray telescope which was used as a flare detector within the spectral range 0.3–6 nm. About 20,000 photographic negatives were obtained with a spatial resolution of 3″. (Group of American and British scientists).

4) Solar images in the light of the L α line of hydrogen were recorded by a TV system. Spatial resolution on the film was 1″. A film with about 50,000 frames was obtained during the mission.

Many results of the Skylab mission have been published (see for example[10]).

Most interesting data about the EUV radiation of prominences[11] have been obtained on board of the satellites OSO-4 and OSO-6. Spectra and heliograms in the spectral range 30–140 nm were recorded with a spatial resolution of 35″ and 60″. The prominences could be seen well in the light of ion lines with ionization temperature less than $3 \cdot 10^5$ K. In the light of 'hotter' lines the prominences are less bright than the surrounding corona.

---

[8] Yang, C.Y., Nicholls, R.W., Morgan, F.J. (1975): Sol. Phys. *45*, 351

[9] Prinz, D.K. (1973): Sol. Phys. *28*, 35

[10] Tousey, R., Bartol, J.-D.F., Bohlin, J.D., Brueckner, G.E., Purcell, J.D., Scherrer, V.E., Sheeley, N.R., Schumacher, R.J., van Hoosur, M.E.: Sol. Phys. *38*, 265; Tousey, R., Sheely, N.R., Bohlin, J.D. (1975): Space Res. *XV*, 651

[11] Noyes, R.W., Dupree, A.K., Huber, M.C.E., Parkinson, W.H., Reeves, E.M., Withbroe, G.L. (1972): Astrophys. J. *178*, 1, 515

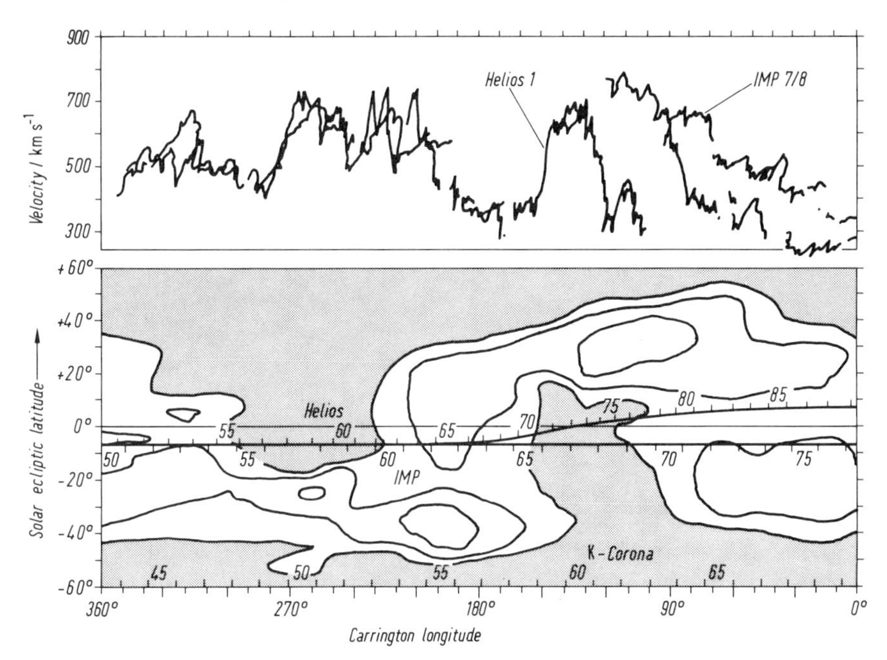

Fig. 13. Top: Solar wind ion bulk velocity for rotation No. 1625 against Carrington longitude. Near-Sun observations (reduced to 1 AU) of space probes HELIOS 1 (West Germany/United States) and IMP 7 and 8 (United States). Bottom: Coronal K-line emission contours from terrestrial observations in white light at the eastern limb (N.C.A.R.-U.S.A.). Low intensity emission ranges are shaded. The high-velocity ranges in the top diagram (around 260° and 140° on the abscissa) coincide rather well with low emission (shaded) ranges at the solar equator[12]

δ) Recently, many papers discuss the so-called *coronal holes*, namely, coronal regions with lower density and temperature. Such coronal structures were known earlier, but they happened to receive the general attention of heliophysicists only after a concise name was given. These regions are seen as dark spots on EUV heliograms. It is probable that the "coronal holes" are the sources of high-energy corpuscular radiation. This was finally proved by the space probes HELIOS A and B, a joint program of West Germany and the United States, see Fig. 13.

9. Photometry of XUV solar images. Methods of XUV solar-image photometry were reviewed by ALLEN[1]. In order to describe such images, he introduced certain terms and expressions and reduced solar-image photographs obtained in the British Skylark experiments over the period 1964–1966. If the EUV emission rate per unit volume is described by a height scale H and a base value I_0 the coronal intensity can be represented by the following expressions:

[12] COSPAR Landesausschuß: Space Research in the Federal Republic of Germany 1976. Bad Godesberg: Deutsche Forschungsgemeinschaft, 1977

[1] Allen, C.W. (1969): Sol. Phys. *8*, 72

a) within the disk ($R \leqq R_\odot - H$):

$$I(R) \approx e_0 I_0 H/(R_\odot - R)^{1/2} \tag{9.1}$$

b) beyond the disk ($R \geqq R_\odot + H$):

$$I(R) \approx 2 e_0 I_0 H \exp\left(\frac{-R_\odot}{H}\right) \frac{R}{H} \cdot \mathrm{K}_1 \left(\frac{R}{H}\right). \tag{9.2}$$

Here K_1 is a modified Bessel function, R radius vector, $R_\odot$ radius of the Sun, $r = R/R_\odot$, e_0 limb darkening.

Polar darkening was considered in accordance with the expression $e = e_0 \cos\phi$, where ϕ is the heliocentric latitude. The results of theoretical calculations were presented in the form of tables and diagrams. For a case of polar darkening, the limb brightening relative to the flux from the entire disk varies slightly over the wide range of limb fluxes: for $H = 0.00$, 0.01, and $0.10 R_\odot$, the relative limb brightening is respectively 0.39, 0.41, and 0.39. After photometry, ALLEN derived the limb, disk, and source emission fluxes.

As an example we can give the results of the Skylark experiment of 9 April 1965 (SL 303). The image studied was obtained within the wavelength ranges 3–10 nm, 17–22 nm and at 30.4 nm yielding 10, 75 and 10% as relative contributions of the different ranges. ALLEN also derived the source emission, the limb emission, and the entire disk emission fluxes in units of flux at the solar disk center; his results are respectively 0.49, 2.1, and 6.7.

The main conclusions are that the emission fluxes from the active regions are generally greater (relative to the disk or the limb) at shorter wavelengths. Also the flux from the limb appears to be greater (relative to the disk) than that accounted for by the decrease in the optical depth of the emerging XUV radiation. The disk brightness is nonregular and undoubtedly associated with the solar activity.

10. XUV solar eclipse experiments. *α)* On 7 March 1970 a combined group from different departments launched an Aerobee-150 rocket into a total solar eclipse and recorded during 180 s a sequence of 50 *slitless spectra* over the range 85–215 nm[1].

Alternate exposures of 0.2 and 1 s were taken throughout the whole frame sequence. An absolutely calibrated $f/10$ one-meter normal-incidence Wadsworth spectrograph was used (dispersion 1.6 nm/mm, solar picture diameter 4.5 mm).

The most interesting feature is that the H Lyman alpha emission (121.6 nm) in the coronal spectrum extended up to 0.3 solar radii above the limb. Lyman α emission in the inner corona has an intensity of the order of 10^{-3} of the underlying chromospheric Lyman α. The authors conclude that the observed coronal H Lyman α radiation stems from the intense chromospheric Lyman α emission scattered by neutral hydrogen atoms in the corona. Although the

[1] Speer, R.J., Garton, W.R.S., Morgan, J.F., Nicholls, R.W., Goldberg, L., Parkinson, W.H., Reeves, E.M., Jones, T.J.L., Paxton, H.J., Shenton, D.B., Wilson, R. (1970): Nature *226*, 249

relative concentration of neutral hydrogen at coronal temperature is only 10^{-7}, the scattering cross section is large. Therefore the intensity distribution in the corona reflects the distribution of neutral hydrogen. (Calculations show that within sufficiently dense coronal regions Lyman α emission can also be due to recombinations.)

On 22 September 1968, 4 h after a total eclipse in Asia, solar disk images were photographed in the far ultraviolet, 17–50 nm, from an Aerobee rocket launched from the White Sands Missile range in New Mexico[2]. The right-hand side of Fig. 14 shows one normally exposed, picture, with bright regions visible on the disk and strong luminosity beyond it, particulartly in the equatorial zone. The overexposed picture on the left demonstrates a large extension of coronal luminosity. Beyond the limb, the corona is brighter than against the disk. Overlapping solar images in 30.4 nm (He II), 28.4 nm (Fe XV), and 35.5 nm (Fe XVI) were also recorded. An example is shown on Fig. 15. The Fe XVI and Fe XV lines were emitted mostly in the corona above the plages but very little on the disk. He II was emitted all over the Sun's surface, very little however in the outer atmosphere, except for a large prominence.

β) The same rocket carried two *white-light coronographs* (with occulting disks in order to record the corona that is distant from the limb by 3 to 9 solar radii, $R_\odot$. The same group observed the solar corona with an apparatus on an Aerobee rocket launched 2 h after the totality on 7 May 1970. Apart from the white-light corona images obtained with the two occulting coronagraphs, the solar images in some lines between 17–50 nm were photographed. For these purposes an off-axis paraboloidal mirror of normal incidence with 250 mm focal length was used. The photographs show the coronal emission off the limb; a bright ring is clearly visible at the limb and is associated with emission originating within the transition layer: at wavelengths 46.5 nm (Ne VII), 36.9 nm (Mg IX), and from Fe XIII–XI at different wavelengths. The coronal emission beyond the limb has a less steep gradient than was expected from estimates of electron-collision excitation. Therefore the authors suppose that a considerable contribution stems from resonance scattering.

In general, the inner EUV corona is similar to the white-light one obtained during the eclipse on 7 March 1970. The depressions in the EUV corona seem to be associated with structures in the H Lyman alpha emission.

γ) During the total eclipse of 7 March 1970, an American group[3] obtained solar x-ray images. A *grazing-incidence telescope* and a sequence of filters covering specific (rather broad) bands in the range 0.3–6 nm were used. The resolution of the telescope was between 3″ and 5″. The images obtained of the solar disk and coronal active regions were compared with white-light corona photographs taken from the ground during the totality and with solar magnetic field records. The authors drew the following conclusions.

1) In the corona there are many large-scale active configurations that are "similar to chromospheric-photospheric complexes", as defined by BUMBA and HOWARD[4].

2) The coronal active regions are interconnected by emitted loops filled with plasma and growing over the solar equator.

3) The diffuse x-ray emission beyond the active regions is identified in

---

[2] Koomen, M.J., Seal, T.R., Purcell, J.D., Tousey, R. (1969): Sky and Telescope *37*, 356

[3] van Spreybroeck, L.P., Krieger, A.S., Vaiana, G.S. (1970): Nature *227*, 818

[4] Bumba, V., Howard, R. (1965): Astrophys. J. *141*, 1502

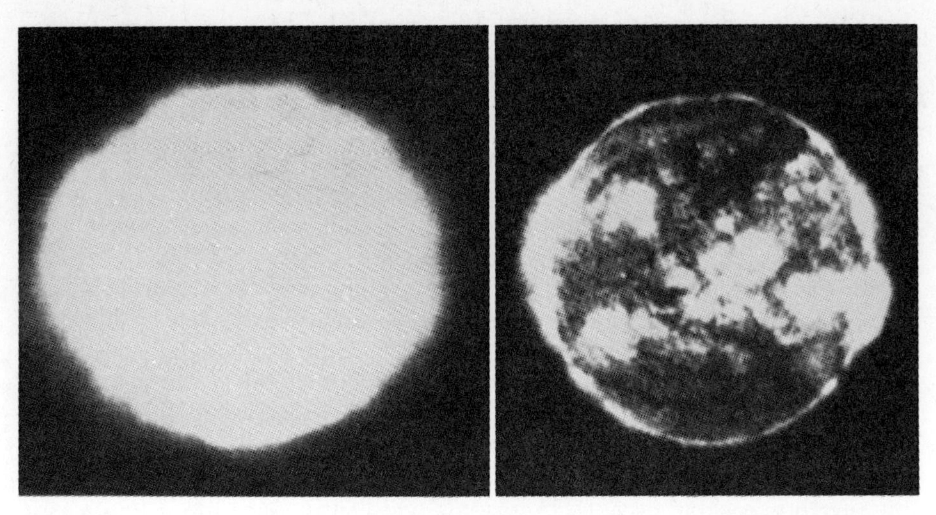

Fig. 14. The Sun photographed with off-axis paraboloidal mirror in the EUV range 17–50 nm on 22 Sept. 1968. Since the picture is overexposed, coronal EUV radiation extending beyond the solar limb can be seen. (NRL: KOOMEN, SEAL, PURCELL, TOUSEY [2])

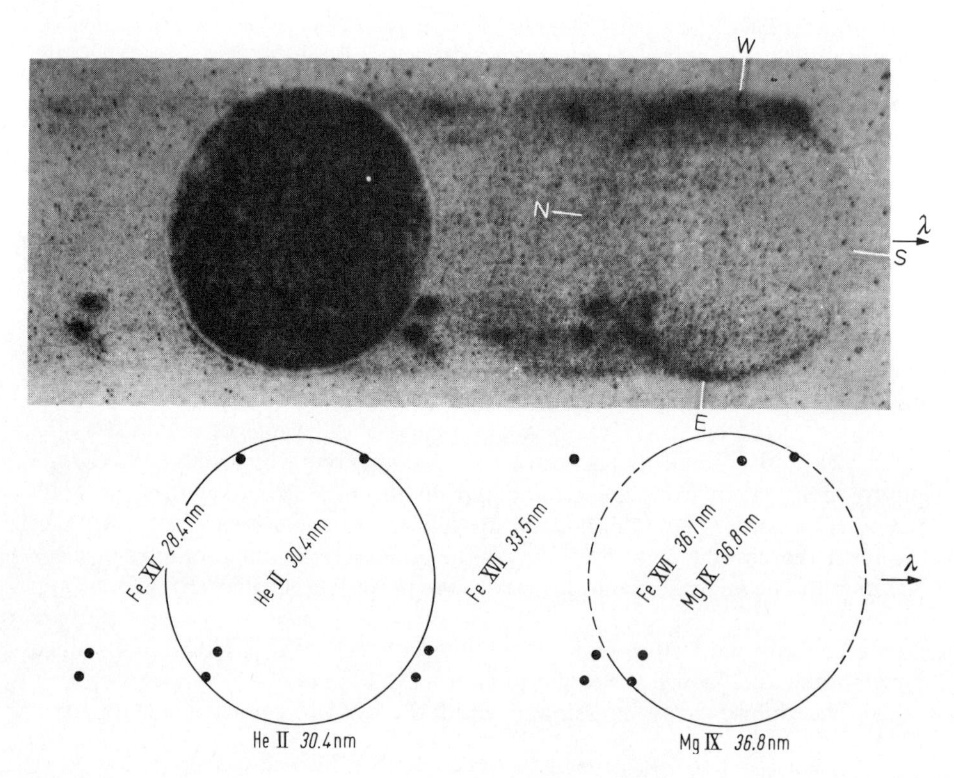

Fig. 15. EUV spectroheliograms photographed on 1 February 1966 with an objective-type single-concavegrating spectrograph (courfesy: TOUSEY, NRL)

some cases with remnants of dissipated active complexes connected with the unipolar magnetic groups and their "ghosts" (such regions are also considered by BUMBA and HOWARD[4]).

4) There are x-ray emitting regions associated with flocculi. Sometimes these are small, bright x-ray regions at such places where a sharp magnetic-field gradient could be identified. Such regions are similar to the x-ray flare features observed earlier by the same authors.

δ) On 20 May 1966 an annular solar eclipse occurred for which *solar radio noise* data are available. It was practically a total eclipse, because the difference between solar and lunar visible radii was only 2″. The NASA Explorer-30 satellite crossed the zone of totality twice; over Africa and over Europe. The satellite carried EUV and x-ray photometers for the following spectral ranges: 0–2 nm, 0.1–0.8 nm, 0.8–1.6 nm, 108–135 nm, and 122.5–135 nm. After reduction of the telemetered data combined with a computation of the trace of the Moon over the Sun's disk, the XUV brightness distribution over the disk was derived[5]. While EUV radiation showed homogeneous distribution, x-ray brightness varied in steps over the disk, originating generally within three active regions where enhanced decimeter radio emission was also observed; 96% of the 0.1–0.8 nm emission was found to be concentrated within the above sources. This concentration decreases for longer wavelengths and is only about 4% in the range 0.8–1.6 nm. Corresponding active regions emitted radio noise at 3.1 and 9.1 cm which was respectively 7% and 20% of the total flux. The active region areas increase with the x-ray wavelength.

For example, the diameter of the active region is respectively 65″ and 70″ at 0.1–0.8 nm and 0.8–1.6 nm. The diameters of the same region within the radiowave range are 97″ and 151″ at a radio wavelength of 3.1 cm and 9.1 cm, respectively.

11. Summary. The general results of the XUV image analysis are as follows.

The contrast between active region and undisturbed background for soft x-rays (wavelengths below 10 nm) increases toward the shorter wavelengths. Radiation in the range 1–2 nm originates almost only in the active regions.

The limb brightening for x-rays and for the emission of highly ionized atoms (Fe XV, XVI, C VI, and even O VI) is so large that the emitting regions extend beyond the limb by about 1′ (50,000 km) and more. It is this emission that causes the remaining ionization of the E layer of the terrestrial ionosphere during totality. According to the data obtained by FRIEDMAN's group during the totality on 12 October 1958, the radiation flux in the ranges 0.8–2 nm and 4.4–6 nm was 10–15% of those from the nonocculted disk.[1] The limb brightening disappears in the heliographic pole regions yet an intensity decrease occurs there. The heliograms obtained in the lines of ions of lower ionization stages (He II, C III, etc.) show the entire solar disk with comparatively low-contrast active regions. The limb brightening is neither absent nor small.

---

[5] Simon, G. (1969): Solar Phys. 7, 295

[1] Chubb, T.A., Friedman, H., Kreplin, R.W., Blacke, R.L., Unzicker, A.E. (1961): Mem. Soc. Roy. Sci. Liege *4*, 228

Thus the contrast of active regions increases with the ionization potential of the emitting ions. This, the principal conclusion of the XUV solar image analysis, allows us to understand the different intensity variations of various EUV lines. For example, the Lyman α flux from the entire disk with a contrast of about 5 shows a variation by a factor of 1.5–2 from maximum to minimum of solar activity, while the x-ray flux with contrast 30–100 varies by a factor of 10–30.

It is important to note that the total area of the active regions decreases for higher-contrast emission. Thus, the higher is the temperature of an active feature, the smaller is the relative surface on the solar disk.

III. Observation of x-ray solar radiation over wide spectral ranges

12. Soft x-ray fluxes. Now we shall consider briefly the results of the solar x-ray emission at wavelengths below 0.1 nm. This radiation is absorbed in the lowest ionosphere. Its influence on the terrestrial atmosphere is less important than that of solar EUV.

α) The first measurements of the solar x-ray radiation flux were made by FRIEDMAN's NRL group, using *rockets* equipped with scintillation counters with and without gas filling and suitably chosen windows of beryllium, aluminum, and other materials. Such equipment enabled radiation to be detected within the following spectral ranges: wavelength less than 0.4 nm, less than 1.0 nm and the two bands with wavelength 4–6 nm and 2–10 nm.

Before the first experiment of this group, photographic observations of the solar x-ray spectrum had been made by BURNIGHT in 1949 [1] and by TOUSEY and coworkers, who used phosphors [2].

FRIEDMAN's group performed a series of launchings during the solar activity cycle. It was found that the x-ray flux from the quiet Sun varies within the activity cycle, decreasing from maximum to minimum by a factor of 40 in the range 0.8–2 nm and even more from 0.2–0.8 nm.[3] During large flares the x-ray flux was enhanced by more than an order of magnitude. X-ray flux enhancement following flares causes an ionospheric disturbance called SID (Sudden Ionospheric Disturbance).[4]

Until the advent of double-stage rockets (before 1957), the Rockoon method had been used for the observation of x-ray emission accompanying flares. A rocket supported by a balloon was kept floating at an altitude of about 20 km until information that a flare was in progress was received by radio. The rocket was then fired. The method was not entirely satisfactory because in

[1] Burnight, T.R. (1949): Phys. Rev. *76*, 165
[2] Purcell, J.D., Tousey, R., Watanabe, K. (1949): Phys. Rev. *76*, 165
[3] Chubb, T.A., Friedman, H., Kreplin, R.W. (1960): Space Res. *1*, 695
[4] Rawer, K., Suchy, R.: This Encyclopedia, vol. 49/2, Sect. 43

Table 5. Intensity variation of soft x-rays from minimum to maximum solar activity. Energy fluxes are given in nW m^{-2} near Earth[8,9]. Annual average flux values were taken from the day-average diagrams

Year	Range			
	4.4–0 nm	0.8–2 nm	0.8–1.2 nm	0–0.8 nm
1964	$(2–3) \cdot 10^{-2}$	10^{-3}	3.10^{-5}	10^{-5}–3.10^{-3}
1965	$(3–10) \cdot 10^{-2}$	10^{-4}–2.10^{-3}	10^{-5}–10^{-4}	10^{-4}–3.10^{-3}
1966	0.03–0.3	0.003–0.03		3.10^{-4}–3.10^{-3}
1967	0.3	0.01–0.1		3.10^{-4}–10^{-2}
1968	0.1–0.5	0.01–0.3		
1969	0.5	0.01–0.1		

most cases the rocket drifted away from the prescribed allowable firing range before a flare occurred.

β) During the next few years x-ray emission was measured during *satellite* space missions carried out by the USSR, the United States, and the United Kingdom. Various types of detectors were used: scintillation counters, multipliers, and photon counters. Since the calibration methods were different, the data obtained differed considerably. The most reliable measurements of x-ray emission were made within three ranges: wavelength below 0.8 nm, below 2 nm, and in the band 4.4–6 nm. A summary is given in Table 5. Such data are often compared with the solar radio noise flux values at the wavelength of 10.7 cm. This is the so-called Covington index of solar activity (primarily measured at the Observatory of Ottawa, Canada).[5]

The correlation of the x-ray flux with other daily observed indices of solar activity is weak, for instance, with the daily Zürich sunspot number[6,7]. However, in spite of the wide dispersion of daily figures, long-term sliding averages through a large part of the solar activity cycle show a clear correlation between x-ray flux and solar activity level; flux enhancement is greater at shorter wavelengths. On the average the x-ray flux increases from minimum to maximum by a factor of 200 in the 0.8–2 nm range and by a factor of 20 in the 4.4–6 nm range (corresponding minimal and maximal Covington indices are 50 and 250).[8]

Soft X-radiation fluxes recorded on different United States satellites have been presented by KREPLIN.[8] They are SOLRAD 7A (1964-01 D), SOLRAD 7B (1965-16 D), SOLRAD 8 (Explorer 30), OGO-IV, and SOLRAD 9 (Explorer 37). The measuring device was always an ionization chamber with suitably chosen gas fillings and windows of different materials. The period described is 1964 (minimum) to 1969 (maximum). The daily average x-ray flux

[5] The numerical value of this dimensionless index is obtained by multiplying the measured flux spectral density at the wavelength of 10.7 cm by 10^{22} W^{-1} m^2 Hz. (Fluxes use to be measured in units of W m^{-2}H^{z-1}).

[6] Ivanov-Holodnyj, G.S. (1965): Geomagn. Aeron. 5, 705

[7] Allen, C.W. (1965): Space Sci. Rev. *4*, 91

[8] Kreplin, R.W. (1970): Ann. Géophys. *26*, 567

shows fluctuations caused by the solar rotation from which the author drew certain conclusions about the altitudes of the x-ray sources. Table 5 gives the annual average x-ray fluxes.

IV. Identification and prediction of extreme ultraviolet lines

13. The identification problem. The most important step in the analysis of any line spectrum is to identify its lines as the emission of known molecules, atoms, or ions. Any subsequent study of the properties of the emitting material depends entirely on the correctness of the line identification. The problem is very complicated when physical conditions within the emitting region are such as to produce a variety of ionization and energy stages, as well as types of ions, and consequently a large number of EUV lines.

In spectroscopy the general criterion of correct line identification is the coincidence of the measured and computed wavelengths. Certainly, some corrections are necessary, i.e., for the Doppler shift and, occasionally, some other effects on the wavelength value. However, because of the large number of lines, often blended, the uncertainties of line position measurement, and the uncertainties of the theoretically computed wavelengths, this criterion is unreliable. Comparison of the relative intensities of the lines within the same multiplets is claimed to be a good identification method but is not always satisfactory. Firstly, the observed intensities are quite often not reliable, even as relative intensities. Secondly, many atoms do not have a sufficiently sharp multiplet structure. EUV solar spectra can also be directly compared with spectra obtained in the laboratory from a plasma under conditions simulating those in the solar atmosphere. However, we cannot accurately reproduce the solar atmosphere conditions - temperature, chemical composition, excitation - since these are the main matters under investigation.

In the EUV spectrum there appears a certain number of bright lines that have been reliably identified and their intensities have been precisely measured. Unfortunately, most of these lines are located in the spectrum at wavelengths below 50 nm, where an absolute intensity calibration is rather difficult. Therefore, it seems reasonable to start with a "theoretical prediction" for the far-ultraviolet spectrum. From a comparison of observed and predicted spectra we can identify most of the lines sufficiently accurately.

The importance of line identification in the study of the solar atmosphere is obvious, but is line identification as important for geophysical and ionospheric problems? It does not seem to matter what sorts of atoms or ions are responsible for EUV radiation. To understand the effects of far-ultraviolet radiation on the terrestrial atmosphere in a general way, we need indeed only to know the energy distribution in the EUV spectrum. However, for the prediction of *fluctuations* in the terrestrial ionosphere it is important to know the possible EUV radiation flux and EUV spectrum variations[1] due to the

[1] See SCHMIDTKE's contribution in volume 49/7

11 years solar activity cycle, the solar rotation of about 27d and some short-term events, e.g., flares, flocculi, and spots. This is impossible without knowing the physical conditions within the regions of the upper solar atmosphere where EUV lines are formed (temperature, density, excitation mechanism, etc.). Therefore, it is clear that without identification of those EUV lines which give the greatest power contribution in the different parts of the EUV spectrum, a deeper investigation of events in the ionosphere will be impossible.

There is another important point. Since at present energy calibration is reliably feasible only within the long-wave part of the EUV spectrum, for a given instrument it may have to be revised and corrected after line identification and comparison of the predicted with measured line intensities.

14. Physical conditions of EUV line formation in the solar atmosphere

α) The problems of predicting numerous EUV lines were considered independently by different authors.[1,2] Their calculation methods were completely different. ALLEN adopted a relation between the so-called emission measure (which is the height integral of the squared electron density in the solar atmosphere) and temperature in accordance with a particular solar-atmosphere model[3], whereas Russian authors derived the *emission measure* as a *function of temperature* on the basis of the intensities of some reliably identified EUV lines. Among the correctly identified EUV lines are those emiited by atoms at various ionization stages: Mg II, C III, He II, O VIII, Si XII, Fe XVI, and many others. The data suggest that large differences in temperature exist within the emitting regions. Therefore, the ions selected for line prediction could be those that exist over the entire solar temperature range from 6000 K to $(2\text{–}3) \cdot 10^6$ K. Many different ionization stages are found over this wide temperature range; from neutral atoms in the lower chromosphere to ions of the Ca XV type within the "hot" region of the solar corona. The corresponding range of ionization potentials is approximately 10–1000 eV.

β) As in the EUV spectrum continua are rather weak, the *line-excitation mechanism* may be ion-electron collisions. To produce a line, collisions must either lift ions to a higher energy state (electron-impact excitation), or bring ions to a lower ionization stage followed by ion excitation (radiative recombination). Under solar atmosphere conditions, ion excitation following radiative recombination is a much less effective mechanism (by a factor of several thousand) than electron impact. Generally, the process that brings excited ions down to lower energy states is spontaneous radiative transition, whose rate in the solar atmosphere, even for forbidden lines, is nearly always higher than the rate of de-excitation transitions caused by electron impact.

All this leads us to the following conclusions: a) Almost all ions are in their lowest (ground) energy state; b) among the various emission lines of any ion, the most intensive ones are generally those associated with the ground

---

[1] Allen, C.W. (1961): Mém. Soc. R. Sci. Liège *4*, 265
[2] Ivanov-Holodnyj, G.S., Nikol'skij, G.M. (1961): Astron. Ž̌h. *38*, 828
[3] Oster, L. (1956): ZS. Astrophys. *40*, 28

level, especially the lines with a lower excitation potential (resonance lines). These conclusions are derived from the steady-state balance equation for the ground (1) and excited (2) levels:

$$N_{i1}^{(j)} N_e W_{12} = N_{i2}^{(j)} A_{21}. \tag{14.1}$$

Here N_e and $N_i^{(j)}$ represent the electron and ion number density, the second index identifies the excitation state, while the upper index (j) is the degree of ionization (i.e., the number of "stripped" electrons). W_{12} is the electron-impact excitation rate (unit, s^{-1}), A_{21} is the Einstein coefficient for spontaneous radiative transition from state 2 into state 1. The right side of Eq. (14.1) is proportional to the photon flux within the considered line $(2 \rightarrow 1)$ per unit volume for any ion of ionization stage (j).

Let us consider the left side of Eq. (14.1), representing the volume excitation rate. We adopt the electron-impact rate after Born's approximation giving

$$W_{12} = \frac{10 f_{12}}{T^{3/2}} \left[\frac{e^{-x}}{x} - \mathrm{Ei}(x) \right] = \frac{f_{12}}{T^{3/2}} W'(x), \tag{14.2}$$

where f_{12} is the oscillator strength, $x = \chi_{12}/kT$, χ_{12} - the excitation potential, and $\mathrm{Ei}(x)$ is the exponential integral function. This expression can be used as a universal description and provides results of limited accuracy. The density of any ion in the ground state is the most important part so that the sum over all possible energy states is not much greater.

If $N_i^{(k)} / \sum_j N_i^{(j)}$ is the ionization degree of any chemical element, we have for a given ionization stage (k)

$$N_{i1}^{(k)} N_e \approx N_e^2 \kappa N_i^{(k)} / \sum_j N_i^{(j)},$$

where κ represents the abundance of a given element relative to hydrogen (ratio of numerical densities). Since electrons are not only supplied from hydrogen, the last equation is only an approximation.

γ) The *degree of ionization* in the corona and upper chromosphere is determined entirely by electron collisions. In this case, the degree of ionization is a function only of temperature and the atomic properties of the ion in question. This function can be represented as a curve with a prominent sharp maximum at a temperature which is particular for each specific kind of ion. It is known as the "ionization temperature". Hence, ions of given ionization state (j) exist within a range of temperatures around the value T_j.

δ) Turning back to Eq. (14.1), we can now carry out an *integration over the height h* within the temperature range where ions of ionization stage (j) exist. Taking the above argument and Eq. (14.2) into account, we obtain

$$2.3 \cdot 10^{13} \frac{\lambda}{\mathrm{nm}} F = \frac{f_{12}}{2} W' \kappa \int_{(j)} N_e^2 (T/K)^{-3/2} \mathrm{d}h. \tag{14.3}$$

Here F is the solar flux of ionizing quanta at 1AU, and λ is the wavelength considered. The numerical coefficient takes a bright ring at the solar limb into account, too. The factor 1/2 was introduced into the right side of Eq. (14.2) because the degree of ionization at $T=T_j$ is not greater than about 1/2.

15. Measuring emission using EUV lines

α) The integral on the right side of Eq. (14.3) describes the radiation output of the solar atmospheric region where the ions in question exist; it is defined by the condition that the ambient temperature T is close to T_j. To distinguish it from the "emission measure" $\int N_e^2 \mathrm{d}h$ usually applied in astronomical treatises, the integral on the right side of Eq. (14.3) is called the "*generalized emission measure*"[1]; the sign $\Delta\varphi_j$ or $\Delta\varphi(T_j)$ will be used for this "generalized emission measure".

Thus, the generalized emission measure can be expressed in terms of the atomic parameters and observational data referring to any solar EUV line[2]:

$$\Delta\varphi(T_j)=\int_{(j)} N_e^2 T^{-3/2} \mathrm{d}h=\frac{4.6\cdot 10^{13}(\lambda/\mathrm{nm})F}{\kappa f_{12}\cdot W'(T_j,\lambda)}. \tag{15.1}$$

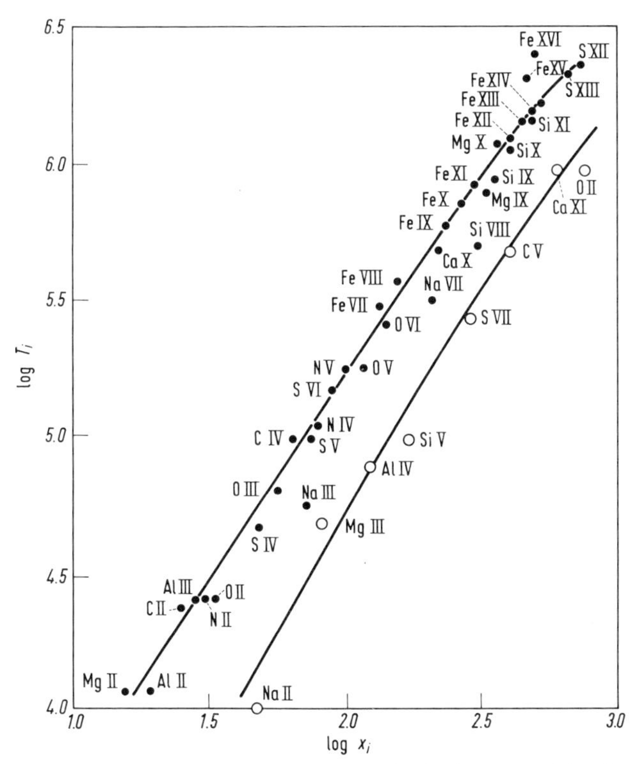

Fig. 16. Relation between ionization temperature T_j and ionization potential χ_j for 50 types of ions (computed by IVANOV-HOLODNYJ and NIKOL'SKIJ [3])

[1] See [8], p. 228

[2] The SI unit for $\Delta\varphi$ is $\mathrm{m}^{-5}\,\mathrm{K}^{3/2}$. Measured data were often expressed in $\mathrm{cm}^{-5}\,\mathrm{K}^{3/2}$ which gives numerical values smaller by a factor of 10^{-10}. See also [3], p. 58.

Table 6. Computation of the generalized emission measure using "basic" EUV lines

Ion	$\log_{10}(T_i/K)$	Wave-length λ [nm]	f_{12}	F [mW m^{-2}]	$\log_{10}(\Delta\varphi_i/m^{-5}K^{-3/2})$
Mg II	4.05	279.6 280.3	0.9	18	34.34
C II	4.25	90.36 90.45	(0.1)	0.003	32.62
Si III	4.40	120.65	1.1	0.085	31.71
Si IV	4.65	139.4 140.3	0.7	0.070	30.33
C III	4.67	97.70	1.38	0.082	29.77
S IV	4.67	106.27 107.33	0.6	0.007	30.13
C IV	5.00	154.8 155.1	0.3	0.170	29.00
O IV	5.05	79.0	0.23	0.015	28.10
N IV	5.08	76.5	0.8	0.007	28.40
S VI	5.20	93.3 94.4	0.6	0.001	28.75
N V	5.28	123.88 124.28	0.23	0.005	27.90
O VI	5.40	103.2 103.8	0.2	0.074	27.70
Ne VII	5.60	89.2	0.1	0.002	27.44
Mg VII	5.75	84.3	(0.1)	0.002	27.08
Ne VIII	5.82	77.04 78.03	0.15	0.015	27.82

For the lines emitted by various ions existing within the entire temperature range of the solar atmosphere, the generalized emission measure can be computed in accordance with Eq. (15.1) as a function of the ionization temperature T_j. The function $\Delta\varphi(T_j)$ was constructed using measured intensities of reliably identified lines that are sufficiently bright. The relation between T_j and the ionization potential computed for 50 ions can be represented by a simple curve [3] (Fig. 16).

The majority of the values are concentrated near the same curve, except for ions with filled outer shells. These ions form another curve slightly shifted from the other one.

β) To determine $\Delta\varphi_j$ we need to know *a few parameters*, namely, the oscillator strength of the index lines and the corresponding elemental abundance in the solar atmosphere. The problem of chemical abundance will be discussed in detail below. Table 6 gives the values of the most abundant elements whose ions can be used for line prediction. For many elements, abundances are close to those usually assumed for the sun and the solar system.[3] Satisfactory estimates of the oscillator strength values f_{12} of some ions and isoelectronic sequences are given in the literature [*1*].[4]

Table 6 contains the initial data and values $\Delta\varphi_j$ computed with Eq. (15.1) for 15 ions with emission lines above 70 nm and whose ionization temperatures are within the range 10,000–600,000 K. Unfortunately, there are no reliably identified lines within this spectral range emitted by ions of higher ionization temperature. Therefore, for T_j above 600,000 K, it is necessary to use other data.

The radiation fluxes in Table 6 are taken from the 1963 photoelectric measurements by HINTEREGGER[5] and from TOUSEY's photographic data[6] (for lines of 5 ions)[5]. The photoelectric data are from the minimum of solar activity. They completely define the behavior of the function $\Delta\varphi(T_j)$. Although the photographic data correspond to a somewhat higher level of solar activity, they fit the curve $\Delta\varphi(T_j)$ rather well. It should be noted that the lines considered in this particular case belong to ions of low-ionization states and do not vary much throughout the solar activity cycle.

16. Additional inputs for the calculation of the generalized emission measure as a function of temperature. As additional inputs, we used the intensities of the forbidden coronal lines within the visible spectrum as measured with ground instruments. The excitation mechanism of these lines is similar to that of EUV emissions and the accuracy of intensity measurements made from ground is better than that reached in rocket experiments.

Observed input observational data concerning the forbidden coronal lines during a minimum of solar activity, and the corresponding values of the generalized emission measure $\Delta\varphi_j$ are presented in Table 7.

Table 7. Computation of the generalized emission measure using forbidden coronal lines

Ion	$\log_{10}(T_i/K)$	Wavelength λ [nm]	F [$nW\,m^{-2}$]	$\log_{10}(\Delta\varphi/m^{-5}\,K^{-3/2})$
Fe X	5.78	637.4	100	174.0
XI	5.90	789.2	100	173.5
XIV	6.11	530.3	600	177.5
Ni XII	6.00	423.1	18	172.4
XIII	6.08	511.6	23	177.8
XV	6.20	670.2	36	175.4
XVI	6.25	360.1	18	169.0
Ca XII	6.28	332.8	9	167.2
XIII	6.42	408.6	4	162.5
XV	6.55	569.4	6	165.0

[3] See [*2*], § 5.9
[4] Veselov, M.G. (1949): Žu. Exp. Teor. Fiz. *19*, 959
[5] Hall, L.A., Damon, K.R., Hinteregger, H.E. (1963): Space Res. *3*, 745
[6] Detwiller, C.R., Purcell, J.D., Garrett, D.L., Tousey, R. (1961): Ann. Géophys. *17*, 263

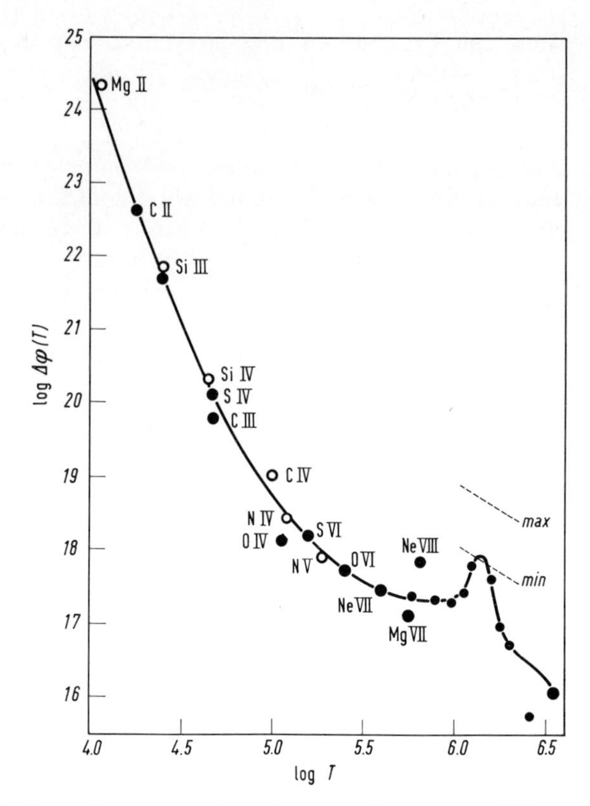

Fig. 17. Generalized emission measure $\Delta\varphi$ as function of temperature T. Intensities of basic lines were taken from observations by DETWILLER et al., 19 April 1960 (open circles), HINTEREGGER et al., 23 August 1961 (dark circles). For coronal data ($\log T > 5.8$) observations from ground were used; points refer to forbidden coronal lines. The short dashed curves refer to the $\Delta\varphi$ as derived by integration of the electron distribution in VAN DE HULST' model

Figure 17 describes $\Delta\varphi$ as a function of $\log T_j$. In spite of the rather coarse analysis of quite a few of the observational data, the ordinate values concentrate near the regular curve shown in this figure. Information obtained from the forbidden coronal lines point to the existance of a maximum occurring near $\log_{10} T_j = 6.15$.

Values derived for comparison from coronal electron-density-distribution data were used as an additional check [3]. They are shown by thin lines in the right-hand-bottom corner of Fig. 17. For an average temperature of $1.4 \cdot 10^6$ K and a degree of inhomogeneity $\overline{N_e^2}/(\overline{N_e})^2 \approx 3$ at minimum solar activity follows a $\Delta\varphi_j$ of about $5 \cdot 10^{17}$ and about $7 \cdot 10^{18}$ at solar activity maximum. These values agree rather well with the coronal determinations of $\Delta\varphi(T_j)$ derived independently from the forbidden coronal line intensities. The general appearance of the curve $\Delta\varphi(T_j)$ demonstrates how the generalized emission measure decreases with increasing temperature.

17. Determining EUV line intensities

α) The *dependence* of the generalized emission measure *on temperature*, as plotted in Fig. 17, was used as a basis for the determination of the EUV line

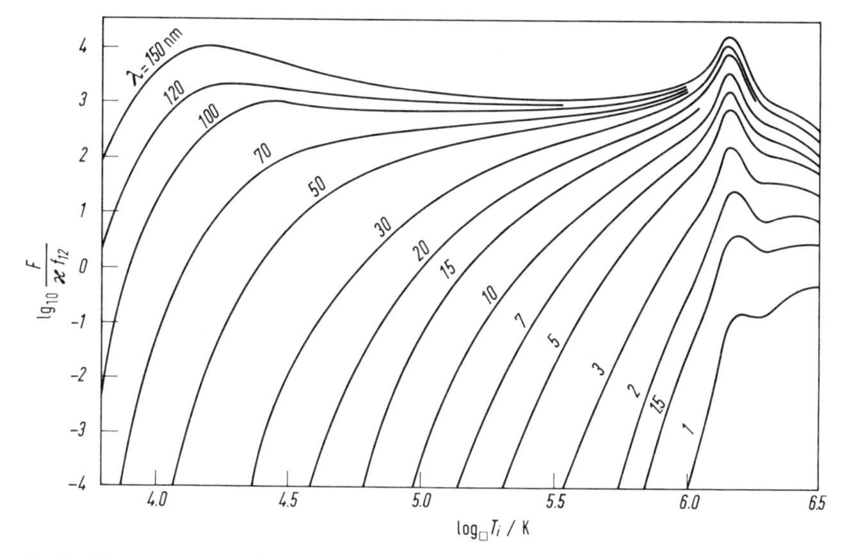

Fig. 18. Auxiliary nomogram for computing EUV line intensity (F). Here F is the energy flux in $\mathrm{mW\,m^{-2}}$ near Earth. κ chemical abundance relative to hydrogen; f_{12}, oscillator strength; T_i, ionization temperature

intensities F. Since the dispersion of the $\Delta\varphi_j$ values does not exceed half an order of magnitude, one may expect to obtain the intensity of any EUV line with the same accuracy. With Eq. (15.1) and according to[1], one has

$$\frac{F}{\kappa f_{12}} = 2.2 \cdot 10^{-14} \frac{W'(\lambda, T_j)}{(\lambda/\mathrm{nm})} \Delta\varphi(T_j). \tag{17.1}$$

All the lines used for calculating $\Delta\varphi_j$ are resonance lines. In the case of a few possible transitions downward from the excited level (3) ($3 \to 2$, $3 \to 1$), the line intensity is

$$\begin{aligned} &F_{31} \approx F A_{31}/\Sigma A && \text{for the main series and} \\ &F_{32} \approx F(\lambda_{31}/\lambda_{32}) A_{32}/\Sigma A && \text{for the subordinate series.} \end{aligned} \tag{17.2}$$

Here F is the quantity defined for the line λ_{31} and ΣA is the sum of the probabilities of all possible spontaneous transitions from the excited level.

β) The *calculating procedure* can be much simplified by the use of a chart giving $F/\kappa f_{12}$ as a function of T_j for various wavelengths λ (see Fig. 18). The selection and calculation of the wavelengths of the various transitions providing sufficiently strong lines is the most labor-consuming part of the EUV line prediction procedure.

Table I in [3] represents the calculated F quantities for 700 multiplets of 146 ions belonging to 21 elements (more than 1000 lines with a flux $F > 10^{-6}\,\mathrm{mW\,m^{-2}}$).

[1] Ivanov-Holodnyj, G.S., Nikol'skij, G.M. (1961): Astron. Žh. *38*, 828

Twenty-six isoelectronic sequences from H I to Fe I were analyzed. The wavelengths of the lines in question are within the range 1–200 nm. The wavelength determination was generally carried out using the table in [*4*]. When, in a few cases, the terms of some ions (especially of high-ionization stages) were unknown, the relevant energy was derived by an extrapolation or interpolation procedure throughout the isoelectronic sequence. The combinations of wave number ν and ionization stage Z considered for the extrapolation were chosen so as to have very small variations along the isoelectronic sequence (type νZ^k).

Generally, the only permitted transitions considered were those that conform to LAPORTE's rule and the selection rules of the orbital and inner quantum numbers. Along with [*1*] and[2] the f_{12} estimates and wavelengths of 205 EUV lines are given by[3]. The f_{12} values for Fe X and Fe XIV were taken from[4], but for most of the lines no f_{12} values are available. With the rule of ALLEN[5] one of four f_{12} quantities can be found through the variation of the principal and azimuthal quantum numbers.

γ) The *list of predicted lines* in [*3*] is considerably extended as compared with previous publications.[6]

The additional information on wavelengths and oscillator strengths was taken from various papers giving line parameter calculations or EUV line predictions. Data on some of the Ni X–XIII lines are contained in[7–10], on Fe VIII–XIV lines in papers[8,10–12], and on Ne IV lines in[13].

The book by STRIGANOV and SVENTICKIJ [*5*] gives precise wavelengths for practically all transitions of relatively low-ionization stages (VI–VIII).

For the resonance lines emitted by the ions of the isoelectronic sequence Ne I to Fe XVII, the oscillator strengths are given by[14]. The wavelengths of the lines of ions of the Ca XII–XIV sequence are listed in[15]. Some of the lines of the A I sequence are reported in[16]. The oscillator strengths of the line series of Fe XVII, O VII–VIII and Ne IX were computed by FROHSE (unpublished data) cited by[17]. The wavelengths of some of the lines 5.7–31 nm are given in[18] where EUV line intensity prediction is discussed.

δ) Let us come back to the *accuracy of predictions of line intensity* in the far ultraviolet. Apart from the method itself, errors can be caused by uncertainty about the intensities of the "reference" lines.

An interesting method for estimating EUV lines was proposed by ŠKLOVSKIJ[19] and used in[20]. In 1945, before the advent of rocket observations and with GROTIAN's[21], identification of the

[2] Veselov, M.G. (1949): Zh. Eksp. Teor. Moskva *19*, 959

[3] Varsavsky, C.M. (1961): Astrophys. J., Suppl. Ser. *6*, N 53, 75

[4] Pecker, Ch., Rörlich, F. (1961): Mém. Soc. R. Sci., Liège *4*, 265

[5] Allen, C.W. (1961): Mém. Soc. R. Sci., Liège *4*, 265

[6] Ivanov-Holodnyj, G.S., Nikol'skij, G.M. (1962): Geomagn. i. Aeronom. *2*, 425

[7] Gabriel, A.H., Fawcett, B.C. (1965): Nature *206*, 808

[8] Feldman, U., Fraenkel, B.S., Hoory, S. (1965): Astrophys. J. *142*, 719

[9] Gabriel, A.H., Fawcett, B.C., Jordan, C. (1965): Nature *206*, 392

[10] Gabriel, A.H., Fawcett, B.C., Jourdan, C. (1965): Nature *206*, 392

[11] Cowan, R.D., Peacock, N.J. (1965): Astrophys. J. *142*, 390

[12] Stockhausen, R. (1965): Astrophys. J. *141*, 277

[13] Tilford, S.G., Giddins jr., L.E. (1965): Astrophys. J. *141*, 1226

[14] Kastner, S.O., Omidvar, K., Underwood, J.H. (1967): Astrophys. J. *148*, 269

[15] Tousey, R., Purcell, J.D., Austin, W.E., Garrett, D.L., Widing, K.G. (1964): Space Res. *4*, 703

[16] Alexander, E., Feldman, U., Fraenkel, B.S., Hoory, S. (1965): Nature *206*, 176

[17] Pottasch, S.R. (1966): Bull. Astron. Inst. Neth. *18*, 443

[18] Zirin, H. (1964): Astrophys. J. *140*, 1332

[19] Šklovskij, J.S. (1945): Astron. J. *22*, 249

[20] Nikol'skij, G.M., Šilova, N.S. (1963): Geomagn. Aeron. *3*, 431

[21] Grotrian, W. (1939): Naturwissenschaften *27*, 214

visible coronal lines as emissions of highly ionized coronal plasma, ŠKLOVSKIJ predicted the far ultraviolet spectrum with wavelengths of a few tens of nanometers. He also estimated the lower limit of the intensity of FeXV 42.4 nm, with the line following the subordinate forbidden line 705.9 nm in cascade fashion. With semiqualitative considerations he estimated the total energy of the solar radiation flux at wavelengths below 50 nm. Although his flux value was too low compared with more recent data ($\approx 0.01\ \mathrm{mW\,m^{-2}}$), the principle he used still seems quite interesting.

If for some atom the transitions between quantum levels $2\rightarrow 1$ and $3\rightarrow 1$ produce visible and ultraviolet lines, respectively, with sufficient accuracy the photon flux ratio will be equal to the ratio of the excitation rates of both upper levels. The latter follows from the steady-state equation, Eq. (14.1). Consequently,

$$\frac{F_{31}}{F_{21}}=\frac{\lambda_{12}}{\lambda_{13}}\cdot\frac{W(\lambda_{13},T_j)}{W(\lambda_{12},T_j)}. \tag{17.3}$$

Such an approach is independent of the abundance of the element in question.

This method was improved and used in [*3*] for the determination of the fluxes of the EUV lines of iron, nickel and calcium ions, since the visible coronal lines of these, ions were used for the calculation of the generalized emission measure in the corona.

18. Identification of EUV lines. Several hundred EUV lines have been identified below 122 nm. At the upper limit of this spectral region the very bright hydrogen Lyman alpha line (121.57 nm) is found which is a most interesting subject for astrophysical and geophysical investigations. In Table II of [*3*] the predicted lines are listed in order of increasing wavelength.

α) For the identification procedure, two features were considered in [*3*], namely, line intensity and numerical coincidence of wavelength values. So as to avoid lines that are too bright or too weak, *relative* intensities were considered.

For example, within a group of recorded lines with almost identical wavelength, the brightest line was compared with the most intensive one from the corresponding group of predicted lines. With this procedure NIKOL'SKIJ [1] identified 180 of 255 lines recorded by [2] and by the Air Force Cambridge Research Laboratory group [3,4]. Later, on the basis of the extended list of predicted lines and additional observational data [5-7] 239 out of 297 recorded lines could be identified. The new prediction list in [*3*] allows the identification to be carried out most accurately.

It was performed on the basis of the most reliable data obtained from American rocket experiments: 1) a record 1.37–2.48 nm obtained by FRIEDMAN and coworkers [8] on 25 July 1963; 2) a photographic spectrum 3.3–7 nm, taken by TOUSEY and coworkers [9] on 20 November 1963; 3) records 5.5–31 nm made by the Air Force Cambridge Research Laboratory group [10] on 2 May 1963; 4) the best of the measurements 2.5–121.5 nm made by the same group [6] on 23 August 1961. All these data provided the wavelengths and intensities of 450 lines below 122 nm.

[1] Nikol'skij, G.M. (1962): Geomagn. Aeron. *2*, 425
[2] Violett, T., Rense, W. (1959): Astrophys. J. *130*, 954
[3] Hinteregger, H.E., Damon, K., Neroux, L., Hall, L. (1960): Space Res. *1*, 615
[4] Hinteregger, H.E. (1960): Astrophys. J. *132*, 801
[5] Tousey, R., Purcell, J.D., Austin, W.E., Garrett, D.L., Widing, K.G. (1964): Space Res. *4*, 703
[6] Hall, L.A., Damon, K.R., Hinteregger, H.E. (1963): Space Res. *3*, 745
[7] Nikol'skij, G.M. (1964): Doctor Thesis, Moskva
[8] Black, R.L., Chubb, T.A., Friedman, H., Unzicker, A.E. (1965): Astrophys. J. *142*, 1
[9] Austin, W., Purcell, J., Tousey, R., Widing, K. (1966): Astrophys. J. *145*, 373
[10] Hinteregger, H.E., Hall, L.A., Schweizer, W. (1964): Astrophys. J. *140*, 319

β) Let us *compare* these *results* with the most extensive identifications of other authors:

159 lines were listed by VIOLETT and RENSE[2]; data for most of these 101 lines were confirmed by other experiments; 74 of these lines were identified and 47 identifications coincide with [*3*]. In 1961 within this spectrum the lines of S VI (94.5 nm) and (93.4 nm), Si XII (49.9 nm) and Ne VII (46.4 nm) were identified for the first time.[11]

HINTEREGGER and ZIRIN[12] identified 88 lines in the range 5.75 to 30.84 nm and 11 lines between 31.3 and 55.8 nm. All their identifications except 16 coincide with [*3*]. Line identifications in the range 3.3–50 nm were indicated by TOUSEY and coworkers[13]; only four of the 58 lines listed differ from ours and one of those was identified in [*3*] as an emission of CV.

About 20 lines were discovered in the 0.95–3.4 nm range by American and Russian experiments. The identification of 10 of 14 lines recorded by FRIEDMAN and coworkers at NRL[14] was confirmed; 15 lines were identified by MANDEL'STAM's group and 12 of these agree with [*3*]; 8 coincide with earlier identifications carried out by the NRL group. Wavelength measurements made by MANDEL'STAM and coworkers[15] seem to have been of lower accuracy than claimed (less than 0.015 nm) since the observed and calculated wavelengths differ by 0.03 nm. The results of the intensity calculation for lines of 96 multiplets (76 multiplets in the range 4.4–121.6 nm) are represented in a paper by ALLEN[16] who identified 31 lines below 122 nm; 24 of his identifications agree with [*3*] and only one line had not been identified earlier (Si II 99.2 nm).

PECKER and ROHRLICH[17] calculated the wavelengths of 4 coronal lines: 105.87 nm (Al XIII), 104.89 nm (Si VII), 95.24 nm (Si IX), and 65.87 nm (Al XIII). The first two lines were photographed by VIOLETT, and RENSE. ZIRIN, HALL, and HINTEREGGER identified a series of lines in the spectrum recorded by the last two of the authors and DAMON. Along with already known lines identified earlier by DETWILLER et al., additional identifications were made. Thus, lines 27.0 nm, 34.4 nm, and 36.5 nm were assigned to Fe XIV emission. In [*3*] the identification of the first two lines[18] is confirmed, but the third agrees better with the bright line of Si XI. The lines 33.2 nm and 36.1 nm that we identified as Fe XVI emission were also identified so independently by EDLEN and TOUSEY[19]. It may be stated that ZIRIN et al., after analysis, concluded that compared with existing estimates, there is a lower abundance of N and Ne and a higher abundance of Fe in the solar atmosphere, in agreement with the findings given below.

The photoelectric spectrometer designed for the OSO experiments recorded the spectrum 12–38 nm. A list of 42 lines with 24 previous identifications is given in [20]. Of these identifications only 7 coincide with those in [*3*]. There is some doubt about the remaining ones. For example, the lines at 31.6 and 3.3 nm should be assigned to the more intensive emissions of Si VIII and Al X instead of Ni XV; the line at 27.4 nm is due to Si VII and not Cu XIX; the line at 26.3 nm is emitted by Fe XVI, not A XIV, and an emission in the 18.4 nm line is hardly due to an element of low abundance but agrees better with O VI radiation.

In the spectrum recorded by HINTEREGGER and coworkers[10] 44 of 77 identifications between 5.5 and 31 nm coincide with those in [*3*].

γ) In the photographs of the *solar limb* spectrum obtained in 1967–1969 by BURTON and RIDGELEY[21] in the range 30.4–122 nm, there are 91 lines, 78 of which are identified. Four of the identifications differ from [*3*], and 5 identified and 3 unidentified lines are absent from this list.

[11] Ivanov-Holodnyj, G.S., Nikol'skij, G.M. (1961): Astron. Žh. *38*, 45
[12] Hinteregger, H.E., Zirin, H. (1964): Astrophys. J. *140*, 1332
[13] Tousey, R., Austin, W.E., Purcell, J.D., Widing, K.G. (1965): Ann. Astrophys. *28*, 755
[14] Blake, R.L., Chubb, T.A., Friedman, H., Unzicker, A.E. (1965): Astrophys. J. *142*, 1
[15] Žitnik, I.A., Krutov, V.V., Malavkin, L.P., Mandel'štam, S.L., Čeremuhin, G.S. (1967): Kosm. Issled. SSSR *5*, 276
[16] Allen, C.W. (1961): Mém. Soc. R. Sci. Liège *4*, 241
[17] Pecker, S., Rohrlich, F. (1961): Mém. Soc. R. Sci. Liège *4*, 253
[18] Ivanov-Holodnyj, G.S., Nikol'skij, G.M. (1962): Geomagn. Aeron. *2*, 425
[19] Zirin, H., Hall, L.A., Hinteregger, H.E. (1963): Space Res. *3*, 760
[20] Behring, W.E., Neupert, W.M., Lindsay, J.C. (1963): Space Res. *3*, 814
[21] Burton, W.M., Ridgeley, A.R. (1970): Sol. Phys. *14*, 3

During the OSO-IV experiment in 1967, spectra of the quiet sun were recorded[22] and 93 of 120 lines between 30.4 and 122 nm were identified (12 of the lines stem from the second order of the grating). Fourteen identifications disagree with [*3*], 12 of 15 unknown lines can be identified with those in Tables I and II of [*3*], and 4 of the lines are listed there.

δ) Summarizing, it can be stated with reference to [*3*] that of 450 lines discovered in the solar spectrum about 90% below 122 nm were identified by means of EUV line intensity prediction technique; only 40 identifications are doubtful.

19. Radiated power in the EUV range. There are two intensity values for every identified line: the observed intensity F_o and that predicted by theory, F_p. Sometimes these values differ considerably, but on the average they are within half an order of magnitude. There are many reasons for such deviations, which can roughly be divided into two classes: "accidental" or "systematic" ones. In the first case there is equal probability for the ratio F_p/F_o to be greater or less than one. To distinguish between these two types one needs a large number of identified lines. "Systematic" differences may be due to errors in the power calibration for a given spectral range; these are then similar for all lines in that range. On the other hand, uncertainties concerning the abundance of elements in the Sun are only important for the lines of these elements.

α) Figure 19a, b shows a *logarithmic plot* of F_p/F_o against wavelength for 350 reliably identified lines below 122 nm. More than half of these lines are found in the region below 30 nm. The dispersion of points in the plots is large, but most of them (about 75%) are clustered near a curve within a belt that is one order of magnitude wide. The method of intensity prediction is independent of wavelength, since the lines emitted by ions at various ionization potential appear within different spectral regions (except for the soft x-ray region below 10 nm). Thus a correction ratio $f(\lambda)$ may be used for calibration of an observed EUV spectrum and allows an estimation of the uncertainty of line intensity measurements.[1] Since the predicted line intensities may differ substantially from their true values and determine only the *average* calibration curve, the real spectrum can be described by $f(\lambda)$ and F_o, better than by $F_p(\lambda)$.

β) Table 8 gives F according to HINTEREGGER's[2] *rocket measurements* (F_H), the correction factors $f(\lambda)$ obtained from Fig. 18 (in Sect. 17) for the energy distribution in the most interesting region of the EUV spectrum below 103 nm, together with corrected values. HINTEREGGER's data refer to the minimum of solar activity, as may be seen from a comparison of his data with the corresponding F_o values; the deviations do not exceed 30%. It should be noted that for the prediction procedure HINTEREGGER's data were smoothed and intensities of individual lines were not considered.

Figure 20 shows the energy distribution in the EUV spectrum in units of $\mathrm{mW\,m^{-2}\,nm^{-1}}$ smoothed over spectral subranges.

[22] Dupree, A.K., Reeves, E.M. (1971): Astrophys. J. *165*, 599
[1] Ivanov-Holodnij, G.S., Nikol'skij, G.M. (1962): Geomagn. Aeron. *2*, 425
[2] Hinteregger, H.E. (1965): Space Sci. Rev. *4*, 461

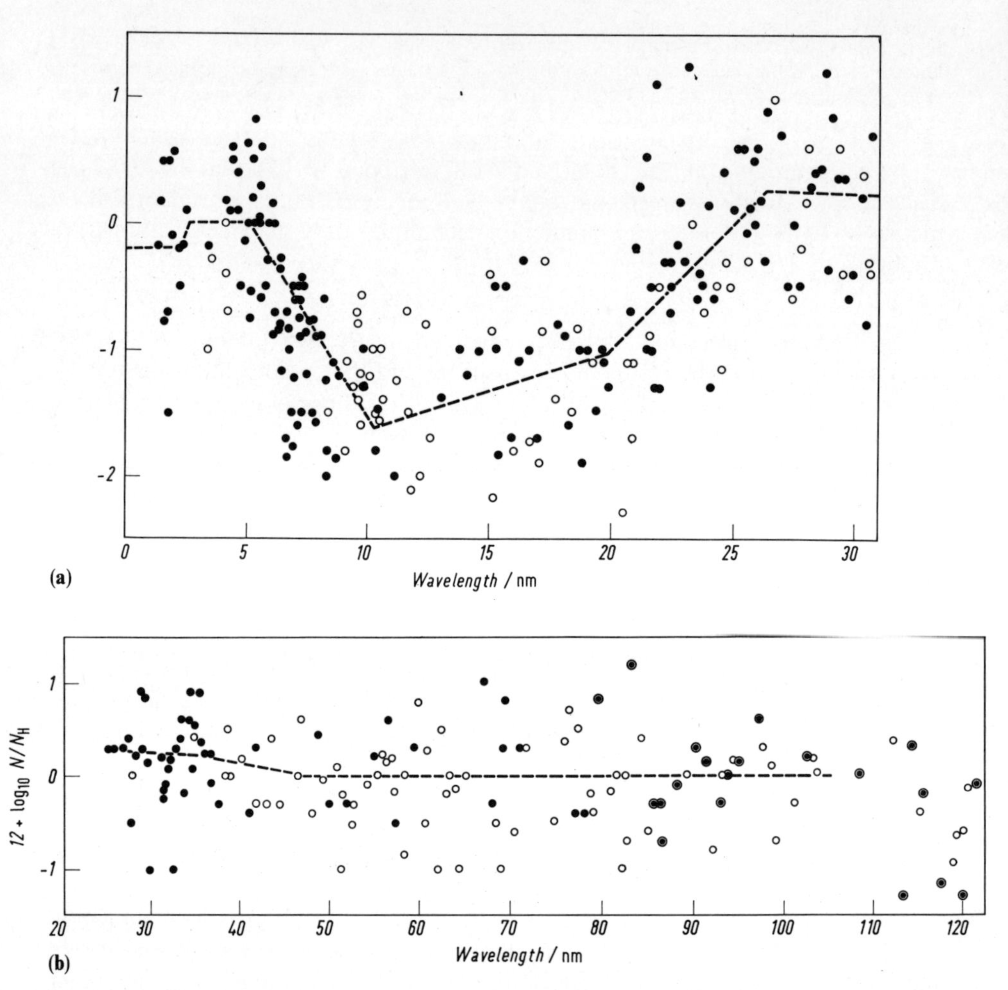

Fig. 19a, b. EUV solar spectrum calibration. Decimal log of the ratio of predicted to observed intensities is shown as a function of the wavelength. Solid circles, lines of coronal ions; open circles, lines of ions in the transition layer; points in circles, chromospheric lines

The appearance of a broad maximum between 9 and 18 nm is understood when the lines are identified: this spectral region is a "peculiar zone" because coronal lines are practically absent. There are only a few weak ones, while the majority of the emission lines within this range originate within the transition layer, where the electron temperature is too low to excite lines of higher excitation energy and thus wavelengths below 10 nm.

Later, more accurate flux density measurements in the wavelength range 6–20 nm however, did, not show significantly lower values, just one order of magnitude.[3] The calibration curve in Fig. 19 can therefore not yet be considered as final. It may be that computations of the excitation probability of high quantum levels of ions should also take account of inelastic collisions, in

[3] Manson, J.E. (1976): J. Geophys. Res. *81*, 1629

Table 8. Energy distribution in the near Earth solar EUV spectrum block distribution F_H according to HINTEREGGER's[2] observations (made before 1965). F: after our correlation

Spectral interval [nm]	F_H [mW m^{-2}]	Correlation factor $f(\lambda)$	F [mW m^{-2}]
0.1– 0.5	10^{-5}	–	–
0.5– 1.5	0.002	–	–
1.5– 3.1	0.007	0.64	0.004
3.1– 4.1	0.083	1	0.083
4.1– 6.2	0.135	1	0.135
6.2– 10.3	0.286	0.1	0.029
10.3– 16.5	0.191	0.04	0.008
16.5– 20.5	0.780	1/16	0.049
20.5– 24.0	0.140	1/3	0.047
24.0– 28.0	0.149	1.6	0.238
30.4	0.250	1.6	0.400
28.0– 32.6[a]	0.160	1.6	0.256
32.5– 37.0	0.125	1.6	0.200
37.0– 46.0	0.098	1.2	0.120
46.0– 50.0	0.072		
50.0– 54.0	0.059		
54.0– 58.0	0.050		
58.0– 63.0	0.161		
63.0– 70.0	0.037		
70.0– 74.0	0.027		
74.0– 78.0	0.049	1	F_H
81.0– 86.0	0.108		
86.0– 91.1	0.185		
91.1– 92.0	0.028		
92.0– 95.0	0.031		
95.0– 99.0	0.113		
99.0–102.7	0.101		

[a] With exception of the He^+ emission at 30.4 nm

particular those leading to recombination. This may be important in solar regions of relatively low temperature.

We must emphasize once more that the predicted line intensities, and therefore the spectrum in Fig. 20 and Table 9, were obtained at minimum solar activity. The dependence of the emission measure on the temperature used for line-intensity prediction (F_0) was obtained from line intensity measurements made on 23 August 1961 by[4]. The solar activity level was very low on that day: the Zürich susnspot number (WOLF's number) was 33, Covington's (10.7 cm radio noise) index 98, the relative spot area was $48 \cdot 10^{-6}$, and the Ca^+ plage relative area was 0.017.

We find from these data[2] that the total flux of radiation on wavelengths below 103 nm, for minimum activity conditions, should roughly be

$$F_{\min}(\lambda < 102.7\text{ nm}) \approx 2.6\text{ mW m}^{-2}. \tag{19.1}$$

[4] Hall, L.A., Damon, K.R., Hinteregger, H.E. (1963): Space Res. *3*, 745

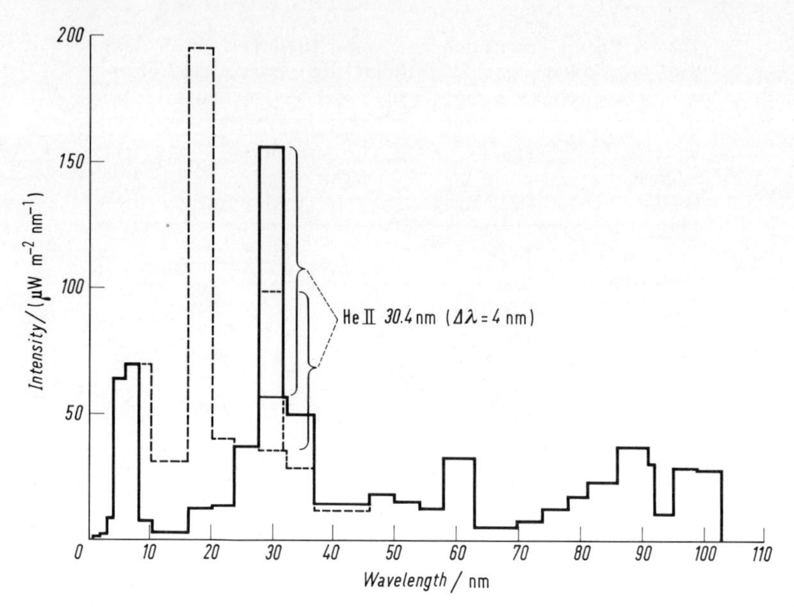

Fig. 20. Block diagram of energy distribution in the solar EUV spectrum. Broken line, observed distribution (HINTEREGGER). Continuous line, spectrum after our correction. The line of He II at 30.4 nm was taken apart and smoothed over a 4 nm interval

Unfortunately, such global indication is not sufficient for aeronomic computations. In the past most of these used HINTEREGGER's data as shown in Sect. 4, Fig. 4. Meanwhile a critical discussion on calibration methods has shed some doubt on the reliability of ground calibration (see Sect. 20β).

γ) The first experiment with in-flight calibration was flown on the *aeronomy* satellites AEROS, a joint program of West Germany and the United States. It was shown that the degradation, in particular of the open multipliers, is very important during the integration tests and during the first mission days. It appears, unfortunately, to be quite different for different multipliers, even if taken from the same production series. Also the multiplier efficiency depends to a nonnegligible extent on the working temperature of the multiplier[5].

The instrument is described by SCHMIDTKE and coworkers.[6,7] The two channel spectrometer with wavelength range 106–31 nm and 57–16 nm was installed on board a spin-stabilized United States Scout satellite (6-s spin period) on a polar orbit. Two replicas with platinum coating (2160 and 3600 mm^{-1}) were used as dispersive elements. The solar light came in at an angle of 84.5° and after diffraction passed through a multiple grid type collimator, its intensity being recorded by a photomultiplier. The Hinteregger-type spectrometer had no entrance slit and reached a spectral resolution of 0.3 to 0.6 nm. Ground calibration was carried out with the EUV radiation of an electron synchrotron. The important step was repeated, in-flight calibration of

[5] Schmidtke, G., Knothe, M., Heidinger, F. (1975): Appl. Opt. *14*, 1645–48

[6] Schmidtke, G., Schweizer, W., Knothe, M. (1974): J. Geophys. *40*, 577

[7] Schmidtke, G. (1970): Appl. Opt. *9*, No. 2

the multipliers by a radioactive Ni_{63} beta source, thus monitoring the effect of multiplier deterioration in space environment. An absolute calibration accuracy of 30–40 % for 106–60 nm, and 30–25 % for $\lambda < 60$ nm was achieved. The relative accuracy over the mission was much better. AEROS was the first satellite aboard which both aeronomic parameters (density and temperature of neutral atoms, ions, and electrons) in the upper thermosphere and solar EUV radiation fluxes were measured simultaneously.

Two satellites were launched: AEROS-A recorded EUV radiation from December 1972 through August 1973, AEROS-B from July 1974 through September 1975.

Solar pointing accuracy was 0.3–3.5°, which is acceptable for a Hinteregger-type spectrometer with plane grating geometry.

The results are presented as energy and quanta flux values for 42 spectral subranges carefully selected after aeronomic considerations.[8,9] Tables are available for both mission periods[10] giving daily median fluxes in these 42 spectral intervals with emission identification where appropriate. As a typical example, the observations of 23 March 1973 ($F_{10.7}=87$) are given in Table 9.[9] They are in good agreement with rocket observation reported by HEROUX et al.[11] for 23 August 1972 (see Table 4 on p. 26 in THOMAS's contribution to this volume).

Flux sums for wider spectral ranges are given in the last 11 lines of Table 9. SCHMIDTKE[8] feels these should be used as EUV indices providing relevant activity measures in the future.[12] The total radiation flux is 2.7 mW m^{-2}, in good agreement with Eq. (19.1) above.

When the variation in the solar activity cycle is taken into account, the AEROS absolute flux values are significantly higher than was concluded from the first quantitative observations (see Sect. 4), before more involved calibration methods were developed. The values of Table 9 are, however, in good agreement with recent measurements which HINTEREGGER[13] obtained from the Atmospheric Explorer-C.[12]

δ) What is the *contribution* of the *different regions* of the *solar atmosphere?* Do identified lines supply nearly all the energy in the EUV spectrum? In order to answer these questions we computed the sum of the power contributed by ions of different ionization potential or T_j; groups of ions with similar T_j were arranged to this end. The following regions were considered: the corona with temperature $T > 600{,}000$ K (hot coronal regions were placed in a special subgroup with $T > 3{,}000{,}000$ K), the transition layer with $T = 30{,}000$–$600{,}000$ K and the chromosphere, $T < 30{,}000$ K. The computed results are given in Table 11. Thus, at minimum solar activity, the solar corona supplies nearly one third of the total solar EUV flux and the transition layer a little more, while one fourth

---

[8] Schmidtke, G. (1976): Geophys. Res. Lett. *3*, No. 10, 573

[9] Schmidtke, G., Rawer, K., Botzek, H., Norbert, D., Holzer, K. (1977): J. Geophys. Res. *82*, 2423

[10] Schmidtke, G.: IPW Sci. Rep. WB3, Freiburg, Inst. phys. Weltraumforschung, Jan. 1978, and WB11, Dec. 1979

[11] Heroux, L., Cohen, M., Higgins, J.E. (1974): J. Geophys. Res. *79*, 5237

[12] See G. SCHMIDTKE's contribution in this volume

[13] Hinteregger, H.E. (1977): Geophys. Res. Lett. *4*, 231

Table 9. Typical solar EUV spectrum as of AEROS-A (SCHMIDTKE [9])*

Range	Wavelength (interval) [nm]	Identification	Energy flux [μW m^{-2}]	Photon flux [10^{12} m^{-2} s^{-1}]
1	103.8+103.2	O VI	78	4.1
2	102.6	H Ly2	66	3.4
3	99.1	N III	13	0.6
4	97.7	C III	88	4.3
5	97.2	H Ly3	14	0.7
6	95.0	H Ly4	8	0.4
7	94.4	S VI	3	0.1
8	102.7–91.1	Unresolved [a]	40	1.9
9	91.1–89.0	H continuum	81	3.7
10	89.0–86.0	H continuum	71	3.1
11	83.4	O II, III	13	0.6
12	86.0–83.0	H continuum	40	1.7
13	83.0–80.0	H continuum	21	0.9
14	91.1–80.0	unresolved [a]	9	0.4
15	79.02–78.77	O IV	12	0.5
16	77.0	Ne VIII	6	0.2
17	80.0–77.0	H continuum	14	0.5
18	76.0	O V	5	0.2
19	77.0–74.0	H continuum	9	0.3
20	70.3	O III	7	0.2
21	80.0–63.0	Unresolved [a]	23	0.8
22	63.0	O V	51	1.6
23	62.5	Mg X	12	0.4
24	61.0	Mg X	25	0.8
25	58.4	He I	54	1.6
26	55.4	O IV	24	0.7
27	52.1	Si XII	12	0.3
28	50.4–47.0	He I continuum	44	1.1
29	49.9	Si XII	14	0.4
30	46.5	Ne VII	9	0.2
31	63.0–46.0	Unresolved [a]	34	0.9
32	46.0–37.0	Unresolved	31	0.6
33	36.8	Mg IX	38	0.7
34	36.1	Fe XVI	15	0.3
35	33.5	Fe XVI	32	0.5
36	30.4	He II	453	6.9
37	28.4	Fe XV	46	0.7
38	37.0–28.0	Unresolved [a]	178	2.9
39	28.0–23.1	Unresolved	263	3.4
40	23.1–20.5	Unresolved	141	1.5
41	20.5–17.6	Unresolved	572	5.5
42	17.6–15.5	Unresolved	112	0.9

* Measurements are for 23 March 1973: $F_{10.7}=87$.

[a] Describes the irradiance of minor emissions in the range indicated. In order to obtain the total irradiance, that of resolved lines in the same range must be added. For example, since line emissions 33–37 must be added, the total irradiance in range 38 is 762 and not 178 μW m^{-2}.

of the flux is contributed by the chromosphere. This description concerns only that part of the solar spectrum which is most effective for reactions in the terrestrial upper atmosphere.

20. Dependence of solar EUV fluxes on solar activity and on solar rotation

α) Flux variations are quite different in the different wavelength groups.

The first observational data were extremely poor: a few isolated rocket observations of the AFCRL group[1,2] and the data obtained with the OSO-I experiment (March-May 1962)[3,4].

Starting with theoretical estimates, it was taken for certain that to a first approximation the ionization temperature of the emitting ions should be of primary importance in this context. Table 10 shows some results of such considerations, namely, the relative importance of the contributions of different solar layers for minimum and maximum solar activity. According to this estimation, the EUV flux at maximum activity should be three times greater than at minimum and the total energy flux should be

$$F_{\max}(\lambda < 102.7\ \mathrm{nm}) \approx 8\ \mathrm{mW\,m^{-2}}. \tag{20.1}$$

β) The limitations of photometric calibrations and of the observational conditions cause serious difficulties when variations of the solar EUV fluxes are to be investigated experimentally. In particular for satellite missions it is necessary to account for changes of the calibration parameters during the flight and the accuracy of solar pointing. Also background corrections for scattered light and for particle effects are needed.

Table 10. The contribution of various regions of the solar atmosphere to the EUV radiation below 102.7 nm, for minimum and maximum solar activity

The region of the solar atmosphere and its temperature	Percentage of total EUV radiation flux		Flux variations	Contrast between active and quiet regions
	Min	Max	Max/Min	
Corona > 3,000,000 K	5 %	15 %	10	30
600,000–3,000,000 K	30 %	50 %	5	14
Transition layer 30,000–600,000 K	40 %	25 %	2	5
Chromosphere below < 30,000 K	25 %	10 %	1.3	2
The solar atmosphere as a whole	2.6 mW m^{-2}	8 mW m^{-2}	3	8

[1] Hall, L.A., Schweizer, W., Hinteregger, H.E. (1965): J. Geophys. Res. *70*, 2241
[2] Hall, L.A., Schweizer, W., Heroux, L., Hinteregger, H.E. (1965): Astrophys. J. *142*, 13
[3] Behring, W.E., Neupert, W.M., Lindsay, J.C. (1964): Space Res. *4*, 719
[4] Neupert, W.M. (1965): Ann. Astrophys. *28*, 446

In order to discuss and overcome these difficulties (according to SCHMIDTKE's advise) an ad hoc working group was organized at the COSPAR XVIIIth meeting in 1975 at Varna (Bulgaria). The first report of this group was presented at COSPAR XIX (Philadelphia, 1976) and came to the following conclusions [7].

1) Solar EUV radiation is variable. However, at present the errors of the flux estimation may often surpass the true solar variation. Therefore the period when the observations were made must be taken into consideration when comparing solar fluxes.

2) The variations caused by the Sun's *rotation* (i.e., due to the inhomogeneity on the solar disk) are estimated to be as follows:

Wavelength/nm	300	200	150	121.6	20	(Fe XV, XVI)	<10	<2
Variation/%	1	5	10	20–30	50–100	200–500	$2\cdot10^3$	$2\cdot10^4$

3) It is very difficult at present to determine the variation of radiation in the spectrum due to the *solar activity cycle*. For cycle N20 a rough estimate for Lyman alpha is a factor of 1.6. Estimates for wavelengths below 100 nm, which were based on correlations with the 10.7-cm radio noise flux of the Sun (the so-called Covington index) are highly speculative.

4) With the exception of intensive emissions like Lyman alpha, optimistic *error estimates* are:

Wavelength/nm	300–200	200–100	100–20	<20
Variation/%	10–200	15–25	20–35	50

5) Since variations with time are large enough in the soft x-ray range, it is recommended to compare various x-ray observations with high timing accuracy.

Careful *in-flight calibration* may provide the final answer to the problem of solar EUV radiation variation with the cycle. An important achievement in this direction were the EUV-flux measurements made on board the aeronomy satellites AEROS (see Sect. 19 β).

γ) The AEROS-A mission produced the first in-flight calibrated EUV spectra as a continuous series through 220 days in 1973.[5] *Comparison* of this with the *classical measures of activity* (the Zürich sunspot number R and Covington's 10.7-cm radio noise index $F_{10.7}$) lead SCHMIDTKE et al.[6] to the following conclusions, which apply at least for the period of decreasing solar activity near minimum conditions.

[5] Schmidtke, G.: Daily solar EUV intensities obtained during the AEROS-A mission. IPW, Sci. Report WB 3. Freiburg: Inst. phys. Weltraumforschung, Jan. 1978

[6] Rawer, K., Emmenegger, G., Schmidtke, G. (1979): Space Res. *XIX*, 199

Table 11. SCHMIDTKE's activity indices N1...N8 with typical values for rather low solar activity[7]

Range	Wavelength (interval) [nm]	Identification		Energy flux [$\mu W\,m^{-2}$]	Photon flux [$10^{12}\,m^{-2}\,s^{-1}$]
N1	102.7–91.1	Integral	*Index:*	232	11.4
N2	91.1–80.0	Integral		235	10.4
N3	80.0–63.0	Integral		76	2.7
N4	63.0–46.0	Integral		279	8.0
N5	46.0–37.0	Integral		31	0.6
N6	37.0–28.0	Integral		762	12.0
N7	28.0–20.5	Integral		404	4.9
N8	20.5–15.5	Integral		684	6.4
	103–80	Integral		467	21.8
	80–15	Integral		2236	34.6
	103–15	Integral		2703	56.4

Measurements of 23 March 1973; $F_{10.7}=87$.

1) The solar rotation provokes a more regular variation of the Covington's index $F_{10.7}$ than of any other index or flux intensity. As for the (logarithmic) amplitude of the short-term variations that of R is largest, that of $F_{10.7}$ smallest. The EUV fluxes lie in between.

2) Similarly, the logarithmic long-term variation with the (then decreasing) solar cycle was greatest for R, smallest for $F_{10.7}$ with the EUV fluxes in between.

3) The EUV fluxes for low-energy solar emissions show smaller variations, nearer to those of $F_{10.7}$. The emissions from higher-energy ions of the Sun behave more like R.

The final conclusion is that neither R nor $F_{10.7}$ can be used to reproduce satisfyingly the variations of the solar EUV fluxes. This is important because many geophysical models use R or $F_{10.7}$ as a decisive index. Examples are models of the neutral atmosphere (like CIRA 1972[8] or OGO-6[9]) but also models of the ionosphere (e.g., the "Pennstate model"[10] and the "International Reference Ionosphere"[11]) and computing procedures for practical applications (e.g., the CCIR electron peak density program[12]).

Since it is obvious that the upper atmospheric processes controlled by solar

[7] Schmidtke, G., Rawer, K., Botzek, H., Norbert, D., Holzner, K. (1977): J. Geophys. Res. *82*, 2423

[8] CIRA 1972=COSPAR International Reference Atmisphere 1972. Berlin: Akademie-Verlag, 1972

[9] Hedin, A.E., Mayr, H.G., Reber, C.A., Spencer, N.W. (1974): J. Geophys. Res. *79*, 215

[10] Nisbet, J. (1971): Radio Sci. *6*, 437

[11] Rawer, K., Bilitza, D., Ramakrishnan, S., Sheihk, N.M. (1978): Intern. Reference Ionosphere. Bruxelles: Union Internationale Radioscientifique (U.R.S.I.)

[12] Atlas C.C.I.R. des caractéristiques ionosphériques (rapport 340). Genève: Union Internationale des Télécommunications. See also Rawer, K.: This Encyclopedia, vol. 49/7 his sect. 14

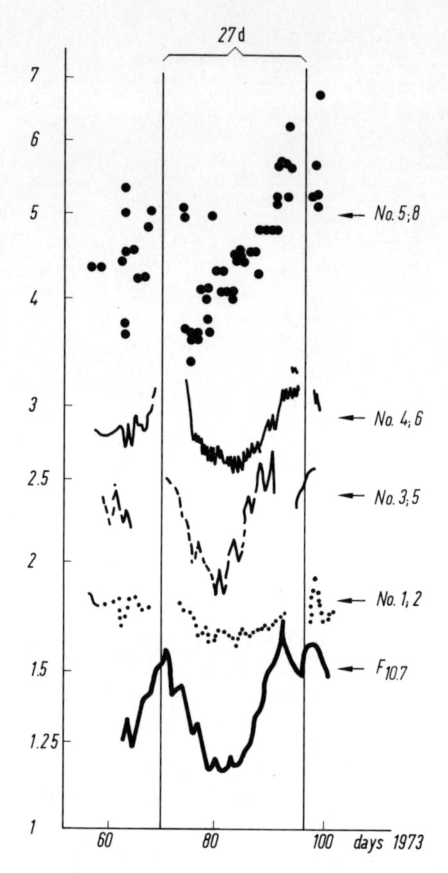

Fig. 21. Variability of solar EUV fluxes on logarithmic scale (relative values) during a period of about 40 d. Wavelength ranges after SCHMIDTKE: N1 ~ 103–91 nm; N2 ~ 91–80 nm; N5 ~ 46–37 nm; N6 ~ 37–28 nm; N8 ~ 20.5–15 nm. Data from AEROS-A satellite mission[5]

EUV should follow the variations of this radiation, SCHMIDTKE[6] proposed eight EUV indices, N1–N8, indicating the total radiation in $\mu W\,m^{-2}$ within the wavelength limits shown in Table 11. Typical indices, i.e., absolute range radiation values in $\mu W\,m^{-2}$, for solar activity minimum conditions are shown there. The numerical values obtained during the AEROS-A mission demonstrate that each EUV index undergoes variations with the solar rotation period and long-term variations which correspond to the sunspot cycle variation.

δ) Let us consider the "solar origin" of the indexes. Each of SCHMIDTKE's indices is linked with particular EUV emissions.

1) In the wavelength range 103–91.1 nm (index N1) the main emissions stem from H I, C III, O VI. Thus, the average radiation is emitted from layers in the solar atmosphere with a temperature of about $5 \cdot 10^4$ K.

2) The main radiation in the interval 91–80 nm (index N2) is caused by the hydrogen Lyman continuum. The temperature at the level of optical depth 1 is $3 \cdot 10^4$ K. Therefore, N2 should have variations rather close to those of N1.

3) In the wavelength range 46–37 nm (index N 5) the most intensive radiations come from Fe XI–XV and O III, C IV, Mg V ions. The last group of rather "cold" ions should decrease the expected stronger variations due to coronal ion emissions of Fe. One may argue that N 4 characterizes the radiation of the transition zone between chromosphere and corona with a temperature of about $3 \cdot 10^5$ K on the average.

4) In the interval 37–28 nm (index N 6) the resonance line 30.4 nm (He II) is dominating. This radiation stems from the upper chromosphere and the transition region.

5) The interval 15–20.5 nm (index N 8) is almost exclusively determined by "hot" coronal radiations of Fe IX–XIII ions. This is confirmed by the significant amplitude of the variations of this index.

An example of the behavior of SCHMIDTKE's indices for more than one 27 d period is shown in Fig. 21. It is clearly seen that as concluded by the AEROS group, the variations of the EUV indices can not be described by those of $F_{10.7}$ or R, neither in phase nor in amplitude.[13] This conclusion is further confirmed by new data from the Atmospheric Explorer missions[14]. EUV activity indices are thus urgently needed.

V. Chemical composition of the solar atmosphere derived from EUV data[0]

21. Emission measure and most abundant elements. The EUV spectrum contains a large number of resonance lines of various elements emitted by a generally optically thin atmosphere. It is therefore a very suitable basis for a quantitative analysis of the chemical composition of the solar atmosphere. In this respect Fraunhofer's absorption spectrum, because of its subordinate lines, provides considerably less information.

α) In 1961, when the generalized emission measure $\Delta\varphi_j$ was first calculated[1], it was found that the plots corresponding to N V and N IV ions deviated substantially from the general curve. From this fact it was concluded that the abundance of emitting nitrogen atoms in the solar atmosphere was lower than had been supposed (about $100\,\mathrm{m}^{-3}$). After the next, more careful analysis (including comparison of F_p with F_o for the following lines: 108.5 nm (N II), 76.44 nm and 99 nm (N III), 76.51 nm (N IV), and 123.88 nm (N V)[2]) a value of $30\,\mathrm{m}^{-3}$ was derived, i.e., 3 to 10 times lower than was then usually assumed. When the abundance of elements is determined from a *comparison of theoretical and observed line intensities*, the lines of such elements must be excluded when calculating the generalized emission measure $\Delta\varphi_j$.

[13] Compare G. SCHMIDTKE's contribution in volume 49/7

[14] Hinteregger, H. (1977): Geophys. Res. Lett. *4*, 231

[0] In this chapter the symbol {element symbol} is used for relative, not absolute abundancies, i.e., a dimensionless number giving the ratio of individual to total number density.

[1] Ivanov-Holodnyj, G.S., Nikol'skij, G.M. (1961): Astron. Žh. *38*, 45

[2] Ivanov-Holodnyj, G.S., Nikol'skij, G.M. (1962): Geomagn. Aeron. *2*, 425

The function $\Delta\varphi(T_j)$ was constructed on the basis of the already corrected abundances. It would be desirable to calculate the emission measure as a function of ionization temperature, using only the lines of a given element of which the abundance is very well known. Such attempts were made by POTTASCH and others. The starting value $\Delta\varphi_j$ (see Fig. 17 in Sect. 16) is defined reliably by the lines of C and O ions for T less than $3\cdot10^5$ K and by the ratio of the abundances of Fe and Ni combined with the coronal emission measure deduced from the coronal continum for T greater than $6\cdot10^5$ K. A value for the ratio {Fe}/{Ni} = 10 is enough reliably defined from the analysis of the forbidden coronal lines of ions from the same isoelectronic sequence, for example, Fe X and Ni XII, Fe XIV, and Ni XVI. At the same time, the C and O abundances are checked by comparing F_p and F_o values obtained from the lines of "cold" ($T<3\cdot10^5$ K) and "hot" ($T>6\cdot10^5$ K) ions.

The best agreement of the plots in Fig. 19 (in Sect. 19) associated with ions of C (12 lines), O (39 lines), Fe (60 lines), and Ni (15 lines) occurs with the following relative abundance values: $2\cdot10^{-4}$, $6\cdot10^{-4}$, $2\cdot10^{-5}$ and $2\cdot10^{-6}$, respectively. The same abundance of Fe was derived as early as 1962 from a consideration of a smaller number of lines.[3] All these data define rather exactly the ratio F_p/F_o as a function of wavelength. Six hydrogen lines may be added for determination ratio. Thus, the abundance of the four elements considered, and hence the average ratio F_p/F_o as a function of λ, is finally defined by the dependence of the coronal emission measure upon a temperature derived from ground observations of the coronal continuum and coronal forbidden lines in the visible spectrum. A curve $\Delta\varphi(T_j)$ for $T>6\cdot10^6$ K is shown in Fig. 17 (in Sect. 16).

β) For *other element* the ions of which are responsible for the lines identified in the EUV spectrum (wavelengths below 122 nm), the following relative abundances, indicated by {brackets} were obtained:

1) *Helium:* Six lines; good agreement for {He} = 0.1.

2) *Nitrogen:* 15 lines; previously derived value[2] $\{N\}=3\cdot10^{-5}$ is definitely confirmed.

3) *Neon:* The solar abundance of neon had not been previously determined because of the absence of Ne lines in the visible spectrum[4] of the Sun. When derived from the observations of nebulae and hot stars, the abundance is $\{Ne\}=5\cdot10^{-4}$. The EUV lines of Ne are rather bright. The solar abundance of Ne derived on the basis of 19 identified lines is $\{Ne\}=3\cdot10^{-5}$.

4) *Sodium:* Only a single line was found. The upper limit of flux was estimated. The accepted sodium abundance, perhaps somewhat overestimated, is $\{Na\}=2\cdot10^{-6}$.

5) *Magnesium:* 23 lines; in plots of the generalized emission measure, the dispersion is comparatively small: $\{Mg\}=4\cdot10^{-5}$.

6) *Aluminum:* 7 lines; relative abundance $\{Al\}=3\cdot10^{-6}$.

7) *Silicon:* This element provides an extremely large number of lines (comparatively bright) over the entire EUV spectrum below 121.6 nm; 65 lines were found. The value $\{Si\}=3\cdot10^{-5}$ is reliable.

---

[3] Ivanov-Holodnyj, G.S., Nikol'skij, G.M. (1962): Astron. Žh. *39*, 777

[4] R.D. DIETZ and F.Q. ORRAL: Astrophys. J. **158**, 1239 (1969) discuss the line 640.20–640.24 nm in the spectrum of RSOph (Nova-like), which is also found in coronal active regions. The observed line intensity agrees with the theoretical abundance value $\{Ne\}=5\cdot10^{-5}$.

Table 12. Relative abundance of elements in the solar atmosphere

H	1	Na	$2 \cdot 10^{-5}$	Cl	$7 \cdot 10^{-8}$	Mn	$8 \cdot 10^{-8}$
He	0.1	Mg	$4 \cdot 10^{-5}$	A	$3 \cdot 10^{-6}$	Fe	$2 \cdot 10^{-5}$
C	$2 \cdot 10^{-4}$	Al	$3 \cdot 10^{-6}$	K	$5 \cdot 10^{-8}$	Co	$2 \cdot 10^{-7}$
N	$3 \cdot 10^{-5}$	Si	$3 \cdot 10^{-5}$	Ca	$4 \cdot 10^{-6}$	Ni	$2 \cdot 10^{-6}$
O	$6 \cdot 10^{-4}$	P	$2 \cdot 10^{-7}$	Ti	$5 \cdot 10^{-8}$		
Ne	$3 \cdot 10^{-5}$	S	$1 \cdot 10^{-5}$	Cr	$2 \cdot 10^{-7}$		

8) *Sulfur:* 25 identified lines provide the reliable value of $\{S\} = 1 \cdot 10^{-5}$.

9) *Argon:* 6 lines; the most probable abundance is $\{Ar\} = 3 \cdot 10^{-6}$.

10) *Calcium:* On the basis of 9 identified lines, the calcium abundance is given as $\{Ca\} = 4 \cdot 10^{-6}$ with very good accuracy.

As a summary Table 12 shows relative abundances for 28 elements. The uncertainty of the abundance determinations for 14 elements, excluding sodium, lies within a factor of 2. The abundances of six of the remaining elements were taken from [*3*]. Ratios of abundances relative to hydrogen deduced in this way are extremely helpful for line intensity prediction.

22. Difference between coronal and photospheric abundances of elements. Let us consider some other investigations concerning the chemical composition of the solar atmosphere. Starting in 1963, POTTASCH published a series of papers on this problem[1-3].

α) Using the ratio of the emission measure to temperature calculated from observed EUV line emissions[4], relative abundances for eight elements were derived; 28 of these were used for establishing the dependance on temperature, $M_E(T)$.

The abundances were determined relative to oxygen by shifting the plots parallel to the $\log M_E$ axis till the best agreement was obtained with the average curve for atomic oxygen. O. All plots associated with the same element have to be shifted equally. Equal shifts along the $\log M_E$ axis correspond to a change of abundance by the same factor. Such a method is not always good, especially in cases of small deviations from the average curve, because the deviations can be caused both by an incorrectly assumed element abundance and by uncertainties in the theory or the line intensities. POTTASCH obtained in this way the abundances of various elements relative to oxygen.

To determine the oxygen abundance relative to hydrogen, solar radio-noise data (radio-noise temperature T_R as a function of the wavelength in the 1–100 cm range) were used. For various values of the relative abundance $\{O\}$ and thus for various $M_E(T)$, POTTASCH calculated $T_R(\lambda)$. Comparison of the computed and observed curves $T_R(\lambda)$ then gives a value for $\{O\}$.

According to[1], the best agreement between theory and observations occurs for $\{O\} = 7 \cdot 10^{-4}$. Some time later the same author[2] derived another value of $\{O\} = 4.5 \cdot 10^{-4}$. In 1966 after an analysis of the EUV lines emitted by Fe VIII–XIV, he[3] found $\{Fe\} = (4\text{–}6) \cdot 10^{-5}$ relative to

[1] Pottasch, S.R. (1963): Astrophys. J. *137*, 945
[2] Pottasch, S.R. (1965): Ann. Astrophys. *28*, 148
[3] Pottasch, S.R. (1966): Bull. Astron. Inst. Neth. *18*, 237, 443
[4] Pottasch, S.R. (1964): Space Sci. Rev. *3*, 816

hydrogen. Later from a consideration of the line intensity ratios of O VIII/Fe XVII and O VII/Fe XIV (according to his own data[3] derived from two active regions), he obtained abundance ratios {O}/{Fe} with the value 10–20 and 6, respectively. This is to be compared with 30 given by[5]. In the same paper POTTASCH obtained {O}/{N} = 6.3 and {O}/{Ne} = 29 (according to Sect. 21, both ratios would be 20).

With twelve elements POTTASCH obtained rather high figures, except for oxygen, where his figure is likely to be too small. He concluded that the coronal abundance of heavy elements is higher than the photospheric one. However, a considerable discrepancy was only found for Fe (within a factor of 5–10). It seems that POTTASCH's photospheric iron abundance is not reliable; first because of the uncertainty of the distribution between excited and ground levels, but mainly because of the lack of accurate data on the degree of photospheric ionization of Fe.

β) On the basis of Fe I lines (visual spectrum), GOLDBERG and coworkers[6] found $\{Fe\} = 4.4 \cdot 10^{-6}$. WITHBROE[7] obtained a value four times lower and suggested some possible explanations for such a discrepancy. Recent determinations of {Fe} in the photosphere showed practically no difference between photospheric and coronal Fe abundance. From an analysis of Fe I lines the value $\{Fe\} = 4 \cdot 10^{-5}$ was obtained, which is close to the coronal value[8].

POTTASCH's hypothesis of the accretion of interplanetary meteoric material as a possible mechanism providing a superabundance of heavy elements in the corona[9] seems doubtful. Van de HULST points out[2] that the particle streams leaving the Sun, now known as "solar wind", make the downward motion of meteoritic material impossible; its density in the vicinity of the Sun is 100-times higher than in interplanetary space.

JORDAN analyzed the intensities of 15 EUV lines of Fe XIII–XV, 8 lines of Ni X–XIII and 17 lines of Si VI–X and derived the relative abundances for these three elements[10]. She used data from the photoelectric measurements of the AFCRL group[11] and ionization temperatures according to the dielectronic recombination process. Her result, {Si}:{Fe}:{Ni} = 8:10:0.8, seems acceptable. She also computed the "absolute" abundances of these elements (relative to hydrogen) and found $\{Fe\} = 6 \cdot 10^{-5}$.

In another paper[12] JORDAN derived from an analysis of forbidden coronal lines in the visible spectrum and of EUV lines emitted by Fe ions $\{Fe\} = 5 \cdot 10^{-5}$. Like POTTASCH, she concluded that there is a tenfold superabundance of iron and nickel in the corona as compared with the photosphere.

γ) WARNER[13] considered the relative element abundances in the corona and photosphere. Using the "*forbidden*" *absorption lines* of Fe II and Ni II, he deduced the photospheric abundances $\{Fe\} = 6 \cdot 10^{-5}$ and $\{Ni\} = 6 \cdot 10^{-6}$. He also noted that the ratio of coronal to photospheric abundances increases parallel to the decrease in the excitation potentials of the lines used for the determination. WARNER pointed out that the conclusion regarding the super-

---

[5] Beigman, I., Vainstein, L. (1968): Astrophys. Lett. *1*, 33

[6] Goldberg, L., Kopp, R.A., Dupree, A.K. (1964): Astrophys. J. *140*, 707

[7] Withbroe, G.L. (1967): Sci. Rep. Shock Tube Spectr. Lab., N. 17. Cambridge, Mass.: Harvard College Obs.

[8] Garz, T., Holweger, H., Kock, M., Richter, J. (1969): Astron. Astrophys. *2*, 446

[9] Pottasch, S.R. (1964): Ann. Astrophys. *27*, 163

[10] Jordan, C. (1966): Mon. Not. R. Astron. Soc. *132*, 463

[11] Hall, L.A., Damon, K.R., Hinteregger, H.E. (1963): Space Res. *3*, 745

[12] Jordan, C. (1966): Mon. Not. R. Astron. Soc. *132*, 515

[13] Warner, B. (1964): Observatory *84*, 14

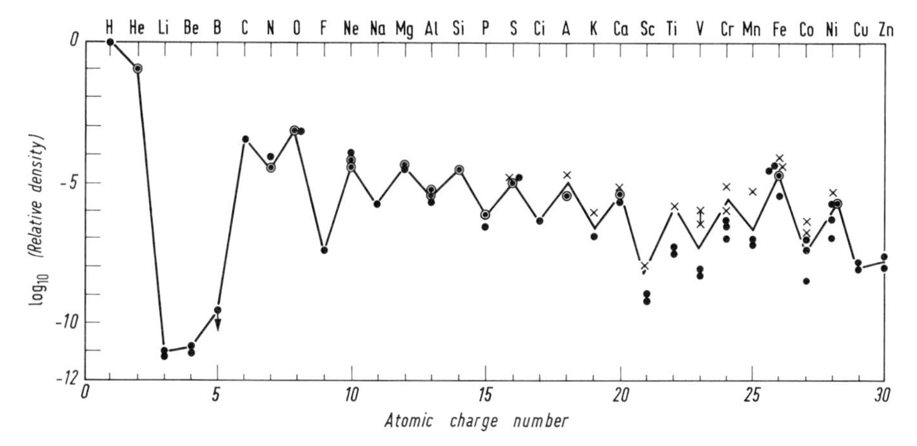

Fig. 22. Chemical composition in the solar atmosphere. Points in circles refer to EUV data, points to photospheric data and crosses to chomospheric and coronal determinations from ground

abundance of iron and nickel in the corona was premature and recommended a critical analysis of the methods used for determining element abundance in the photosphere.

δ) DUPREE and GOLDBERG[14] deduced the abundances of Si, O, and Fe. Their height profile $T(h)$ was calculated from observed *fluxes in the resonance EUV lines* O II–VI, Si II–IV, VI, VIII, X–XII and Fe XV, XVI (generally using HINTEREGGER's measurements). The abundances of Fe and O relative to Si were taken in accordance with data of GOLDBERG et al.[15] for the photosphere. The abundance of Si was found by comparing the computed and observed radio temperatures as a function of wavelength. The value of $\{Si\}=3.2\cdot 10^{-5}$ agrees very well with [*3*]. For $\{O\}$ and $\{Fe\}$ the values $9.3\cdot 10^{-4}$ and $3.8\cdot 10^{-6}$, respectively, were obtained.

The latter value seems not to be well founded. The abundance value $\{Fe\}=2\cdot 10^{-5}$ obtained above from an analysis of 60 lines of various ions is probably more accurate. It should be observed that the method used for the $T_R(\lambda)$ calculation by DUPREE and GOLDBERG is open to criticisim since in the calculation of the optical depth for radio waves the emission measure values within each temperature range were summed up without taking account possible overlapping. As a consequence, the optical depth values for all temperature ranges are probably overestimated.

ε) Figure 22 *summarizes* the pecularities of solar chemical composition: it shows the solar elemental abundances found by different methods as a function of the atomic number up to 50 (Sn). The numerical values were taken from the review[16]. Apparently, the elemental abundances of the iron group derived from the photospheric lines are lower than those derived from chromospheric and coronal emission lines.

Such a discrepany may be due to the uncertainty of the procedure. More recent determinations have shown that the photospheric abundance of iron (the most abundant element of the

[14] Dupree, A.K., Goldberg, L. (1967): Sol. Phys. *1*, 229

[15] Goldberg, L., Müller, E., Aller, L.H. (1960): Astrophys. J., Suppl. Ser. 5, N45

[16] Hauge, D., Engvold, O. (1970): Inst. of Theor. Astrophys., Blindern-Oslo, Rept. N31

iron group) coincides very well with the coronal abundance.[17] However, in a recent survey[18] WALKER presents an abundance value {Fe} of $3.1 \cdot 10^{-5}$, which is twice the value given above. His abundance value for nitrogen is even three times larger. With these exceptions his values (which are mainly due to WITHBROE's[19] determinations) agree with those given above, within a factor of 2. We feel that this is the best attainable accuracy which can be reached at present.

23. "Extraordinary" identifications in the EUV spectrum

α) In 1964 HELLMANN and ZWEIG independently postulated the existence of three elementary particles named *quarks*[1], which, being responsible for meson and baryon multiplets, complete the general theory of elementary particles. Quarks have rather large masses about 5 to 10 times that of the proton, m_p, and a fractional charge of $+2/3$ to $-1/2$ of an elementary charge. ZEL'DOVIČ and coworkers[2] considered various possible effects of quarks on EUV lines emitted by ions of the most abundant nuclei (N, O, C, Fe) containing a quark of charge $-1/3$. Such particles would change the effective charge of ions and consequently the energy of their quantum-level systems. TOUSEY and coworkers[3] calculated the wavelength of the most intensive ion lines possibly influenced by quarks: C III–IV, N II–V, O III–VI.

The calculations were performed by interpolation in isoelectronic sequences (Edlen method). For an ionic charge of about 2 to 5, the wavelengths would be approximately 10% larger. If any such lines were discovered from a comparison of its intensity with that of the corresponding line emitted by an "ordinary" ion, the concentration of "quark" elements could be estimated. An attempt to discover quark lines in the EUV solar spectrum was reported[3]. For the above-mentioned ions the wavelengths of 30 lines in the range 20–170 nm assigned to twelve multiplets were calculated with an uncertainty of 0.005 to 0.02 nm. The transitions responsible for the sufficiently strong ordinary lines were taken into account. A search for the predicted quark lines in an observed solar EUV spectrum showed that eleven of them were overlapping with intensive ordinary lines, and there was no emission at the position of the other 16 lines within experimental error.

From this an upper limit for the relative concentration of quark elements (symbol ′) can be derived: $\{C'\}/\{C\} < 10^{-4}$; $\{N'\}/\{N\} < 10^{-3}$; $\{O'\}/\{O\} < 3 \cdot 10^{-5}$. These limits are considerably greater than the cosmological estimates of quark concentration made by ZEL'DOVIČ and coworkers[2] (10^{-9} to 10^{-18}). At the position of three quark lines, generally N II and C IV multiplets, weak lines appeared.[3] All the data discussed above are reproduced in Table 13.

β) There is a more recent paper by BENNETT[4] containing a list of 36 lines due to *subordinate transitions* between the levels of the ordinary ions O, N, C. BENNET noted that the wavelengths of these lines coincided with the quark lines within experimental error. Thus, any attempt to discover quark lines in the solar spectrum is virtually hopeless. Indeed, the emissions in the solar

---

[17] Garz, T., Holweger, H., Kock, M., Richter, J. (1969): Astron. Astrophys. 2, 446

[18] Walker Jr., A.B.C. (1972): Space Sci. Rev. *13*, 672

[19] Viethbroe: Withbroe, G.L. (1971): The chemical composition of the photosphere and corona. N.B.S. Special Publ. No. 353. Washington DC: National Bureau of Standards

[1] Zel'dovič, J.B. (1965): Usp. fiz. Nauk *86*, 303

[2] Zel'dovič, J.B., Okun', L.B., Pikel'ner, S.B. (1965): Usp. fiz. Nauk *87*, 113

[3] Sinanoglu, O., Skutnik, B., Tousey, R. (1966): Phys. Rev. Lett. *17*, 785

[4] Bennet, W.R. (1966): Phys. Rev. Lett. *17*, 1196

Table 13. Identification of solar EUV lines emitted by ions with quarks

Ion	Transition	Ordinary	With quarks[3] Wavelength λ [nm]	Observed Wavelength λ [nm]	Possible identification[4]		
C IV	$2^2S_{1/2}-2^2P^0_{3/2}$	154.92	168.90 ± 0.02	Weak lines	C II	$4p\,^2P^0_{3/2}-4f\,^4D_{5/2}$	168.91
N II	$2^3P_2-2^3D^0_3$	108.57	120.76 ± 0.02	120.77	O III	$2p^4\,^1D_2-2d'\,P^0_1$	120.77
N II	$2^3P_2-2^3D^0_2$	108.55	120.74 ± 0.02	120.72	O III	$3p\,^3D^0_3-3d\,^3P_2$	120.72

spectrum with wavelengths coinciding with the expected quark lines can be more simply identified as subordinate ordinary transitions.

Table 11, column 5, gives wavelengths of quark lines that coincide with faint unidentified observed lines in the solar spectrum[3]. Columns 1–3 give the ions responsible for quark lines and the transitions and wavelengths of ordinary lines, and column 6 shows the possible identification of observed lines as ordinary emissions according to Bennett[4]. It may be that not all of the given identifications are final.

γ) Let us consider next the conclusions of Tousey et al.[3]. From the 120.78 nm line they found the abundance of "quark" nitrogen: $\{N'\}/\{N\}=5\cdot 10^{-4}$. This means that the difference between the intensities of observed and ordinary N II 108.554 nm line are of the same order. The intensity of the ordinary line is 0.01 mW m^{-2} according to the AFCRL group[5]. Consequently, the intensity of 120.78 nm is $5\cdot 10^{-6}$ mW m^{-2}. On the other hand, we can estimate the intensity of the 120.778 nm O III line, the probable identification of which was suggested by Bennet[4]. Using an oscillator strength value of about 0.1 and taking the excitation energy of the upper level (26 nm) into consideration, we derive a line intensity of $F=3\cdot 10^{-6}$ mW m^{-2}. Thus, the identification of this line is rather reliable.

In conclusion, the problem of the spectroscopic detection of "quark" elements on the Sun is a very difficult one. Spectral methods are effective only when the relative concentration of such elements is higher than 10^{-5} to 10^{-4}.

General references

[*1*] Allen, C.W. (1955): Astrophysical quantities. London: Athlone Press

[*2*] Aller, L.H. (1961): The abundance of the elements. New York: Interscience Publ.

[*3*] Ivanov-Holodnyj, G.S., Nikol'skij, G.M. (1969): Solnce i ionosfera (Sun and ionosphere). Moskva: Izdatel'stvo "Nauka"

[*4*] Moor, S.E. (1948; 1952): Atomic energy levels. Washington: National Bureau of Standards, vol. I: 1948, vol. II: 1952

[*5*] White, O.R. (ed.) (1977): Boulder: Col. Ass. Univ. Press. The solar output and its variation.

[*6*] Striganov, A.R., Sventickij, N.S. (1966): Tablicy spectral-nyh linij nejtral'nyh i ionizovannyh atomov (Tables of spectral lines of neutral and ionized atoms). Moskva: Atomizdat

[*7*] Delaboudiniere, J.P., Donnelly, R.F., Hinteregger, H.E., Schmidtke, G., Simon, P.C. (1978): Intercomparison/compilation of relevant solar flux data related to aeronomy. COSPAR Techn. Manual No. 7. Paris: Committee on Space Research (COSPAR)

[*8*] Šklovskij, I.S. (1962): Fizika solnečnoj korony (Physics of solar corona). Moskva: GFML

[5] Hall, L.A., Damon, K.R., Hinteregger, H.E. (1963): Space Res. *3*, 745

Subject Index

Physics and Chemistry in Space

Editors:
J.G. Roederer, J.T. Wasson

The series "Physics and Chemistry in Space" will cover the following main topics: Solar Physics, Cosmic Rays, Interplanetary Plasma and Fields, Physics of the Magnetosphere, Radiation Belts, Aurora and Airglow, Ionosphere, Aeronomy, Moon, Planets, and Interplanetary Condensed Matter, Satellite Geodesy, Satellite Meteorology, Satellite and Rocket Astronomy Techniques, Exobiology.

Volume 1: A.J. Jacobs
Geomagnetic Micropulsations
1970. 81 figures. VIII, 179 pages. ISBN 3-540-04986-X

Volume 2: J.G. Roederer
Dynamics of Geomagnetically Trapped Radiation
1970. 94 figures. XIV, 166 pages. ISBN 3-540-04987-8

Volume 3: I. Adler, J.I. Trombka
Geochemical Exploration of the Moon and Planets
1970. 129 figures. X, 243 pages. ISBN 3-540-05228-3

Volume 4: A. Omholt
The Optical Aurora
1971. 54 figures. XIII, 198 pages. ISBN 3-540-05486-3

Volume 5: A.J. Hundhausen
Coronal Expansion and Solar Wind
1972. 101 figures. XII, 238 pages. ISBN 3-540-05875-3

Volume 6: S.J. Bauer
Physics of Planetary Ionospheres
1973. 89 figures. VIII, 230 pages. ISBN 3-540-06173-8

Volume 7: M. Schulz, L.J. Lanzerotti
Particle Diffusion in the Radiation Belts
1974. 83 figures. IX, 215 pages. ISBN 3-540-06398-6

Volume 8: A. Hasegawa
Plasma Instabilities and Nonlinear Effects
1975. 48 figures. XI, 217 pages. ISBN 3-540-06947-X

Volume 9: A. Nishida
Geomagnetic Diagnosis of the Magnetosphere
1978. 119 figures. 1 table. VIII, 256 pages. ISBN 3-540-08297-2

Volume 10: A.V. Gurevich
Nonlinear Phenomena in the Ionosphere
Translated from the Russian by J.G. Adashko
1978. 76 figures, 17 tables. X, 370 pages. ISBN 3-540-08605-6

Springer-Verlag
Berlin
Heidelberg
New York